Optical Fiber Sensors

NATO ASI Series

Advanced Science Institutes Series

A Series presenting the results of activities sponsored by the NATO Science Committee, which aims at the dissemination of advanced scientific and technological knowledge, with a view to strengthening links between scientific communities.

The Series is published by an international board of publishers in conjunction with the NATO Scientific Affairs Division

A	Life Sciences	Plenum Publishing Corporation
B	Physics	London and New York
C	Mathematical and Physical Sciences	D. Reidel Publishing Company Dordrecht, Boston, Lancaster and Tokyo
D	Behavioural and Social Sciences	Martinus Nijhoff Publishers Boston, Dordrecht and Lancaster
E	Applied Sciences	
F	Computer and Systems Sciences	Springer-Verlag Berlin, Heidelberg, New York
G	Ecological Sciences	London, Paris, Tokyo
H	Cell Biology	

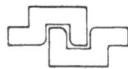

Series E: Applied Sciences – No. 132

Optical Fiber Sensors

edited by:

A.N. Chester
Hughes Aircraft Company
El Segundo
California
USA

S. Martellucci
The Second University of Rome
Rome
Italy

A.M. Verga Scheggi
National Research Council
Institute of Research on Electromagnetic Waves
Florence
Italy

Springer-Science+Business Media, B.V.

Proceedings of the NATO Advanced Study Institute on "Optical Fiber Sensors",
Erice, Italy, May 2-10, 1986

Library of Congress Cataloging in Publication Data

NATO Advanced Study Institute on Optical Fiber Sensors
    (1986 : Erice, Sicily)
    Optical fiber sensors.

    (NATO ASI series. Series E, Applied sciences ;
no. 132)
    "Proceedings of the NATO Advanced Study Institute on
Optical Fiber Sensors, Erice, Italy, May 2-10, 1986"--
T.p. verso.
    "Published in cooperation with NATO Scientific
Affairs Division."
    Includes index.
    1. Fiber optics--Congresses. 2. Optical detectors--
Congresses. I. Chester, A. N. II. Martellucci, S.
III. Verga Scheggi, A. M. (Anna Maria) IV. North
Atlantic Treaty Organization. Scientific Affairs
Division. V. Title. VI. Series.
TA1800.N39 1986   621.36'7   87-7819

ISBN 978-94-010-8116-0   ISBN 978-94-009-3611-9 (eBook)
DOI 10.1007/978-94-009-3611-9

Distributors for the United States and Canada: Kluwer Academic Publishers,
P.O. Box 358, Accord-Station, Hingham, MA 02018-0358, USA

Distributors for the UK and Ireland: Kluwer Academic Publishers, MTP Press Ltd,
Falcon House, Queen Square, Lancaster LA1 1RN, UK

Distributors for all other countries: Kluwer Academic Publishers Group, Distribution
Center, P.O. Box 322, 3300 AH Dordrecht, The Netherlands

This volume presents the proceedings of a summer course on "Optical
Fiber Sensors", which was held from May 2 to 10, 1986, in Erice, Italy. This
is the 11th in a series of courses conducted by the International School
of Quantum Electronics, on behalf of the "Ettore Majorana" Center for
Scientific Culture.

Although optical systems have long played an essential role in the
fields of instrumentation and sensors, the development of optical fiber communications and related technologies has greatly enlarged the possibility for
sensor systems. The successful applications of optical fibers have stimulated additional creative work in other guided-wave technologies, both in fibers and in the planar geometries characteristic of integrated optics.

Among this newer work, fiber optic and integrated optic sensors play
a particularly promising role. The propagation of light in guided-wave
structures is sensitive to a number of phenomena which can change the physical geometry or refractive properties of the material. By using ingenui-
ty and good design to isolate the desired effects, researchers have been
able to construct a variety of compact and useful sensor devices, as are
described in these papers. In many cases, a sensitive optical device involving no electrical connections to the phenomena being sensed provides a
unique and necessary non-interfering sensing technique.

The papers published here include discussions of the basic principles of optical fiber sensors and their major applications in measuring
rotation, acoustic vibration, intensity, temperature, strain, and chemical
concentration. In addition, there are discussions of integrated optic structures as applied to sensing, and of the engineering technologies underlying
both fiber and integrated optic sensor devices.

This course was initiated at the suggestion of one of us and was
greatly successful, bringing together a hundred people among the leading

VI

researchers in this field coming from all over the world (16 NATO countries
and 9 non NATO countries). We appreciate the hard work of the lecturers
in providing their manuscripts for rapid publication.
Owing to their exceptional level a few papers have been also included
from lecturers who had prepared their papers for presentation in Erice
but at the last minute were unable to attend the course.

Due to the severe time requests we have been obliged to leave out
the contributions of very busy authors (B.Crosignani, D.N.Payne, G.Tangonan).
The level of the course has been so high that it might be considered more
similar to a workshop than a summer school; accordingly contributed papers
by attendees have been included in the proceedings. The articles in this
volume have been judged and accepted on their scientific quality, and
language corrections may have been sacrificed in order to allow quick
dissemination of knowledge to prevail.

We would like to acknowledge with thanks the financial support for
this course provided by NATO, the North Atlantic Treaty Organization, the
Italian Ministry of Scientific and Technological Research, the Sicilian
Regional Government and the National Research Council (C.N.R.). We also
thank the U.S.National Science Foundation for providing travel grants
for three graduate students to attend the course, and the U.S. Army Research
Development and Standardization Group, U.K. We also welcome Martinus
Nijhoff as the Publishers of our proceedings.

Finally, but most importantly, we are glad to take the opportunity
to acknowledge the skillfull assistance of Mrs. Vanna Cammelli and
Mrs. Mary Schram as regards both the course and its proceedings.

The Editors

Arthur N. Chester   Sergio Martellucci   Anna Maria Verga Scheggi
Hughes Aircraft Company   The Second University   IROE - C.N.R.
El Segundo,California,USA   Rome, Italy   Florence, Italy

TABLE OF CONTENTS

MONOMODE FIBRE OPTIC INTERFEROMETERS AND THEIR APPLICATION IN SENSING
SYSTEMS

D.A. JACKSON

PHYSICS LABORATORY, UNIVERSITY OF KENT, CANTERBURY, KENT CT2 7NR, UK.

1. INTRODUCTION
    In this chapter we review the Michelson, Mach Zehnder and Fabry-Perot
interferometers and introduce their fibre optic equivalents, together with
signal processing techniques which enable these devices to operate over a
large dynamic range with constant sensitivity to induced optical phase
changes. The lower resolution polarimetric (differential) interferometer
and associated signal processing to enable it to be operated remotely is
also considered. The application of these novel fibre optic interferometers to a variety of measurements such as temperature, magnetic field,
displacement and sound waves is described.

2. OPTICAL INTERFEROMETRY
2.1. Introduction
    In the laboratory, optical interferometers are mainly used either: (i)
to determine the fundamental parameters of an optical source such as its
wavelength or coherence length; or (ii) in high precision optical path
difference (OPD) measurements, where the change in the OPD may occur
because of a physical displacement, or a change in the optical constants of
the light transporting medium (e.g. refractive index) in part of the interferometer.
    Optical interferometer is a relatively old subject, and most of these
classic instruments were introduced well over 70 years ago; indeed the
first interferometers are associated with the founders of modern optics
such as Newton, Young and Michelson.
    Before discussing fibre optic interferometers, it is appropriate to
summarise the basic operation of a conventional interferometer and to
evaluate its potential as a general purpose displacement sensor. For
simplicity we choose the Michelson two-beam interferometer shown in
figure 1(a). Light from the optical source, ideally a single-frequency
laser, is amplitude divided at the beam splitter to produce reference and
signal beams which propagate in the arms of the interferometer. These
beams may be represented by

$$\text{Reference} \equiv A_R \exp[i(\omega_L t + 2kx_R)] \tag{1.1}$$

$$\text{Signal} \equiv A_S \exp[i(\omega_L t + 2kx_S)] \tag{1.2}$$

A_R and A_S are the amplitudes of the reference and signal beams; x_R and
x_S are the distances the light travels between the reference (M_R) and
signal (M_S) mirrors respectively; $k = 2\pi n/\lambda$, the propagation constant,
where λ is the vacuum wavelength of the light, n is the refractive index
of the air path and ω_L is the angular frequency of the light source.
After traversing the interferometer arms the beams coherently recombine at

the beam splitter. The current I_D of the photodetector used to record the irradiance output of the interferometer is given by

$$I_D = \varepsilon(1 + K \cos \phi(t)) \qquad (1.3)$$

$\phi(t)$, the time-dependent phase difference between the arms of the inter-ferometer, is equal to $2k|x_S(t) - x_R|$ where $x_R \neq x_R(t)$, K is the fringe visibility and equals 1 when $A_R = A_S$ and ε is related to the input optical power. The transfer function of the interferometer, I_D, plotted as a function of ϕ is shown in figure 1(b), where we have assumed that it is operating with the following restrictions, which can be closely approxi-

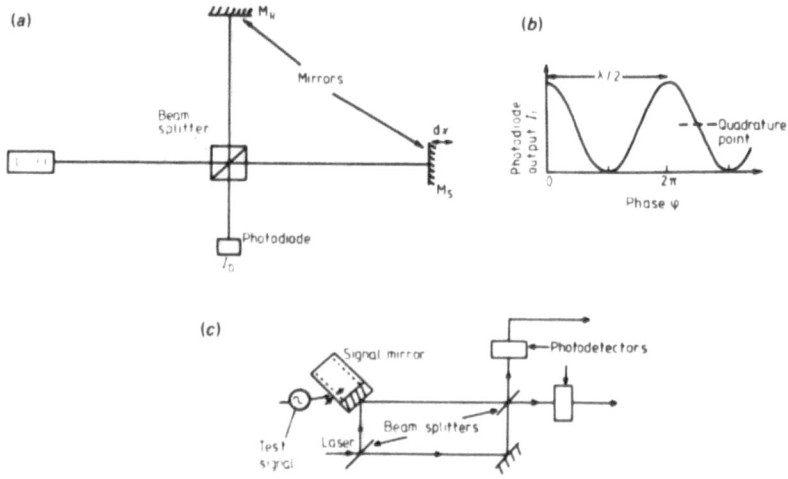

FIGURE 1. (a) Schematic diagram showing a conventional Michelson inter-ferometer. (b) Michelson transfer function, photodiode current I_D as a function of the relative (static) phase ϕ. (c) Schematic diagram showing a conventional Mach–Zehnder interferometer.

mated to under normal laboratory conditions:
(i) $A_R = A_S$, i.e. the beams are of equal intensity and state of polarisation;
(ii)   the bandwidth of the optical source is extremely narrow;
(iii)   the absolute optical frequency of the source is constant;
(iv)   the optical alignment is 'perfect' and stable.
    Under these operating conditions it is possible to detect extremely small changes in the OPD of the interferometer, for example Moss et al (1971) have determined periodic displacement amplitudes of the order of 10^{-14} m. The ultimate limit of detectability is defined by a signal-to-

noise ratio of unity, the so called 'shot-noise' limit (set by the photo-detector); this concept is discussed in §6.2.1.

It is evident that the performance of many primary sensors based upon a measurand-induced dimensional change could be significantly improved if it were possible to use interferometric techniques. Although interferometers offer tremendous resolution (or sensitivity), it is only very recently that any significant effort has been devoted to developing them into general purpose displacement sensors. This is primarily because their operation is dependent on the relative alignment of: (i) the internal optical beams and (ii) the input light beam with the interferometer itself. In a typical instrument where conventional beam splitters and mirrors are used to control the amplitude division and recombination of the light beams, the output will fluctuate due to random noise perturbing the alignment of the optical components. To a very large extent, both the internal and external alignment problems are eliminated if the interferometer is constructed entirely from optical fibres allowing the possibility of a novel high resolution primary sensing element capable of remote operation.

Although the optical fibre approach solves the stability problem associated with all interferometers and has encouraged the development of guided wave systems for sensor applications, it does so at the cost of introducing a potentially major source of error, namely the variability of the optical constants of the fibre itself.

There are several other fundamental properties of all interferometers, whether implemented in a fibre optic or conventional form, which tend to restrict their use in sensor applications and are discussed further below.

2.2. Periodicity and dynamic range

From figure 1(b) we see that the output of the Michelson interfero-meter - as for all interferometers - varies with a periodicity of 2π rad, equivalent in this case to a change in relative mirror separation of $\lambda/2$. If it is used as a sensor and we require that its output is a unique function of mirror displacement, then the sensor must be operated in a 'zero pathlength difference' mode with an operational range of only $\lambda/4$, corresponding to 2×10^{-7} m for a typical solid-state laser source, with a wavelength of 8×10^{-7} m. Although it is possible in principle to construct an interferometer without any imbalance in its optical path it would require special techniques; consequently most interferometers are operated with an arbitrary optical imbalance between their arms, and displacement measurements are made relative to an initial unspecified OPD. Another consideration is that if reproducible relative optical displacement measurements are to be made then the optical pathlength of the reference (arm) must not change. Assuming that this can be achieved and the instrument is operated at the shot-noise limit, which typically corresponds to a displacement sensitivity of about 10^{-5} - 10^{-6} of an optical radian, then an optical sensor with a dynamic range (ratio of measurement range to resolution) of greater than 10^5 could be realised with a maximum displacement of only 2×10^{-7} m. If a specific application requires that the sensor operates over a much larger displacement range then it is necessary to retain both the current fringe number N (where a 'fringe' corresponds to an optical phase change of 2π rad, such that the total excursion of the Michelson interferometer is approximately $\frac{1}{2}N\lambda$) and also ascertain the 'direction' of the fringe motion to determine whether the optical path difference is increasing or decreasing. Various signal processing concepts (§6) have been developed which enable this information to be retained unambiguously *from an arbitrary starting point*; however,

4

this information is usually lost when the system is switched off. If the
application for the interferometric sensor is to obtain the amplitude of a
periodic measurand, such as in surface vibration studies, then this problem does not arise. However, if the sensor is to be used for relatively
slowly varying measurands, such as temperature, then this loss of information is unacceptable and the system will require initialisation every time
it is switched on, see §8.3.3.

2.3. Linearised output
    Again referring to figure 1(b) we see that the displacement sensitivity
of the interferometer $(dI_D/d\phi) \propto \sin\phi$ varies periodically, being at a
maximum when $\phi = \pm n\pi \pm \pi/2$ and zero when $\phi = \pm n\pi$ $(n = 0,1,2,...)$. The point
of maximum sensitivity corresponds to an OPD of $(2n + 1)\lambda/4$; this is the
so called 'quadrature position'. In sensor applications it is important
that the interferometer is effectively operated at constant linear
sensitivity. Various signal processing systems to achieve this requirement are discussed in §6.

3. OPTICAL FIBRE WAVEGUIDES AND COMPONENTS
The complete analysis of waveguide propagation in optical fibres has been
extensively treated by several authors (see, for example, Gloge 1971,
Midwinter 1979) and only those fundamental aspects of the waveguide
relevant to the practical realisation of optical fibre interferometers
will be considered here.
    At the most basic level, the waveguide can be considered to be an
infinitely long two-component coaxial transparent medium. The central
core which guides the light has a larger refractive index (n_1) than that
of the outer cladding (n_2). In the ray-optics limit corresponding to
fibre core diameters greater than 100 µm, the light will propagate without
loss through the optical fibre provided it is launched into the central
core such that it meets the core-cladding interface at an angle greater
than the critical angle, $\theta_1 = \sin^{-1} (n_2/n_1)$. The guiding nature of the
fibre is primarily determined by: (i) the core radius, a; (ii) its
numerical aperture (NA), equal to $n_1(2\Delta)^{\frac{1}{2}}$ where $\Delta = [1 - (n_2/n_1)]$; and
(iii) the propagation constants of the light in the fibre.
    Gloge (1971) has shown that Maxwell's equations, which describe the
propagation of the light in the fibre, may be solved in the limit $\Delta \to 0$,
corresponding to 'weak' guiding of the light. Gloge's analysis shows that
in general there is not a unique 'optical path' for the injected light to
follow as it propagates through the optical fibre, but a large number of
paths, termed 'modes', which have different propagation constants. The
propagation constant β of any guided mode in the fibre is defined by
$\beta_2 < \beta < \beta_1$, where $\beta_2 = 2\pi n_2/\lambda$ and $\beta_1 = 2\pi n_1/\lambda$. The number of modes in
the fibre, N, has been shown (Midwinter 1979) to be equal to $4V^2/\pi$, where
V is the normalised frequency of the fibre equal to $(2\pi a/\lambda)(n_1^2 - n_2^2)^{\frac{1}{2}}$.
When V < 2.405 then only the lowest order spatial mode, the LP_{01} mode, can
propagate; and the fibre is classified as monomode. The LP_{01} mode
comprises two independent orthogonal linearly polarised modes with propa-
gation constants β_f and β_s equal to $2\pi n_f/\lambda$ and $2\pi n_s/\lambda$ respectively, where
n_f and n_s are the refractive indices of the modes. Thus a monomode fibre
element is birefringent and can be thought of as a general optical phase
plate. The increases in the optical phase of two orthogonal linearly
polarised light beams propagating in the eigenmodes of a birefringent
fibre of physical length L are $\beta_f L$ and $\beta_s L$. The length L_p of the fibre
over which these two orthogonal modes change their relative phase by 2π

rad is known as the 'beat length' of the fibre.

A typical optical fibre designed to be monomode at 633 nm will have a NA of about 6° and a core radius of 2.5 μm with a cladding radius typically between 40 to 60 μm; the whole fibre is usually coated in a soft polymer jacket (primary coating) to both protect ahd outer surface of the fibre and to give it mechanical strength.

If the V number of the fibre is greater than 2.405 then the fibre is multimode; in figure 2 the far-field patterns produced when a laser beam is propagated through separate multimode and monomode fibres are shown. The complex nature of the far-field pattern of the multimode fibre, figure 2(a), is readily distinguished from that of the simple gaussian profile of the monomode fibre, figure 2(b). Clearly it is not possible to fabricate a high contrast interferometer from multimode fibres as the

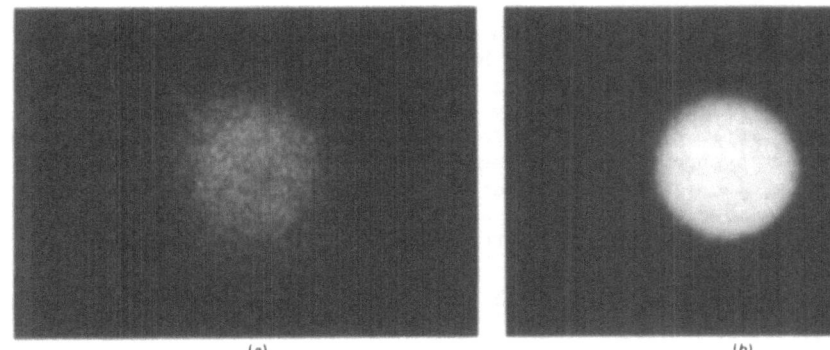

(a)   (b)

FIGURE 2.   Far-field patterns observed when a laser beam is propagated through (a) a multimode fibre, (b) a monomode fibre.

multiplicity of guided 'light paths' will produce a similar multiplicity of interferometers and, as there will be arbitrary pathlength differences in each of these interferometers, the overall contrast will generally be both very poor and unstable. Even when the interferometer is made from monomode fibre, its transfer function will tend to be more complex than that of the conventional interferometer because of the birefringent nature of the medium guiding the light beams.

3.1. Specialised monomode fibres

As was stated above, monomode fibre supports the propagation of two orthogonally polarised modes; if the fibre core has perfect rotational symmetry then $n_s = n_f$, the modes are degenerate and the fibre shows no birefringence (Rashleigh 1983a); hence the polarisation state of the light remains constant as it propagates. As most commercial monomode optical fibres (primarily supplied for communications applications) are produced by pulling the fibre from a preform with a high degree of rotational symmetry in its refractive index profile, they usually exhibit relatively low birefringence, and are therefore also suitable for incorporation into fibre optic interferometers. Unfortunately, random levels of birefringence can be induced into the optical fibre by local twists or strains which are difficult to avoid when it is deployed as an interferometer. In addition to induced linear birefringence it is also possible that bends and twists in the fibre will cause random coupling between the orthogonal modes of the monomode fibre (circular birefringence); both these effects can combine to deleteriously affect the fringe visibility in the interferometer.

To overcome these problems, considerable research effort has recently been devoted to the design and development of special fibres for interferometric sensor applications. One approach has been to produce optical fibres with very high levels of 'built-in' birefringence as this tends to inhibit both externally induced mode coupling and bend birefringence. Highly birefringent fibre has been produced by several methods (Payne et al 1982); the fibre with the best polarisation holding parameter 'h' has been fabricated with internal stress lobes (Birch et al 1982). Highly birefringent fibres are often incorrectly called 'polarisation state maintaining fibres' although in general the polarisation state of the injected light is not maintained, in fact it is only preserved under special conditions - when the azimuth of an incident linearly polarised light beam is coincident with one of the eigenaxes of the fibre.

Monomode optical fibres with very low linear birefringence have also been developed (Norman et al 1979) to exploit magnetically induced circular birefringence (Faraday effect) (Smith 1978) for current measurement. Although in principle this particular fibre would be useful for interferometric sensors, its high susceptibility to externally induced birefringence has generally restricted its use in these applications.

3.2. Monomode fibre optic components

In order to implement all-fibre optic equivalents of the classical interferometers, it has been necessary to develop monomode fibre optic components to replace the conventional optical components such as beam splitters, mirrors, polarisers, etc. used in these instruments. Optical phase and frequency modulators compatible with the fibre optic waveguides are also required in order to recover the measurand-induced optical phase shift from the interferometer.

3.3. Directional couplers

Beam splitters - often called 'directional couplers' when implemented in optical fibre - have been developed at several laboratories and are discussed by D.N. Payne in this volume.

3.4. Mirrors

High quality, but low reflectivity mirrors can be readily implemented at the end of a monomode fibre by accurately cleaving the fibre at the desired position. Here the reflection coefficient of the mirror is determined by the Fresnel reflection coefficient and will be approximately 4%. Higher reflectivities can be obtained either by vacuum deposition of a dielectric or metal coating or, as has proved very successful in our laboratory, using a silvering solution commonly used to coat vacuum flasks. Several authors have attempted to attach conventional mirrors to monomode fibres; however, considerable difficulties are experienced with long term stability in the relative alignment between the mirror and the fibre end (Petuchowski et al 1981).

3.5 Joints

Permanent low loss joints can be made using a fusion process based upon a high voltage arc (Kato et al 1982). In the laboratory, however, recently introduced inexpensive commercial monomode connectors have been used very successfully in demanding applications such as the fibre optic gyroscope (Burns et al 1984).

3.6. Polarisers (wavelength selective attenuators)
Optical fibre polarisers have been made by selectively increasing the attenuation of one of the independent polarisation mode's propagation in the monomode fibre. This has been achieved by: (i) selective etching of the cladding to allow a metal coating to be deposited sufficiently close to the guiding core that it distorts the electric field of one of the propagating modes thereby greatly increasing the loss (Eickhoff 1980); and (ii) coiling the fibre at specific radius so that the bending loss for one of the propagating modes is greatly increased. Extinction ratios of greater than 25 dB have been obtained (Varnham et al 1983b) by this approach.

3.7. Phase modulators
Fibre optic phase modulators were first introduced by Davies and Kingsley (1974) for use in multimode fibre optic sensors; here the fibre is tightly wound around a large piezoelectric (PZ) cylinder. Applying a voltage to the PZ cylinder produces a redistribution of the light amongst the guided modes enabling a signal related to the movement of the cylinder to be superimposed on the far-field pattern of the output light emanating from the fibre. A similar concept has been used to produce a phase modulator for monomode fibre (Jackson et al 1980b). The monomode fibre is similarly tightly wound around a PZ cylinder but now the voltage-induced dimensional change of the PZ is used to alter the optical pathlength of the fibre. This type of phase modulator has virtually no insertion loss as the light is always contained in the fibre, and if many turns are wrapped around the cylinder only moderate drive voltages are required to produce optical pathlength changes of greater than 100 μm.

3.8. Frequency shifter
Bragg cells are commonly used to change the absolute optical frequency of a light beam; however, these components are not compatible with monomode optical fibre as the light has to be reinserted into the fibre after passing through the Bragg cell. Recently there have been reports (Nosu et al 1983, Risk et al 1984) of fibre optic based frequency shifters using acousto-optic modal coupling in highly birefringent monomode fibres. At the present time the reported efficiencies are very low, about 5%, and further development to increase this efficiency is necessary before they can be used for signal processing applications in monomode fibre interferometers.

4. CLASSICAL INTERFEROMETERS AND THEIR FIBRE OPTIC EQUIVALENTS
4.1. Two-beam: Michelson and Mach-Zehnder
The conventional Mach-Zehnder interferometer is shown schematically in figure 1(c) and although this configuration is slightly more complex than that of the Michelson interferometer, as it requires an extra beam splitter and mirror, it does offer two significant advantages over the Michelson interferometer. (i) Optical feedback to the light source is at a minimum - this is very important when laser sources are used as any out of phase light fed back into the laser tends to induce random changes in its optical output frequency (Kanada and Nawata 1979). (ii) There are two

antiphase outputs from the interferometer which are equal in magnitude
only at the quadrature point and therefore can be used as differential
inputs for a servo to maintain the interferometer at maximum sensitivity;
this approach also has the advantage that it tends to reduce the effects
of intensity noise in the laser's output (Dandridge et al 1980a).

The all-monomode-fibre equivalents of both the Michelson and Mach-
Zehnder interferometers are shown in figures3(a) and (b) respectively. In
the case of the Michelson the conventional beam splitter is replaced with
a monomode fibre directional coupler and the mirrors are formed by
chemically coating the normally cleaved ends of the fibre (see §3.4). In
the all-fibre Mach-Zehnder the directional couplers replace both the conventional beam splitters and mirrors. Indeed, it is this component which
makes these interferometers unique, as it virtually eliminates all the
effects of random mechanical disturbances which normally induce misalign-
ment of the optical components in a conventional interferometer. Piezoelectric fibre phase modulators are readily incorporated into each configuration and can be used either to induce test signals or to form part
of a 'quadrature' maintaining servo, as indicated in figure 3(c).

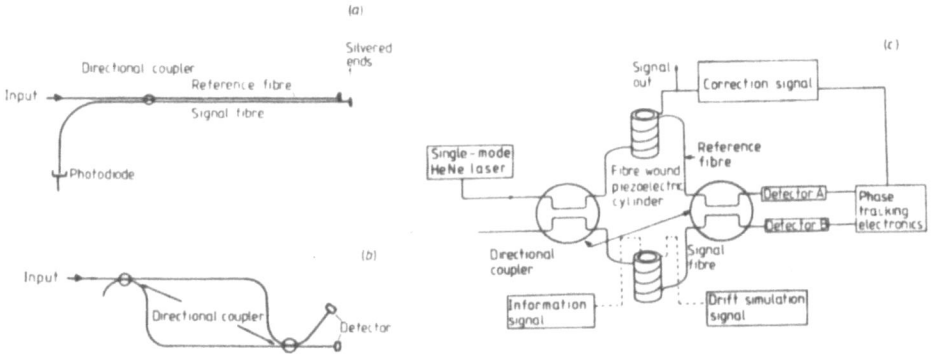

FIGURE 3. Fibre optic equivalents of the classical interferometers:
(a) Michelson, (b) Mach-Zehnder, (c) fibre optic Mach-Zehnder
test system incorporating piezoelectric phase modulators.

4.2. Multiple beam: Fabry-Perot
The classical Fabry-Perot interferometer (FPI) has the simplest configuration of all the interferometers as it is essentially an optical
cavity formed by two ideal mirrors adjusted to be perfectly parallel with
each other; again a requirement which is difficult to satisfy. The FPI
can be used either in a reflection or transmission mode and its transfer
function is described by the well known Airy function.

Fibre optic equivalents of the FPI are shown schematically in figures
4(a) and (b); in figure 4(a) the instrument is used in the conventional
transmission mode and mirrors have been attached to the fibre ends to
enhance the finesse (Petuchowski 1981). The alternative configuration,
figure 4(b) is essentially a single section of monomode fibre with
normally cleaved uncoated ends. The finesse will be low and the contrast
of this FPI in transmission will be very poor; however, in back reflection,
the transfer function and contrast are very similar to those of the

Michelson interferometer. The inherent simplicity of this fibre FPI, combined with the lack of ancillary optical components, tend to make the back reflection mode the most appropriate for sensing applications (Kersey et al 1983b).

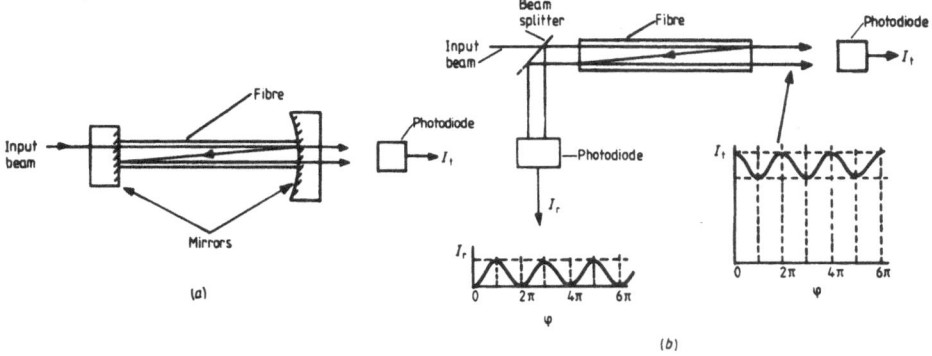

FIGURE 4. Fibre Fabry-Perot configurations: (a) mirrors attached to the fibre to enhance the reflectivity of the optical cavity, (b) FPI cavity formed by accurately cleaving the fibre ends; as indicated the FPI can be used either in reflection on transmission. The inserts show the transfer functions for reflection I_r, again ϕ, and transmission I_t against ϕ.

4.3. Differential interferometer (polarimetric interferometer)

Induced optical birefringence has been exploited for many years in the analysis of both static and dynamically induced strains in optical structures; for example, glass vacuum systems are usually examined between crossed polaroids to determine if the component has been properly annealed.

Differential interferometers are based upon the same fundamental principle and have been implemented using both monomode (Rashleigh 1980) and multimode (Jones and Spooncer 1983) optical fibres; both induced linear and circular birefringence have been exploited for sensor applications.

4.3.1. Induced linear birefringence interferometer. As was stated in §3, monomode optical fibres have two orthogonal polarisation modes which are degenerate in the limit that the fibre has rotational symmetry; however, if this rotational symmetry is broken, either by design in manufacture or through some extraneous perturbation such that the constituent atoms (or molecules) are subject to an anisotropic stress, the modes are no longer degenerate and have different propagation constants. The basic polarimetric sensor is shown in figure 5.

The polarised input beam is injected into the fibre such that it equally excites both eigenmodes of the fibre and can therefore be either linearly polarised with its azimuth at ±45° to the horizontal,

$$\frac{1}{\sqrt{2}} \begin{bmatrix} 1 \\ 1 \end{bmatrix} \quad \text{or} \quad \frac{1}{\sqrt{2}} \begin{bmatrix} 1 \\ -1 \end{bmatrix}$$

or left or right circularly polarised,

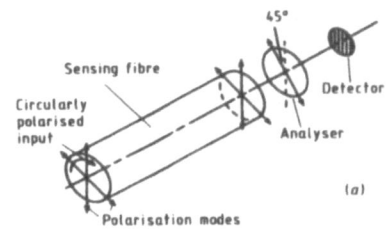

FIGURE 5. Polarimetric fibre
optic interferometer: (a) basic
configuration, (b) with
additional optical components
to enable limited range linearised
operation with $\delta(\phi_{ft} - \phi_{sl}) = (I_1 - I_2)/(I_1 + I_2)$.

SBC, Soleil-Babinet compensator;
WP, wollaston prism.

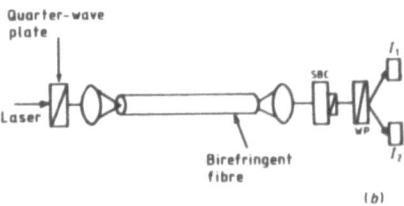

$$\frac{1}{\sqrt{2}} \begin{bmatrix} 1 \\ i \end{bmatrix} \quad \text{or} \quad \frac{1}{\sqrt{2}} \begin{bmatrix} 1 \\ -i \end{bmatrix}$$

(or elliptically polarised with azimuth at ±45°); here Jones vector
notation is used to represent the polarisation state of the light. When
these two orthogonal components exit the fibre, they can be made to
interfere by passing them through a polarisation analyser with its eigenaxes at ± 45° to the horizontal (see figure 5(a)).

The need to use additional optical components in the vicinity of the
fibre to recover the phase information can severely restrict the practical
use of polarimetric sensors; a method which overcomes this restriction
enabling the sensor to operate remotely is described in §8.3.2.

As the birefringent fibre may be considered to be an optical phase
plate it is convenient to use Jones matrix notation (Jones 1941) and
represent it by

$$\bar{F} = \begin{bmatrix} \exp(i\phi_{ft}) & 0 \\ 0 & \exp(i\phi_{sl}) \end{bmatrix}$$

where ϕ_{ft} and ϕ_{sl} are equal to $(2\pi/\lambda)n_f L$ and $(2\pi/\lambda)n_s L$ respectively; L is
the length of the sensor and the fast eigenaxis is vertical. The ampli-
tude of the transmitted light for the complete system when illuminated
with right circularly polarised light with Jones vector E = (1,-i) can
then be written

$$A_t = \bar{P}\bar{F}E$$

where

$$\bar{P} = \frac{1}{2} \begin{bmatrix} 1 & 1 \\ 1 & 1 \end{bmatrix}$$

and is the Jones matrix of the analyser.

The intensity is thus given by

$$I_t = A_t \cdot A_t^* \propto [1 + \sin(\phi_{ft} - \phi_{sl})]$$

where the asterisk indicates complex conjugation.

Hence the transfer function of the differential interferometer is identical to that of the conventional interferometer with its basic sensitivity to induced phase changes $\delta\phi_i$ reduced in the ratio $2\delta(\phi_{ft} - \phi_{sl})/\delta(\phi_{ft} + \phi_{sl})$ when compared with that of a 'standard' interferometric configuration.

4.3.2. <u>Induced circular birefringence interferometer</u>. It is well known that a linearly polarised beam of light can be represented as the summation of two coherent right- and left-handed circularly polarised light beams. For example,

$$\begin{bmatrix} 1 \\ 0 \end{bmatrix} = \frac{1}{2} \begin{bmatrix} 1 \\ -i \end{bmatrix} + \frac{1}{2} \begin{bmatrix} 1 \\ i \end{bmatrix} \qquad (4.1)$$

When this beam propagates through a monomode optical fibre with cylindrical symmetry, the two orthogonal states will propagate at the same rate and the azimuth of the input beam will remain constant.

If the fibre is subjected to a measurand field (such as a magnetic field) which interacts directly on the constituent atoms (or molecules) of the fibre so as to modify its susceptibility tensor producing complex off-diagonal elements in the tensor (Faraday effect) then the degeneracy between the orthogonal circular states will be removed. The propagation constants of the orthogonal circular states will no longer be equal and the relative phase between the two states will increase linearly. It is readily shown that the azimuth of the propagating beam will rotate through an angle equal to half the induced phase delay between the beams.

In the case of Faraday rotation, the change in the azimuth is

$$\theta = V \int_L H.dL$$

where V is the Verdet constant of the fibre, approximately 4 μrad A^{-1}, H is the magnetic intensity and L the length of the fibre (Smith 1978).

5. OPTICAL SOURCE

In the laboratory the ideal source to use for most (prototype) interferometric sensors is a commercial single-mode HeNe laser operating at a visible wavelength of 633 nm. The optical frequency of these lasers can be very precisely controlled to better than ±10 MHz and they exhibit very little short term frequency jitter and intensity noise. Unfortunately, these lasers are both expensive and insufficiently rugged for sensor applications, and a single-mode solid-state laser is a much more appropriate source as it is both extremely rugged and can be driven from a very low voltage power supply.

Another of the unique features of these lasers, which makes them particularly suited for sensors, is that it is possible to rapidly alter their absolute optical oscillation frequency by varying their injection current (Dandridge and Goldberg 1982). This property can be exploited to recover the input modulating phase signal in an interferometric sensor if the interferometer is operated with a finite optical path difference between its signal and reference arms. The phase change $\Delta\phi_L$ induced in the interferometer for an absolute optical frequency shift $\Delta\nu$ of the laser is given by

$$\Delta\phi_L = \frac{2\pi n\ell}{c} \Delta\nu \qquad (5.1)$$

where $n\ell$ is the optical path imbalance in the interferometer.

Unfortunately most single-mode solid-state lasers suffer from both intensity noise and random fluctuations in their emission frequency (jitter) (Dandridge et al 1981). The effect of this frequency jitter is to produce phase noise in the output of the interferometer which is directly proportional to $n\ell$ from (5.1).

6. SIGNAL PROCESSING

In §2 it was shown that the sensitivity of an interferometer to an induced optical phase change is not constant, due to the periodic nature of the transfer function. In most sensor applications this variable sensitivity is not acceptable as it causes signal fading, and inter-ferometric sensors are usually operated such that the output is linearly related to the induced optical phase change. Linearisation of the inter-ferometric sensor's transfer function has been achieved by a variety of different approaches which may be classified as: (i) active homodyne, (ii) passive homodyne, or (iii) heterodyne. Although these signal processing techniques have been primarily developed for the classical interferometer configurations, they can also be used to linearise the polarimetric sensors discussed in §§4.3 and 8.3.2.

6.1. Periodic phase modulation

When the optical path length of one arm of the interferometer is modu-lated by a signal Φ_s of frequency ω_s, the optical phase difference between the optical beams $\phi(t)$ may be written as (Jackson et al 1980a)

$$\phi(t) = \phi_d + \Phi_s$$

where $\Phi_s = \phi_s \sin\omega_s t$, ϕ_d is the 'static' phase difference and ϕ_s is the amplitude of the phase change induced in the interferometer by the perturb-ing signal. The output current of the optical detector has the following form from equation (1.3):

$$I_D = \varepsilon[1 + K \cos(\phi_d + \phi_s \sin\omega_s t)] \qquad (6.1)$$

where

$$\cos(\phi_d + \phi_s \sin\omega_s t)$$

$$= \left[\cos\phi_d\left(J_0(\phi_s) + 2 \sum_{n=1}^{\infty} J_{2n}(\phi_s)\cos(2n\omega_s t)\right)\right.$$

$$\left. - \sin\phi_d\left(2 \sum_{n=0}^{\infty} J_{2n+1}(\phi_s)\sin[(2n + 1)\omega_s t]\right)\right] \qquad (6.2)$$

and J_n is the Bessel function of order n.

The spectrum of I_D is essentially the spectrum of a phase modulated carrier with the carrier frequency equal to zero. Under normal operating conditions ϕ_d will fluctuate randomly in time due to temperature changes, causing the amplitude of the Bessel components to fluctuate in a similar manner.

6.2. Active homodyne

One of the most effective techniques available to recover the input signal modulating the relative phase in the interferometer ($\phi_s \sin\omega_s t$) is to use a piezoelectric based fibre optic phase modulator (Jackson et al 1980b) as shown in figure 3(c); in this technique the phase modulator is incorporated in the reference arm and forms part of a servo feedback loop to maintain the interferometer locked at its point of maximum sensitivity (quadrature point) where $\phi_d = (2n + 1)\pi/2$.

Although the amplitude of the peak phase deviation of ω_s could be obtained from I_D by band pass filtering at ω_s, this would not be appropriate if the spectrum of the input signal were complex.

If the servo system used to lock the interferometer at quadrature had an infinite gain bandwidth product, then I_D would remain constant, and contain no spectral information. However, as the complete servo-controll-ed interferometer shown in figure 3(c) is effectively a phase-locked loop with a $\pi/2$ phase difference between the signal and reference arms (when locked), the modulating signal $\phi_s \sin\omega_s t$ can be directly recovered from the feedback signal V_F, provided ω_s is less than the bandwidth of the servo. In practice, the servo will have a finite gain bandwidth product and the spectrum of I_D for a frequency ω_s above the servo cut-off frequency is

$$I_D \propto J_1(\phi_s)\sin\omega_s t + J_3(\phi_s)\sin 3\omega_s t + \ldots \qquad (6.3)$$

provided ϕ_s is relatively small.

A similar approach for active homodyne signal recovery has been demon-strated by Dandridge and Tveten (1982), where the quadrature condition is maintained in an unbalanced interferometer by active wavelength tuning of the solid-state laser used as the source. This approach has the advantage that no electrically active element is deployed in the interferometer. The tracking range of both of these active homodyne signal processing schemes is limited by the maximum (allowable) voltage generated in the feedback loop. Although the tracking range can be extended by increasing the path imbalance, when active wavelength tuning of the laser is used, the attendant increase in the phase noise in the interferometer (equation (5.1)) makes this unattractive.

6.2.1. Calibration and minimum detectable displacement. There are various methods available to determine the minimum amplitude of a periodic signal detectable from the phase change it induces in the interferometer. The most straightforward approach is to utilise equation (6.2) where the arguments of the Bessel function (ϕ_s) are seen to be directly related to the vibration amplitude. If the amplitude of the signal is relatively large, greater than 10^{-7} m, then the ratio of J_1/J_3 or J_2/J_4 will give ϕ_s directly. If the signal amplitude is smaller than 10^{-10} m then the only significant term in the spectrum of I_D is the first harmonic at ω_s. As the recorded amplitude of this component depends upon $\sin\phi_d$ it must be measured when $\sin\phi_d = 1$. If the interferometer is operated in a thermally stable environment then $\sin\phi_d$ can be set to unity manually; alternatively it can be locked at the quadrature point using a servo. When operated in this manner it is useful if the system can be independently calibrated from a unique feature in the spectrum - for example J_{1max} corresponds to $\phi_s = 1.84$ rad. In a Michelson configuration $\phi_s \equiv 2kx$, hence the displacement amplitude corresponding to $J_{1max} \cong 9 \times 10^{-8}$ m for λ equals 600 nm.

If we assume that the only noise in the system is due to the shot-noise current at the detector then we can make an estimate of the minimum detectable value of the induced phase. From equation (6.2) the recovered signal at ω_m is

$$S = \epsilon K(2J_1(\phi_s)).$$

For a fringe visibility equal to unity, $K = 1$ and $\epsilon \equiv I_D$

$$S = I_D(2J_1(\phi_s)) = 2I_DJ_1(2kx)$$

as $x \to 0$, $J_1(2kx) \to kx$ and $S = 2I_Dkx$.
The shot-noise current of the detector

$$= (2eI_D\Delta_f)^{\frac{1}{2}}$$

where e is the charge on the electron and Δ_f is the bandwidth of the system. The signal-to-noise ratio, S/N, is thus given by

$$S/N = \frac{2I_Dkx}{(2eI_D\Delta_f)^{\frac{1}{2}}}$$

for $S/N = 1$, with $\Delta_f = 1$ Hz, $I_D = 10^6$ A and $\lambda = 600$ nm, $x_{min} \cong 10^{-13}$ m (or $\phi_s < 10^{-6}$ rad).

In figure 6 the variation of the minimum detectable displacement (and phase) are plotted as a function of the frequency of an induced periodic displacement for the first linearised all-fibre interferometer (Jackson et al 1980a). The figure clearly shows that phaseshifts below 10^{-6} rad can be detected at frequencies above 100 Hz, and that detectability falls very rapidly at low frequencies. The apparent reduction in low frequency sensitivity is due to random environmental noise modulating the phase of the interferometer.

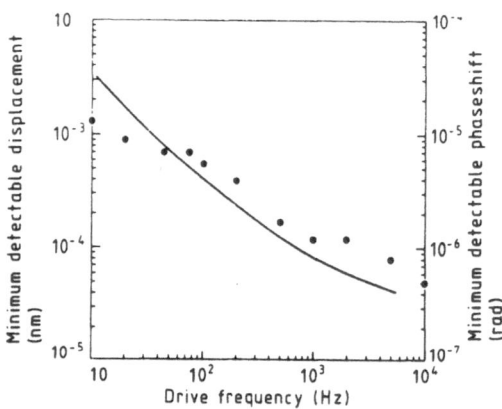

FIGURE 6. Variation of the minimum detectable displacement and equivalent optic phase shift as a function of frequency.

————, without compensator in drift free environment;

●, with compensator in laboratory environment, after Jackson et al (1980b).

6.3. Passive homodyne

Passive homodyne demodulation can be used if the interferometer is configured to produce two outputs which vary as quadrature functions of the
phase difference term $(\phi_d + \phi_s)$ (Sheem et al 1982), i.e. in addition to
the normal output, $I_c = \epsilon[1 + K \cos(\phi_d + \phi_s)]$, a second output $I_s =$
$\epsilon[1 + K \sin(\phi_d + \phi_s)]$ can be generated. The constant terms can be electronically nulled to produce the following terms:

$$I_c' = \epsilon K \cos(\phi_d + \phi_s) \quad \text{and} \quad I_s' = \epsilon K \sin(\phi_d + \phi_s) \quad (6.4)$$

before final signal processing.

These quadrature components have been produced by: (i) replacing the
four-port output directional coupler of the all-fibre Mach-Zehnder interferometer with a six-port directional coupler, usually referred to as a
3 × 3 coupler (Koo et al 1982); (ii) switching the absolute frequency
(Kersey et al 1983c) of the laser source at a high rate between two
optical frequencies ν_1 and $\nu 2$ such that the effective phase change induced
in the (unbalanced) interferometer is

$$(2\pi n\ell/c)(\nu_1 - \nu_2) = \tfrac{1}{2}\pi \; ;$$

(iii) generating orthogonal polarisation states at the output of the interferometer which are then (simultaneously) combined, (a) directly to produce I_c' and (b) indirectly via a quarter-wave plate to produce I_s' (Kersey
et al 1982a).

If $\phi_s \ll 1$ then the output signals I_c' and I_s' can be high pass filtered
to produce new signals

$$I_{c_f}' = \epsilon K \sin\phi_d J_1(\phi_s)\sin\omega_s t$$

and

$$I_{s_f}' = \epsilon K \cos\phi_d J_1(\phi_s)\sin\omega_s t \; .$$

These signals may be combined to produce an output signal S which does not
suffer from fading:

$$S = [(I_{c_f}')^2 + (I_{s_f}')^2]^{\tfrac{1}{2}} = \epsilon K J_1(\phi_s)\sin\omega_s t \; .$$

The requirement that $\phi_s \ll 1$ is too restrictive for most applications and
rather more complex non-linear signal processing has been used with the
3 × 3 coupler configuration (Dandridge et al 1982). As is shown in §6.6
the restriction placed on the magnitude of ϕ_s can be eliminated by an
alternative method of processing I_c' and I_s'.

6.4. Heterodyne and pseudo-heterodyne

Heterodyne methods have been used in conventional optical interferometers for more than 15 years (Eberhardt and Andrews 1970). This method
of signal recovery is very attractive as: (i) the phase detection
sensitivity is constant and does not depend on the value of ϕ_d (i.e. the
recovered signal does not suffer from fading), (ii) the phase tracking
range is effectively infinite, (iii) the dynamic range of the modulating
signal can be very large and (iv) the 'direction' of fringe motion is
readily determined. These features can be used very effectively when the
interferometric sensor is used for quasi-steady-state measurands.

Heterodyne signal processing requires that the light in one of the arms in the interferometer is frequency shifted with respect to the other. This may be accomplished with an acousto-optic modulator such as a Bragg cell. The output signal of the interferometer is then given by

$$I_D = \varepsilon K \cos(\omega_B t + \phi_d + \phi_s \sin\omega_s t)$$

where ω_B is the offset frequency. This signal represents a carrier at frequency ω_B which is phase modulated by $(\phi_d + \phi_s)$. Demodulation of this signal can be accomplished in a variety of ways as described below. Unfortunately, the conventional Bragg cell is not readily integrated into a fibre optic interferometer and a considerable amount of effort has been directed to new methods for producing a phase modulated carrier. These new methods may be classified as (i) synthetic heterodyne, (ii) pseudo-heterodyne and (iii) quadrature recombination heterodyne.

6.4.1. <u>Synthetic heterodyne</u>. If the fibre optic interferometer is phase modulated at a frequency ω_m, which is much higher than the maximum frequency of the signal of interest, then a phase modulated carrier can be produced if the output signal is conditioned by multiplying it by phase related harmonics of ω_m. Several schemes of varying complexity have been described in the literature based on this principle (Cole et al 1982, Green and Cable 1982, Kingsley 1983, Dandridge et al 1982). Recently an electronically simpler approach for generating the carrier has been demonstrated by Kersey et al (1984c) which is briefly outlined here. From equation (6.1)

$$I_D \propto 1 + K \cos(\phi(t) + \phi_E \sin\omega_m t)$$

where ϕ_E is the amplitude of the induced high frequency (ω_m) phase modulation, hence

$$I_D \propto 1 + K \cos\phi(t)[J_0(\phi_E) + 2(J_2(\phi_E)\cos2\omega_m t + ...)]$$

$$- K \sin\phi(t)[2(J_1(\phi_E)\sin\omega_m t + J_3(\phi_E)\sin3\omega_m t + ...)] \ .$$

If I_D is gated (effectively multiplied) by a square-wave $G(t)$ with an equal mark space ratio derived from the oscillator driving the phase modulator, then

$$G(t) = \frac{1}{2} + \frac{2}{\pi} (\cos\omega_m t - \frac{1}{3}\cos3\omega_m t + \frac{1}{5}\cos5\omega_m t + ...) \ .$$

If the signal resulting from this multiplication is band pass filtered at $2\omega_m$ it will be composed of two sets of terms:

(i) dependent on $\cos\phi(t)\cos2\omega t$

$$B(\phi_E) = KJ_2(\phi_E) \ ;$$

(ii) dependent on $\sin\phi(t)\sin2\omega t$:

$$A(\phi_E) = -\frac{8K}{\pi}\left[\frac{J_1(\phi_E)}{3} + \frac{J_3(\phi_E)}{5} - \frac{J_5(\phi_E)}{21} + ...\right] \ .$$

Hence the complete signal R generated by this technique at $2\omega_m$ is

$$R_{(2\omega_m)} = A(\phi_E)\sin\phi(t)\sin2\omega t + B(\phi_E)\cos\phi(t)\cos2\omega t$$

if

$$|A(\phi_E)| = |B(\phi_E)|$$

then

$$R_{(2\omega_m)} = A(\phi_E)\cos(2\omega_m t \mp \phi(t))$$

$$= A(\phi_E)\cos[2\omega_m t \mp (\phi_d + \phi_s \sin\omega_s t)]$$

i.e. a phase modulated carrier.

6.4.2. Pseudo-heterodyne. As shown in §5 it is possible to change the relative optical phase difference between the arms of an unbalanced interferometer by varying the absolute frequency of the injected laser light. If this emission frequency increases linearly in time, then a linearly moving fringe pattern will be created at the optical output of the interferometer, where from equation (5.1)

$$\frac{d\phi_L}{dt} = \frac{2\pi n\ell}{c} \frac{d\nu}{dt} .$$

The spectrum of the photodetector current is complex and contains components at the fundamental harmonics of the ramp repetition frequency f_r (Voges et al 1982). In the limit when the interferometer output is only driven over one complete fringe per cycle, most of the optical power will be concentrated in the first harmonic. Band pass filtering at f_r produces a distortion-free carrier at f_r (Jackson et al 1982), with a phase modulation identical to that of the interferometer, i.e. the output is identical to that produced by conventional heterodyne modulation. This pseudo-heterodyne technique has features in common with FMCW radar and is sometimes termed Serrodyne modulation (Giles et al 1983).

6.4.3. Quadrature recombination. As was discussed in §6.2, it is possible with an appropriate configuration of the interferometer to produce quadrature outputs, equation (6.4). If these signals are translated in frequency by multiplying them with quadrature components of a high frequency oscillator at ω_c, then signals of the form

$$\sin\omega_c t\cos(\phi_d + \phi_s) \qquad \text{and} \qquad \cos\omega_c t\sin(\phi_d + \phi_s)$$

are produced which can be linearly combined to produce a phase modulated carrier, $A\sin(\omega_c t + \phi_d + \phi_s)$, identical to the two previous examples (Kersey et al 1984a, Jackson et al 1984b, Kim and Shaw 1984).

6.4.4. Demodulation of the phase modulated carrier. Demodulation of a phase modulated carrier may be accomplished by several methods which include the use of FM discriminators, phase-locked loop (PLL), or carrier phase tracking. Conventional FM discriminators and PLL are not ideally suited for optical sensor applications because the dynamic range of the signal is generally large and its modulation index is small ($\sim 10^{-6} - 1$). Several applications of the interferometric sensor require that the

information related to the total phase excursion be retained; again a
conventional PLL based upon analogue techniques cannot be used for these
applications, and digital phase tracking circuits have been designed
specifically for this purpose (Nokes et al 1978, Jackson 1981).

7. SENSOR CLASSIFICATION

The basis of the fibre optic interferometric sensor (FOIS) is the
measurement of a physical parameter through the phase modulation it induces in the guided optical beam in a sensing element; in many cases the
sensing element will be the optical fibre itself. FOIS can be conveniently classified into two main categories.
(i) Internal, which may be subdivided into:
    (a) direct: the measurand modulates the optical path directly;
examples are temperature, strain and pressure (Kingsley and Davies 1976);
    (b) indirect: the measurand modulates the optical path length via
an auxiliary sensing element; the fibre is bonded to some form of conventional sensing element or coated with a material to locally modify the
physical properties of the fibre; for example, magnetic fields can be
detected via magnetostriction (Dandridge 1980b).
(ii) External, the optical beams are guided by the fibre to a remote
measurement area where the beams are then projected into the measurement
volume either directly or via ancillary optics; here the phase modulation
is induced external to the fibres. External sensors have been developed
for holography (Jones et al 1984), laser Doppler velocimetry (Jackson et
al 1984a), and non-contact vibration analysis (Lewin et al 1985, Nokes et
al 1978).

7.1. Sensitivity: direct interactions

Following the treatments of Kingsley and Davies (1976) and Hocker
(1979a) we may make estimates of the phase change induced in the sensing
fibre for a range of measurands. As stated in §3 the optical phase change
ϕ in the light propagating through a fibre of length L is equal to βL; if
the fibre is subject to an external stimulus then ϕ changes by an amount

$$\Delta\phi = \beta\Delta L + L(\Delta\beta_n + \Delta\beta_a)$$

where $\beta\Delta L$ corresponds to a physical change in the fibre length and
$L(\Delta\beta_n + \Delta\beta_a)$ corresponds to a change in its propagation constants.

If we assume that the strain optic coefficients (Pockells coefficients)
of the fibre are equal to those of pure fused silica, then the induced
phase change sensitivities for strain, force and pressure can be directly
calculated. The sensitivity of thermally induced phase changes may be
similarly obtained from the known thermal expansion and temperature
dependent refractive index of quartz. Typical phase sensitivities for
these measurands are summarised in table 1.

7.2. Indirect interactions

If a measurand-induced dimensional change in a conventional sensor can
be coupled into the optical waveguide so as to produce a change in its
optical pathlength, then this can form the basis of an indirect sensor.
Clearly the coupling between the sensing element and the fibre must be
very efficient if the full sensitivity of the primary sensor is to be
exploited. To date, the most successful example of an indirect sensor is
that of the fibre optic magnetometer (Giallorenzi et al 1982) based upon
the magnetostrictive effect. Typical sensitivities are given in table 1

TABLE 1. Induced phase sensitivities for typical 100 μm diameter mono-mode optical fibres for a range of measurands. Parameters assumed for the fibre are: refractive index = 1.5, Pockel's coefficients P_{11} = 0.12 and P_{12} = 0.27, Poisson's ratio = 0.17, and Young's modulus = 70 GPa. The wavelength of the source is taken as 850 nm.

Internal: direct	
Strain	$\Delta\varphi/\varepsilon L = 10^7 \text{ rad m}^{-1}$
Force	$\Delta\varphi/FL = 2 \times 10^4 \text{ rad N}^{-1} \text{ m}^{-1}$
Pressure	$\Delta\varphi/L\Delta P = -5 \times 10^{-5} \text{ rad Pa}^{-1} \text{ m}^{-1}$
Temperature	$\Delta\varphi/L\Delta T = 100 \text{ rad K}^{-1} \text{ m}^{-1}$
Internal: indirect	
Magnetic field	Magnetostrictive effect $\Delta\varphi/L\Delta H$
	nickel $\sim 1.3 \times 10^{-2}$
	metallic glass ~ 1 $\Big\}$ rad A

(where the coupling coefficient is assumed to be unity).

7.3. Differential interferometer sensitivities

The action of the measurand field, F, on the sensing fibre produces a relative phase change, $\Delta\phi$, which may be detected by a change in the polarisation state of the recombined emergent beams in the case of induced linear birefringence or a relative change in the azimuth of a linearly polarised beam in the case of induced circular birefringence. For the case of induced linear birefringence we may write

$$\frac{1}{L}\frac{d(\Delta\phi)}{dF} = \frac{k}{L}(\Delta n_s - \Delta n_f)\frac{dL}{dF} + k\frac{d}{dF}(\Delta n_s - \Delta n_f)$$

where again the first term corresponds to a physical change in length of the fibre and the second term represents the change in the propagation constants of the fibre's eigenmodes. The sensitivity of the differential sensor is strongly dependent on the type of fibre used, as shown in table 2.

TABLE 2. Induced phase sensitivities for typical highly birefringent fibres used in the 'polarimetric' mode.

	Internally stressed $\Delta n = 2 \times 10^{-4}$ $L_p = 3.2$ mm	Elliptical core $\Delta n = 8 \times 10^{-4}$ $L_p = 0.75$ mm
Temperature	2.4 rad K^{-1} m^{-1}	0.0024 rad K^{-1} m^{-1}
Pressure	1.61×10^{-8} rad Pa^{-1} m^{-1}	3.9×10^{-8} rad Pa^{-1} m^{-1}
Strain	1.43×10^6 rad m^{-1}	—

8. APPLICATIONS OF FIBRE OPTIC INTERFEROMETRIC SENSORS

8.1.   **Pressure sensors.** Currently little attention has been paid to
the development of static pressure sensors based upon the FOI. A possible
approach would be to use the interferometer to measure the displacement of
a pressure responsive element – displacement sensors are discussed in §8.3.

8.1.1. **Acoustic sensors.** The first demonstrations of an optical interferometer incorporating monomode optical fibres as the sensing and reference arms to produce a sensitive acoustic detector were reported independently by Bucaro et al (1977) and Cole et al (1977). These first systems
were very much laboratory prototypes with relatively low pressure detectability of 120 dB re: 1 μPa at about 400 Hz. For the fibre optic hydrophone to be competitive with conventional (piezoelectric-based) hydrophones its pressure detectability must be better than 20 dB re: 1 μPa (sea
state zero at 10 kHz). If we assume that an FOI can be operated outside
the laboratory with a minimum phase detectability of about 10^{-6} optical
radians then approximately 1 km of monomode fibre is required in the
sensor arm to achieve the desired pressure sensitivity. Such long lengths
of fibre would preclude the construction of compact inexpensive fibre optic
hydrophones; however, the pressure sensitivity of the fibre can be increased by one to two orders of magnitude by embedding it in a material of
lower elastic modulus such as a silicone-rubber casting (Hocket 1979b),
hence only about 10 m of coated sensing fibre is necessary for the construction of a high performance hydrophone. The data presented in
figure 7 shows the dramatic improvements made in the performance of laboratory prototypes from the initial hybrid fibre interferometer
implementation to one of the first all-fibre configurations (Cole and
Bucaro 1980). The first practical hydrophone capable of undersea
operation was demonstrated by the NRL group (Giallorenzi et al 1982). This
device is essentially a ruggedised all-fibre Mach-Zehnder interferometer
using a single-mode solid-state laser source with active homodyne signal
recovery.

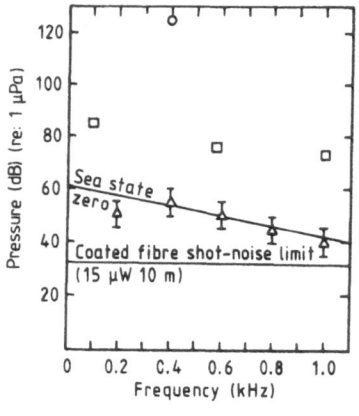

FIGURE 7.   Improvements in the
demonstrated pressure detectability of
laboratory fibre optic based hydrophones.
O, □, hybrid systems; Δ, all monomode
FOI, after Cole and Bucaro (1980).

To obtain the required optical phase sensitivity of approximately 10^{-6} rad
the contribution from the laser's phase noise (Dandridge et al 1981) to
the output signal (equation (5.1)) had to be minimised by very carefully

matching the lengths of the fibres in the arms of the interferometer.

The first all-fibre hydrophone was designed as a 'point' acoustic wave sensor, although the 'distributed' nature of the sensing fibre allows a large variety of geometrical configurations for a hydrophone which are not possible with competing technologies, for example if the fibre is wound in a helical fashion on a long rod then both the magnitude and direction of the sound wave can be detected. Further possibilities such as apodisation of the hydrophone's effective acoustic aperture can be achieved by varying the distribution of the fibre windings. Henning et al (1983) have demonstrated side lobe reductions of greater than 30 dB by adapting this technique. To ensure that the hydrophone can be operated at maximum sensitivity it is important to acoustically de-sensitise the reference arm of the interferometer. Lagakos and Bucaro (1981) have shown that if the fibre is coated with a thin metal film of appropriate thickness its pressure sensitivity is greatly reduced. A further consideration in the design of the fibre optic hydrophone (as in the case of all hydrophones) is that the sensitivity will become both frequency- and direction-dependent as the wavelength of the sound begins to approach the dimensions of the hydrophone; further frequency-dependent effects can also be expected as the sound wavelength approaches the dimensions of the fibre coil itself (Jarzynski et al 1981).

Other forms of fibre optic hydrophone have recently been demonstrated and are discussed by Dakin in this volume.

8.2. Magnetic field sensors

8.2.1. <u>Fibre optic magnetometers</u>. Yariv and Winsor (1980) proposed that high sensitivity magnetometers could be realised by 'bonding' or 'coating' the sensing fibre of an FOI with a magnetostrictive element. The basic concept is that a magnetically induced change in the dimensions of the magnetostrictive element can be used to directly strain the optical fibre causing the relative optical phase between the reference and sensing fibres to be magnetic field dependent. The first practical fibre optic magnetometer was reported by Dandridge et al (1980b). In these first experiments the magnetostrictive material used was either a nickel strip bonded directly to the fibre or a film of nickel deposited on the fibre by direct vacuum deposition. The detection sensitivity of these devices show a very strong dependence on the frequency of the applied magnetic field, as shown in figure 8, varying from $\sim 8 \times 10^{-4}$ A m^{-1} at frequencies below 100 Hz to better than 6×10^{-6} A m^{-1} at frequencies above 10^3 Hz for 1 m of coated sensing fibre. The apparent decrease in sensitivity as the

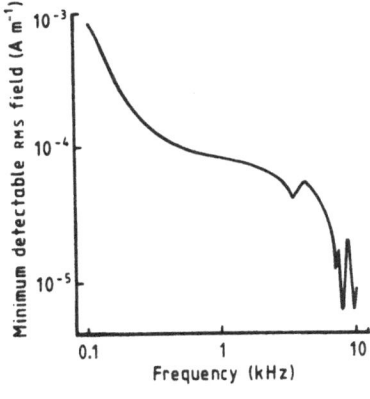

FIGURE 8. Variation of the minimum detectable magnetic field as a function of frequency, after Dandridge et al (1980b).

frequency of the applied field is reduced is because the optical phase changes induced by the magnetic signals are smaller in magnitude than those induced in the interferometer by low frequency random thermal or mechanical disturbances. Subsequent experiments have shown that certain magnetic glasses have greatly enhanced magnetostrictive coefficients (Giallorenzi et al 1982, Koo and Sigel 1982), and sensitivities as high as 4×10^{-7} A m^{-1} with 1 m of sensing fibre have been reported for periodic fields at frequencies about 1 kHz. In all cases the response of the magnetostrictive elements show a strong dependence on the value of the local steady-state magnetic field, H_{dc}, and a static magnetic bias field is usually required in order that the sensitivity can be optimised (figure 9(a)).

8.2.2. Enhanced low frequency sensitivity. Several important applications for magnetometers, geophysical surveying for example, require magnetic field sensitivities of 8×10^{-5} A m^{-1} at frequencies around 1 Hz. Such low frequency sensitivities cannot be obtained if the magnetometer is operated conventionally by attempting to recover the magnetically induced signal directly as it will be masked by the environmentally induced low frequency noise signals. Significant improvements in the sensitivity of the magnetometer in the low frequency range have been obtained by utilising the dependence of the high frequency magnetostrictive responsitivity on the static bias field (Kersey et al 1983a, Koo et al 1983).

In this approach (Kersey et al 1983a) the fibre optic magnetometer is operated with an active servo to maintain the interferometer at quadrature, i.e. its sensitivity to induced optical phase shifts is constant. The DC magnetostrictive response, i.e. the interferometer output (ϕ) as a function of local magnetising field H_{dc}, will then be as shown in figure 9(a). The dynamic magnetostrictive response is determined by applying a calibration field H_{ac} at a high frequency f_c to the magnetometer via a second coil and measuring the change in the interferometer output as the value of H_{dc} is varied. The resulting dynamic transfer function ($\Delta\phi$ as a function of H_{dc} at the calibration frequency), essentially the derivative of the curve in figure 9(a), is shown in figure 9(b). The magnetometer can now be operated in either an open-loop

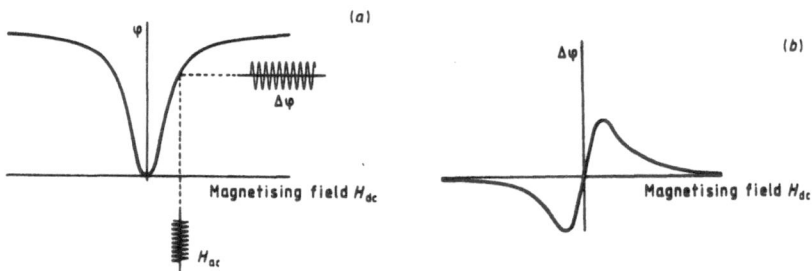

FIGURE 9. (a) Variation of the DC magnetostrictive response ϕ as a function of the local magnetising field H_{dc} for a typical magnetostrictive element bonded to the sensing arm of an FOI. (b) Variation of the AC magnetostrictive response $\Delta\phi$ when the magnetometer is subjected to a constant frequency test field H_{ac} (at constant amplitude) as the local magnetic field H_{dc} is varied.

or closed-loop mode as shown in figure 10 (Kersey et al 1984b). In the
open-loop mode the value of the static field H_{dc} is adjusted until the
operating point of the magnetometer corresponds to the maximum in the
dynamic response curve. The output signal at the photodetector will be a
periodic signal composed primarily of odd harmonics of f_c as described by
equation (6.3), provided that both f_c is much higher than the 'cut-off'
frequency of the servo and the amplitude of the modulating magnetic field
is small. Low frequency magnetic field fluctuations will contribute to
the total bias field causing the operating point of the magnetometer to
vary with the same spectral density; hence the Fourier components of the
output signal will be subject to amplitude modulation, and the spectrum of
the perturbing magnetic field can be recovered by standard AM techniques.
Using metallic glass sensing elements (Vitro-vac 40/40) sensitivities of
greater than 8×10^{-5} A m^{-1} with 1 m of sensing fibre have been reported
at frequencies below 10 Hz. Although this approach enables very weak low
frequency magnetic fields to be determined, there are problems associated
with (i) the stability of the calibration signal at f_c, (ii) the linear
range of the technique and (iii) variation in sensitivity caused by
hysteresis effects. These problems are essentially eliminated when the
magnetometer is operated in a closed-loop mode (Kersey et al 1984b, 1985).
The modifications required to enable the magnetometer to operate in the
closed-loop mode are indicated in figure 10. As before, a high frequency

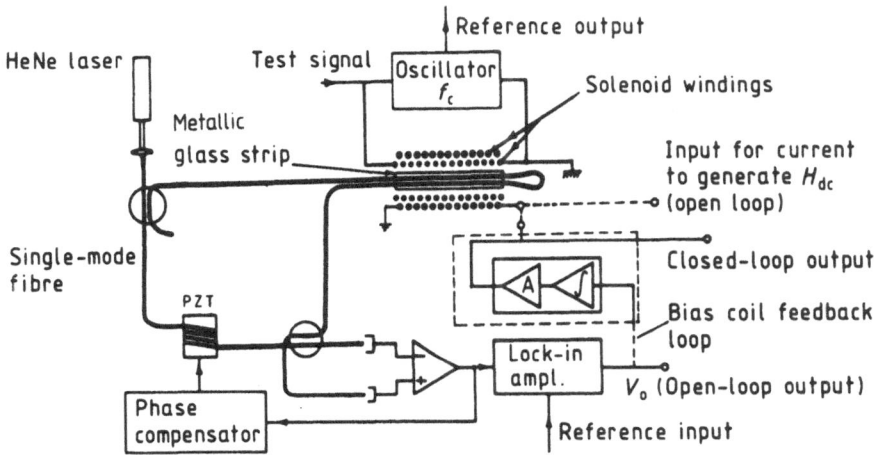

FIGURE 10. Fibre optic magnetometers; as discussed in the text it can be
operated in either an 'open-loop' or 'closed-loop' mode, after Kersey et
al (1984b).

calibration field is applied to the fibre optic magnetometer, again
maintained at quadrature by an active servo with a relatively low 'cut-
off' frequency, and as in the open-loop case the output signal from the
difference amplifier will primarily be composed of odd harmonics of the
calibration signal. The amplitude of the first harmonic of this signal
is recovered using a 'lock-in' amplifier referenced to the calibration
frequency. The 'lock-in' output is then applied after integration to the

second coil generating a magnetic field which now acts as the bias field H_{dc} for the magnetometer. If this second servo-loop is operated such that the phase of the generated field opposes that of the local magnetic field then its effect will be to reduce the magnetic field at the magneto-strictive element to zero; the error current supplied to the bias field coil is directly proportional to the variation in local magnetic field. Operated in this closed-loop mode the magnetostrictive element is always maintained at a constant magnetic field level, unlike the open-loop mode when the element is continuously driven over a hysteresis cycle.

Clearly the linear range of operation of the 'open-loop' magnetometer is very limited because of the nature of the DC magnetostrictive response, however in the closed-loop mode a linear dynamic range of greater than 10^5 has been demonstrated with a detection sensitivity of 8×10^{-5} A m^{-1} at frequencies below 2 Hz (Kersey et al 1985).

8.2.3. <u>Magnet gradiometer</u>. Some applications for magnetic sensing require the detection of small local magnetic perturbations in the presence of strong spatially uniform background magnetic fields. In this case it is necessary to measure the gradient rather than the magnitude of the field. A magnetic gradiometer based upon a Mach-Zehnder interfero-meter with magnetostrictive elements incorporated into both fibre arms has been demonstrated by Koo and Sigel (1983). Again a high frequency calibration field f_c is applied to the magnetometer to allow low frequency open-loop operation of the magnetometer (Koo et al 1983). In this case the relative phase and amplitudes of the calibration signals are adjusted such that when the magnetometer is in a constant magnetic field the inter-ferometer output at f_c is zero. When the system is subject to a field gradient, the output will contain a component at f_c which is directly proportional to the gradient.

8.3. Temperature and displacement

As both the propagation constants and length of an optical fibre vary when it is subject to either a changing temperature or force, it is possible to use a FOI as a sensor for quasi-steady-state measurands such as strain, displacement and temperature. Inspection of tables 1 and 2 shows that high resolution devices can be realised if the FOI is imple-mented either in a conventional or differential configuration. Surprisingly few practical systems for these measurands have been described in the literature, possibly because the initialisation problem has yet to be satisfactorily solved (see §8.3.3.).

8.3.1. <u>Conventional interferometric configurations</u>. The initial experiments of Hocker (1979a) demonstrated that both temperature and strain measurement could be made with optical fibres. The first high resolution monomode fibre optic temperature sensor incorporating some form of signal processing was described by Musha et al (1982) in which a Mach-Zehnder hybrid interferometer was used with a 10 m coiled fibre arm. This system demonstrated a sensitivity of approximately 1200 rad K^{-1}, however an expensive single-mode gas laser with very good frequency stability was required due to the large path imbalance in the interferometer.

Recently a fibre optic thermometer based on an unbalanced all-fibre Michelson configuration, using a solid-state laser for the source, has been reported (Corke et al 1983). This system utilises pseudo-heterodyne signal processing (§6.4) with a digital phase tracker to enable the thermometer to operate over a large dynamic range.

The Michelson configuration is particularly attractive for high resolution thermal measurements because of the ease with which the signal and reference arms may be placed in good mechanical and thermal contact throughout their common length. Excellent common mode rejection is thus achieved so that the sensitive region is restricted to the small additional length of the sensing arm. Differencial temperature measurements are also possible if the signal and reference arms are separated. The demonstrated temperature range of this thermometer was 0 - 250°C which can be extended since the working range is set by the mechanical and thermal properties of the fibre. The resolution of the thermometer with a sensing element of about 2 cm was better than 10^{-3} K. The signal processing developed for this system had an equivalent resolution of about 10^{-3} rad with an effectively infinite tracking range. The relative change in optical pathlength caused by the 250 K temperature change is equivalent to a change in the OPD of about 60 μm. Consequently this particular FOI can also be used as a *large dynamic range displacement* sensor provided the system can be made thermally insensitive.

8.3.2. <u>Polarimetric configurations</u>. The polarimetric sensor has been used to measure both temperature (Eickhoff 1981) and displacement (Rashleigh 1983b) with significantly reduced sensitivities when compared with the conventional interferometric configurations of the FOI (see tables 1 and 2). Most of the polarimetric sensors described to date may only be considered to be laboratory prototypes because (i) ancillary optical components are required close to the sensing element to recover the induced phase shift (as indicated in figure 5(b)), (ii) the linear dynamic range is very small and (iii) the total length of the fibre acts as a sensor, including the sections used for the input and output leads, which generally precludes remote operation of this device.

These problems have been solved by taking advantage of the unique properties of highly birefringent monomode fibre enabling both remote operation of the polarimetric sensor (Varnham et al 1983b, Corke et al 1984) coupled with large dynamic range pseudo-heterodyne signal recovery (Kersey et al 1984a).

The basic optical arrangement used to enable remote operation of the polarimetric sensor is shown in figure 11(a). The sensing element is a short length of birefringent fibre fusion-spliced to a similar long length of fibre such that the relative orientation of their eigenaxes is 45°. The long length of fibre serves as both the input and output lead for the system; the far end of the sensing fibre is silvered.

Light from a linearly polarised source is launched into the input fibre such that it only excites one of its eigenmodes. After propagating through this fibre the light on arriving at the fusion splice will equally populate both eigenmodes of the sensing fibre. After traversing the sensing fibre these orthogonally polarised beams are reflected back towards the splice, where they recombine to produce polarised light of arbitrary ellipticity determined by the birefringence in the sensing fibre. Any change in this birefringence is monitored by detecting the light propagating in the unused eigenmode of the input lead as indicated in figure 11(b). The electric field vector E' of the light measured at the output is $\alpha\{\bar{P}_R\bar{L}(\bar{R}^-\bar{S}\bar{R}^+)\bar{L}\bar{P}_T E\}$ (Jones and Jackson 1985) where E, the electric field vector for the input light, is $(1,0)$, $\bar{P}_T$ and $\bar{P}_R$ are the Jones matrices for transmission and reflection of the polarising beam splitter, $\bar{S}$ and $\bar{L}$ are the Jones matrices for the sensing element and input lead and $\bar{R}^+$ and $\bar{R}^-$ are the matrices for rotation of +45° and −45°

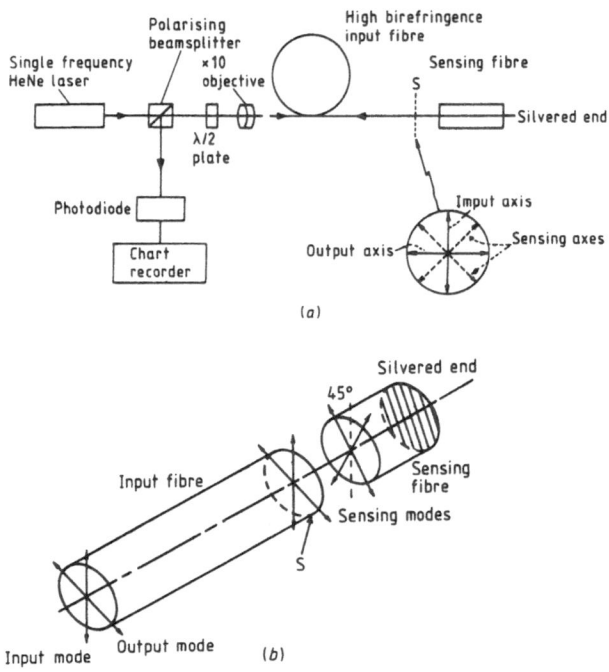

FIGURE 11. Remote polarimetric sensor: (a) details of sensing region
which is fusion-spliced at 'S' to the input lead, (b) optical arrangement
to enable access to the remote sensing element, after Corke et al (1984).

respectively. The output irradiance I is $\alpha(1 - \cos(\phi_{ft} - \phi_{sl}))$ and is
similar to that of the conventional Michelson interferometer (equation
(1.3)). Hence this sensor will suffer from the same problems of variable
sensitivity, signal fading, etc. as the conventional two-beam interferometers. A heterodyne-type signal processing technique has been
developed to overcome this problem (Kersey et al 1984a) which utilises
the wavelength dependence of the linear birefringence. It can readily be
shown that the relative phase change $\Delta\phi_p$ in the output of a polarimetric
sensor (whether operated locally or remotely) caused by a change in the
frequency of the source $\Delta\nu_L$ is

$$\Delta\phi_p = \frac{2\pi\Delta\ell\Delta\nu_L}{c} \tag{8.1}$$

essentially the same relationship as described by equation (5.1), except
$\Delta\ell = L\lambda/L_p$, where L is the physical length of the fibre and L_p its beat
length. If the optical frequency of the light source is switched such
that $\Delta\phi_p \equiv \pi/2$ rad, then signals of the form $(1 + \cos\Delta\phi_p)$ and $(1 + \sin\Delta\phi_p)$
can be recovered at the output of the sensor. These signals can then be
recombined, as described in §6.4.3, to produce a carrier phase modulated
by changes in the birefringence of the sensing element. As in the case of
the all-fibre Michelson, this system has been used for both *temperature*
and *displacement* sensing.

It is important to note that the displacement sensitivity is very much lower for this sensor, for example it is necessary to extend a 10 cm length of birefringent fibre by about 50 μm to change the modal phase delay by 2π. The phase resolution obtained with the signal processing system described above was approximately 10^{-5} rad (of the modal phase delay) equivalent to an extension of the fibre of approximately 5×10^{-1}nm. This measurement range is well matched to many conventional transducers; it is therefore possible that the polarimetric sensor could be used as a direct replacement for several conventional sensing elements with the added advantage of an enhanced range to resolution of greater than 10^5.

8.3.3. Initialisation. Clearly the performance of the FOI temperature and displacement sensors makes them attractive alternatives to many conventional primary transduction elements. However, there still remains the one major problem associated with all interferometric measurement systems, in that the absolute relative phase difference is lost when the system is switched off. Until this data loss problem is solved, inter-ferometric fibre optic sensors are likely to be used only in specialised sensor applications.

There are several methods available to solve this problem which have their origins in conventional optical interferometry. One possible approach is to use 'white light' interferometry, with an additional local reference interferometer (Al-Chalabi et al 1983, Bosselman and Ulrich 1984). A possible all-fibre implementation for such a system is shown in figure 12; light from the broad-band source (i.e. very short temporal coherence) is transferred to the remote sensor where it is amplitude-divided into wave packets of extremely short duration. These two separate wave packets then propagate in the sensor interferometer and arrive back at the beam splitter with a time difference $\Delta t_B = (\ell_1 - \ell_2)/c$. These wave packets now separated in time are coupled into the output fibre and transferred to the variable path reference interferometer. When the relative path difference between the arms of this interferometer is scanned, interference effects will only be observed for two conditions, (i) when the path difference is around zero - here we observe the auto-correlation of the two individual wave packets, and (ii) when the path difference is identically equal to the path imbalance of the sensing interferometer, i.e. $|\ell_1 - \ell_2|$. An accuracy of better than 0.2 μm has been reported using this approach; this is sufficient to enable the current integral fringe number of the sensing interferometer to be deter-mined. If the sensor incorporates both a 'white light' and laser source, the system could be initialised using the 'white light approach' to determine the current fringe number of the sensor and then switched to a second mode to enable rapid high precision measurements to be made over a large tracking range.

As stated in §2.2, it is possible to operate an interferometer with a unique output provided the change in the path imbalance does not exceed one half cycle of the transfer function. This restriction applies to both the interferometric and polarimetric configurations of the FOI. Recently it has been demonstrated (Corke et al 1985) that it is possible to greatly extent the 'unique measurement range' of an FOI by combining the polari-metric and interferometric configurations in a single device. Highly birefringent monomode fibre is used for the sensor and two independent fibre Fabry-Perot interferometers (FFP) are implemented in parallel along a common length of fibre. The resolution of the sensor is equal to either one of the two interferometers. The outputs from the two FFP interfero-

28

meters are combined creating a dependent differential interferometer which
defines the unique operating range of the complete sensor.

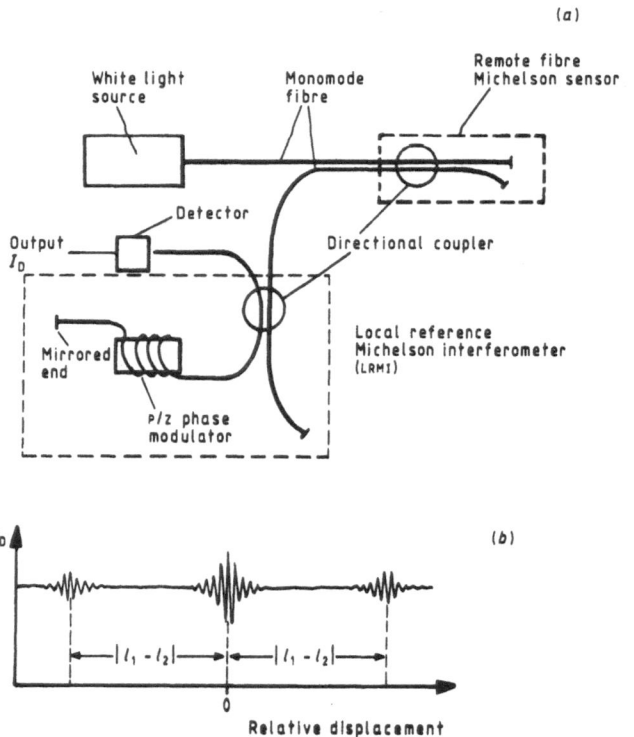

FIGURE 12. (a) Possible all-fibre implementation of the 'white light'
initialisation scheme; accuracy depends on the long term stability of the
PZ phase modulator. (b) Output signals obtained by Bosselman and Ulrich
(1984) for a hybrid system based upon this principle, where $|\ell_1 - \ell_2|$ is
the path imbalance in the remote sensor.

ACKNOWLEDGEMENTS
    The material contained in this chapter is a shortened version of a
paper entitled "Monomode optical fibre interferometers for precision
measurement" published in J. Phys. E. Vol. 18, 981, 1985. The Institute
of Physics (U.K.) retain the copyright of this material and the author is
grateful for their permission to republish.

REFERENCES
Al-Chalabi, S.A., Culshaw, B. and Davies, D.E.N., 1983, Partially
coherent sources in interferometric sensors, Proc. 1st Int. Conf. on
Optical Fibre Sensors (London: IEE), pp.132-5.

Bergh, R.A., Kobler, G. and Shaw, H.J., 1980, Single-mode fibre optic
directional coupler, Electron. Lett. 16, 260-1.

Birch, R.D., Payne, D.N. and Varnham, M.P., 1982, Fabrication of polarisation maintaining fibres using gas phase etching, Electron. Lett. 18, 1056-8.

Bosselman, T. and Ulrich, R., 1984, High accuracy position-sensing with fibre coupled white light interferometers, Proc. 2nd Int. Conf. on Optical Fibre Sensors, Stuttgart (Berlin: VDE), pp.361-4.

Bucaro, J.A., Dardy, H.D. and Carome, E.F., 1977, Fibre optic hydrophone, J. Acoust. Soc. Am. 62, 1302-5.

Burns, W.K., Moeller, R.P., Villarruel, C.A. and Abebe, M., 1984, All fibre gyroscope with polarisation-holding fibre, Opt. Lett. 9, 520-72.

Cole, J.H. and Bucaro, J.A., 1980, Measured noise levels for a laboratory fibre interferometric hydrophone, J. Acoust. Soc. Am. 67, 2108-10.

Cole, J.H., Danver, B.A. and Bucaro, J.A., 1982, Synthetic heterodyne interferometric demodulation, IEEE J. Quantum Electron. QE-18, 694-7.

Cole, J.H., Johnson, R.L. and Bhuta, P.G., 1977, Fibre optic detection of sound, J. Acoust. Soc. Am. 62, 1136-8.

Corke, M., Jones, J.D.C., Kersey, A.D. and Jackson, D.A., 1985, Dual Fabry-Perot interferometer implemented in parallel on a single monomode optical fibre, Technical Digest, 3rd Int. Conf. on Optical Fibre Sensors, San Diego (New York: OSA), p.128.

Corke, M., Kersey, A.D., Jackson, D.A. and Jones, J.D.C., 1983, All fibre 'Michelson' thermometer, Electron. Lett. 19, 471-3.

Corke, M., Kersey, A.D., Liu, K. and Jackson, D.A., 1984, Remote temperature sensing using polarisation preserving fibre, Electron. Lett. 20, 67-9.

Dandridge, A. and Goldberg, L., 1982, Current induced frequency modulation in diode laser, Electron. Lett. 18, 302.

Dandridge, A. and Tveten, A.B., 1982, Phase compensation in interferometric fibre optic sensors, Opt. Lett. 7, 279.

Dandridge, A., Tveten, A.B. and Giallorenzi, T.G., 1982, Homodyne demodulation scheme for fibre optic sensors using phase generated carrier, IEEE J. Quantum Electron. QE-18, 1647-53.

Dandridge, A., Tveten, A.B., Miles, R.O. and Giallorenzi, T.G., 1980a, Laser noise in fibre optic interferometer systems, Appl. Phys. Lett. 37, 526-8.

Dandridge, A., Tveten, A.B., Miles, R.O., Jackson, D.A. and Giallorenzi, T.G., 1981, Single-mode diode phase noise, Appl. Phys. Lett. 38, 77-8.

Dandridge, A., Tveten, A.B., Sigel, G.H., West E.J. and Giallorenzi, T.G., 1980b, Optical fibre magnetic field sensors, Electron. Lett. 16, 408.

Davies, D.E.N. and Kingley, S.A., 1974, Method of phase modulating signals in optical fibres: application to optical-telemetry systems, Electron. Lett. 10, 21-2.

Eberhardt, F.J. and Andrews, F.A., 1970, Laser heterodyne systems for measurements and analysis of vibration, J. Acoust. Soc. Am. 48, 603-9.

Eickhoff, W., 1981, Temperature sensing by mode-mode interference in birefringent optical fibres, Opt. Lett. 6, 204.

Giallorenzi, T.G., Bucaro, J.A., Dandridge, A., Sigel, G.H., Cole, J.H., Rashleigh, S.C. and Priest, R.G., 1982, Optical fibre sensor technology, J. Quantum Electron. QE-18, 626-65.

Giles, I.D., Uttam, D., Culshaw, B. and Davies, D.E.N., 1983, Coherent optical fibre sensors with modulated laser sources, Electron. Lett. 19, 14-5.

Gloge, D., 1971, Weakly guiding fibres, Appl. Opt. 10, 2252.

Green, E.L. and Cable, P.G., 1982, Passive demodulation of optical inter-ferometric sensors, IEEE J. Quantum Electron. QE-18, 1639-44.

Henning, M.K., Thornton, S.N., Carpenter, R., Stewart, N.J., Dakin, J.P. and Wade, C.A., 1983, Optical fibre hydrophone with down lead insensi-tivity, Proc. 1st Int. Conf. on Optical Fibre Sensors (London: IEE), p.23.

Hocker, G.B., 1979a, Fibre optic sensing of pressure and temperature, Appl. Opt. 18, 1445.

Hocker, G.B., 1979b, Fibre optic acoustic sensors with composite structure: an analysis, Appl. Opt. 18, 3679-83.

Jackson, D.A., 1981, A prototype digital phase tracker for the fibre interferometer, J. Phys. E: Sci. Instrum. 14, 1274-8.

Jackson, D.A., Dandridge, A. and Sheem, S.K., 1980a, Measurement of small phase shifts using a single-mode optical fibre interferometer, Opt. Lett. 5, 139-41.

Jackson, D.A., Jones, J.D.C. and Chan, R.K.Y., 1984a, A high-power fibre optic laser Doppler velocimeter, J. Phys. E: Sci. Instrum. 17, 977-80.

Jackson, D.A., Kersey, A.D., Corke, M. and Jones, J.D.C., 1982, Pseudo-heterodyne detection scheme for optical interferometers, Electron. Lett. 18, 1081-3.

Jackson, D.A., Kersey, A.D. and Lewin, A., 1984b, Fibre gyroscope with passive quadrature demodulation, Electron. Lett. 20, 399-401.

Jackson, D.A., Priest, R., Dandridge, A. and Tveten, A.B., 1980b, Elimination of drift in a single-mode optical fibre interferometer using a piezoelectrically stretched coiled fibre, Appl. Opt. 2926-9.

Jarzynski, J., Hughes, R., Hickman, T.R. and Bucaro, J.A., 1981, Frequency response of interferometric fibre optic coil hydrophones, J. Acoust. Soc. Am. 69, 1709-808.

Jones, B.E. and Spooncer, R.C., 1983, Photoelastic pressure sensor with optical fibre links using wavelength characterisation, Proc. 1st Int. Conf. on Optical Fibre Sensors (London: IEE), pp.173-7.

Jones, J.D.C., Corke, M., Kersey, A.D. and Jackson, D.A., 1984, Single-mode fibre optic holography, J. Phys. E: Sci. Instrum. 17, 271.

Jones, J.D.C. and Jackson, D.A., 1985, Monomode fibre optic temperature sensor, Anal. Proc. Royal Society of Chemistry, 22, 207-10.

Jones, R.C., 1941, A new calculus for the treatment of optical systems Parts I-III, J. Opt. Soc. Am. 31, 488.

Kanada, T. and Nawata, K., 1979, Injection laser characteristics due to reflected optical power, IEEE J. Quantum Electron. QE-15, 559-65.

Kato, Y., Seikai, S. and Teteda, M., 1982, Arc-fusion splicing of single-mode fibres. 1: Optimum splice conditions, Appl. Opt. 21, 1332.

Kersey, A.D., Corke, M. and Jackson, D.A., 1984a, Linearised polarimetric optical sensor using a 'heterodyne-type' signal recovery scheme, Electron. Lett. 20, 209-11.

Kersey, A.D., Corke, M. and Jackson, D.A., 1984b, Phase nulling DC field fibre optic magnetometer, Electron. Lett. 20, 573-4.

Kersey, A.D., Corke, M., Jackson, D.A. and Jones, J.D.C., 1983a, Detection of DC and low frequency AC magnetic fields using an all single-mode fibre magnetometer, Electron. Lett. 19, 469-71.

Kersey, A.D., Jackson, D.A. and Corke, M., 1982a, Passive compensation scheme suitable for use in the single-mode fibre interferometer, Electron. Lett. 18, 392-3.

Kersey, A.D., Jackson, D.A. and Corke, M., 1983b, A simple Fabry-Perot sensor, Opt. Commun, 45, 71-4.

Kersey, A.D., Jackson, D.A. and Corke, M., 1983c, Demodulation scheme for interferometric sensors employing laser frequency switching, Electron. Lett. 19, 102-3.

Kersey, A.D., Jackson, D.A. and Corke, M., 1985, Single-mode fibre optic magnetometer with DC bias field stabilisation, J. Lightwave Tech. 3, 836-40.

Kersey, A.D., Lewin, A.C. and Jackson, D.A., 1984c, Pseudo-heterodyne detection scheme for the fibre gyroscope, Electron. Lett. 20, 368-9.

Kingsley, S.A., 1983, Fibre optic interferometric signal processor employing a differential and cross-multiply frequency discriminator, Proc. 1st Int. Conf. on Optical Fibre Sensors (London: IEE), pp.205-9.

Kingsley, S.A. and Davies, D.E.N., 1976, Use of optical fibres as instrumentation transducers, Proc. CLEOS, Conf. (New York: OSA).

Koo, K.P., Dandridge, A., Tveten, A.B. and Sigel, G.H., 1983, A fibre optic DC magnetometer, J. Lightwave Tech. 1, 524-5.

Koo, K.P. and Sigel, G.H., 1982, Characteristics of fibre optic magnetic field sensors employing metallic glasses, Opt. lett. 7, 334-6.

Koo, K.P. and Sigel, G.H., 1983, A fibre optic magnetic gradiometer, J. Lightwave Tech. 1, 509-13.

Koo, K.P., Tveten, A.B. and Dandridge, A., 1982, Passive stabilisation scheme for fibre interferometers using (3 × 3) fibre directional couplers, Appl. Phys. Lett. 41, 616.

Lagakos, N. and Bucaro, J.A., 1981, Pressure desensitisation of optical fibres, Appl. Opt. 20, 2716-20.

Lewin, A.C., Kersey, A.D. and Jackson, D.A., 1985, Non-contact surface vibration analysis using a monomode fibre optic interferometer incorporating an open air path, J. Phys. E: Sci. Instrum. 18, 604-8.

Midwinter, J., 1979, Optical Fibres for Transmission (Chichester: Wiley).

Moss, G.E., Miller, L.R. and Forward, R.L., 1971, Photon-noise-limited laser transducer for gravitational antenna, Appl. Opt. 10, 2495.

Musha, T., Kamimura, J. and Kakazawa, M., 1982, Optical phase fluctuations thermally induced in a single-mode optical fibre, Appl. Opt. 21, 694.

Nokes, A.M., Hill, B.C. and Barilli, 1978, Fibre optic heterodyne interferometer for vibration measurements in biological systems, Rev. Sci. Instrum. 49, 722-8.

Norman, S.R., Payne, D.N., Adams, M.T. and Smith, A.M., 1979, Fabrication of single-mode fibres exhibiting extremely low polarisation birefringence, Electron. Lett. 15, 309-10.

Nosu, K., Rashleigh, S.C., Taylor, H.F. and Weller, J.F., 1983, Acousto-optic frequency shifter for single-mode fibres, Electron. Lett. 19, 816.

Petuchowski, S.J., Giallorenzi, T.G. and Sheem, S.K., 1981, Sensitive fibre optic Fabry-Perot interferometer, IEEE J. Quantum Electron. QE-17, 2168.

Payne, D.N., Barlow, A.J. and Ramskov Hansen, J.J., 1982, Development of low and high birefringence optical fibres, IEEE J. Quantum. Electron. QE-18, 477.

Rashleigh, S.C., 1980, Acoustic sensing with a single coiled monomode fibre, Opt. Lett. 5, 392-4.

Rashleigh, S.C., 1983a, Origins and control of polarisation effects in single-mode fibre, J. Lightwave Tech. 2, 312-32.

Rashleigh, S.C., 1983b, Polarimetric sensors: exploiting the axial stress in high birefringence fibres, Proc. 1st Int. Conf. on Optical Fibre Sensors (London: IEE), pp.210-3.

Risk, N.P., Youngquist, R.C., Kino, G.S. and Shaw, H.J., 1984, Acousto-optic frequency shifting in birefringent fibre, Opt. Lett. 9, 309.

Sheem, S.K., Giallorenzi, T.G. and Koo, K.P., 1982, Optical techniques to solve the fading problem in fibre interferometers, Appl. Opt. 21, 689.

Smith, A.M., 1978, Polarisation and magneto-optic properties of single-mode optical fibre, Appl. Opt. 17, 52-6.

Varnham, M.P., Barlow, A.J., Payne, D.N. and Okamoto, K., 1983a, Polarimetric strain gauges using high birefringence fibre, Electron. Lett. 19, 699-700.

Varnham, M.P., Payne, D.N., Birch, R.D. and Tarbox, E.S., 1983b, Single polarisation operation of highly birefringent bow-tie optical fibres, Electron. Lett. 19, 245-7.

Voges, E., Ostwald, O., Schiek, B. and Neyer, A., 1982, Optical phase and amplitude measurements by single sideband homodyne detection, IEEE J. Quantum Electron. QE-18, 124-9.

Yariv, A. and winsor, H.V., 1980, Proposal for detection of magneto-strictive perturbation of optical fibres, Opt. Lett. 5, 87.

OPTICAL FIBER INTERFEROMETER TECHNOLOGY AND HYDROPHONES

Thomas G. Giallorenzi

Naval Research Laboratory
Washington, DC 20375-5000

1. INTRODUCTION

Because of the extremely high sensitivities possible with inter-fermetric fiber sensors many laboratories have conducted research on these devices. Substantial progress has been realized, and interferometric sensors are in use in laboratory instrumentation and are undergoing field trials to determine their reliability and performance under uncontrolled conditions.[1,2]

By varying the coating on the fiber in the interferometric sensor, many different environmental parameters such as pressure, temperature, magnetic/electric fields, etc., could be measured. The application of these coatings usually is straightforward and does not affect the optical performance of the optical fiber waveguide. The optical phase shift induced by a particular environmental parameter and transduced to the fiber via its coating is detected by configuiring the fiber in the appropriate interferometric configuration. The most widely utilized interferometric configuration is the Mach-Zehnder configuration in which one arm of the fiber interferometer (signal arm) is exposed to the perturbation whereas the other arm (reference arm) is either desensitized to the perturbation via appropriate coatings or is shielded from the perturbation. Critical to the successful operation of fiber interferometers is the demodulation technique used to process the output of the interferometer.

The purpose of the demodulation scheme[3] in fiber optic inter-ferometers is to transform the optical output of the interferometer into an electrical signal proportional to the amplitude of the relative phase shift. The design of the demodulation scheme is made non-trivial by the presence of low frequency random temperature and pressure fluctuations which the arms of the interferometer experience. These fluctuations produce differential drifts between the arms of the interferometer resulting in changes in the amplitude of the detected signal (signal fading), as well as distortion of the signal. Typical requirements of the detection scheme are to be able to resolve signals corresponding to 10^{-6} rad phase shift, to have a linear response and to have a large dynamic range ($10^6 \rightarrow 10^7$). Specific requirements such as packaging and low power consumption are also important. In this paper we will discuss a) the active homodyne approach in which the interferometer is locked at maximum sensitivity (quadrature) with feedback circuitry and a phase shifter and b) on two newer techniques using laser tuning: passive homodyne techniques and (3x3) directional couplers. Each of these schemes have achieved rad performance; however, each scheme has specific advantages and disadvantages which will be briefly be detailed.

The homodyne scheme (referred to as phase swept) uses a large phase modulation (produced by laser tuning) applied to the interferometer, such that by appropriate beating and filtering, two signals in quadrature are produced. The 3x3 directional coupler scheme achieves the required

quadrature condition between the two outputs through the use of an all
optical 3x3 fiber directional coupler. In both of these schemes the output
signals contain sine and cosine functions which are electrically manipulated
(demodulated) to produce a final output which is proportional to the signal
to be detected. The various demodulation techniques are in wide use.

The performance of interferometric sensors is limited in part by noise
in the detection process. The typical single mode fiber interferometer
constructed with high quality components (e.g. couplers) is an intrinsically
'quiet' device. Most noise problems are associated with a) perturbations
causing unwanted $\Delta\ell$ and Δn contributions to $\Delta\phi$ and b) noise associated with
the optical source which is used to measure $\Delta\phi$ induced by the signal field.
The first noise source is basically related to fiber coatings and packaging
issues. Properly designed coatings can be used in the signal arm to
maximize   due to the signal field while reducing the effects from other
fields.

Optical source noises which result in a loss of sensor sensitivity may
be split into three basic groups: 1) coherence length; 2) intensity noise;
and 3) phase noise frequency jitter. All three parameters are strongly
dependent on light fed back into the laser cavity. The free running laser
has a coherence length of only a few meters, requiring the sensing and
reference fiber to be balanced to within ±1m. The amplitude of the laser's
intensity noise is similar to that of gas lasers, but has a 1/f
characteristic which may limit performance at low frequencies; common mode
rejection techniques are therefore desirable. The interferometer is
extremely sensitive to low frequency (i.e., in the signal band) jitter of
the laser's emission. Different schemes to stabilize the laser output using
optical feedback or electronic feedback and reduce sensor noise are being
considered.

As an example of an interferometric sensor, the acoustic sensor which
has demonstrated state of the art performance will be described. Coatings
to highly sensitize interferometer signal arm optical fibers to acoustic
fields and to desensitize reference arm fibers to acoustic fields have been
perfected and will be discussed in a companion paper by J. Bucaro in this
book. Acoustic interfermetric sensors now have been packaged and are
capable of reliable at sea operation. Critical to this success was the
development of rugged optical components for these interferometers.

2. MACH-ZEHNDER INTERFEROMETRIC SENSORS

Single mode fiber optic interferometers may be configured in a number
of different ways--similar to their bulk analogues. Common configurations
are: Mach-Zehnder, Michelson, Sagnac, and polarimetric (birefringent). The
Sagnac (used to fabricate fiber gyros) and Mach-Zehnder are by far the most
common forms currently in use. Two beam interferometers, i.e. Mach-Zehnder,
Michelson and polarimetric; have their own advantages and disadvantages in
their use as fiber sensors. Referenced to the Mach-Zehnder, for the same
length of signal arm, the Michelson has twice the sensitivity (owing to the
double path); however, the phase noise limitation also doubles for an
equivalent fiber length mismatch. The main disadvantage of the Michelson is
that the complementary output is launched back into the laser, which for a
low loss system means (when laser/fiber coupling is included) that the
optical feedback term is $\sim 10^{-1}$ which is far above the $\sim 10^{-6}$ ratio required
for stable operation of semiconductor sources. The polarimetric type of
sensor is sensitive to changes in birefringence, caused by the signal field,
of a single fiber. This approach may be considered to be a single fiber
Mach-Zehnder in the sense that the signal and reference arms are the two

othogonal polarization modes of the fiber. This configuration has a number of advantages in terms of common mode noise rejection; however, the sensitivity is somewhat less than for the two fiber Mach-Zehnder interferometer.

A schematic of a Mach-Zehnder is shown in Fig. 1. A laser beam is split, part being sent through a "reference" fiber arm and part through a "sensing" fiber arm which is immersed in, and sensitive to, its environment. After passing through the fibers, the two beams are recombined and allowed to interfere on the surface of a photodetector. There results in the photodetector current a signal which is directly related to the environmentally induced phase shift, $\Delta\phi$.

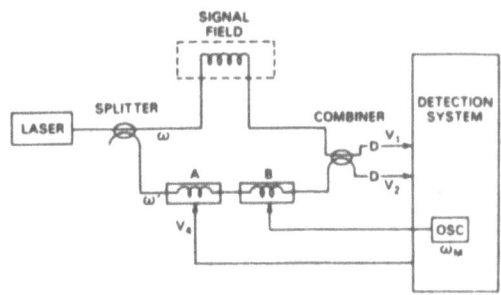

Fig. 1. Generalized fiber interferometer system.

Interferometers allow the measurement of extremely small phase shifts generated by the field to be detected (e.g., magnetic or acoustic) in the optical fiber. The optical phase delay (in radians) of light passing through a fiber is given by

$$\phi = nk\ell \tag{1}$$

where n is the refractive index of the fiber core, k is the optical wave number in vacuum ($2\pi/\lambda$) and   is the length of the fiber. Small variations in the phase delay are found by differentiation:

$$\frac{\Delta\phi}{\phi} = \frac{\Delta\ell}{\ell} + \frac{\Delta n}{n} + \frac{\Delta k}{k}. \tag{2}$$

The first two terms are related to physical changes in the fiber caused by the perturbation to be measured. Accordingly, they describe the transduction mechanism whereby the fibers can act as sensors. Generally, changes in pressure, temperature, electric field, etc. produce different contributions to $\Delta\phi$ via the $\Delta\ell$ and Δn terms. The last term in Eq. (2) takes into account any wavelength (or frenquency) variation associated with the laser source. This term has been specifically included to facilitate our discussion of certain demodulation issues, e.g., phase noise.

If we consider the interference of two light waves as in an interferometer, the amplitude of the two fields are

$$E_1 = E_{10}\sin(\omega t - \phi_1) \text{ and } E_2 = E_{20}\sin(\omega t - \phi_2) \tag{3}$$

where ϕ_1 and ϕ_2 are the total path length in each arm of the interferometer and the optical frequency. The resultant is given by

$$E_1 + E_2 = \sin\omega t(E_{10}\cos\phi_1 + E_{20}\cos\phi_2) - \cos\omega t(E_{10}\sin\phi_1 + E_{20}\sin\phi_2) \tag{4}$$

Rewriting this expression, the electric field is

$$E = E_0\sin(\omega t - \phi) \tag{5}$$

where

$$E_0^2 = E_{10}^2 + E_{10}^2 + E_{10}E_{20}\cos(\phi_2 - \phi_1); \quad \tan\phi = \frac{E_{10}\sin\phi_1 + E_{20}\sin\phi_2}{E_{10}\cos\phi_1 + E_{20}\cos\phi_2} \tag{6}$$

If $E_{10} = E_{20}$

$$E_0^2 = E_{10}^2 [1 + \cos(\phi_2 - \phi_1)] \tag{7}$$

and as a consequence, the interferometer is only sensitive to differential changes in ϕ_2 and ϕ_1. Defining $\Delta\phi = \phi_2 - \phi_1 = nk$ we have the classical interference equation; the resultant intensity is given by

$$E_0^2 = 2E_{10}^2 [1 + \cos\Delta\phi] \tag{8}$$

To make a sensitive interferometric sensor, the phase shift $\Delta\phi$ for the particular field to be measured must be maximized, the nonlinear output of the interferometer demodulated, and the interferometer operated such that noise contributions which would mask the signal are minimized. The phase shift may be maximized by making the signal arm sensitive to the field, while desensitizing the reference fiber.

3. DEMODULATION TECHNIQUES

The purpose of the demodulation scheme in fiber optic interferometers is to transform the optical output of the interferometer into an electrical signal proportional to the amplitude of the relative phases shift. The design of the demodulation scheme is made non-trivial by the presence of low frequency random temperature and pressure fluctuations which the arms of the interferometer experience. These fluctuations produce differential drifts between the arms of the interferometer. The drift causes changes in the amplitude of the detected signal (signal fading), as well as distortion of the signal. Typical requirements of the detection scheme are: to be able to resolve signals corresponding to 10^{-6} rad phase shift, to have a linear response and to have a large dynamic range ($10^6 - >10^7$).

3.1. Phase Tracking Homodyne Detection (PTDC)

In equation 8, $\Delta\phi$ may be expressed as the sum of many phase terms; ϕ_1, ϕ_2, $S(t)$, $A(t)$, $B(t)$. $S(t)$ is the phase shift which is proportional to the signal to be detected and $A(t)$ and $B(t)$ are signals intentionally introduced into the interferometer by for example a piezoelectric fiber stretcher. This stretcher is used to maintain the quadrature condition $\phi_1 - \phi_2 - A(t) \approx \pi/2 + m\pi$. In this case, the detector output voltage from the detector becomes

$$V = \alpha V_o \cos(\phi_1 - \phi_2 - A(t) + S(t) - \pi/2) \qquad (9)$$
$$\approx V_o \alpha(\varepsilon - A)$$

In this expression we have imiplicitly taken the phase modulo 2π. Since V vanishes at the desired quadrature condition, it is an ideal error signal from the view point of linear control theory. If an appropriate feedback voltage can be produced from V and applied to the piezoelectric element, the phase A can be made to exactly cancel ε, thus driving the error signal to zero. Such a feedback signal is the integral of V.

$$V' = g \int_o^t V(t')dt' \qquad (10)$$

The differential equation for the feedback signal V is

$$V' + ghV_o \,\alpha V' = g\alpha V_o E \qquad (11)$$

where h is the volts to radians constant of the piezoelectric stretcher. The combination $ghV_o\alpha$ is the gain bandwidth product of the feedback circuit. If the variation in $S(t)$ corresponds to frequencies much less than the gain bandwidth product, (11) ensures that $A=V'h=\varepsilon$, the quadrature condition. Since $hV' = \varepsilon = S(t) + \phi_s - \phi_r - \pi/2$, V' is linear in the signal S. Thus, $S(t)$ can be separated from ϕ_s and ϕ_r by appropriate filtering.

The strong point of the PTDC detection system is that it involves only linear operations. This is possible because the phase shift A produced by the piezoelectric element is a very linear function of the feedback voltage V'. A drawback to the PTDC scheme is that the voltage range of the integrator producing V' is limited in a practical system to ±10V. Once this voltage is approached, V' must be reset to zero. This reset produces a transient in the output which must be appropriately smoothed out. The frequency of these resets can be reduced by making the piezoelectric element bigger and associated fiber longer. However, the larger the cylinder the more troublesome are cylinder resonance problems. An advantage of the PTDC scheme is that the feedback and balanced mixer hold V' near zero, thus ensuring that amplitude fluctuations are common mode to both arms and thus cancel out. More elaborate implementations of the PTDC scheme are possible. Various filters can be incorporated to allow high gain bandwidth ratios to be used. Some gain control can be used to stabilize the gain bandwidth product. In the simple version described here, the output signal V is independent of the optical power and mixing efficiency, but the gain bandwidth product depends on these parameters.

A variant of the PTDC system uses an oscillating phase term (B) as well as a feedback phase (A). The oscillating phase causes the output signals to oscillate at ω_m (approximately 100 kHz) and its harmonics. The oscillating signal V is mixed with the local oscillator which drives the "B" piezoelectric element. This mixing produces a slowly varying signal proportional to $\cos(S(t) + \phi_s - \phi_r - A)$. This will be recognized as being of the same form as V in the PTDC scheme. The difference between these two schemes is that the PTDC scheme uses the low frequency signal V directly as the error signal, while the ω_m scheme uses the amplitude of a high frequency "dither" signal as the error signal. In the appropriate small signal analysis, the error signal in both cases is proportional to $S(t) + \phi_s - \phi_r - A - \pi/2$.

3.2 Passive (phase swept) homodyne detection

The homodyne scheme (referred to as phase swept) uses a large phase modulation (produced by laser tuning) applied to the interferometer, such that by appropriate beating and filtering, two signals with a /2 relative phase shift are produced. The variation in the light intensity detected at the output of an interferometer may be written as

$$I = A + B \cos \Delta\phi(t) \tag{12}$$

where $\Delta\phi(t)$ is the phase difference between the arms of the interferometer. The constants A and B are proportional to the input optical power, but B also depends on the mixing efficiency of the interferometer. If a sinusoidal modulation with a frequency ω_0 and amplitude C is imposed on the interferometer, then (12) becomes

$$I = A + B \cos (C \cos \omega_0 t + \Phi(t)) \tag{13}$$

where (t) includes not only the signal of interest, but environmental effects as well. Expanding (13) in terms of Bessel functions produces.

$$
\begin{aligned}
I = A + B\Bigg\{ &\left[J_0(C) + 2\sum_{k=1}^{\infty}(-1)^k J_{2k}(C)\cos 2k\omega_0 t \right]\cos\Phi(t) \\
&- \left[2\sum_{k=0}^{\infty} (-1)^k J_{2k+1}(C) \cos (2k+1)\omega_0 t \right] \sin\phi(t)\Bigg\}
\end{aligned}
\tag{14}
$$

From this expression it is clear that when $\Phi(t)=0$, only even multiples of ω_0 are present in the output signal, whereas for $\Phi(t)=\pi/2$ rad (quadrature condition), only the odd multiples of ω_0 survive.

In a similar fashion the phase angle $\Phi(t)$ can be separated into a signal component of frequency ω and the environmental drifts $\Psi(t)$, ($\Phi(t) = D \cos \omega t + \Psi(t)$) and expanded

$$
\begin{aligned}
\cos \Phi(t) = &\left[J_0(D) + 2 \sum_{k=1}^{\infty} (-1)^k J_{2k}(D) \cos 2k\omega t \right]\cos \Psi(t) \\
&- \left[2 \sum_{k=0}^{\infty} (-1)^k K_{2k-1}(D) \cos (2k + 1) \omega t \right] \sin \Psi(t)
\end{aligned}
\tag{15}
$$

$$\sin(t) = \left[2 \sum_{k=0}^{\infty} (-1)^k J_{2k+1}(D) \cos(2k+1) \omega t \right] \cos \Psi(t)$$

$$+ \left[J_0(D) + \sum_{k=1}^{\infty} (-1)^k J_{2k}(D) \cos 2k\omega t \right] \sin \Psi(t). \tag{16}$$

These equations show that when $\Psi(t)=0$, even (odd) multiples of ω are present in the output signal centered about the even (odd) multiples of ω_0. For the case when $\Psi(t) = \pi/2$ rad, even (odd) multiples of ω are present about the odd (even) multiples of ω_0. The sidebands contain the signal of interest and are either present about the even or the odd multiples of ω_0. The signal is obtained by mixing the total output signal with the proper multiple of ω_0 and low-pass filtering to remove the terms above the highest frequency of interest. For the carrier frequencies considered in the experiment, namely 0, ω_0, and $2\omega_0$, the output signals after mixing and filtering are

$$A + BJ_0(C) \cos \phi(t)$$

$$BGJ_1(C) \sin \phi(t) \tag{17}$$

$$-BHJJ_2(C) \cos \phi(t),$$

respectively, and where G and H are the amplitude of the mixing signals for ω_0 and $2\omega_0$. In order to obtain a signal that does not fade as a function of undesired fluctuations, two signals, one containing the sine $\phi(t)$ and the other cosine $\phi(t)$ are utilized. The time derivative of the sine and cosine terms are cross multiplied with the cosine and sine terms, respectively, to yield the desired sine and cosine squared terms. The process will be illustrated by considering the output signals for $_0$ and 2_0. The time derivative of these are obtained from (17) and are given by

$$BGJ_1(C) \dot{\phi}(t) \cos(t)$$

$$BHJ_2(C) \dot{\phi}(t). \tag{18}$$

Multiplying this by the signal for the other frequency produces

$$B_2 GHJ_1(C) J_2(C) \dot{\phi}(t) \cos_2(t)$$

and

$$-B^2 GHJ_1(C) J_2(C) \dot{\phi}(t) \sin^2 \phi(t). \tag{19}$$

Subtracting gives

$$B^2 GHJ_1(C) J_2(C) \dot{\phi}(t) (\sin^2 \phi(t) + \cos^2 \phi(t))$$

$$= B^2 GHJ_1(C) J_2(C) \dot{\phi}(t). \tag{20}$$

This output can then be integrated to produce the signal $\Phi(t)$ which includes all of the drift information in addition to the actual signal. This scheme has the advantage of having no reset problem. This scheme requires an AGC (owing to polarization fading) and may have a somewhat limited frequency range (two orders of magnitude), however, the tuning range may be varied over a fairly wide frequency range (~ 0.1 Hz $\rightarrow$ 10kHz).

3.3 (3x3) Passive Coupler Detection

The (3x3) passive coupler scheme is potentially a true passive scheme, where the required $\pi/2$ phase shift between the two outputs is achieved by replacing the conventional final (2x2) directional coupler, by a (3x3) device. The optical and electronic configuration is shown in Fig. 2.

$$P_1 = -2B_2(1 + \cos\theta),$$

$$P_{II,III} = B_1 + B_2 \cos\theta \pm B_3 \sin\Phi, \tag{21}$$

where P_{II} (P_{III}) takes the plus (minus) sign and B_i (i=1,2,3) are constants dependent on the coupling coefficients of the fiber coupler. To obtain the required $\pi/2$ phase difference required P_{II} and P_{III} were processed to form the sum and difference

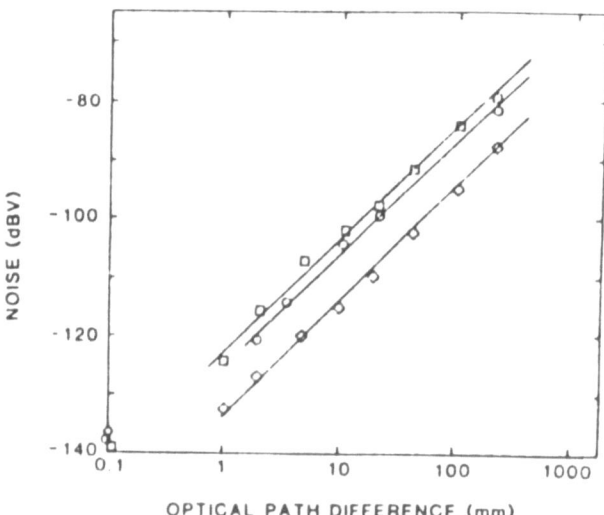

Fig. 2. Schematic of the optical and electronic circuitry for a (3x3) passive detection scheme.

$$P_{II} = P_{II} + P_{III} = 2(B_1 + B_2 \cos\theta),$$

$$P_{III} = P_{II} - P_{III} = 2B_3 \sin\theta. \tag{22}$$

After signal processing one obtains

$$D = P_{II} \dot{P}_{III} - \dot{P}_{II} P_{III}$$

$$= 4B_1 B_3 \dot{\theta} \cos\theta + 4B_2 B_3 \dot{\theta},$$

$$E = \int D\, dt = +4B_1 B_3 \sin\theta + 4B_2 B_3 \theta. \tag{23}$$

Note that B_1 is the dc output of P_{II}, P_{III}. By introducing offsets in the processing electronics, B_1 can be set to zero. Then

$$D = 4B_2 B_3 \dot{\theta},$$

thus

$$E = \int D\, dt = 4B_2 B_3 \theta, \tag{24}$$

which is directly proportional to the interferometer's phase shift. It can be seen from Eq. (23) that unless the electronics is correctly balanced (i.e., $B_1 = 0$) distortions will be present in the output. These equations, although derived for a single polarization system, serve to demonstrate the principle of operation. In practice, small deviations from the $\pi/2$ phase shift (between P_{II} and P_{III}) were observed as different input polarizations were used. Consequently, Eq. (24) should be multiplied by a factor cos Δ (where Δ is the angular deviation from $\pi/2$) to allow for this effect. As before, an AGC is required. In general, the properties of the (3x3) directional coupler determine the stability and fidelity of the detection scheme.

4. NOISE LIMITATION IN INTERFEROMETRIC SENSORS

Noise sources which result in a loss of sensor sensitivity may be split into three basic groups, 1) coherence length, 2) intensity noise, and 3) phase noise frequency jitter. All three parameters[4] are strongly dependent on light fed back into the laser cavity. The amplitude and phase of the light fed back into the laser cavity as well as the effective external cavity length are all important in determining the laser's properties[4]. The free running laser has a coherence length of only a few meters, requiring the sensing and reference fiber to be balanced to within ±1 m. The amplitude of the laser's intensity noise is similar to that of gas lasers[4], but has a 1/f characteristic which may limit performance at low frequencies; common mode rejection techniques are therefore desirable. The interferometer is extremely sensitive to low frequency (i.e., in the signal band) jitter of the laser's emission. This quasi-random noise source (also with a 1/f characteristic) results in noise in the interferometer which has an amplitude proportional to the path length difference in the interferometer. Typical results are shown in Fig. 3 for a number of different laser structures. The measures were made with an unbalanced Michelson interferometer (this noise is often termed phase noise). As can be seen from this figure for μradian performance, path differences of less than $\sim$ mm are required. Different schemes to stabilize the laser output using optical feedback or electronic feedback are being considered; they will be discussed briefly below.

4.1 Intensity Noise

If we consider the output of the interferometer after the demodulator, the output voltage, V_0, may be represented by

$$V_0 = V_1 (1 + \cos\phi) + V_2 \tag{25}$$

Here V_1 is the amplitude of the output dependent on the optical phase of the interferometer and V_2 a DC term due to either unbalanced splitting ratios of the couplers (or uneven losses in the fiber arms) or lack of coherence of the optical source resulting in a loss of fringe visibility $V(= V_{max} - V_{min}/V_{max} + V_{min})$. Changes in $\Delta\phi$ due to perturbation of the signal field results in an output ΔV_0

$$\Delta V_0 = -V_1 \sin\phi\Delta\phi \tag{26}$$

which at the quadrature (maximum sensitivity) point ($\phi = \pi/2$) is equal to

$$\Delta\phi_{min} = \frac{\Delta V_0}{V_1} \quad \text{and} \quad \frac{\Delta I}{I} = \frac{\Delta V_0}{V_1 + V_2} \tag{27}$$

Thus

$$\frac{\Delta V_0}{V_1} = \frac{\Delta I}{I}\left(1 + \frac{V_2}{V_1}\right) \quad \text{and} \quad \Delta\phi_{min} = \frac{\Delta I}{I}\ \frac{1}{V} \tag{28}$$

PASSIVE HOMODYNE

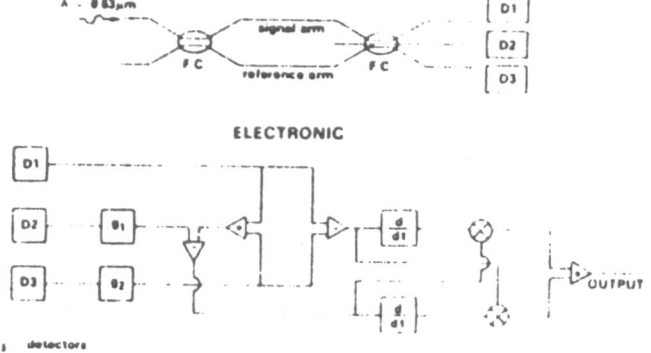

OPTICAL

ELECTRONIC

$D_{1,2,3}$ detectors
F C Fiber Couplers

Fig. 3. Noise output of a Michelson interferometer as a function of path length diference for three types of GaAlAs laser stuctures: ◊, TJS: O, Bh; □, CSP.

In terms of laser properties, the coherence length (and path difference D), as well as the intensity noise itself, determines the sensor noise floor. The relative intensity noise is given in terms of dB ($20 \log (\Delta I/I)$). Thus, typical values of $\Delta I/I$ and V indicate values of $\Delta\phi_{min} \sim 10^{-5}$ rad. However, this noise term can be reduced either by using electronic feedback schemes. For example, by subtracting the complementary outputs of the Mach-Zehnder, as much as 40 dB intensity noise rejection may be achieved. The two output signals of the interferometer V_A and V_B are 180° out of phase and may be written as

$$V_A = I(1 + \Delta I) + VI (1 + \Delta I)\Delta\phi \text{ and } V_B = I(V + \Delta I) - VI(1 + \Delta I)\Delta\phi \tag{29}$$

for the case where one (or two) couplers has a 50:50 splitting ratio. Here V indicates the fringe visibility with the interferometer at quadrature so that

$$V_A \simeq I(1 + \Delta I + V\Delta\phi) \text{ and } V_B = I(1 + \Delta I - V\Delta\phi) \tag{30}$$

Consequently,

$$V_A - V_B = 2VI\Delta\phi + 2VI\Delta I\Delta d \approx 2VI\Delta\phi \tag{31}$$

and the first order laser intensity noise term is eliminated.

4.2 Phase Noise

As indicated earlier, the phase delay in the optical fiber is sensitive to the source wavelength. This can result in a noise source if the frequency ν is unstable ($\Delta\nu$). However, light travels in both arms of the interferometer, and only pavelength difference, $\Delta\ell$, is important.

The phase shift $\Delta\phi$ resulting from frequency instability $\Delta\nu$ is

$$\Delta\phi = \frac{\phi}{k} \Delta k = n\ell\Delta k \tag{32}$$

If the two arms of the interferometer have path lengths ℓ_1 and ℓ_2 such that $\ell_1 - \ell_2 = \Delta\ell_{12}$, then from Eq. (32), the minimum detectable phase shift $\Delta\phi\min$ is

$$\Delta\phi\min = \frac{2\pi n}{C} \Delta\ell_{12} \Delta\nu \tag{33}$$

where C is the speed of light in vacuum.

Shown in Fig. 3 is the variation of $\Delta\phi_m$ with path difference for a number of different GaAlAs lasers. Similar behavior has been observed with lasers operating at 1.3 μm and 1.55 μm. To ensure high performance operation, it is necessary to operate the interferometer with less than 1 mm path difference.

A method to increase the frequency stability of the laser is to take the noise output and form a feedback loop to the laser's constant current supply. The experimental arrangement is shown in Fig. 4. The feedback circuit is adjusted such that when the voltage output of the photodiode is equal to the voltage that corresponds to the maximum slope of the Fabry-Perot response, the error signal applied to the diode laser is zero. However, when the laser's frequency shifts, the feedback circuit imposes a current on the constant bias current of the laser proportional to the laser's frequency deviation. The frequency response of the feedback loop is determined by the overall loop gain of the system and is adjusted to a maximum for stable operation.

The results of the phase noise measurement for the free running laser are shown in Fig. 5a (upper trace). The characteristic $f^{-1/2}$ frequency dependence is observed; it should be noted that the phase noise contribution was ~ 2.5 orders of magnitude larger than the contribution due to amplitude noise, which is shown in Fig. 5b. The lower trace in Fig. 5a shows the phase noise output when the feedback circuit is switched on; a substantial reduction in the phase noise is observed. Below 5 Hz, the noise is reduced by 60 dB (electrical noise power); at 250 Hz a ~30 dB reduction was observed. Despite the large reduction in phase noise, no change in the laser's amplitude noise was observed. This is similar to the effect of optical feedback stabilization. Similar results may of course be obtained by using a Mach–Zehnder (bulk or fiber) to provide the stabilization feedback.

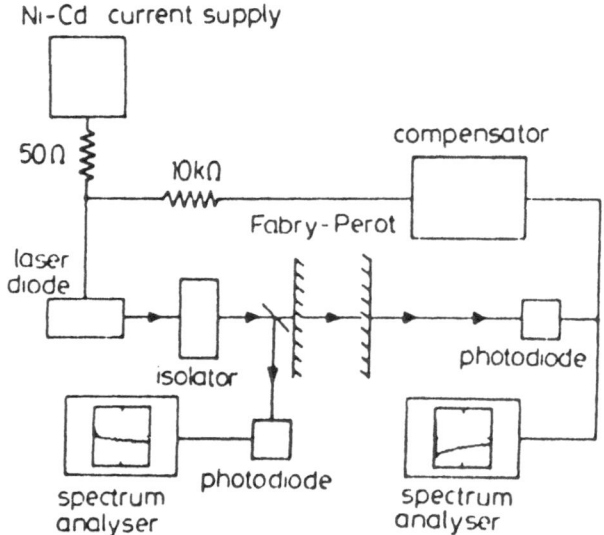

Fig. 4. Schematic diagram of an electronic laser stabilization.

It has been shown[5] that the frequency and intensity instabilities of GaAlAs lasers are partly correlated. However, the correlation is too low to, for example, stabilize the frequency by stabilizing the intensity.[6] However, the intensity noise emitted by the laser's front facet may be substantially reduced by stabilizing the output of the rear facet.

5. ACOUSTIC SENSOR

One of the earliest interferometric sensors[1,3] to be demonstrated was the acoustic sensor, and since that time, a significant amount of work has gone into optimizing the response of fibers for acoustic applications. This has involved the choice of the right elastomeric materials regarding the dynamic behavior of their elastic moduli. For fibers with typical coating thicknesses (<1 mm), the acoustically induced strains — and therefore the

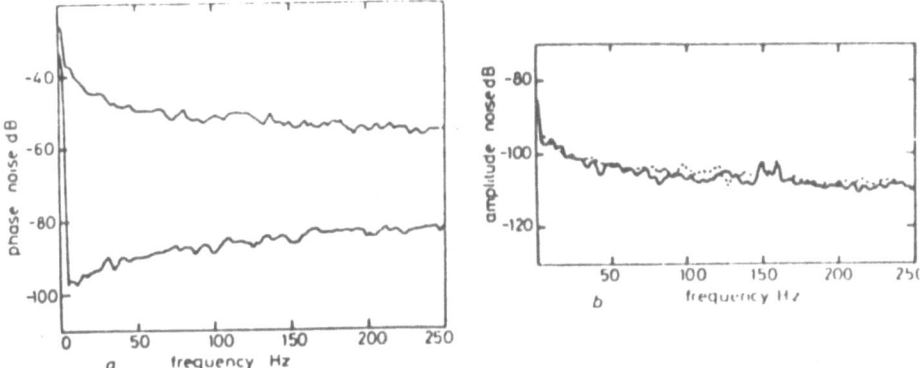

Fig. 5. Frequency dependence of the laser diodes phase noise (1 Hz band-width) obtained using output of a Fabry-Perot interferometer. Upper trace: free running laser; lower trace; with current feedback stabilization. (b) Frequency dependence of intensity noise (1 Hz bandwidth) with and without frequency stabilization.

acoustic sensitivity -- are a complicated function of the elastic moduli. For such a composite fiber geometry, the axial stress carried by a particular layer is governed by the product of the cross-sectional area and the Young's modulus of that layer. In Figure 6, the HYTREL and U.V. ACRYLATE indicate the role of the Young's and bulk moduli of various material types play in determining the fiber sensitivity for typical fiber thicknesses. In the case of any fiber optic sensor, the dynamic response of the sensor must be understood, and even if interested in only the low frequency response as addressed here, the position of the first mechanical resonance must be determined to insure a uniform frequency response. If higher frequency response is desired, then a more complete mechanical approach must be pursued.

As the spatial wavelength of the strains developed by perturbing field approach the fiber length, a length resonance is observed beyond which the fiber can no longer dynamically respond in axial strain[1]. Thus, the fiber response is constrained axially ($e_z = 0$) and if the spatial wavelength of the developed radial strains are still large compared to the fiber radius which guarantees isotropic radial strains, the phase response is reduced to the fiber response of the axially constrained radially isotropic condition:

$$(\phi_c / \phi)_{fiber} = -(n^2/2)(P_{11} + P_{12})e_r \tag{34}$$

Assuming a fiber coating thickness large enough to dominate the acoustic response, the isotropic radial strains are given by:

$$e_r = -(P/3B)(3B-E)/2E \tag{35}$$

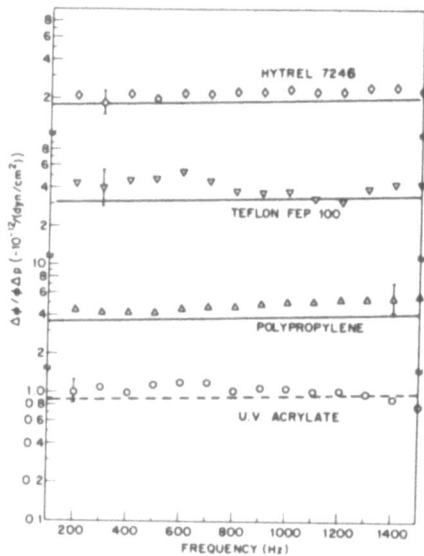

Fig. 6. Pressure sensitivity (points: experimental; lines: analytical) versus frequency of fibers with different OD's and outer coatings at 27°C. Upper 0.5 mm Od Hytrel 7246; second from top: 0.69 mm OD Teflon FEP 100; second from bottom: 0.62 OD polypropylene 6523; bottom: 0.5 mm OD UV acrylate (multimode fiber).

where the relationships between the Poisson's ratio of the fiber and the Young's (E) and bulk (B) moduli have been utilized. Substituting Eq. (35) into Eq. (34) and using $e_r=e_z=P/3B$, the ratio $(\Delta\varphi c/\varphi)$ $(\Delta\varphi u/\varphi)= .18$ for Hytrel coated fiber is determined. This ratio is in good agreement with the Hytrel coated fiber response, which has been more accurately determined. The above analysis is not intended to be rigorous but attempts to illustrate simplifications which can be utilized in estimating sensor performance. The reader is referred to appropriate references if interested in more rigorous analysis (1,7).

As the frequency of the perturbing field is increased further (7), and the spatial wavelength of the corresponding strains approach the fiber radius, the induced radial strain field becomes anisotropic and the fiber becomes birefringent supporting two phase velocities associated with the two principal indices of refraction n_1 and n_2.

$$(\Delta\phi_1/\phi)_{fiber} = -(n^2/2)(P_{11}e_1 + P_{12}e_1) \text{ and} \qquad (36)$$

$$(\Delta\phi_2/\phi)_{fiber} = -(n^2/2)(P_{12}e_1 + P_{12}e_2)$$

The acoustic response of an uncoated fiber and a fiber coated with a silicone inner buffer and an outer Hytrel 7246 jacket is plotted in Figure 7. In the low frequency (Unconstrained Isotropic) regime, the e_z strain level in the core is substantially enhanced by the Hytrel elastomer over the uncoated fiber. Theory and experimental data on an uncoated fiber below 1 MHz demonstrate the Axially Constrained Radially Isotropic regime corresponding to Eq. (34). Finally, the two phase responses $\Delta\phi_1$ and $\Delta\phi_2$ associated with the principal indices of refraction are illustrated above 1 MHz. Here, the strains e_1 and e_2 are complicated functions of frequency and elastic and mechanical parameters. For example, the resonance peak corresponds to the n=2 mode or the quadrupole resonance where two wave lengths are equal to the circumference of the fiber.

The discussion to this point has been in terms of normalized field sensitivity. An example is provided here to estimate actual sensor performance for the normalized sensitivity. The acoustic field sensitivity $\Delta\phi/\phi\Delta P = 6.0 \times 10^{-12}$ cm²/dyne for a 410 m coating thickness of Teflon PFA 340. The actual phase shift of a sensor is: $\Delta\phi/\Delta P = 2.3 \times 10^{-3}$ radians-cm²/dyne where $\phi = 2\pi n \ell/\lambda$, and the sensor parameters are ℓ = 50 meters, n = 1.458 and λ = 0.83 μm have been assumed. Given a minimum detectable phase shift of the demodulator of $\phi_{mm} = 1 \times 10^{-6}$ radians, the threshold of pressure detection, Pmin = 4.3×10^{-4} dynes/cm^{-2}. Such a detection level compares favorably with conventional acoustic sensors.

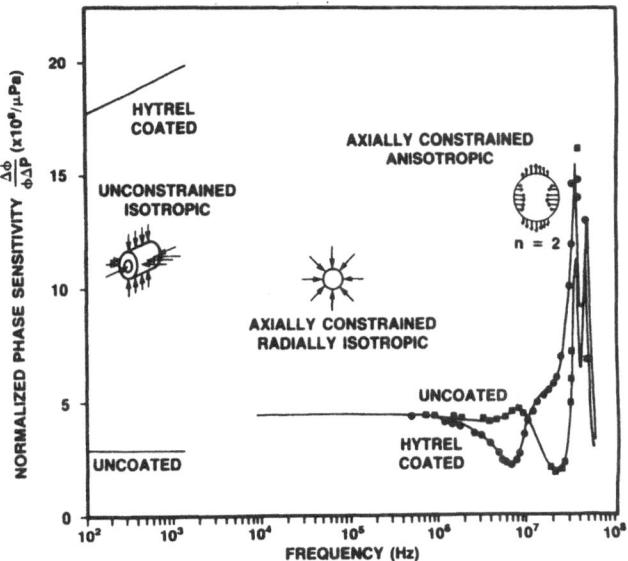

Fig. 7. Frequency response and different regimes of response of optical fiber sensors.

5.1 Sensor Packaging

Since the first reports of laboratory fiber interferometric acoustic sensors, considerable progress has been made in packaging of such devices (1). A significant improvement in S/N can be realized by employing an all-fiber interferometer. Noise levels for these all-fiber interferometers are considerably reduced and very low acoustic signals have been detected (approximately 10^{-4} dyn/cm² = 100 pa = 10^{-9} atm) down to freqencies as low as a few hundred hertz.

One of the first attempts at packaging a fiber optic acoustic sensor for field use is shown in Fig. 8. This hydrophone employed a phase stabilized demodulation scheme to maintain the sensing and reference fiber signals in quadrature. This fiber acoustic sensor was completely self-contained and included the sensing and compensator coils, a stable low noise, single-mode diode laser, two silicon photodectors (the Pair discriminates against optical intensity noise), and the demodulator electronics (not shown) which was placed in the space between the compensator and detector pair. The massive metal flange behind the sensor coil has been included to permit coupling of the sensor to a high pressure acoustic testing facility and normally would not exist in most hydrophone (1) successfully demonstrated the feasibility of taking laboratory acoustic sensors into actual field deployment. Many other fiber acoustic sensors have successfully been tested in ocean environments.

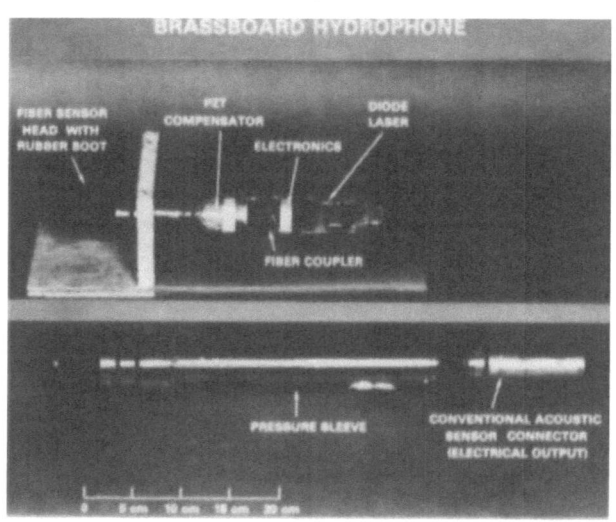

Fig. 8. Hydrophone demonstrating ability to package interferometric all fiber acoustic sensor for use in stressing environments.

REFERENCES

1. T.G. Giallorenzi, J.A. Bucaro, A. Dandridge, G. Sigel, J.H. Cole, S.C. Rashleigh and R.G. Priest, "Optical Fiber Sensor Technology," IEEE J. Quantum Elect., 18, 626, 1982.

2. T.G. Giallorenzi, "Progress in Optical Fiber Sensor," NATO' AGARD Book CCP-383, August 1983.

3. A. Dandridge, J. Cole, T.G. Giallorenzi, J.A. Bucaro, "Optical Fiber Sensors," Opto-Electronics for the Information Age, edited by Chinion Lin, Van Nostrand, New York, 1986 (in press).

4. L. Goldberg, H.F. Taylor, A. Dandridge, J.F. Weller and R.O. Miles, "Spectral Characteristics of Semiconductor Lasers with Optical Feedback," IEEE J. Quantum Elect., 18, 55, 1982.

5. A. Dandridge and H.F. Taylor, "Intensity and Frequency Instabilities and GaAlAs Diode Lasers," IEEE J. Quantum Elect., 18, 1738, 1982.

6. A. Dandridge and A.B. Tveten, "Properties of Diode Lasers with Intensity Noise Control," Appl. Opt., 12, 311, 1983.

7. J.A. Bucaro, N. Lagakos, J.H. Cole, T.G.Giallorenzi, "Fiber Optic Acoustic Transduction," Physical Acoustics Vol XVI, edited by W.P. Mason and R.N. Thurston, Academic Press, p. 385, New York, 1982.

OPTICAL FIBRE HYDROPHONES AND HYDROPHONE ARRAYS

Dr J P Dakin

Plessey Electronic Systems Research Limited
Roke Manor, Romsey, Hampshire SO51 0ZN

Preliminary Note

This contribution consists of two discrete sections. The first discusses the basic concepts of, and engineering progress with, a reflectometric method of polling the acoustically-induced length variations of a passive array of optical fibre hydrophones. The second part of the tutorial paper describes an alternative sensor, which, in its simple uncompensated form has been considered for use as a single hydrophone element by the U.S. Naval Research Laboratories. The present paper does not give any account of its use for this purpose, but describes how such a sensor may in general be compensated for ambient common-mode pressure or temperature changes and how it may be configured for practical sensors of pressure or magnetic field. The latter technique, therefore, in addition to its possible relevance to hydrophone systems, has a more general application to the field of optical fibre sensors.

1. PROGRESS WITH MULTIPLEXED SENSOR ARRAYS BASED ON REFLECTION AT SPLICED JOINTS BETWEEN SENSORS

1.1 Summary

This section of the paper reports recent progress made in developing a time-division-multiplexed, fibre optic hydrophone array using optical-time-domain-reflectometry (OTDR) techniques.

1.2 Introduction

Interferometric fibre optic sensors are attractive because of their high sensitivity relative to other types, such as those based on intensity modulation. Most reported to date have been single sensors, but there is currently considerable interest in using multiplexing techniques to drive an array of passive sensors from one source and detector. Such a system was recently reported by the author's laboratory (Reference 1.8.1).

The basic operating principle is shown in Figure 1. The system consists of a concatenated series of identical optical fibre sensors (which in the simplest case would be coils of fibre), each joined to the next by a partially-reflecting joint. Pairs of optical pulses, generated by applying pulses of RF to the Bragg cell, are launched into one end of the array. The first and second pulses of each pair have slightly different frequencies $f(1)$ and $f(2)$ respectively. As the transmitted pulses propagate down the array, a small proportion of light is reflected back from each partially-reflecting joint, and a series of reflections is received on a photodiode (Figure 2). The delay between the two transmitted pulses is chosen to be equal to the two-way propagation time through each sensing section, so that the reflection of the first pulse from a particular joint is received simultaneously with the reflection of the second pulse from the preceding joint. The two therefore mix on the photodiode and generate a heterodyne signal, whose phase depends on the difference in optical paths. These paths only differ by twice the length of the sensor that separates the two relevant reflecting joints, and therefore changes in the length of this sensor, caused for example by acoustic signals, modulate the phase of the heterodyne signal. Changes in the length of the downleads and previous sensors have little effect. The photodiode output consists of a sequence of short bursts of phase-modulated heterodyne signal, each corresponding to a particular sensor in the array. If the whole cycle is repeated continuously, the photodiode output consists of a set of phase-modulated carriers, time-division-multiplexed together. The acoustic signal on a particular sensor can then be recovered by time demultiplexing the photodiode output followed by phase-demodulation to recover the acoustic information.

Recently, improvements have been made in four main areas: development of partially reflecting splices; development of a high power single mode gas laser for use in the present system; development of a balanced-optical-path arrangement to allow shorter coherence length sources to be used in future systems; and development of an all-fibre frequency shifter, with the ultimate aim of producing a complete all-fibre system.

1.3 Partially reflecting splices

The fabrication of suitable low-loss, partially reflecting joints is crucial to the system. There is an optimum value for the size of reflection required, which depends on the number of sensors in the array.

For an array of ten sensors the optimum reflectivity is calculated to be in the range 0.1% to 2.0%, depending on the source power and system losses. A convenient way of producing a small reflection with low excess loss, is to introduce a refractive index mismatch into the optical beam. Fresnel reflection then occurs at each interface. Early breadboard systems, constructed to show the feasibility of the technique, used fibre joints with a small air gap between the fibre ends. This produced a pair of silica/air interfaces, each having a reflectivity of approximately 4%, in effect forming a low-finesse Fabry Perot cavity. The total reflectivity of such a splice is approximately a sinusoidal function of the fibre end separation, having maxima approaching 16% and minima close to zero. Splices fabricated in this way were rather lossy, and the reflection coefficient was highly sensitive to environmental effects such as temperature and strain on the splice support. This latter effect was due to the difficulty of adequately supporting the fibre ends and maintaining their separation.

Improved splices, having greater stability and lower transmission losses, have recently been produced by setting the fibre ends in transparent media having a different refractive index from silica (Figure 3). This technique still produces a double reflection and splice reflectivity is therefore still a function of end separation, but once the potting medium has set the splice is quite stable and relatively insensitive to environmental effects.

Various different potting media have been tried, including UV-setting cement (ref index = 1.51), visible-light-cured adhesive (ref index = 1.59) and polystyrene (ref index = 1.6). The optically cured adhesives are particularly convenient to use and give maximum theoretical reflectivities of 0.16% and 1% for the types mentioned above. Polystyrene gives a theoretical maximum reflectivity of 1%, but must be melted in order to make the joint, although this can be done easily with a small electrical heater element. In this way polystyrene splices have been made with reflection coefficients of 0.5% to 0.8% and a transmission loss of 0.5dB.

Using potted splices we have recently been successful in constructing an engineered array of 7 potted hydrophones with protected splices. The average loss per sensor, including splices was less than 0.5dB for this system.

1.4 High power single mode gas laser

The laser used for initial experiments on the sensor array was a 1mW HeNe device operating at a wavelength of 1152nm. This had a multi-longitudinal mode output which was responsible for a certain amount of fading of the received heterodyne signal, due to drifting of the modes within the laser gain curve. Additionally, the 1mW output power was only sufficient to drive an array of up to two sensors plus a downlead. An improved source for early experiments was therefore developed, with the aim of producing 10mW output in one longitudinal mode.

Conventional gas lasers have an output spectrum consisting of a set of discrete lines or modes, whose frequency separation is dependent on the length of the cavity. The total number of these modes in the spectrum depends on the Doppler-broadened linewidth.

Lasing can be confined to a single longitudinal mode by replacing one of the cavity mirrors with a pair, thus forming a secondary cavity (Figure 4). The secondary cavity behaves like a single mirror with wavelength-dependent reflectivity, maxima occuring when the cavity length is an integral number of half wavelengths. Laser modes having wavelengths that do not coincide with these reflection maxima will be suppressed. Thus, by judicious choice of mirror reflectivities and spacings, a three-mirror single-mode laser can be constructed.

Such a laser was constructed to operate at 1152 nm. A rigid structure of Invar rods was used to support the gas discharge tube and the three mirrors. The low thermal expansion coefficient of Invar helped to minimize the effect of temperature fluctuations on the mirror separations, but a piezoelectric support for one of the mirrors was still necessary to control the length of the secondary cavity and maintain single mode operation.

Figure 5 shows the emission spectra of the laser with two and three mirrors: an interference filter was used to remove low intensity lines at wavelengths other than 1152 nm. The laser had a total output power of 13mW, with 9mW concentrated in a single mode at 1152 nm.

1.5 Balanced optical path interferometer

In the basic system described in the introduction, the two light beams that interfere on the photodiode to produce a signal originate in the laser consecutively with a time difference t. This means that the coherence time of the laser must exceed t, or the beams will not interfere. On current systems, t is of the order of 1us and the coherence requirement has therefore precluded the use of short coherence length sources such as semiconductor or solid state lasers, both of which are more compact and rugged than gas lasers. A second consequence of the time difference between the generation of the two interfering pulses, is that any fluctuations in laser frequency will cause fluctuations in the heterodyne frequency, which are indistinguishable from phase modulation due to sound on the sensor. The balanced optical path configuration, described below, avoids both these problems and is expected to reduce laser microphony in current systems and enable Nd:YAG or semiconductor lasers with phase-noise reduction to be used in future systems.

A balanced optical path is achieved by generating the transmitted optical pulse pair from a single initial pulse, which is split into two paths, one of which contains a delay and a frequency shifter. The two paths are then combined, producing the required pairs of pulses. This "pre-delay" is matched to the differential delay in the sensor array, thus forming a balanced system, with the minimum source coherence length being determined by the error in matching the delays.

Figure 6 shows an all-fibre configuration of the balanced system using a fibre serrodyne frequency shifter currently under development. This device is described in the next section.

1.6 All-fibre frequency shifter

It is well known that an optical phase shifter can be driven with a ramp waveform to produce a frequency shift (ref 1.8.2). A simple optical fibre phase shifter can be constructed by winding fibre on to a PZT cylinder (ref 1.8.3); voltages applied to the PZT strain the fibre and produce a phase shift in the propagating light. However, if such a PZT is driven with a sawtooth waveform to produce a frequency shift, severe ringing occurs, particularly on the flyback, resulting in a badly distorted phase ramp. We have devised three modifications to this simple idea, for reducing the resonances and improving the fidelity of the response.

A small strain gauge was attached to the PZT cylinder in order to monitor its response. The basic intention was to apply feedback to force the PZT to follow the applied drive signal accurately. Initially the transfer function of the PZT was found to have large phase shifts coinciding with its resonances, and these had to be reduced by mechanical damping before feedback could be applied, or the system would have oscillated. This was achieved by packing the PZT in a proprietary plastic damping compound. Modification of the ramp waveform used to drive the PZT, so as to remove the sharp edges without affecting the linear portion also helped. Negative feedback was then applied (Figure 7) and resulted in a considerable improvement, the output of the strain gauge accurately following the PZT drive signal at ramp repetition rates up to 10kHz.

The device was incorporated into one arm of a fibre Mach-Zehnder interferometer, so that its optical performance could be investigated. The heterodyne signal generated at the output of the interferometer during the linear ramp sections was found to be slightly distorted, indicating that the optical fibre was not experiencing quite the same strain as the strain gauge. Further modifications are therefore required before the device will be of practical use. However these early results are highly encouraging and subsequent devices are expected to give improved performance.

1.7 Conclusion

We have described a fibre optic hydrophone array, consisting of a chain of passive sensors addressed by coherent optical time domain reflectometry, and recent progress made in improving the system. The areas in which improvements were described are: fabrication of reflective splices between sensors; the construction of a high performance gas laser source; modifications to the optical system to enable short coherence length sources to be used in subsequent systems; and an all-fibre serrodyne frequency shifter.

1.8 References

1.8.1 Dakin J P, Wade C A, and Henning M L, "Novel optical fibre hydrophone array using a single laser source and detector." Electronics Letters, vol 20, No 1, pp 53-54, Jan 1984.

1.8.2 Wong K K, and Wright S, "An optical serrodyne frequency translator", Proceedings of First European Conference on Integrated Optics, London, pp 63-65, September 1981.

1.8.3 Jackson D A, Priest R, Dandridge A, Tveten A B, "Elimination of drift in a single mode optical fibre inteferometer using a piezoelectrically stretched coiled fibre", Applied optics, Vol. 19, No 17, pp 2926-2929, Sepember 1980.

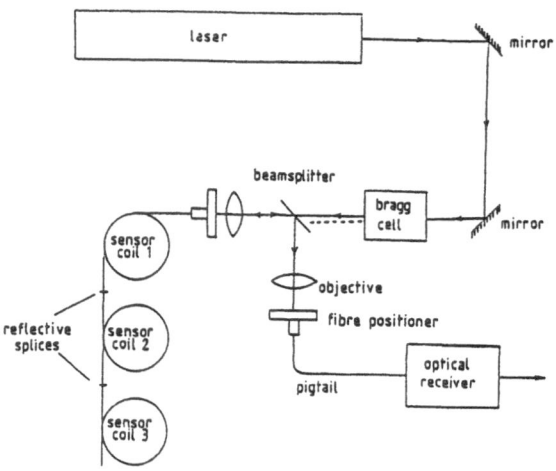

Figure 1. The basic reflectometric sensor array.

57

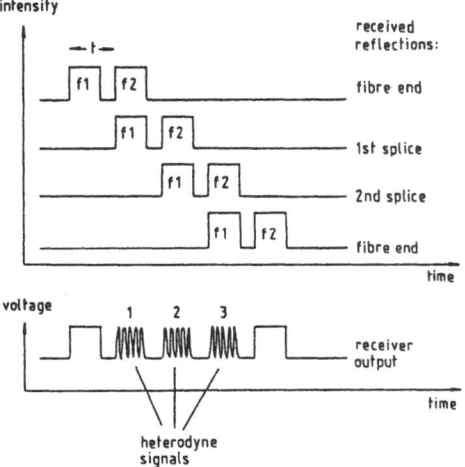

Figure 2. Timing diagram of reflections received, and corresponding receiver output for the array shown in Figure 1.

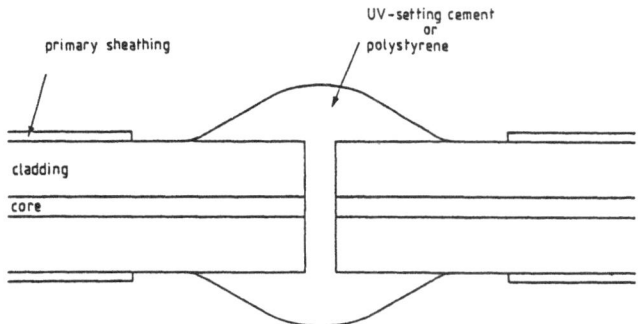

Figure 3. Construction of partially-reflecting splices using UV-setting cement or polystyrene.

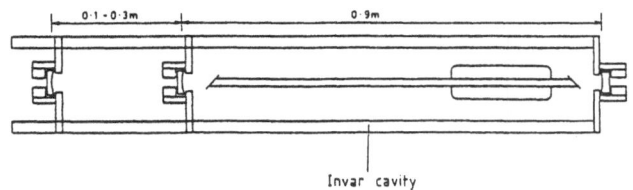

Figure 4. Construction of single-mode, 9mW, 1152nm HeNe laser.

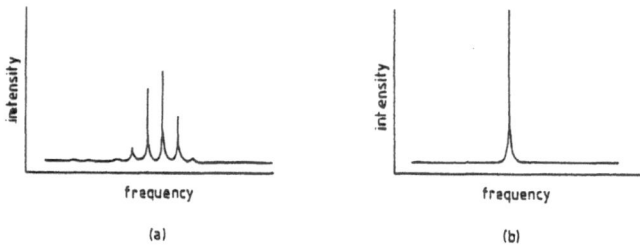

Figure 5. Fabry Perot interferometer scans of laser output:
(a) Multimode, two mirrors, 0.9m cavity
(b) Monomode, three mirrors, 0.9m & 0.18m cavities

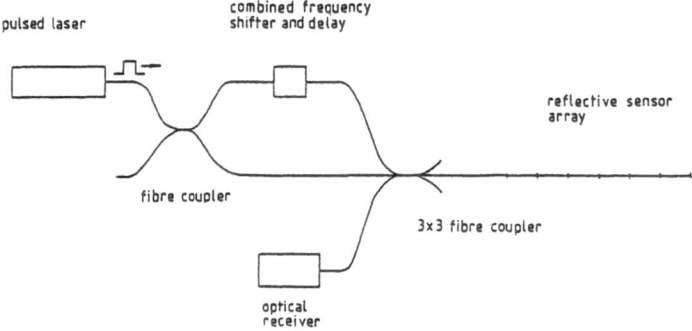

Figure 6. All-fibre sensor array, using short coherence laser and combined
frequency-shifter and delay.

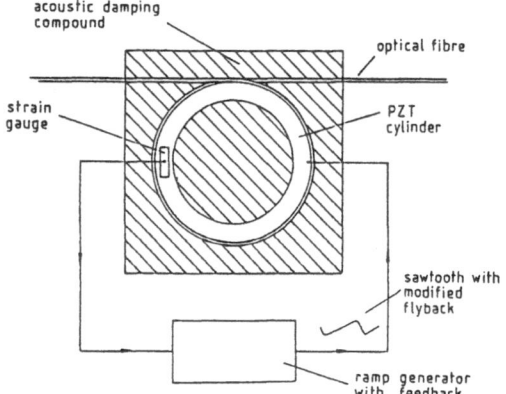

Figure 7. Optical fibre serrodyne frequency shifter.

2. MAGNETIC AND PRESSURE SENSORS USING THE COMPENSATED POLARIMETRIC SENSOR CONFIGURATION

2.1 Summary

Two identical lengths of high birefringence fibre spliced together with a 90° axial rotation to couple their orthogonal birefringent axes form the basis of a compensated polarimetric optical fibre sensor. This can be used to detect differential strains in the two fibre lengths caused by such phenomena as hydrostatic pressure or magnetic fields (the strain being introduced by magnetostrictive material bonded to the fibre). Magnetic and pressure sensors have been constructed using this principle and have shown good sensitivity yet better stability than Mach-Zehnder arrangements.

2.2 Introduction

The polarimetric optical fibre sensor was first proposed by Rashleigh of the U.S. Naval Research Laboratory and has been used to detect acoustic and magnetic fields, temperature, electrical current and strain (Ref. 2.9.3). The compensated version developed at Plessey ESR (Ref. 2.9.1) offers two main advantages over the original polarimetric sensor. Firstly, the original sensor of reference (2.9.3) could show an undesirable response to changes in its ambient conditions, as well as detecting the parameter required, whereas the compensated polarimetric sensor (CPS) is insensitive to any common-mode change of ambient conditions affecting both halves of the fibre. Secondly, the CPS is much less sensitive to phase noise in its light source, and can even employ a relatively incoherent LED source when well balanced (Ref 2.9.2).

We will first describe the theory of operation and the construction of a compensated polarimetric optical fibre sensor system, and then the application of this system to the measurement of AC and DC magnetic fields, and to the measurement of differential hydrostatic pressure. Finally, we will briefly describe a method for conferring downlead insensitivity to these sensors.

2.3 Theory of Operation of the Compensated Polarimetric Sensor System

A conventional polarimetric sensor (Fig. 1) uses a length of polarisation-maintaining optical fibre, which transmits light in two orthogonally polarised modes which have different propagation velocities. Plane-polarised light launched into the fibre at 45° to its polarisation axes will couple equal optical power into each of the two modes. The fibre introduces a phase delay between the two modes propagating along it. Changes in this delay can be observed simply by using an analyser at 45° to the polarisation axes. The light intensity passing through the analyser is proportional to (1 + cos ∅),where ∅ is the relative phase delay between the two polarisation modes. External physical fields can be made to change this phase delay, and are hence detected as cyclic variations of the analyser output. More sophisticated polarisation analysers can provide compensation for intensity changes and even allow tracking of phase changes over more than 2π radians. The simple polarimetric sensor has two disadvantages. Firstly its phase delay and hence output signal respond to changes both in temperature and strain in the fibre. Secondly, the differential phase delay introduced by the fibre may be as great as 3.10^6 radians in 100 metres, so that light launched into the fibre must be highly coherent in order to give high fringe contrast and low phase noise at the output from the analyser.

The compensated polarimetric sensor (Fig. 2) overcomes both these disadvantages. It consists of two identical lengths of polarisation-maintaining fibre which are spliced together with a 90° axial rotation in order to couple orthogonal polarisation axes. Light is input as before, but the phase delay between modes introduced in the first fibre length is exactly cancelled in the second. The fast mode in the first length couples into the slow mode of the second, and vice versa, so there is no net phase delay at the output.

Any uniform change in the physical conditions of the whole sensor will not affect the output signal, because the resulting change in the first length's phase delay is cancelled by an identical change in the second fibre length. The parameter to be sensed is allowed to change the phase delay in one of the fibre lengths relative to the other, causing an overall phase delay which is detected at the analyser. Hence the CPS is sensitive to differential-mode but not common-mode changes.

As the overall phase delay in a well compensated sensor will always be small, a light source of relatively high phase noise will normally give insignificant excess noise at the analyser. Reference 2 describes the successful operation of a CPS with an LED source showing that high fringe contrast may be achieved even using relatively broad linewidth sources.

2.4 Construction of Basic System

The layout of the basic CPS system is shown in Fig. 3.

The polarisation-maintaining optical fibre used in this study was type HB 800/1 made by York Technology Ltd., with core diameter 8um and cladding diameter 125um. It has a "bowtie" pattern of borosilicate glass within the cladding region our particular example having a beat length of 2.4mm at a wavelength of 633nm.

To make the CPS fibre, two equal lengths of polarisation-maintaining fibre were cut and their ends bared of coating. The polarisation axes at the fibre ends could then be located by viewing the bowtie pattern on the fibre endface under a microscope. Using this technique, the fibre lengths were aligned end to end with a relative 90° twist. The ends were then laterally adjusted to bring the fibre cores into line, using maximisation of light transmission between the lengths. Finally the ends were butted together and fusion spliced.

The CPS fibre requires an input light beam either circularly polarised or plane-polarised at 45° to each polarisation axis. This ensures that an equal intensity of light is launched into each polarisation mode of the fibre. The light source used was a Hitachi HL 7801E laser diode which emitted infrared light at a wavelength of 780nm, and was plane-polarised horizontally with an extinction ratio of 40:1.

The polarisation axes of the fibre were aligned at 45° to the laser junction. For initial experiments, discrete optical components were used to efficiently launch the laser into the fibre. Naturally in a practical sensor the launching system may be greatly simplified by using a packaged laser with a factory-aligned polarisation- maintaining fibre tail.

The polarisation state leaving the CPS fibre may be most simply analysed by a linear polariser at 45° to the polarisation axes, and the transmitted light measured to give the sensor's output signal. This was performed by a silicon PIN photodiode covered with an infrared polariser, while a second uncovered silicon photodiode within the same package measured the total light intensity leaving the fibre to allow compensation for source intensity fluctuation. The fibre end was fixed in a ferrule and held firmly in place so that its (unfocussed) output cone of light illuminated both photodiodes, with its polarisation axes aligned at 45° to the axis of the linear polariser (Fig. 4). Although this arrangement has higher losses than a polarisation dependent beam splitter, it has significant savings in both cost and complexity, and hence is a more practical proposition for many low cost sensor applications.

The photodiode photocurrents are converted to proportionate voltages by transimpedance amplifiers. Thus the "analysed" output, V_p, is proportional to $1 + \cos \phi$ (ϕ = phase delay introduced in the fibre as before) and the bare photodiode output, V_T, is proportional to the total output intensity. These two signals may be electronically divided, and as V_p is also proportional to total output intensity, the divider output is therefore again proportional to the factor (1 + cos ∅), but is now independent of the total light intensity in the fibre and amplitude noise is eliminated.

2.5 <u>Application of the Compensated Polarimetric System to Magnetic Sensing</u>

A magnetic field can be made to strain an optical fibre simply by bonding the fibre to a magnetostrictive material.

Magnetostrictive material deforms parallel to an applied magnetic field, with the deformation approximating to a parabolic function of field strength. The magnetic sensor was constructed by bonding strips of magnetostrictive metallic glass on opposite sides of a polarisation-maintaining fibre, forming a "sandwich" as illustrated in Fig. 5. Metglass 2605SC metallic glass supplied by Allied Corporation was used, and the best bonding agent was found to be a rubber-based contact adhesive.

The magnetostrictive element was placed in a solenoid to produce a magnetic field parallel to the fibre. Measurements were made of the change of phase delay, ∅, resulting from the application of magnetic field, B. The DC response is plotted in Fig. 6; the curve was a reasonable approximation to the expected parabolic shape, except it did not pass through the origin, due to an offset in the readings. The ordinate offset was believed to be due to magnetic remanence in the magnetostrictive material, which had not been annealed, whereas the abscissa offset was probably caused by a systematic phase offset from the point of quadrature of the CPS. From the shape of the curve, it is apparent that the small-signal AC sensitivity improves with increasing DC bias field as anticipated. Earlier workers have already reported the possibility of annealing the magnetostrictive material to reduce remanence but the long-term effectiveness of this process during magnetic "cycling" must remain in question. The frequency response of the magnetic sensor is shown in Fig. 7; the amplitude of alternating phase delay resulting from an AC field amplitude of 57uT, plus a DC bias field of 339uT, is plotted against frequency. There are several peaks of response due to mechanical resonances of the magnetostrictive element, and a region of high sensitivity around 20kHz to 50kHz. Here the sensitivity per unit length (with a DC bias of 339uT) is given by:-

$$\frac{\Delta\phi}{L\Delta B} = 1900 \text{ radians. } T^{-1}.m^{-1}$$

where L is the length of fibre bonded to the magnetostrictive strips (10cm) and B is the AC field amplitude. The minimum detectable field of the magnetic sensor is 25nT. $Hz^{-0.5}$ (based upon the minimum detectable phase change of the detector system of 5×10^{-6} radians. $Hz^{-0.5}$).

Clearly, the magnetic sensor's response could be increased by bonding greater lengths of fibre to the magnetostrictive element, using configurations such as illustrated in Fig. 8.

2.6 The application of the compensated polarimetric system for sensing pressure

Isotropic pressure will cause strain in polarisation–maintaining optical fibre, and hence a differential mode delay. Therefore a CPS for sensing pressure can be made simply by enclosing one of the fibre lengths in a chamber, which can be pressurised with the second length in close thermal contact, but unpressurised. For this study, both fibre lengths were enclosed (Fig. 9) so that they could be independently pressurised in order to demonstrate the differential and common–mode response of a compensated sensor. The sensor was arranged in the basic CPS configuration, and the changes of phase delay, $\Delta\phi$, resulting from step changes in the differential pressure between the chambers, P, were measured. When the chambers were air–filled, the sensor was found to respond more to the transient temperature rise caused by adiabatic compression than to the pressure rise itself. However, temporarily covering the fibre with water virtually eliminated the transient component of the response due to this temperature rise. (A practical sensor may utilise a compliant silicone rubber filling). The sensitivity per unit length was found to be

$$\frac{\Delta\phi}{L\Delta P} = 4.96 \times 10^{-6} \text{ radians. } Pa^{-1}.m^{-1}$$

where L is the length of the fibre pressurised (1.5m). Fig. 10 compares the differential and common mode responses of the pressure sensor. Both photographs show a trace of output voltage V_R, against time over a 40kPa step decrease in pressure. The common mode response is less than 3% of the differential response. This common mode response may have been due to a slight imbalance between the lengths of fibre in the two chambers. The pressure sensor's response may be increased simply by increasing the length of fibre pressurised, as its sensitivity is proportional to this length. In addition, this is likely to improve the common mode rejection ratio as small errors in matching their lengths, or positioning of the fibre within the chambers will be less significant.

2.7 A Compensated Polarimetric Sensor with Insensitive Downleads

The compensated polarimetric sensors described so far have been sensitive to localised strains or temperature changes over their whole lengths from source to detector. This was exploited in bringing them to a quadrature point by localised heating, but it would not be desirable in a practical sensor. A configuration for a CPS with insensitive downleads from source to sensing element and sensing element to detector is shown in Fig. 11. This is an extension of a previously published method for remoting a conventional polarimetric sensor (Ref. 2.9.4).

The sensing element consists of two identical fibre lengths joined by a 90° splice as before, whereas the extension leads are further lengths of polarisation-maintaining fibre spliced to the sensing fibre, with a 45° relative rotation about their axes at each splice. The light source system is now arranged to launch light into only one plane-polarised mode of the first downlead. Thus the light transmitted to the sensing pair of fibres is plane-polarised at 45° to their polarisation axes, as required. The light leaving the sensing element would be analysed into components at 45° to the element's polarisation axes by the second downlead. The intensity of either of the plane-polarised modes leaving the second downlead would be measured, to give an output signal proportional to $(1 + \cos \emptyset)$. The total light intensity leaving the return lead would also be measured, and used as a reference to confer insensitivity to intensity changes on the system.

Since light in only one mode of each of the extension leads is used to determine the phase delay in the sensor element, changes in the phase delays of the downleads cannot affect the output. Hence the sensor system is insensitive to any changes in temperature, pressure etc. that may be experienced by the downleads, provided that this is not significant enough to cause differential attenuation between the two modes.

2.8 Conclusions

A basic compensated polarimetric sensor system has been constructed and adapted to measure magnetic fields and pressure giving sensitivities per unit fibre length of 1900 $radians.T^{-1}.m^{-1}$ and 4.96×10^{-6} $radians.Pa^{-1}.m^{-1}$ respectively. With the detector system capable of resolving phase changes of 5×10^{-6} $radians.Hz^{-1/2}$, the minimum detectable magnetic field was $25nT.Hz^{-1/2}$, and the minimum detectable pressure change was 0.67 $Pa.Hz^{-1/2}$. These are both quite sufficient for many practical sensing applications, and greater sensitivities can be achieved simply by using higher lengths of optical fibre. In addition, a method of making the sensing element remote from the source and detector system has been presented.

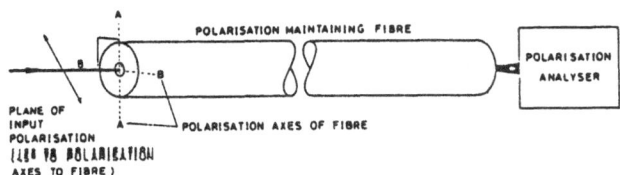

FIGURE 1: THE CONVENTIONAL POLARIMETRIC SENSOR

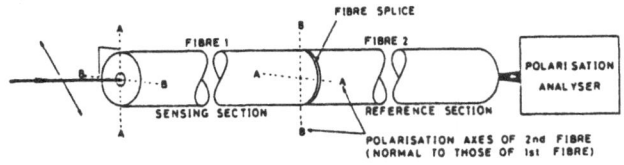

FIGURE 2: THE COMPENSATED POLARIMETRIC SENSOR

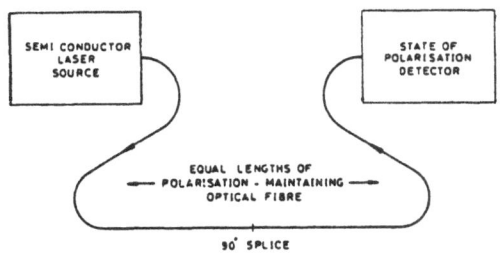

FIGURE 3: BASIC CPS CONFIGURATION

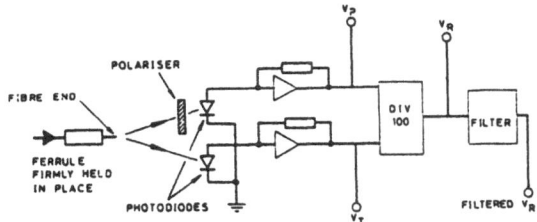

FIGURE 4: STATE OF POLARISATION DETECTOR

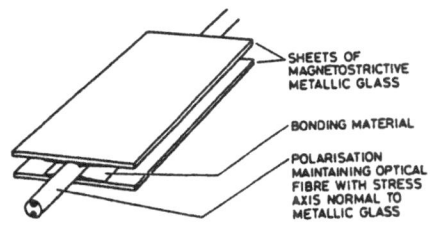

FIGURE 5: EXPERIMENTAL FIBRE OPTIC MAGNETOMETER

66

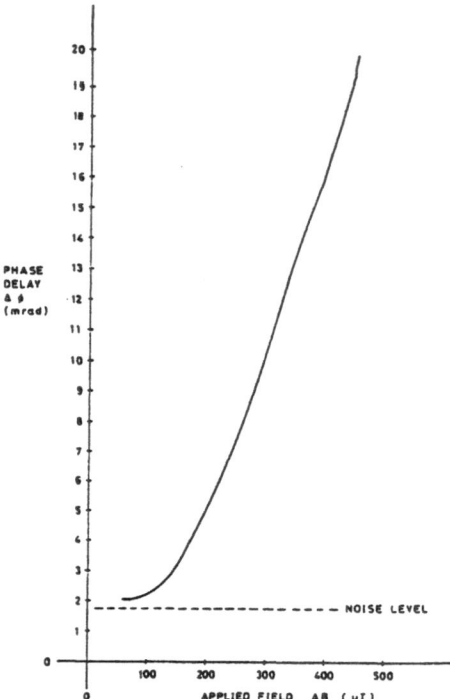

FIGURE 6: DC RESPONSE OF MAGNETIC SENSOR

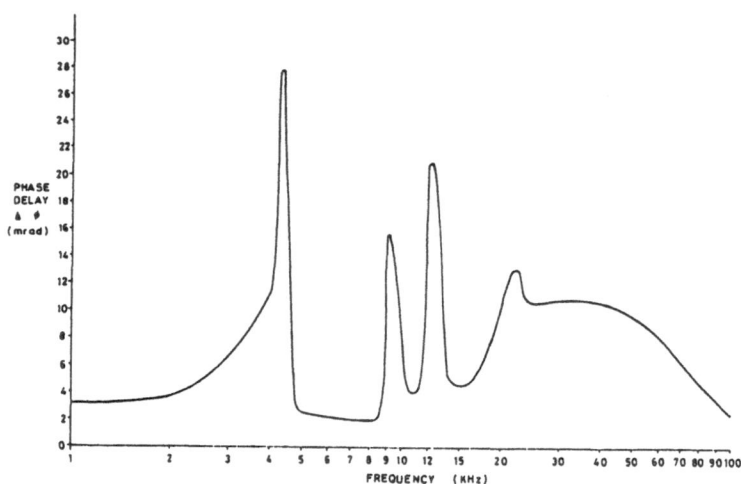

FIGURE 7: FREQUENCY RESPONSE OF MAGNETIC SENSOR

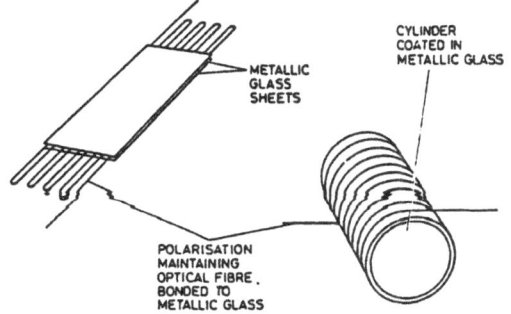

FIGURE 8: MAGNETOMETERS WITH IMPROVED SENSITIVITY

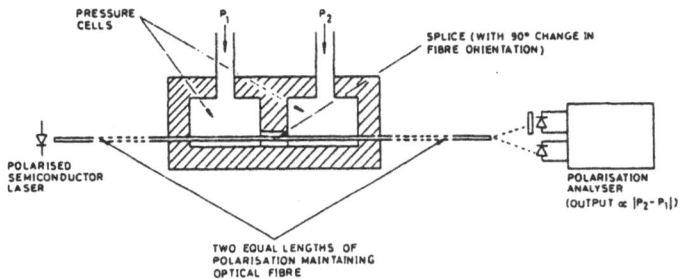

FIGURE 9: THE PRESSURE SENSOR

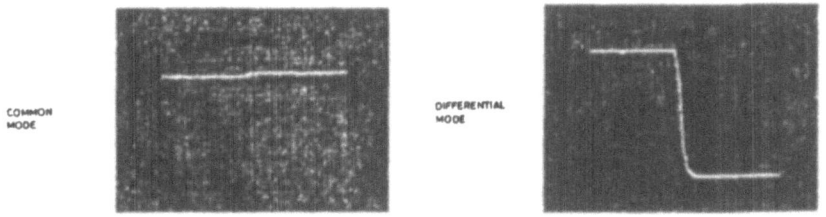

FIGURE 10: PRESSURE SENSOR RESPONSE IN DIFFERENTIAL
AND COMMON MODES

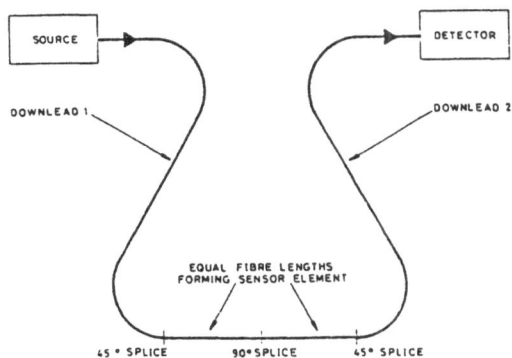

FIGURE 11: CPS WITH INSENSITIVE DOWNLEADS

68

2.9 References

2.9.1 Dakin, J.P., Wade, C.A. "Compensated Polarimetric sensor using polarisation-maintaining fibre in a differential configuration". Electron. Letts., Vol. 20, No. 1 Jan. 1984 pp 51-53.

2.9.2 Dakin, J.P., Broderick, S., Carless, D.C., Wade, C.A. "Operation of a compensated polarimetric sensor with semiconductor light source". Proc. 2nd Int. Conf. on Optical Fiber Sensors, Stuttgart 1984, pp 241-245.

2.9.3 Rashleigh S.C. "Polarimetric Sensors: Exploiting the axial stress in high birefringence fibers". 1st International Conference on Optical Fibre Sensors, London, April 1983.

2.9.4 Varnham M.P. et al. "Polarimetric strain gauges using high birefringence fibre". Electronics Letters, Vol. 19. No. 17, pp 699-700.

Acknowledgements

The author wishes to thank his colleagues, C A Wade, P B Withers, C R Batchellor and J A Rex for their contribution to the work described in this tutorial.

FIBER OPTIC GYROSCOPES

Hervé C. LEFEVRE

THOMSON-CSF, Central Research Laboratory
Domaine de Corbeville
91401 ORSAY (FRANCE)

Inertial guidance and navigation systems are known to be very useful for aircrafts, missiles, land vehicles, robots... For the moment, they make use of mechanical gyroscopes or gas laser gyroscopes. Now the development of low-loss optical fibers and compact and reliable semi-conductor light sources for the telecommunication industry has made possible a new approach : the fiber-optic gyroscope [1, 2, 3]. This device is only composed of low-mass solid-state components and does not require any mechanical movement (this is still the case with laser gyros which need dithering to solve their lock-in problem about zero rotation). This provides several specific advantages :
- withstanding of vibrations, accelerations and shocks
- operating and shelf life time
- high dynamical range (more than 1000°/s)
- fast response time (less than 1 ms)
- fast turn-on
- reasonable power consumption
- low weight
- potential low cost because the components could be massfabricated
- good engineering can make the system insensitive to temperature.

After ten years of research, yielding the demonstration of compact high performance laboratory brassboards, the fiber optic gyroscope is entering its development phase and will take a reasonable share of the market during the nineties [2].

This device is basically a Sagnac loop interferometer with an enhanced sensitivity to inertial rotation because of the use of a multiturn single-mode fiber coil. Research has been focused on designs insensitive to other mesurands but rotation.. This was possible because of the fundamental law of reciprocity of wave propagation in linear medium. It was found that the so-called "reciprocal "configuration is a sufficient condition to ensure this insensitivity to environment. In this configuration (figure 1) light coming from the source is fed through a truly single-mode spatial and polarization filter in the interferometer, where it is splitted in two waves which propagate in opposite directions along the coil. It has been shown that the interference wave which returns through the filter of the common input-output port, carries a phase shift signal which is proportional to rotation, being identically zero at rest. A second splitter is needed to send light back to a detector. Signal processing using phase modulators and the fiber coil as a delay line, allows one to get the theoretical sensitivity and bias stability limited by photon noise.

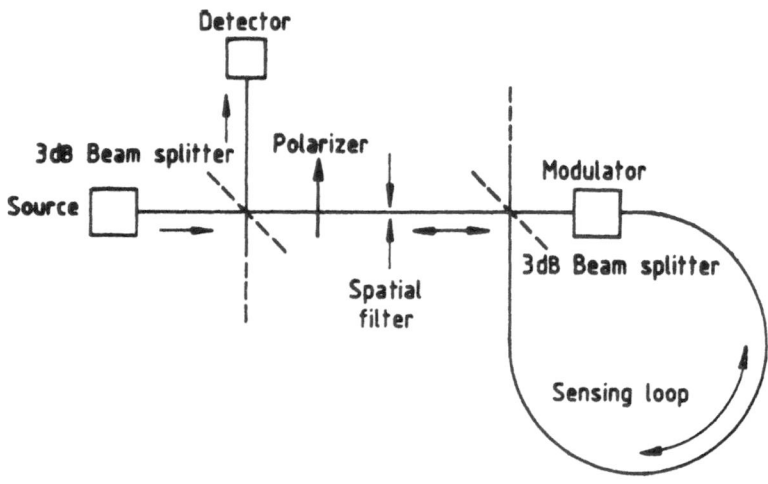

FIGURE 1."Reciprocal" configuration of a fiber gyroscope

However these performances require the use of broadband sources as superluminescent (also called superradiant) diodes. This can be explained, considering the fiber gyroscope as an interferometer where both primary waves interfere with a phase difference proportional to rotation and a continuous set of randomly coupled waves which have propagated along different optical paths. The limited coherence length of a broadband source avoids an efficient coherent detection of these low power spurious waves with the primary waves acting as a local oscillator. Furthermore the use of polarization maintaining fiber relaxes the required rejection of the polarizer and makes practical devices adequate to obtain theoretical bias stability.

Now gyroscopes measure rotation rate when users are actualy interested by its integrated value, the change in angular orientation. This calls, of course, for a good sensitivity by also for a high accuracy of the scale factor (100 to 1 ppm) over the whole dynamical range which extends over 6 to 10 orders of magnitude, depending on the application. This requires a closed-loop operation with a frequency reading ouptut. Two approaches are competing : use of acousto-optic frequency shifters (Bragg cells) and serrodyne modulation (phase ramp). The first one introduces a basic non reciprocity and requires high RF driving power. We have prefered the second solution in its digital form [4] which has none of these drawbacks. The use of integrated optics in the so-called "Y-tap" configuration (figure 2) [4] , fullfilling the conditions of the "reciprocal" configuration,allows one to use this signal processing technique with an optimal technological simplicity. Photograph 1 shows an example of a compact fiber gyro brassboard (65 mm diameter, 30 mm height) which uses this approach and which has been realized in our Central Research Laboratory with in-house components (GaAlAs

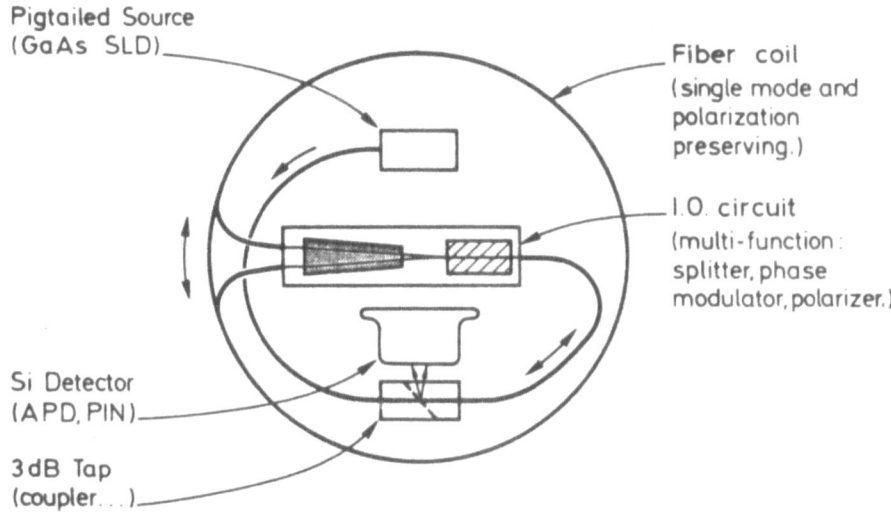

Pigtailed Source
(GaAs SLD)

Fiber coil
(single mode and
polarization
preserving.)

I.O. circuit
(multi-function :
splitter, phase
modulator, polarizer.)

Si Detector
(APD, PIN)

3 dB Tap
(coupler...)

FIGURE 2. "Y-tap" configuration of fiber gyroscope

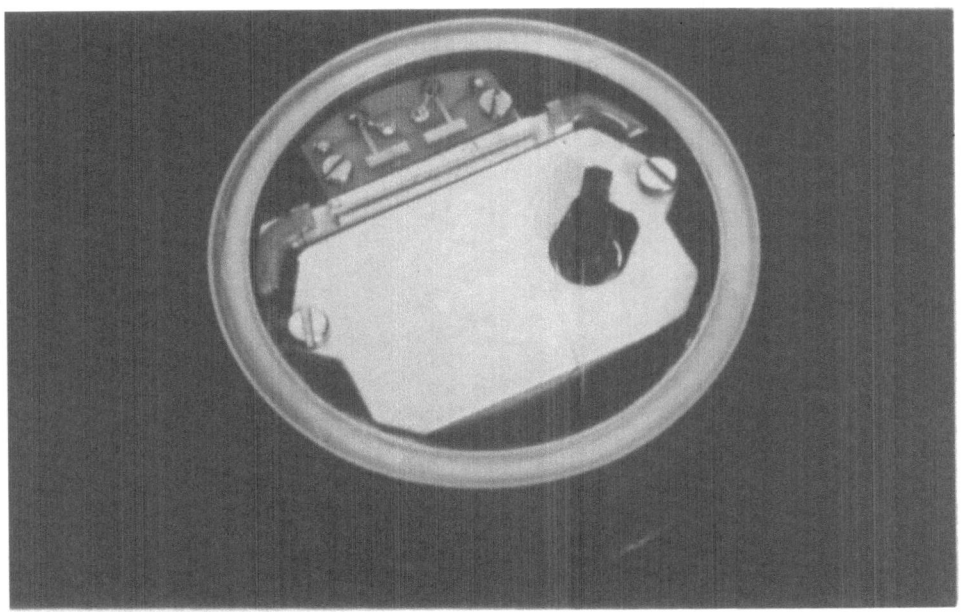

PHOTOGRAPH 1. Compact fiber gyro brassboard (65 mm diameter, 30 mm height)
using THOMSON-CSF components in a "Y-tap" configuration

72

superluminescent diode, multi-function integrated optics circuit, 3 dB tap and 250 meters of polarization maintaining fiber). Bias stability is in the μrad range for phaseshift (0.5°/h) and scale factor accuracy in the 100 ppm range using digital phase ramp.

These performances and those of other laboratories working on this subject [2] clearly show that the fibergyro is not a laboratory curiosity anymore. After going through engineering development it will find broad applications in existing markets but also widen the domain of inertial guidance to new fields because of its specific advantages with in particular its potential cost effectiveness.

REFERENCES

1. S.Ezekiel and H.J.Arditty : Proceedings of the First International Conference of Fiber-Optic Rotation Sensors and Related Technologies : published by Springer-Verlag, Springer Series In Optical Sciences, 32, 1982.
2. Proceedings of the Tenth Anniversary Conference of Fiber Optic Gyros : to be published by SPIE, vol.719, 1986.
3. For a review paper, see for example : R.A.Bergh, H.C.Lefevre and H.J.Shaw : An overview of Fiber-Optic Gyroscopes : Journal of Lightwave Technology, LT-2, 91-107, 1984.
4. H.C.Lefèvre, S.Vatoux, M.Papuchon, C.Puech : Integrated Optics : a practical solution for the fiber-optic gyroscope : SPIE Proceeding, vol.719, to be published 1986.

THEORY OF SPECTRAL ENCODING FOR FIBER-OPTIC SENSORS

R. ULRICH

TECHNISCHE UNIVERSITÄT HAMBURG-HARBURG, D2100 HAMBURG 90,
FEDERAL REPUBLIC OF GERMANY

SUMMARY

A variety of fiber-optic sensors for industrial applications use the scheme of spectral encoding. They represent the measurand by spectral features of the light that is transmitted by a fiber link from the transducer to a receiver unit. Their general advantages are a reduced sensitivity to variations of the link losses and an interchangeability of transducers without recalibration. An analysis of this encoding scheme is given from a system's point of view. Transmission of information from the measurand to the evaluation unit is treated as occurring on a number of discrete spectral channels. Decoding is performed by channel selection and by channel interpolation. The maximum attainable resolution is determined from the power budget of the sensor, distinguishing between scanning and tracking operation and systems with array detection. Finally, the immunity to variation of losses is discussed.

1. INTRODUCTION

Optical fibers offer specific advantages for use with industrial sensors in adverse environments. Electrical insulation, compatibility with extreme temperatures and ionizing radiation, and low optical power levels make the fiber an ideal medium for the transmission of information from sensors in many applications of industrial instrumentation. In order to utilize these possibilities, numerous concepts for fiber-optic sensors have been proposed in recent years [1,2,3].

Ideally with such sensors, the fiber linking transducer and receiver should be of the multimode type that operates with low optical power density and permits convenient installation with low-cost fiber-optic connectors. After evaluation, the output signal of the sensor should be independent (within limits) of the losses of the fiber-optic link. The length, type, curvature, and temperature of the fiber, as well as the quality and perfect insertion of the connectors should not affect the final sensor output. Moreover it is desirable that transducers and receivers of a sensor are interchangeable without the need for recalibration, comparable to the situation with well-designed electrical sensors. To satisfy all these requirements, an important class of proposed fiber-optic sensors operate by analog-encoding the measurand into certain features of a broadband optical spectrum that is guided by the fiber from the transducer to the receiver (see Fig. 1). Representing the measurand by the *position* of lines in that spectrum or by their power *ratio* rather than by the total transmitted power, the required independence from link losses and the interchangeability can be achieved more or less perfectly, generally at the price of a more complicated construction of these sensors with spectral encoding as compared to the simpler power-encoding sensors which typically do not satisfy the mentioned requirements.

In this paper some general properties of sensors with spectral encoding are discussed from a system's point of view. In Sec. II some typical sensors are reviewd so as to introduce here the known variants of spectral encoding. It is shown that each of them uses a number M of spectral channels, with M $\geq$ 2. In Sec. III the central idea of the paper is explained, that the operation of all these sensors may be understood to be based on two principles, i.e. on spectral channel selection and on channel interpolation. A simple digital transducer may encode the measurand into the channel number, the receiver then deciding which channel had been activated. In more sophisticated systems, two neighboring channels are excited simultaneously, and the receiver interpolates between them by evaluating their power ratio. In this way the number R of resolvable values of the measurand may far exceed the number of channels. Based on this view, in Sec. IV the attainable resolution

R is determined from the power budget of the sensor, taking into account the source power, detector noise, link losses, and filter efficiency. It is found that for low resolutions R the two encoding schemes 'fiber-coupled spectrometers' and 'fiber-coupled interferometers' show roughly comparable performance, whereas for high resolutions the coupled interferometers become advantageous. It is further shown that with either scheme the operation in a tracking mode or with an array of M detectors can permit roughly M times faster response or $M^{1/2}$ times larger link losses than the alternative, repetitive scanning operation. Finally, in Sec. V the problem is discussed how variations of the link losses may affect the sensor output. In this respect, too, fiber-coupled interferometers are found advantageous in comparison to fiber-coupled spectrometers.

2. SENSORS USING SPECTRAL ENCODING

2.1 General Arrangement

All fiber-optic sensors to be discussed here can be represented by the optical system shown in Fig. 1. A light-source in the source/receiver unit feeds light with a broad optical spectrum $W_o(\nu)$ through an optical fiber to the remotely located transducer. There, the value x of the measurand determines the spectral transmission $F_1(x,\nu)$ of some kind of filter F_1, where x may represent a position, angle, pressure, temperature, current, or any other variable that is to be measured and ν is the optical frequency in wave numbers. Light with the filtered spectrum $F_1 W_o$, containing encoded the information on x, is guided by a fiber back to the source/receiver unit, where it is decoded to recover the value of x. To this end, the light is passed through a filter F_2 with spectral transmission $F_2(y,\nu)$ onto a detector. Its electrical output signal S is proportional to the total optical power received. Suppressing that factor of proportionality,

$$S(x,y) = \int_o^\infty F_1(x,\nu) \, F_2(y,\nu) \, W(\nu) d\nu \quad . \tag{1}$$

Here the source spectrum $W_o(\nu)$, the spectral sensitivity of the

detector, and other transmission losses have been lumped together into the spectral function $W(\nu)$. It would determine the
detector signal in the absence of both filters. For simplicity
of notation, in the following S is measured directly in units
of the optical power.

According to Eq. (1), the detector signal S depends both on
the value x of the measurand and on the spectral tuning parameter y of the receiving filter. This scheme can be operated
as a sensor if the function $S(x,y)$ is such that by observing
the dependence of S on y an electronic evaluation circuit can
determine the unknown value x of the measurand. In particular,

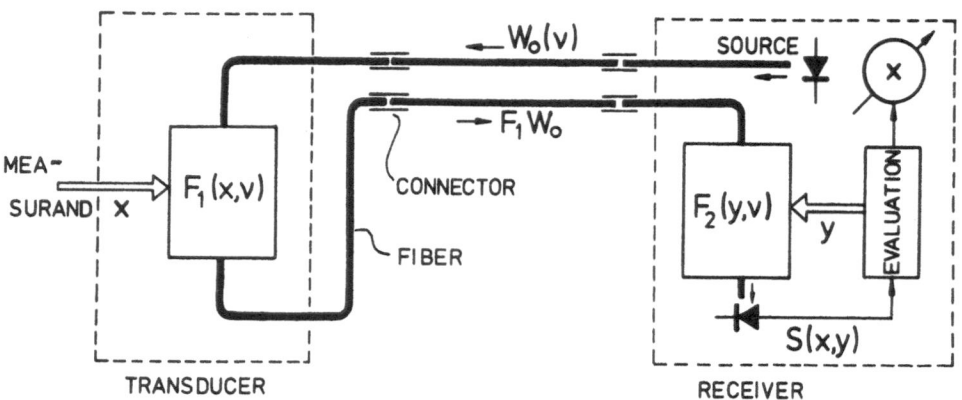

FIGURE 1. General arrangement of fiber-optic sensor system using spectral
encoding. The transducer is characterized by a spectral transmission $F_1(x,\nu)$ that depends in some characteristic way on the
value x of the measurand. In the receiver this spectrum is
decoded by a filter function $F_2(y,\nu)$.

this scheme may be used to analyze the following types of sensors.

(i)   fiber-optic sensors employing two discrete receiving
       channels with different spectral characteristics, de-
       fined e.g. by interference filters, and employing a
       transducer in which the transmission in one channel
       is modulated by the measurand in a different way as
       in the other channel. This includes, in particular,
       transducers that may be described as having a single

transducing channel, continuously tunable by the measurand.

(ii)   sensors based more generally on two fiber-coupled
       spectrometers, i.e. in which the filters F_1 and F_2
       are realized as two dispersive spectrometers of some
       kind, with F_1 being 'tuned' by the measurand, so that
       only light within a narrow spectrum centered at a
       wave number $\nu_1(x)$ is passed. The filter F_2 measures
       this ν_1, operating either in a scanning mode, i.e.
       searching for that value of y for which the detector
       signal S is maximum, thus indicating y = x, or F_2
       operates in a tracking mode in which a servo loop
       maintains S at its maximum, so that y will follow all
       variations of x.

(iii) sensors based on two fiber-coupled interferometers,
       in which the path difference in one interferometer F_1
       is being set by the measurand x, while the path dif-
       ference y in the receiving interferometer is varied
       systematically either in a scanning or in a tracking
       mode, as mentioned, to recover x.

In actual sensors, the arrangement and design of the filters
may differ considerably from those indicated in Fig. 1. For example, the sequence of F_1 and F_2 in the optical path may be inverted, so that the light of the source is passed first through
the receiving filter F_2, and is then guided by fiber to a remote transducing filter F_1 and back to the detector in the
source/receiver unit. It is obvious that the detector signal
S(x,y) in this passive optical system is not affected by such
an interchange of F_1 and F_2 in the optical path. In other possible modifications, one filter or both may be operating in reflection rather than in transmission. This requires at least
one directional coupler, but it permits to guide the light between transducer and source/receiver unit along a single fiber
instead of two separate ones.

The filter F_2 may be a spectrometer or an interferometer, as mentioned, with y denoting e.g. the continuously variable tuning angle of a prism or grating, or the displacement of a mirror, respectively. Alternatively, however, the angle or path difference in F_2 may be tuned in a number M of steps, so that y is a discrete variable, which can assume the values $y_1, y_2, \ldots \ldots y_M$. An equivalent situation exists when the receiving spectrometer or interferometer is equipped with a detector array of M elements. The variation of y then corresponds to the read-out procedure of the array. In a limiting case (M=2), the receiver may contain only two fixed spectral filters, each followed by a detector. In that case of only two spectral channels, which is characteristic of sensors of the type (i), the parameter y simply distinguishes the two channels, denoted e.g. by A and B. Thus, $F_2(A,\nu)$ and $F_2(B,\nu)$ describe their spectral transmission characteristics. Alternatively, it is possible to define M spectral channels by employing a number M of spectrally discrete light-sources, e.g. light-emitting diodes (LEDs) of different colors. Their emission spectra, weighted with the response functions of the detector and the other optics, form the spectral channels $F_2(m,\nu)$ with m = 1,2,...M. Finally it should be pointed out that the mentioned types (i)-(iii) of sensors represent those specific examples that have been investigated so far. It is conceivable, though, to operate a sensor system like Fig. 1 with much more general types of filters F_1 and F_2.

In order to illustrate these various possibilities and to form a basis for the subsequent comparison of various encoding and operating procedures, a number of examples are described in the following, see Figs. 2-5.

2.2 Examples

A simple and typical example of a fiber-optic sensor employing the two-channel spectral encoding type (i) is the fiber-optic thermometer [4] in which the filter function $F_1(x,\nu)$ is given by the strongly temperature-dependent transmission of a Cobalt-Chloride solution, see Fig. 2. The measurand is the temperature of the solution. The spectra $F_2(A,\nu)$ and $F_2(B,\nu)$ of the two receiving channels are defined by interference filters for $\nu_A \simeq 12500$ cm^{-1} and $\nu_B \simeq 15270$ cm^{-1}. The transmission F_1 of the solution varies strongly with temperature near ν_A. Near ν_B, however, transmission is high and independent of temperature, thus providing a reference for the link losses. For a determination of the unknown temperature, the ratio r=S(T,A)/S(T,B) of the two detector signals is formed electronically and evaluated by comparison with a stored calibration table.

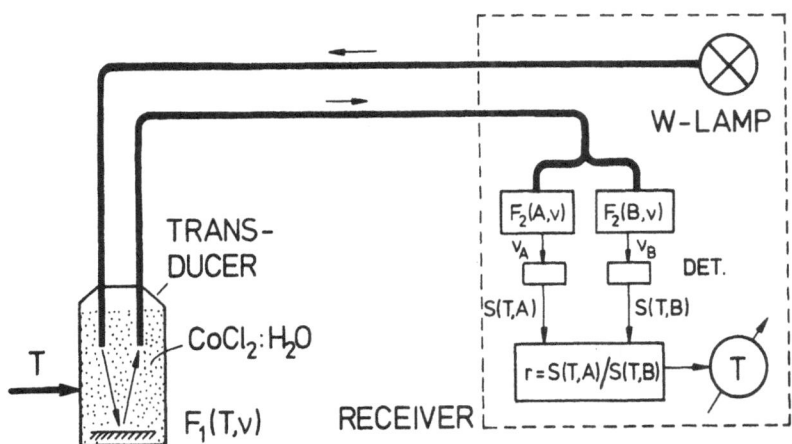

FIGURE 2. Fiber-optic thermometer [4] employing two spectral transmission channels (schematic). The encoding filter function $F_1(T,\nu)$ is the temperature-dependent transmission of the transducer cell, filled with cobalt chloride solution.

Numerous similar fiber-optic sensors emplying two spectral channels have been demonstrated. Some of them measure the concentration x of certain chemical species that change the 'coloration' $F_1(x,\nu)$ of an indicator dye [5,6]. Another one [7] measures the temperature T of a fiber with liquid cladding from its temperature-dependent transmission $F_1(T,\nu)$ near the cutoff where $n_{CLAD}(T) \longrightarrow n_{CORE}$. In these and other cases, the two spectra consist of essentially two isolated, separate bands, with as little overlap or crosstalk as possible, see Fig. 6(a).

In order to reduce the remaining dependence of the sensor

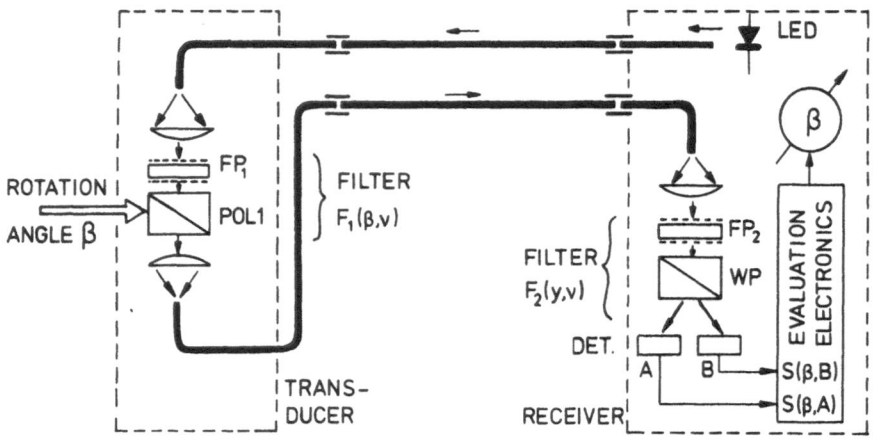

FIGURE 3. Fiber-optic angular sensor employing two interleaved spectral transmission channels [8]. The channels are defined by the birefringent Fabry-Perot filters FP_1 and FP_2. The polarizer POL1 is rotated about the direction of light-propagation through the angle β to be measured.

on variable link losses, the two spectra may also be finely interleaved, as indicated in Fig. 6(d). This type of encoding by 'fiber-coupled comb filters' has been demonstrated [8] in an angular sensor depicted schematically in Fig. 3. Here, either channel consists of a set of equidistantly spaced narrow spectral lines. They are defined by birefringent Fabry-Perot filters (FP1 and FP2). The lines of one channel can pass the filters in one linear state of polarization, those of the other channel in the orthogonal state. With the help of birefringence, the interference orders of the two polarizations are designed to differ by 1/2 or an odd multiple thereof, and therefore the lines of the two channels are interleaved. When the polarizer POL1 is rotated to an azimuth ß, the filter F_1 has a transmission proportional to $\cos^2 ß$ for the set of lines in the first channel, and to $\sin^2 ß$ for those of the second set. Therefore, this arrangement can be used as sensor for the angular position ß of the polarizer POL1. In the receiver, an identical filter F_2 separates the two channels again, and two detectors produce the signals $S(ß,A) \sim \cos^2 ß$ and $S(ß,B) \sim \sin^2 ß$. Their ratio may be formed electronically, and evaluation by inverse trigonometric functions yields the value of ß.

An example of a sensor of group (ii) is the arrangement of fiber-coupled sepctrometers [9] shown schematically in Fig. 4. Here, the measurand is again an angle, denoted as $ß_x$. It determines the rotation of the grating in a grating spectrometer, which serves as the transducing filter, $F_1(ß_x, \nu)$. Consequently, $ß_x$ is encoded into the optical frequency $\nu_1(ß_x)$ of the light that can pass from the source through F_1 and back to the re-

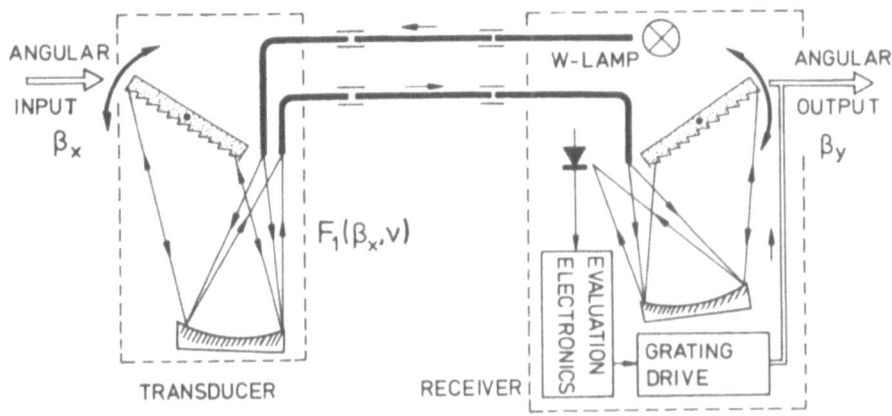

FIGURE 4. Fiber-optic angular transmission system based on two fiber-coupled grating spectrometers and a broad-band source [9]. By the addition of a conventional sensor measuring the output angle $ß_y$, this system becomes a complete fiber-optic angular sensor.

ceiver unit. There, a second grating spectrometer (filter F_2) is tuned by rotation of its grating through an angle $ß_y$. If this second spectrometer is of sufficiently similar design as the first one, the light of wave number $\nu_1(ß_x)$ arriving at F_2 can pass to the detector and produce an output signal S only if $ß_y = ß_x$. This condition can be reached momentarily during a scan of $ß_y$, or it can be maintained by a servo-loop, as mentioned earlier. Alternatively, the mechanical movement may be replaced by the electronic scanning in an array of detectors. In any case, a maximum of the detector signal S(x,y) indicates that $ß_x \simeq ß_y$. In that situation, the measurand $ß_x$ is transmitted fiber-optically from the transducing spectrometer to the receiving spectrometer, where it is reproduced as β_y. Strictly speaking, then, this system is not a sensor, but a transmission system. It becomes a sensor, however, if some conventional angular sensor is added to the receiver unit to read out β_y when S is maximum.

From a system's point of view, this arrangement of two fiber-coupled spectrometers may be viewed as employing a number M of spectral transmission channels. Each channel may be taken to have a spectral width $\Delta\nu$ that is equal to the spectral resolution of the receiving spectrometer. Typically (when input slit and output slit have equal widths) the spectral transmission of a prism- or grating spectrometer has the triangular shape depicted in Fig. 6(a). It is reasonable then to choose $\Delta\nu$ as the full width of this triangle at half maximum height (FWHM). The number of distinguishable transmission channels in this case is of the order of $M = B_\nu / \Delta\nu$, where B_ν is the bandwidth of the effective source spectrum $W(\nu)$.

An illustration of a sensor of group (iii) is given in Fig. 5(a). The arrangement of two fiber-coupled Michelson-interferometers [10,11,12] with a broadband source represents a trans-

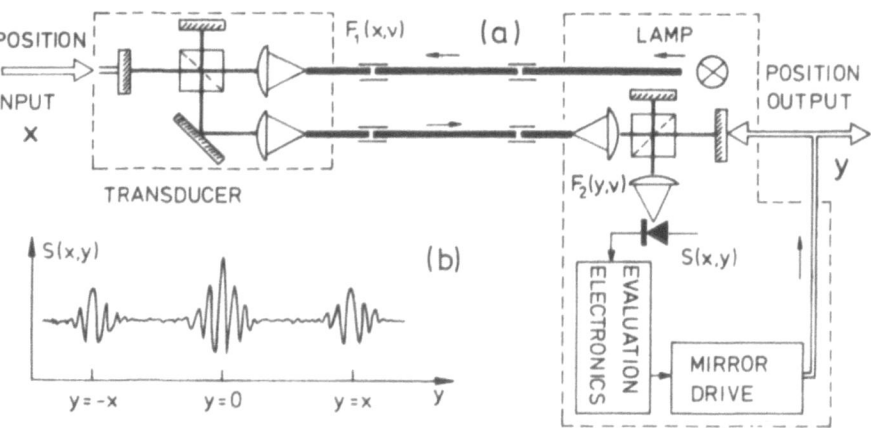

FIGURE 5. Fiber-optic transmission system for position and displacement, consisting of two fiber-coupled Michelson interferometers and a broad-band source [10]. By the addition of a conventional position sensor measuring the output position y, this system becomes a complete fiber-optic sensor. (a) arrangement of optics, (b) detector signal during one scan of the receiving interferometer.

mission system for displacements or positions, in the sense just mentioned. The measurand moves one mirror in the first interferometer, which serves as the transducing filter F_1. It has a periodic transmission function, $F_1(x,\nu) \simeq \cos^2(\pi x \nu)$, where x denotes the interferometric path difference, i.e. half the mirror displacement. To avoid ambiguities we postulate here that x > 0. In this interferometric sensor arrangement, the measurand is encoded into the period $\delta\nu = 1/x$ of the optical spectrum that is guided from the transducer back to the receiver. There, a second interferometer analyzes this spectrum by variation of its path difference y, with a filter function $F_2(y,\nu) \simeq \cos^2(\pi y \nu)$. When $y = \pm x$, both filter functions F_1 and F_2 agree perfectly, and the detector signal assumes a maximum, as indicated in Fig. 5(b). Actually, next to that main maximum there exist a number of smaller maxima, spaced nearly regularly in the y scale with a period $1/\bar{\nu}$, where $\bar{\nu}$ is the mean wavenumber of the spectrum $W(\nu)$. They extend over a distance roughly equal to the coherence length l_c of $W(\nu)$.

When the receiver is tuned to the main maximum y = x, it means that the value of x is reproduced in the receiver. To convert this transmission system to a true displacement sensor, a conventional absolute displacement sensor must be added at the receiver to determine y. This can be done conveniently here by feeding a laser beam of known wavelength (e.g. HeNe-laser) through the receiver interferometer, along with the light from the transmitter, and evaluating the laser interference fringes [10,11,13].

Again, a number of variations of this basic scheme are possible. The system may be operated in a mechanical scanning [10,12] or tracking mode [11]. Electronic scanning by an array of detectors is also conceivable [14]. The interferometers employed as the transducer and receiver may both be of the Michelson type [10,12] or both of the Fabry-Perot type [15], or both types may be combined [11]. With two Michelson interferometers, the evaluation of the situation y = x can be done in a unique way. When a Fabry-Perot interferometer is involved, however, subsidiary maxima of the detector signal may

appear whenever y/x is equal to the ratio of two (small) in-
tegers [11], and the evaluation of x may become ambiguous. For
the identification of the maximum of the detector signal, two
basically different schemes have been demonstrated. With carefully balanced interferometers and a source of large bandwidth
(e.g. tungsten lamp [10,13]) the zero-order fringe at y = x of
the interferogram can be clearly identified, because its amplitude well exceeds that of the higher-order fringes. This case
corresponds to the situation in Fourier-spectroscopy, and the
system may be characterized as being a multi-channel transmission system in which two neighboring channels have a separation $\Delta x = 0.5/\hat{v}$. This fact follows from Nyquist's sampling

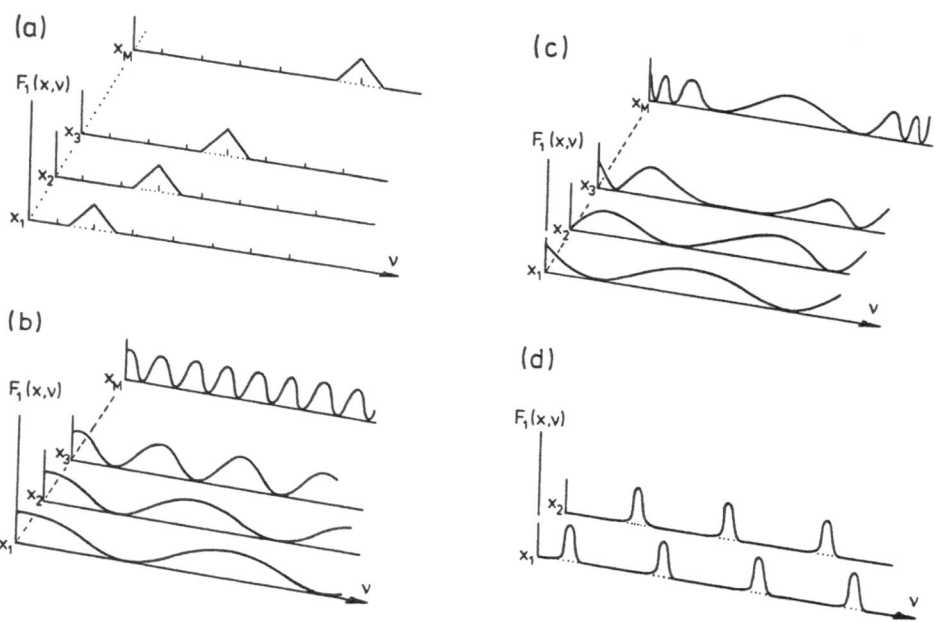

FIGURE 6. Typical transmission spectra of transducing filters used for
multi-channel spectral encoding. Each of these spectra defines
one "channel" of signal transmission. (a) spectra of a grating
spectrometer as used in Fig. 4; (b) spectra of a Michelson interferometer as used in Fig. 5; (c) general spectra which (except
for their DC-part) are "orthogonal"; (d) spectra of the two-
channel sensor used in Fig. 3, which may be generalized to a
larger number of channels.

theorem, with $\hat{\nu}$ representing the highest optical wavenumber contained in the spectrum $W(\nu)$. According to another procedure of operating the system of coupled interferometers, less demanding with respect to the interferometer balance, the envelope of the interferogram [i.e. the 'visibility curve' of the spectrum $W(\nu)$] is evaluated in the receiver [12]. Here, the accuracy of determining the condition y = x is reduced, as individual fringe orders are not identified. In that case, the separation of neighboring channels may be taken as $\Delta x = l_c$ where l_c is the coherence length, i.e. that path difference at which the fringe visibility drops to half of its maximum value. For either possibility of channel definition, the spectra $F_2(y,\nu)$ of the channels have the form indicated in Fig. 6(b). The total number of channels is M = X/Δx where X is the range of variation of x.

3. EVALUATION PROCEDURES

For the receiver in a spectrally encoding sensor, the problem exists of retrieving the value of x from the arriving spectrum $F_1(x,\nu)W(\nu)$. Formally, this problem can be solved [16,17] on the basis of a systematic analysis of that spectrum, combined with the knowledge of the transducer characteristics $F_1(x,\nu)$. Practically, however, it is desirable to perform such an analysis in the most economic way, by using a particularly simple receiver that is 'matched' to the possible spectral characteristics produced by the transducer. Yet, some sophistication in the encoding and resultant complexity of transducer and receiver are necessary so that the spectral features characteristic for the measurand can be distinguished from those modifications of the spectrum that are caused by variations of the link losses. This complexity increases with the required degree of immunity against such variations, and also with the required accuracy of the sensor, expressed e.g. by the number R of resolvable values of the measurand.

It will be shown now that the evaluation procedures used in the receivers may be understood as being based on two principles, i.e. channel selection and channel interpolation. The first principle is employed in sensors of type (ii) and (iii), which have a large number M of spectral transmission channels, as considered earlier. In such sensors, the transducing filter passes light in essentially only one particular spectral channel $F_1(x,\nu)$ that is determined by the value x of the measurand. The receiving filter $F_2(y,\nu)$ is 'matched' in its design to the transducer and is scanned or dithered so that it can recognize which channel is activated. Consequently, this principle of transmitting information may be considered as an encoding of x into the channel number m. In this sense, spectral encoding may be seen as a *binary digital* modulation scheme [18]. Like other digital schemes, it is fairly immune against variations of the link losses. This digital nature would be most obvious in a system of fiber-coupled spectrometers equipped with a detector array in the receiver. There, the evaluation procedure could be reduced essentially to the decision which channel m of the array produces the highest signal. When used in this simple digital fashion, the number R of resolved x values in such a system is equal to the total number M of its channels. It is interesting to note that this resolution represents a fairly poor utilization of the maximum resolution capacity, $R_{max} = 2^M$, however, which a binary M-channel system could theoretically provide, if all channels can operate simultaneously and independently.

If the signal/noise ratio is high in each channel, the resolution of the above-described system may be increased by invoking the second evaluation principle mentioned, i.e. the interpolation between neighboring channels in the receiver. When the transducer is tuned continuously by the measurand, the spectrum $F_1(x,\nu)W(\nu)$ arriving at the receiver may extend over more than one channel of the receiving filter F_2. Thus, the receiver generally detects optical power in at least two of its channels. In sensors of the coupled spectrometer type (ii) with ideally square filter characteristics in transducer and receiver, no other channels than those two would be excited. With real filters, however, and especially in sensors of type (iii), many or all of the other channels will also receive some power. In any case, interpolation between the signals $S(x,y_m)$ and $S(x,y_{m+1})$ of the two channels (m and m+1) excited most strongly can yield x with a higher resolution than a simple evaluation from the channel number alone. This encoding by channel number with subsequent interpolation represents a mixed digital/analog modulation. In the case of the simple 2-channel sensors listed earlier as type (i), however, only the analog interpolation part exists.

In the following, these principles of channel selection and channel interpolation are discussed in more detail. The emphasis that will be put here on the concept of discrete transmission channels may appear unnecessary for some basically analog/continuous systems like the fiber-coupled interferometers. The channel concept forms a convenient basis, however, for the subsequent treatment of errors caused by detector noise and by variations of the link losses.

3.1 Channel Selection

Sensors with multi-channel spectral encoding were illustrated above - and have been demonstrated so far - only with spectrometers and interferometers, i.e. with those kinds of filters which are well known from spectroscopy. Multi-channel sensors in the arrangement of Fig. 1 are conceivable, however, also with filters of much more general spectral transmission functions than those used in spectroscopy. One hypothetical example is given in Fig. 6(e). In fact, under fairly general conditions any spectral filter whose transmission $F(x,\nu)$ depends continuously on some tuning parameter x may be used - at least in principle - as a transducer for that parameter x in a sensor with spectral encoding. It defines a transducer channel that is characteristic of x. An identical or sufficiently similar filter could serve as a matched receiver. It has a number $M \geq 2$ of receiving channels, characterized by different values of the tuning parameter y. Out of these channels, the scanning or tracking process selects that one which best matches the transducer channel. In practice, of course, not all tunable filters are suited. The dependence on x or y should be a strong one, so that a large number of channels and, therefore, a high resolution results.

To explain these possibilities and their preconditions, a real spectral amplitude distribution $w(\nu)$ is formally introduced by

$$w(\nu) = +[W(\nu)]^{1/2} . \tag{2}$$

This spectrum is written in the Dirac notation as a spectral vector $|w>$ in an abstract vector space. Introducing also the scalar product of two spectral vectors $|u>$ and $|v>$,

$$<u|v> = \int_{0}^{\infty} u(\nu) \ v(\nu)d\nu \ , \tag{3}$$

the detector signal in the absence of both filters can be expressed as

$$P = <w|w> . \tag{4}$$

When both filters $F_1(x,\nu)$ and $F_2(y,\nu)$ are inserted, the detector signal is given by Eq. (1) and may be interpreted as the scalar product

$$S(x,y) = <wF_1(x)|F_2(y)w> \tag{5}$$

$$= <u_1(x)|u_2(y)> \tag{6}$$

of two spectral vectors $|u_1(x)>$ and $|u_2(y)>$ which result when the filters F_1 and F_2, respectively, 'operate' on the fundamental spectrum $|w>$,

$$|u_1(x)> = F_1(x)|w> = |F_1(x)w>$$

$$|u_2(y)> = F_2(y)|w> = |F_2(y)w> . \tag{7}$$

In these expressions (5)-(7) the argument ν has been suppressed in the filter functions F_1 and F_2. The vector $|u_1>$ characterizes the transducer channel, and $|u_2>$ one of the receiver channels.

Using this formalism, it can now be shown that the value of x can be determined in a general sensor (Fig. 1) with two sufficiently similar filters by variation of y and subsequent evaluation of that value $\hat{y}$ at which the *normalized* detector function

$$s(x,y) = S(x,y)/[S(y,y)]^{1/2} \qquad (8)$$

assumes its maximum. The value of the unknown measurand is x = $\hat{y}$.

This procedure is a consequence of Schwarz's inequality [18] for integrals or for scalar products of the form (6). It states that

$$|<u_1|u_2>|^2 \leq <u_1|u_1> <u_2|u_2>$$

or

$$S(x,y) \leq [S(x,x) \ S(y,y)]^{1/2} \qquad . \qquad (9)$$

The = sign holds only when x = y. Therefore, the ratio

$$\sigma(x,y) = \frac{S(x,y)}{[S(x,x) \ S(y,y)]^{1/2}} = \cos\beta_{12} \approx 1 - \left(\frac{x-y}{b}\right)^2 \pm \ldots \qquad (10)$$

has its maximum at y = x. In the terminology of linear vector spaces, this ratio σ is the cosine of the 'angle' β_{12} between the vectors $|u_1>$ and $|u_2>$. This angle vanishes only when the two vectors are parallel, i.e. when x = y occurs in a sensor with identical filters. When the ratio $\sigma(x,y)$ is expanded in a power series about the position of the maximum, the term linear in (x-y) vanishes. Among the remaining terms, the leading two describe an approximating parabola whose width is characterized by a parameter b, as expressed in (10).

For the evaluation of x in the receiver the existence of this maximum is used. The function s(x,y), defined in (8), differs from the ratio σ only by the factor $[S(x,x)]^{1/2}$, which is constant during the variation of y. Therefore, the maximum of

$s(x,y)$ is found at the same position $\hat{y} = x$ as the maximum of the ratio σ.

The direct (i.e. non-normalized) detector signal $S(x,y)$ has its maximum generally at a different position. For a correct determination of x, therefore, the detector signal must be normalized with the denominator function of (8), i.e. with

$$Q(y) \equiv [S(y,y)]^{1/2} = [\int F_1(y,\nu)F_2(y,\nu)W(\nu)d\nu]^{1/2} \quad (11)$$

This function can be obtained in practice by a synchronous variation of x and y over the entire range of the tuning parameters, maintaining $x = y$. To be used in (8), then, the shape of $Q(y)$ would have to be recorded once and taken into account later in determining the maximum of $s(x,y)$. In this sense, the function $Q(y)$ forms a part of the calibration of a sensor. However, only the functional shape of $Q(y)$ matters. A constant factor in $Q(y)$ is irrelevant as it does not affect the position of the maximum.

It is interesting to evaluate the shape of this function $Q(y)$ for the two types (ii) and (iii) of sensors. For fiber-coupled spectrometers of moderate or high resolution, the factor $F_1 F_2$ in (11) extends over a narrow spectral interval only. Hence, unless the fundamental spectrum $W(\nu)$ has sharp structures, it may be considered as approximately constant in the integration (11). Assuming further a flat response of the spectrometers, (11) yields $Q(y) = w(\nu_1(y))$, apart from an irrelevant factor. Here $\nu_1(y)$ denotes the tuning characteristic of the spectrometers, i.e. the wavenumber ν_1 of light that passes at a given tuning y.

In the case of two fiber-coupled Michelson interferometers with not too small path differences and with a flat overall response, the factor $F_1(y,\nu)F_2(y,\nu) \simeq \cos^4(\pi y \nu)$ is a rapidly oscillating function whose average value is independent of y. Therefore, the integral (11) is a constant in very good approximation, and $Q(y) = 1$. The measurand x may be found in this case simply from the maximum of $S(x,y)$.

91

3.2 Channel Width

Spectrally encoding sensors with a number of discrete detectors in their receiver, e.g. with an array of photodiodes,
would reasonably be so designed that each detector corresponds
to approximately one resolution element, and each detector
would then define one channel. In receivers with continuous
tuning of y, however, the definition of channels is somewhat
arbitrary. The narrower the channels are chosen, the more
crosstalk exists among them. A simple and general possibility
is to choose the channel width, i.e. the separation of adjacent channel centers, in the y scale equal to the width parameter b of the normalized detector signal $s(x,y)$. Actually, b
had been defined with the ratio σ instead of s in (10). Concerning the y dependence, however, the function $s(x,y)$ is proportional to the ratio σ. Therefore, both have maxima of equal
widths. With b thus specified the channels are simply arranged
next to each other, filling the entire tuning range X. Their
number is $M = X/b$.

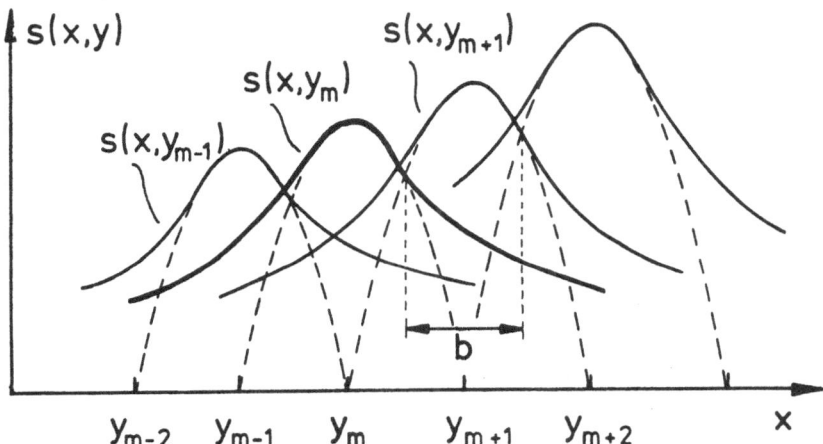

FIGURE 7. Schematic representation of the channel responses [normalized
detector signals $s(x,y_m)$] versus the value x of the measurand
in a sensor with multi-channel encoding. The parameter b of
the approximating parabola is used to define the channel width
and channel separation.

The response of the receiving channels so defined is shown
schematically in Fig. 7 as a function of the value x of the
measurand. By the definition given, the parabolas approximating

s(x,y) have the widths b between their points of intersection,
i.e. at 0.75 of their peak height, and the widths 2b at their
bases.

The definition of b as the channel width measured in the y
scale agrees well with the values given earlier on the basis
of spectroscopic practice. In particular, for a sensor with
two fiber-coupled spectrometers, each having a triangular filter function of FWHM resolution $\Delta\nu$, the width b of the detector signal in the y scale corresponds to a width $(2/3)^{1/2}\Delta\nu \simeq$
$0.82\Delta\nu$ in the ν scale, as can be shown from the overlap integral of two triangular filter functions. For a sensor with two
fiber-coupled interferometers, the width of the detector sig-
nal in the y scale is $b = 2/(\bar{\nu}\sqrt{3}) \simeq 0.37/\bar{\nu}$, where $\bar{\nu}$ denotes
the r.m.s. average wavenumber in the spectrum $W(\nu)$. This fig-
ure is in reasonable agreement with the channel width $0.5/\hat{\nu}$
given earlier from Nyquist's theorem.

The spectral transmission functions of the channels were
illustrated in Fig. 6 for the typical cases. The curves of
Fig. 6(a)-(d) actually refer to the transducing filters, operating at a number of discrete values x_m of the measurand.
Likewise, however, these curves may also be interpreted to represent the spectral transmission $F_2(y_m,\nu)$ of individual receiving channels.

3.3 Channel Interpolation

3.3.1 Two-Channel Sensors

In sensors with two spectral channels, A and B, the evaluation of x from the *power ratio* of the two receiving channels,
$r = S(x,A)/S(x,B)$, is the standard procedure to achieve immunity against variations of the link losses. This ratio-evaluation is closely related to the evaluation by *interpolation* between two adjacent channels that may be used in a multi-channel
sensor. Actually, the two procedures are indistinguishable from
the view-point of the receiver which analyzes the entire arriving light flux. A difference may be construed from the encoding process in the transducer. In sensors of the type (i)

specified above, the transducer may operate either with two discrete, fixed channels and individually modulate their power transmissions. This could be called 'ratio-encoding'. Alternatively, the transducer may encode x by setting a characteristic frequency ν_1 of its filter F_1 at some value $\nu_1(x)$ between the characteristic frequencies of the two channels, thus performing an interpolation. In the other sensors, type (ii) and (iii), the situation is similar. A continuous variation of x from one channel to the adjacent one causes the power transmission of the transducer to decrease in the first channel and to increase in the second one. On the other hand, the transducer could also be described to pass only one channel that is continuously tunable. In view of this ambiguity, both methods of evaluation are discussed now under the common heading "interpolation", because this is the broader technique. The two-channel case is treated first, providing also a further illustration of the sensors of type (i).

1		$F_1 = xT_A(\nu) + T_B(\nu)$	$x = \dfrac{1 - 2\eta}{1 + 2\eta}$
2		$F_1 = (1-x)T_A(\nu) + T_B(\nu)$	$x = \dfrac{4\eta}{1 + 2\eta}$
3		$F_1 = (1-x)T_A(\nu) + xT_B(\nu)$	$x = \dfrac{1}{2} + \eta$
4		$F_1 = \dfrac{1}{2}(1-x)T_A(\nu) + \dfrac{1}{2}(1+x)T_B(\nu)$	$x = 2\eta$
5		$F_1 = T_A(\nu)\cos^2\pi x/2 + T_B(\nu)\sin^2\pi x/2$	$x = \dfrac{1}{\pi}\arcos(-2\eta)$

TABLE I. Modulation techniques for two-channel sensors. For η see Eqs.(17)-(19).

The two transducer channels are defined by spectral filters of transmission $T_{1A}(\nu)$ and $T_{1B}(\nu)$, both overlapping with the fundamental spectrum $W(\nu)$. By these filters, the channels are separated in the transducer, and their transmissions are modulated then in different ways by the measurand. The channels are combined again and are transmitted jointly to the receiver. Some possibilities of modulation in the transducer are listed in Table I, where it is assumed for simplicity that the measurand varies in the range $0 \leq x \leq 1$. In the examples 1 and 2, only channel A is modulated by x, whereas channel B is left unmodulated and serves as a reference [4]. In all other examples, both channels are modulated in opposite senses. The examples 1-4 in Table I, having linear x-dependencies of the channel transmission, are representative of displacement or angular sensors in which a shutter is used as a modulator. The example 3, in particular represents simple *linear* interpolation between two channels, see Fig. 8. This example is also the basis of the graphical representation of two-channel encoding/decoding in Fig. 9. Here $|A\rangle$ and $|B\rangle$ denote the spectral vectors $|F_1(A)w\rangle$ and $|F_1(B)w\rangle$ of the two limiting channels, respectively, and $|x\rangle$ stands for the spectral vector $|F_1(x)w\rangle$ that represents the value of the measurand.

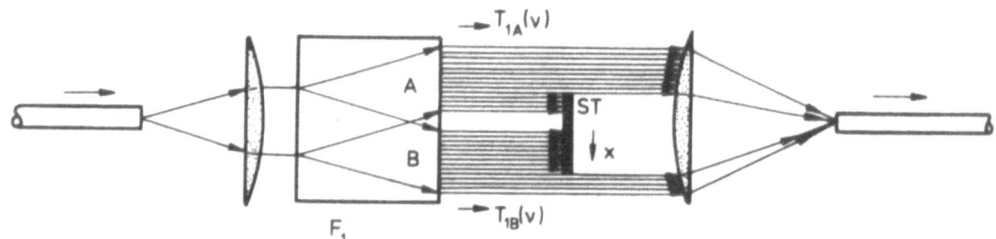

FIGURE 8. Schematic representation of the transducer in a sensor with two-channel spectral encoding, using linear interpolation between the two channels A and B, illustrating the example 2 of Table I. Modulation is achieved by means of a moving stop ST whose position represents the measurand x.

The last example of Table I applies to sensors in which the two channels are modulated in opposite senses by a rotating

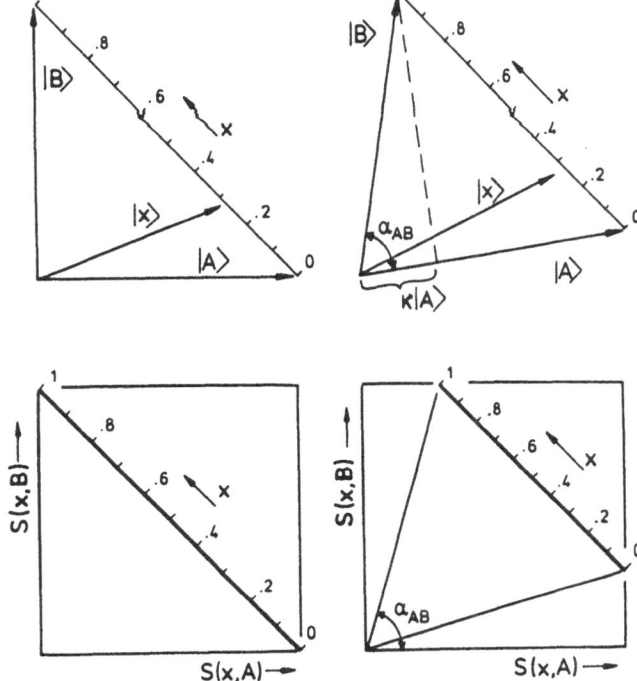

FIGURE 9.   Schematic representation of the spectral vectors (top) and of
the variation of the signals (below) in sensors with two-channel
linear interpolation (see e.g. Fig. 8), shown for idealized filters without crosstalk (left) and for real filters with crosstalk
(right).

polarizer [8] or by measurand-induced birefringence. Illustrating this example further, the transducing filter function is

$$F_1(x,\nu) = T_{1A}(\nu)\cos^2 \pi x/2 + T_{1B}(\nu)\sin^2 \pi x/2 \quad . \qquad (12)$$

This function represents both filters T_{1A} and T_{1B}, as well as
the modulator and the channel recombination. The corresponding
receiver is assumed to be equipped with filters of spectral
characteristics $T_{2A}(\nu)$ and $T_{2B}(\nu)$, each followed by a detector.
Hence

$$F_2(y,\nu) = \begin{cases} T_{2A}(\nu) & \text{if} \quad y=A \quad \text{(channel A)} \\ \\ T_{2B}(\nu) & \text{if} \quad y=B \quad \text{(channel B)} \end{cases} \qquad (13)$$

The resulting general detector signal $S(x,y)$ is

$$S(x,A) = P \left(t_{AA}\cos^2 \pi x/2 + t_{BA}\sin^2 \pi x/2\right)$$

$$S(x,B) = P \left(t_{AB}\cos^2 \pi x/2 + t_{BB}\sin^2 \pi x/2\right) \tag{14}$$

where

$$t_{ij}P = \int_0^\infty T_{1i}(\nu)\, T_{2j}(\nu)W(\nu)d\nu \tag{15}$$

for i,j=A,B and with P denoting the power (4) of the fundamental spectrum. According to (15), then, t_{AA} is the apparent power transmission through the two corresponding filters T_{1A} and T_{2A} in series, measured with the spectrum $W(\nu)$. Likewise, t_{BB} holds for the corresponding filters T_{1B} and T_{2B} in series, and the off-diagonal coefficients t_{AB} and t_{BA} describe the "crosstalk" transmission through two *different* filters in series. Straightforward evaluation of (14) yields the value of x from the two detector signals

$$x = (1/\pi)\ \text{arcos}\ 2\eta \tag{16}$$

where

$$\eta = \frac{1}{2}\ \frac{(t_{AA}+t_{BA})S(x,B) - (t_{BB}+t_{AB})S(x,A)}{(t_{AA}-t_{BA})S(x,B) + (t_{BB}-t_{AB})S(x,A)}\ . \tag{17}$$

Exactly the same combination η of parameters appears in the evaluation formulas of the other examples of Table I. It is apparent, then, that this combination η is characteristic of two-channel interpolation in general, assuming equal statistical weights of both channels. The actual dependence $x(\eta)$, on the other hand, depends on the details of the modulation process employed.

In any case, the evaluation becomes particularly simple in the special situation when $t_{AA} = t_{BB}$ and $t_{AB} = t_{BA}$ holds, which exists e.g. in the two-channel encoding with interleaved spectra and identical filters in transducer and receiver [8]. Abbreviating the relative crosstalk of the filters T_A and T_B by $\varkappa = t_{AB}/t_{BB} = t_{BA}/t_{AA}$, the value of η becomes

$$\eta = \frac{1+\varkappa}{1-\varkappa} \left[\frac{1}{2} \frac{S(x,B) - S(x,A)}{S(x,B) + S(x,A)} \right] \tag{18}$$

$$= \frac{1+\varkappa}{1-\varkappa} \left[\frac{1}{2} \frac{1 - r}{1 + r} \right] \tag{19}$$

These expressions show clearly the relationship between interpolation and ratio-evaluation. The bracket [] in (18) is an
interpolation factor. It ranges from $-1/2$ to $+1/2$. The value
$-1/2$ is assumed when $S(X,B)$ vanishes completely, i.e. when
only the first channel is activated ($x \to 0$) and crosstalk is absent. The other extreme value, $+1/2$, is assumed for $x \to 1$ when
only the second channel is active, i.e. when $S(x,A) \to 0$ and
crosstalk is zero. Hence, this interpolation factor [] varies
by one unit from channel A to channel B. In practice, with finite crosstalk, this factor varies over only a part of the
$-\frac{1}{2} \ldots +\frac{1}{2}$ range. It is essential now that this interpolation
factor depends only on the *ratio* r of the two signals. Dividing
numerator and denominator of (18) by $S(x,B)$ yields line (19),
in which this dependence on r is expressed explicitly. The
same considerations apply to the factor η in its more general
form (17). Consequently, interpolation between the signals of
two channels in the form (17) or (18) always includes ratio-
evaluation.

In this context it is interesting to note the role of channel crosstalk which is apparent from (18) and (19). With all
t_{ij} being positive, it follows from (14) that crosstalk pre-
vents both signals $S(x,A)$ and $S(x,B)$ from reaching zero. Thus,
crosstalk reduces the range over which $(1-r)/(1+r)$ varies, as
had been mentioned. These signal variations without and with
crosstalk are shown schematically in the lower half of Fig. 9.
In the evaluation of (18) and (19), then, the role of the
crosstalk term $(1+\varkappa)/(1-\varkappa)$ is to formally expand again that
range of variation to the full interval $(-\frac{1}{2} \ldots +\frac{1}{2})$ that would
exist without crosstalk. The inverse of that term,

$$\gamma = (1-\varkappa)/(1+\varkappa) < 1 \tag{20}$$

is recognized as the factor by which the modulation is reduced
due to crosstalk between the two channels.

Furthermore it should be noted that the described interpolation procedure for two-channel sensors may also be viewed as
a special case of the general procedure of determining x from
the maximum of s(x,y). If y is imagined to vary continuously
between the channels A and B, the normalized signal s(x,y)
would assume a maximum somewhere between A and B. The role of
interpolation is simply to locate the position $\hat{y}$ = x of that
maximum, using the relation between s and x as given by the
transducer. Actually, of course, s(x,y) is discrete, and only
the two values s(x,A) = S(x,A)/Q(A) and s(x,B) = S(x,B)/Q(B)
are available for evaluation, where Q(A) = $(Pt_{AA})^{1/2}$ and
Q(B) = $(Pt_{BB})^{1/2}$. It is possible, in fact, to rewrite the basic interpolation formula (17) in terms of these s instead of
S.

3.3.2 Multi-Channel Sensors

In sensors with a larger number of spectral channels, operating primarily on the digital principle of channel selection,
increased accuracy is possible by evaluating the position $\hat{y}$ of
the maximum of s(x,y) by analog interpolation between two or
more of the channels excited most strongly in the transducer,
i.e. whose channel parameters y_m are near $\hat{y}$.

For interpolation between two adjacent channels, similar
conditions apply as in the two-channel sensors discussed before. This holds in particular for sensors with discrete receiver channels. The two channels concerned may be characterized by the tuning parameters y_m and y_{m+1}. The corresponding
values of the (continuously variable) measurand are called x_m
and x_{m+1}. A situation is considered now where the value of x
falls between those two channels,

$$x = x_m + \xi b \tag{21}$$

where b = $x_{m+1} - x_m$ is the channel separation, and 0 < ξ < 1.
When the construction of the transducer is such that it oper-

ates sufficiently continuously and if the channels are not spaced too widely, the actual spectral transmission function $F_1(x,\nu)$ may be found by linear interpolation between the functions belonging to x_m and x_{m+1}, just as had been illustrated in Fig. 9 and in example 3 of Table I.

$$F_1(x,\nu) = (1-\xi)F_1(x_m,\nu) + \xi F_1(x_{m+1},\nu) \quad . \tag{22}$$

Using the receiving filter functions $F_2(y_m,\nu)$ and $F_2(y_{m+1},\nu)$, the detector signals of the two channels of interest are calculated from (1).

$$\begin{aligned}
S(x,y_m) &= (1-\xi)t_{AA}P + \xi t_{BA}P \\
S(x,y_{m+1}) &= (1-\xi)t_{AB}P + \xi t_{BB}P
\end{aligned} \tag{23}$$

where

$$\begin{aligned}
t_{AA}P &= S(x_m,y_m) & t_{BB}P &= S(x_{m+1},y_{m+1}) \\
t_{AB}P &= S(x_m,y_{m+1}) & t_{BA}P &= S(x_{m+1},y_m)
\end{aligned} \tag{24}$$

The solution of the two linear equations (23) may be written in the form

$$\xi = \frac{1}{2} + \eta \tag{25}$$

where η is given by the expression (17), with $S(x,y_m)$ and $S(x,y_{m+1})$ substituted for $S(x,A)$ and $S(x,B)$, respectively. With this ξ, the value of the unknown measurand follows from (21).

In Eq. (22), *linear* interpolation had been chosen for simplicity. Depending upon the detailed behavior of $S(x,y_m)$ as a function of x, two-point interpolation with other than linear characteristics may be advantageous for improved accuracy. The last line of Table I was an example of such a nonlinear interpolation. Instead of (25), a more complex evaluation formula would result, containing e.g. higher powers of η or some function $f(\eta)$ which represents the inversion of the interpolation characteristic.

The preceding discussion was formulated so as to show the close analogy of interpolation in a two-channel sensor and interpolation between two adjacent channels in a multi-channel sensor. It may also serve to further interprete spectral encoding in geometrical terms in an abstract vector space. In (10) the "angle" β_{12} between spectral vectors $|F_1(x)w\rangle$ of the transducer and $|F_2(x)w\rangle$ of the receiver had been defined. For two-channel interpolation, the angles β_{AB} and β_{BA} between the transducing and receiving channels are of special interest. In the simplest system, with identical filters in transducer and receiver and symmetric channels, $t_{AA} = t_{BB}$ and $t_{AB} = t_{BA}$. In that case $\cos \beta_{AB} = t_{AB}/t_{AA} = \varkappa$. It is recognized, then, that the spectral vectors $|A\rangle$ and $|B\rangle$ of the two channels are orthogonal in the absence of crosstalk, as indicated on the left side of Fig. 9. With crosstalk present, β_{AB} is always less than $\pi/2$, but the interpolation process remains qualitatively unchanged (see right side of Fig. 9). It is obvious that this geometric interpretation can readily be generalized to cases of two-channel interpolation with non-identical filters and non-symmetric channels, or to nonlinear interpolation.

For multi-channel systems, the discussion of the angles β between channels show that each channel constitutes its own dimension in the vector space of the spectra used for encoding or decoding. The total vector space which is needed to represent any possible spectrum $F_1(x,\nu)$ or $F_2(y,\nu)$ has M dimensions when both filters F_1 and F_2 are of similar design, but it may require 2M dimensions when they are different.

3.3.3 Generalizations

The interpolation procedure that was employed to determine the position $\hat{y}$ of the maximum of $s(x,y)$ can be generalized to involve more than just two adjacent channels. Mathematical procedures are well established [19] for fitting a higher-order polynomial to a number N of given data points, i.e. to a set of detector signals, $S(x,y_{m+\rho})$ with $\rho = 1,2,...N$, obtained in the receiver when y is varied. Taking into account the normalizing denominator (11), this polynomial may then be differentiated to find $\hat{y}$. Rather than performing these steps sequentially, however, they may all be combined into a single evaluation formula

$$\hat{y} = y_m + bf(\eta_N) \tag{26}$$

with

$$\eta_N = \sum_\rho c_\rho S(x, y_{m+\rho}) \Big/ \sum_\rho d_\rho S(x, y_{m+\rho}) \; . \tag{27}$$

This η_N is a generalization of the two-point interpolation parameter η defined earlier in (17). The summations in (27) extend over N adjacent channels. The coefficients c_ρ and d_ρ correspond to the t_{ij}, they are linear combinations of the signals from all possible transducer/receiver channels $S(x_i, y_j)$ falling into the range of the N channels involved. In calculating these coefficients, different statistical weights for the N channels may be taken into account, emphasizing channels which ly close to the central one that is of interest in (26).

The decision what number N of channels are reasonably to be included in such an evaluation depends on the actual structure of $S(x,y)$ as a function of either x or y. Clearly, all channels should be included that contribute to the information on the position $\hat{y}$ of the maximum near y_m. These channels are found only among those that have non-vanishing crosstalk with y_m. However, not all of those channels do necessarily contain relevant information on $\hat{y}$, because the crosstalk may happen to be non-zero but essentially independent of the channel number [e.g. with the spectra of Fig. 6(b)].

Carrying this generalization (27) to its ultimate limit, N → M is reached. Then, the necessary investments in receiver hardware and software become equivalent to those required for a complete spectral analysis of the spectrum $F_1(x,\nu)W(\nu)$ arriving at the receiver, as it is typically performed in Fourier spectroscopy. With such an effort, however, it appears no longer advantageous to use transducing and receiving filters of rather similar designs, necessary for obtaining a well pronounced maximum of $S(x,y)$ near x = y. Rather, the receiver may then be any kind of dispersive or Fourier spectrometer, and the evaluation of x is reduced to the standard spectroscopic task of determining the spectral transmission of a filter F_1, followed by an estimation of that x which fits best that measured transmission [16,17]. The distinction between channel selection and channel interpolation may become obsolete in terms of hardware in that case, but the concepts discussed above must then be carried over to the software of evaluation.

The foregoing discussions did not fully cover multi-channel sensors with continuous tuning of the receiver parameter y. Here, evaluation of x is done by locating the position $\hat{y}$ of the maximum of s(x,y) in the y scale, as had been explained, either by scanning or by tracking. Although these are obviously continuous procedures, they contain hidden the concept of channels and the principles of channel selection and interpolation. This may be recognized by considering how the maximum of s is actually being located. In the scanning mode, for example, some threshold level $\bar{s}$ slightly below the peak value of s(x,y) may be specified by hardware or software, and the two argument values $y^{(+)}$ and $y^{(-)}$ be determined at which s passes through $\bar{s}$. From the existence of the maximum in (10) it follows that $\hat{y}$ lies centrally between those arguments, and the evaluation $\hat{y} = [y^{(+)} + y^{(-)}]/2$ may be regarded as a particular form of interpolation. For this procedure to be optimum, the threshold $\bar{s}$ must be chosen so that the separation $[y^{(+)}-y^{(-)}]$ is approximately equal to the channel width b defined in (10). If that separation were much smaller, the decision process at the threshold would become uncertain because the signal slope ds/dy vanishes. On the other hand, if that separation exceeds b too far, asymmetries of s(x,y) may deteriorate the accuracy of $\hat{y}$. Another technique of locating $\hat{y}$ is differentiation of s(x,y) with respect to y, for example by dithering the receiver. This technique may be used in the scanning mode as well as with tracking. In either case it is necessary for optimum accuracy to choose the dither amplitude roughly equal to the channel width b, for the same reason as just explained. Hence,

dithering may also be viewed as an interpolation procedure between two adjacent channels in the sense defined earlier.
These channels may "slide" continuously in the y scale, however. The evaluation or tracking procedure shifts them until
their signals $s(x,y^{(-)})$ and $s(x,y^{(+)})$ become equal, and $\hat{y}$ lies
centrally between them.

Like the interpolation procedure among discrete channels,
the described averaging of the positions of "sliding channels"
in continuous receivers may advantageously be extended to include a larger number (N>2) of channels that contribute to the
accuracy of locating $\hat{y}$. This possibility had been used in a
system of fiber-coupled interferometers [13]. With a tungsten
lamp as the source and a Si detector, approximately 5-10 side-
maxima and minima exist near the main maximum at $\hat{y}$, spaced approximately one mean wavelength $1/\bar{\nu}$ or one channel width b
apart. Hence a total of N = 10-20 arguments $y^{(\rho)}$ could be measured, using the average interferogram signal $\bar{s}$ as the threshold. These arguments, together with the amplitudes of the maxima and minima, were digitized and stored during each scan. In
the evaluation, first the highest maximum was identified, giving a rough estimate of $\hat{y}$. This step clearly represents a channel selection. A more accurate $\hat{y}$ was calculated then as the
average

$$\hat{y} = \sum_{\rho} g_{\rho} y^{(\rho)} \qquad (28)$$

of an even number of the measured arguments, taken symmetrically from either side of the estimated $\hat{y}$. This second step should
be seen as a channel interpolation. The g_{ρ} in (28) are weighting factors, chosen symmetrically about $\hat{y}$ and normalized so
that $\sum g_{\rho} = 1$. A closer investigation shows that the g_{ρ} are
preferably taken proportional to the square of the peak height
of their neighboring maxima, $[S(x,y_{m+\rho+0.5})-\bar{s}]^2$, where $\hat{y}$ denotes the threshold crossing argument just below $\hat{y}$. The reason
for this choice of the g_{ρ} is that the statistical error expected for $y_{m+\rho}$ is inversely proportional to the slope of the
threshold crossing, i.e. to the height of the neighboring max-

imum.

It is recognized that the concepts of channels and of their
selection and interpolation may be applied to systems with continuous receivers in rather similar ways as with discrete systems.

4. SIGNAL/NOISE RATIO, RESOLUTION

Among the various sources of errors and instabilities that
limit the performance of fiber-optic sensors with spectral encoding, two effects are closely related to the particular type
of encoding employed, and they will be discussed next. The
first error results from the limited signal/noise ratio at the
detector, the other error effect is the residual influence of
variations of the link losses. While the former one represents
essentially additive white noise, the latter one is multipli-
cative and typically has very low frequencies. It will be
treated separately in Sec. 5. In the following, then, the noise
is characterized, and the signal levels are estimated for the
various types of spectral encoding. Here the number M of channels and the integration time are the relevant parameters. As
a result, for each type of encoding a kind of power budget is
established. It shows that the dynamic range, available between
the input power of the fiber and the detector noise at a given
integration time τ, must be shared by

o the maximum possible resolution RQ_{SN} of the sensor

o the link transmission factor T_L, describing the minimum
link loss plus all possible variations and any system
reserve

o the power efficiency factor T_F of the channel filters,
which is related to the sophistication of their design.

4.1 Noise

Noise in the detector output of sensors as discussed here
consists of noise from the electronic circuit (including the
dark current) and of shot noise. The first-mentioned contribu-

tion is signal-independent (stationary). When the r.m.s. deviation δS of the detector signal due to this electronic noise is referred to the detector input, it can be expressed as a noise equivalent power $\delta S = N_0 \tau_i^{-1/2}$. Here τ_i denotes the electronic integration time, and N_0 characterizes the circuit. For a small-area PIN silicon detector and a good preamplifier, $N_0 \simeq 10^{-12} \ldots 10^{-13}$ W/$\sqrt{\text{Hz}}$.

The shot noise depends on the detected optical power, expressed here directly as the signal S. Its noise equivalent power can be written as $\delta S = N_1 \tau_i^{-1/2}$ with a noise spectral density $N_1 = (SQ_0)^{1/2}$. This relation results from the shot-noise spectrum $2Ie\Delta f$ of a current I, but is expressed here with the signal power S, and $\frac{1}{2}Q_0 = hc\bar{\nu}/\eta_e$ is the average optical energy required to produce one photo-electron. For a silicon photo-diode and 0.83 μm wave-length, typically $Q_0 \simeq 6 \cdot 10^{-19}$ Ws.

As the two noise contributions N_0 and N_1 are independent, the total noise is characterized by their r.m.s. sum N,

$$\delta S = N\tau_i^{-1/2} = (N_0^2 + SQ_0)^{1/2}\tau_i^{-1/2} . \tag{29}$$

This noise spectral density is plotted vs. S in Fig. 10 for two values of electronic noise N_0. It can be seen that for typical power levels S occurring in fiber optic sensors the noise N may range from the amplifier-limited (at low S) to the shot-noise limited regime (at large S). For each practical type of sensor it must be analyzed which noise is dominant.

Both kinds of noise are treated here as being "white" over the frequency range 10^2-10^6 Hz which is of primary interest, corresponding to integration times $\tau_i \simeq 10^{-3} - 10^{-7}$ s. At lower frequencies, the influence of flicker noise can be avoided in most applications by a modulation of the source at a suitably high frequency and subsequent phase-sensitive detection. Fluctuations of the detector output due to variations of the link losses cannot be avoided. It is the very basis of the spectral encoding technique, however, that these fluctuations generally have very low frequencies, so that they affect the signals of

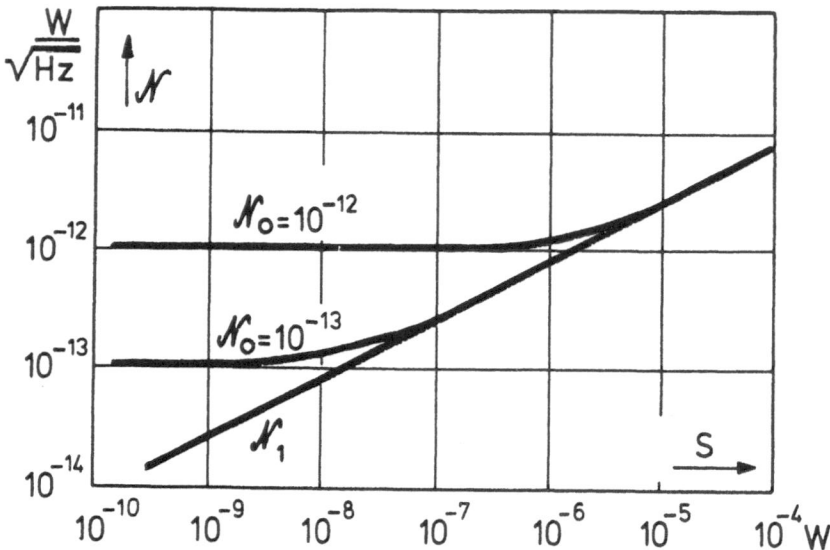

FIGURE 10. Noise-equivalent power (NEP) of electronic noise N_o, shot noise N_1, and total noise N for a silicon photodiode. Quantum effi- ciency $\eta_e = 0.8$; optical wavelength $1/\nu = 0.8$ μm .

all channels of the sensor in the same way and therefore, can- cel out in the evaluation (17) or (27). It is obvious, though, that such a cancellation requires that the various signals are all recorded within a total measuring time τ that is shorter than the characteristic fluctuation times of the fiber link. In sensors of type (i) this τ is of the order of the integra- tion time τ_i, and in sensors of types (ii) and (iii) this τ is the time of one full scan of y or the response time of the tracking servo-loop.

4.2 Resolution

For the comparison of sensors that use different kinds of spectral encoding, resolution R is generally defined here as the maximum number of values x of the measurand that can be distinguished in the sensor output. Hence, $R = X/\delta x_{res}$, where X is the full-scale range of the sensor and δx_{res} is the small- est resolvable variation of x, which for simplicity is assumed here to be independent of x.

For a meaningful definition of δx_{res} it is necessary to

specify also the confidence coefficient p or its complement,
the error rate (1-p), by which one resolvable element δx_{res}
can be distinguished from its neighbors. For analog systems,
in which the sensor output is a continuous variable, δx will
denote in the following the standard or r.m.s. deviation of the
sensor output y from its mean value $\bar{y}$. Consequently, if a normal probability distribution of y is assumed, y will fall with
a probability $p_i(Q_{SN}) = \text{erf}(Q_{SN}/\sqrt{2})$ into an interval $\Delta x = \pm Q_{SN} \delta x$,
centered at $\bar{y}$. Here, erf denotes the error function [20], and
the confidence margin Q_{SN} measures the width of the resolution
element δx_{res} in terms of the r.m.s. noise δx. For the interval $\bar{y} \pm \delta x$, for example, $Q_{SN} = 1$ and a confidence coefficient
$p_i \simeq 0.68$ result . The same relation can be applied, conversely, if it is required to distinguish one value of x with near
'certainty', e.g. an error rate of $(1-p_i) \leq 10^{-9}$, from its
neighbors, situated at least one resolution element $\delta x_{res} = \Delta x$
away. For that case the relation given shows that the r.m.s.
deviation of x must be $\delta x/\delta x_{res} \leq 1/\bar{Q}_{SN}$ with $\bar{Q}_{SN} \simeq 6.1$. Stated
more generally, a change δx_{res} of the measurand can be recog-
nized in the presence of noise δx with a confidence $p_i(Q_{SN})$
where $Q_{SN} = \delta x_{res}/\delta x$ denotes the signal/noise ratio in the x
scale.

This result can be applied now directly to the procedure of
channel interpolation described earlier. It is assumed that
the deviations δS of the two activated channels A and B have
equal r.m.s. amplitudes δS and are statistically independent.
This requires that either channel is observed for at least the
time τ_i. Then, the r.m.s. deviation $\delta\eta$ of the interpolation
parameter η due to detector noise δS is found from (17) or,
with the simplifying assumptions made there, from (18). At mid-
range $\delta\eta \simeq \delta S/(\gamma S\sqrt{2})$, where $S = [S(x,A) + S(x,B)]/2$ now denotes the average detector signal of the channel activated,
and γ was defined in (20). When x or y is tuned over one channel width, η varies by one unit. Therefore, the number R_1 of
elements $\delta\eta_{res}$ resolvable with confidence $p_i(Q_{SN})$ by interpo-
lation of two channels is $R_1 = 1/\delta\eta_{res} = 1/Q_{SN}\delta\eta = \sqrt{2}\gamma S/Q_{SN}\delta S$.
In a sensor with M adjacent channels, there are (M-1) intervals

available for interpolation, and the total number of x values resolvable is $R = (M - 1)R_1$. Inserting for δS the noise (29) yields

$$RQ_{SN} = \sqrt{2}\gamma\tau_i^{1/2} (M - 1) \ S/N \ . \tag{30}$$

This product RQ_{SN} is a reasonable measure of the attainable resolution because it takes into account the error rate $(1 - p_i)$ of the evaluation. The fact that R and Q_{SN} appear here as a product shows that one factor may be traded freely for the other.

Before evaluating (30) for sensors with various types of spectral encoding it remains to be discussed, however, how the noise affects the process of channel selection that may precede the cahnnel interpolation for $M > 2$. In that process, the receiver must find the number m of the channel that has the highest normalized signal $s(x,y_m)$. When $x = y_m$, the signals $s(x,y_{m\pm 1})$ of the neighboring channels are, in the average, smaller than $s(x,y_m)$. In the presence of noise, however, they may occasionally exceed $s(x,y_m)$, then causing an erroneous selection. Assuming the normalizing function $Q(y)$ to vary little between y_m and $y_{m\pm 1}$, the detector signals of the neighbor channels are approximately equal, $S_{m-1} \simeq S_{m+1} \simeq \varkappa\, S_m$, where $\varkappa$ is again the crosstalk ratio. The signal values in channel m have a Gaussian distribution $P_m = P([S - S_m]/\delta S\sqrt{2})/\delta S\sqrt{2}$, and $P_{m+1} = P([S - S_{m+1}]/\delta S\sqrt{2})/\delta S\sqrt{2}$ in channels $(m \pm 1)$, where $P(z) = (2\pi)^{-1/2}\exp(-z^2)$. Counting only errors from these two nearest neighbor channels, the error rate in channel selection is

$$1 - p_s(Q_{SN}) = 2 \int_{-\infty}^{+\infty} P_m(S) \int_{S}^{\infty} P_{m-1}(S') \ dSdS'$$

$$= \pi^{-1/2} \int_{-\infty}^{+\infty} \exp(-z^2) \ \text{erfc}(z + Q_{SN})dz \tag{31}$$

Here $\text{erfc}(z) = 1 - \text{erf}(z)$ is the complementary error function [20], and the term

$$Q_{SN} = 2^{-1/2}(1 - \varkappa) \ S_m/\delta S \tag{32}$$

is the signal/noise ratio that is actually available for the

decision process. It is determined by the difference $(S_m - S_{m\pm1})$
$\simeq (1-\varkappa)S_m$ of adjacent detector signals, relative to the combined noise $\sqrt{2}\delta S$ of the two signals. Here again δS has been
assumed as independent of the signal.

It is revealing now to compare the resolution of the interpolation procedure with that of channel selection. An approximate evaluation of (31) shows that the error rates of interpolation $(1 - p_i)$ and of selection $(1 - p_s)$ agree within a factor close to unity if the argument is not too small, $Q_{SN} \gtrsim 3$.
For channel selection at a specified error rate, a signal/noise
ratio $S_m/\delta S = \sqrt{2}\, Q_{SN}/(1 - \varkappa)$ is required according to (32). If
that ratio and the average signal $S = S_m(1 + \varkappa)/2$ are used to
calculate the resolution, $R_1 = (Q_{SN}\delta\eta)^{-1}$. In a two-channel interpolation procedure, $R_1 = 1$ is found. This result is quite
satisfactory since it means that the sensor can just distinguish one channel from the other with the error rate previously specified. Consequently, whenever the signal/noise ratio $S/\delta S$ is high enough to give an appreciable resolution
$R_1 \gg 1$ in the interpolation process, it is certainly high
enough to make the error rate $(1 - p_s)$ of the selection process negligibly small. Consequently, the selection process
need not be considered further when using the interpolation
result (30). Moreover, that equation (30) is recognized to be
applicable also to sensors that operate by channel selection
only, i.e. those with discrete receiver channels.

Evaluating (30) now for non-tracking sensors, i.e. two-
channel sensors of type (i) and scanning sensors of types (ii)
and (iii) with a single detector, the optimum integration time
is $\tau_i = \tau/M$, where τ is the total measuring time. To make the
resolutions of different sensors comparable, they are all referred to the same τ. From (30)

$$R_{scan}\, Q_{SN} = \sqrt{2}\gamma\; \tau^{1/2}(1 - 1/M)\; M^{1/2}\; S/N \qquad (33)$$

However, if the same sensor were using an array of M detectors,
i.e. one detector per channel, each one may integrate over the

full time, $\tau_i = \tau$, and its resolution would be

$$R_{array}\, Q_{SN} = \sqrt{2}\,\tau^{1/2}\, (M - 1)\, S/N \qquad (34)$$

where again S is the average signal in those channels that are employed for interpolation, and N is the noise of a single channel.

For the sensors operating in a tracking mode with a servo-loop of response time τ, dithering between two adjacent channels, an effective integration time $\tau_i \simeq \tau/2$ per channel results. Consequently, their resolution is

$$R_{track}\, Q_{SN} = \gamma\, \tau^{1/2}\, (M - 1)\, S/N \qquad (35)$$

On the basis of these equations (33), (34), and (35) it is possible now to compare the various types of sensors, all operating according to Fig. 1.

Firstly it is noted that the values RQ_{SN} calculated here are theoretical maximum values. Actual sensors should come close to them if designed optimally. As in other measuring processes with white noise, the resolution increases here in proportion to $\tau^{1/2}$. The factor γ accounts for the reduction of resolution due to crosstalk between adjacent channels, as had been shown earlier.

In all cases covered by (33) - (35), the possible resolution R increases with the signal/noise ratio S/N. Clearly, an increase of the power P_o coupled from the source into the fiber, as well as any improvement in the detector that reduces N, will always increase the possible R. However, when comparing sensors of different types and with different M, the factors S/N and those containing M in (33) - (35) must be discussed jointly, because they are not independent.

4.2.1 Power Budget

A convenient benchmark for the detector signal S is the value P given in (4). It is the signal that would be received in the absence of both filters. This P is related to the input power P_o into the fiber by

$$P = T_L P_o \ , \tag{36}$$

where T_L is the power transmission factor of the fiber link for the spectrum $W(\nu)$ employed. If the link losses vary, T_L is assumed to represent the worst case, i.e. the highest loss. Moreover, T_L may also contain a factor for some system reserve which could account for e.g. tolerances or aging of components.

The absolute power P_o that can be coupled into a high-NA fiber from a light-emitting diode is typically $P_o \simeq 0.1 - 1$ mW. The power P_o available from a W-halogen lamp in the spectral range of a silicon photo-diode has the same order of magnitude.

In a two-channel sensor [type (i)] with a single detector the receiving filter is switched between channels A and B. Assuming that the total time τ available for the measurement is split evenly between both channels, $\tau_i = \tau/2$. This had already been used in (33). Simultaneously, however, the power P is split between A and B by the channel filters. For simplicity they are assumed to have equal transmission coefficients, $t_{AA} \simeq t_{BB} = T_F/2$. Here, the factor 1/2 is inserted to account for the unavoidable loss associated with splitting the power into two channels. Therefore, the other factor defined here, T_F, denotes the *excess loss* – apart from that splitting loss – of the two filters T_{A1} and T_{A2} in series (or of T_{B1} and T_{B2} in series, respectively). In fact, T_F is a measure of how efficiently that filter combination utilizes the available optical power. Ideally, the filters should have square transmission spectra just bordering each other, and peak transmissions of unity. In that case $T_F \rightarrow 1$. For real filters, T_F indicates how close they come to that ideal. According to (14) and (15) the average signal falling on the detector is $S = T_F T_L P_o / 2$. Consequently from either (33) or (35)

$$RQ_{SN}/(\gamma \ T_F \ T_L) = \tfrac{1}{2} D \tag{37}$$

with

$$D = \tau^{1/2} P_O/N \qquad (38)$$

This result (37) may be viewed as a form of power budget. The quantity D is the gross dynamic range, available between the input power P_O into the fiber and the noise equivalent power $N\tau^{-1/2}$ of the detector in the measuring time τ. On the left-hand-side of (37), the various factors must make up, in any combination, the 'effective' dynamic range which is 0.5 D here. Thus, equation (37) shows how link losses, filter losses, maximum possible resolution, and confidence margin may be traded for each other.

A similar result is obtained for two-channel sensors with two separate detectors, like in Fig. 2. In that case, each detector receives its signal for the full time, $\tau_i = \tau$. According to (34) the power budget may also be written in the form (37), but with a dynamic range $D/\sqrt{2}$ being effective on the right-hand-side.

In sensors of type (ii), fiber-coupled spectrometers, the transducing filter splits the available power into M channels and then passes on only one of them. In systems with a single detector, hence, the average detector power is $S = T_F T_L P_O / M$, where the factor T_F describes again the loss of the two filters in series (tuned to y = x) *in excess* of their fundamental 1/M splitting loss. The resulting power budget for M >> 1 can again be written in a form like (37)

$$R Q_{SN} / (\gamma T_F T_L) = q D = D_{eff} \qquad (39)$$

where $q = \sqrt{2}/M^{1/2}$ for scanning operation, and $q = 1\sqrt{2}$ for tracking. For sensors of type (ii) equipped with a detector array, $q = \sqrt{2}$ due to the increased integration time.

In sensors of type (iii), fiber-coupled interferometers with a single detector, the transducing and the receiving filters have $\cos^2 (\pi x \nu)$ and $\cos^2 (\pi y \nu)$ transmission characteristics, respectively, i.e. an average power transmission of 1/2.

When measured in series and $x = y$, however, $\cos^4(\pi\nu x)$ results with an average of 3/8, regardless of $W(\nu)$. Therefore $S = (3/8) \, T_F T_L \, P_0$, independent of the number M of channels ($M \gg 1$). Accordingly, the factor T_F describes here the combined power transmission of both interferometers apart from their basic 3 dB insertion losses. For these sensors, the power budget (39) holds with $q = (3\sqrt{2}/8)M^{1/2}$ for scanning and $q = M$ for tracking operation. When this sensor were operating with a detector array [14], the same factor $q = (3\sqrt{2}/8)M^{1/2}$ would apply as in the scanning mode. The reason is that in the array each element would receive only the fraction 1/M of the total power, whereas in the scanning mode the detector receives the full power nearly all the time. Yet, the array detector may be advantageous because the lower signal level per element may result in reduced shot-noise. For either kind of interferometric sensor, it is also possible to evaluate directly the crosstalk. Using (15) and $x \neq y$ it can be shown that $\varkappa = 2/3$, which yields $\gamma = 1/5$.

The coefficients q for the various types of sensors are summarized in Table II. As the effective dynamic range of a sensor is $D_{eff} = qD$, these coefficients q may be regarded as a measure of the possible performance of the various sensor types.

Sensor Type:	(i)	(ii)	(iii)
No. of Channels	M = 2	M $\gg$ 1	M $\gg$ 1
Interpolating	0.5		
Scanning		$1.41 \, M^{-1/2}$	$0.53 \, M^{1/2}$
Tracking		1	M
Array Detection	0.71	1.41	$0.53 \, M^{1/2}$

TABLE II. Coefficients $q = D_{eff}/D$ of various types of sensors employing spectral encoding, where $D_{eff} = R \, Q_{SN} / (\gamma \, T_F \, T_L)$ denotes the dynamic range that would be possible in the absence of crosstalk ($\gamma \to 1$) and of transmission losses ($T_F \, T_L \to 1$), and D is the ratio of optical input power to the detector noise, see Eqs. (38) and (39).

In this sense, Table II shows that fiber-coupled interferometers are superior by a factor of the order M or $M^{1/2}$ to fiber-coupled spectrometers of the same number M of channels.
This can be a very large factor, for example $M \simeq 5 \cdot 10^4$ in
[8]. Interpreting Table II in another way it is recognized
that for a high theoretical resolution R, fiber-coupled spectrometers should operate preferably in the tracking mode or
with a detector array. If for some reason a scanning mode is
used, these sensors should be designed to have only a small
number of channels, i.e. relatively poor spectral resolution.
Expressed differently this means that a scanning spectrometric
sensor should derive its resolution $R \simeq MR_1$ preferably from
the principle of channel interpolation (large R_1) rather than
by channel selection (large M). Ultimately, then, the case
M = 2 appears as the optimum. It will be seen below, however,
that the requirement of immunity to variations of T_L may favor
the opposite preference, i.e. channel selection rather than
interpolation.

While the q values for sensors of types (ii) and (iii) are
not too different at small M, the coupled interferometers gain
an increasing advantage for large M simply because they have a
higher average spectral transmission than dispersive spectrometers. This is also evident from Fig. 6. Therefore, they produce higher detector signals S in (33) and (35). With all kinds
of coupled interferometers, the effective dynamic range qD increases continuously with the number of channels and may far
exceed the gross range D. In the tracking mode of operation,
in particular, q is proportional to the number M of channels.
This behavior may be understood from the fact that the *absolute*
resolution δx of path difference in an interferometric system
is essentially independent of the value x of the path difference. Therefore, the *relative* resolution $R = X/\delta x$ increases with
the range X and therefore with the number of channels.

A high average transmission could be achieved also in dispersive spectrometers if their narrow slits were replaced by
multi-slit masks [21]. However, the resulting so-called

'throughput advantage' is hardly useful in practice with fiber-coupled spectrometers because of the very limited throughput of the fibers themselves.

Comparing the scanning and the tracking modes of operation, Table II shows for both types (ii) and (iii) of sensors the superiority of the tracking mode by roughly a factor $M^{1/2}$. This advantage results from the fact that a tracking sensor, that has locked onto a measurand x, has a maximum *a-priori* knowledge about the value of x. This knowledge permits the sensor to determine a new value of x in a time τ by comparing only two channels, as explained, rather than scanning through M channels. The resulting advantage may be translated into a resolution increased by a factor $M^{1/2}$ or into a response time reduced by the factor M. This advantage is lost, however, when the loop gets unlocked.

In order to illustrate equation (39) and to mention some typical power budgets, three examples are considered now. In the first one, an optical power P_o = 100 µW into the fiber and a detector with a moderate amplifier-limited $N = 10^{-12}$ W/√Hz are assumed. For $\tau = 10^{-2}$ s the gross dynamic range is D = 10^7 or 70 dB. For a two-channel sensor, the effective dynamic range according to (37) is 67 dB. This may be made up, for example, by a resolution R = 10^3 at Q_{SN} = 1, corresponding to 30 dB, and link losses of 20 dB, still leaving a reasonable 17 dB margin for the filter efficiency T_F and the crosstalk factor γ.

With the same input power and single detector, a sensor with fiber-coupled spectrometers with M = 100 channels has a factor q = 0.14 in the scanning mode, leaving an effective dynamic range of 61.5 dB. A total resolution of RQ_{SN} = 10^3 or 30 dB would require only a very modest $R_1 Q_{SN}$ = 10 in the interpolation procedure, thus providing good immunity to variations of the link losses (see next section). It can be realized e.g. with 15 dB link losses and a realistic 16.5 dB for the combination of spectrometer efficiency T_F and crosstalk γ.

In the third example a sensor with two fiber-coupled Michelson interferometers in a scanning mode is considered [13] that has demonstrated a displacement resolution $\delta x = 10^{-2}$ μm (at $Q_{SN} = 1$) over a range of $x_{max} = 20$ mm. This means $RQ_{SN} = x_{max}/\delta x = 2 \cdot 10^6$ or 63 dB and $M = 2 \ x_{max} \ \bar{\nu} \simeq 50,000$ channels. Here, $P_O = 100$ μW, $N = 6 \cdot 10^{-13}$ W/$\sqrt{Hz}$, and $\tau = 1$ s were used, corresponding to $D \triangleq 82.2$ dB. With $q \triangleq 119$ or 20.7 dB the effective dynamic range is 102.9 dB. From this value, 63 dB is used for R and 7 dB for the cross-talk factor γ, leaving 32.9 dB for the combination of T_F and T_L. This value is in good agreement with the observed maximum permissible link loss of 15 dB.

In the last example, actually an even higher effective dynamic range was available, because the evaluation of the peak position $\hat{y}$ in the receiver had not been done by two-point interpolation [as assumed in deriving (39)], but by an N-point interpolation procedure line (28) with $N \simeq 10 - 20$. It can be shown that the uncertainty $\delta\hat{y}$ is reduced by the factor $N^{1/2}$ if the weighting factors g/ρ are chosen optimally. Consequently, N-point interpolation increases the effective dynamic range D by this factor $N^{1/2}$ above the value given in (38). In the last example given, this factor corresponds to an improvement by ~ 6 dB.

5. IMMUNITY TO VARIATION OF LINK LOSSES

The very purpose of employing spectral encoding in fiber-optic sensors is to make the output signal of the sensor independent of variations of the loss of the fiber link. It should be possible, for example, to use fiber-connectors for convenient installation of the transducer, or to insert or remove some extra length of fiber, or to use fibers of different quality, all without affecting the output signal. Moreover, insensitivity to variations of the reveived absolute power is also a precondition for the interchangeability of similar transducers without the need for recalibration.

In first order approximation, the principle of spectral encoding does provide this immunity that is crucial for a wider acceptance of fiber-optic sensors for industrial applications. Encoding the measurand into the ratio or position of spectral

features and evaluation of x by channel interpolation or selection guarantees indeed that the recovered value $\hat{y}$ is independent of the *mean* transmission factor T_L of the fiber link.
It is necessary, of course, that T_L does not fall below some
minimum. In fact, the discussion of resolution and power budget has shown that with decreasing T_L the maximum attainable
resolution R is reduced, too. Therefore, a minimum T_L can be
specified that is required to achieve a desired R.

In second order of analysis, however, it is obvious that a
change in the sensor output $\hat{y}$ may occur if a variation of the
link attenuation has pronounced spectral features in that part
of the spectrum that is used by the two filters F_1 and F_2, even
if the mean transmission T_L is unchanged. In the following,
then, it is discussed how this residual dependence of $\hat{y}$ is related to the spectrum of a loss variation, to the properties
of F_1 and F_2, and to the procedures of evaluation.

Any variation of the link loss acts as a third spectral filter in the optical path in addition to F_1 and F_2. Its spectral
transmission is denoted by $V(\nu)$. Both possibilities of $V(\nu) < 1$
and $V(\nu) > 1$ may occur, e.g. with increased or reduced fiber
length, respectively. The influence of V may be regarded from
two seemingly contradictory points of view. From the first one,
the filter V acts on the spectrum F_1W coming from the transducer, so that the receiver will interpret VF_1W instead of F_1W
and may arrive at an erroneous result. In the linear vector
space of spectral vectors, the vector $|u_1> = |F_1w>$ is trans-
formed by the operator V, changing its length and direction, so
that the receiver would select a modified $|u_2>$ as a best match
to $V|u_1>$.

Alternatively, however, it may be argued that V simply modifies the fundamental spectrum $W(\nu)$ of the sensor, and therefore
should not have any effect on the operation of the sensor except perhaps a change of R, as mentioned above. In particular,
in a channel-selecting sensor with two filters F_1 and F_2 of
similar design, the receiver searches for the situation of
'parallel' spectral vectors $|u_1>$ and $|u_2>$. This situation ex-

ists for y = x, regardless of whether the fundamental spectrum
is $W(\nu)$ or $V(\nu)W(\nu)$.

Both points of view are correct, in fact, but they are
based on different assumptions about the evaluation procedures
involved. The first one applies when the receiver has no information about the variation and evaluates y with the 'old'
normalizing function $Q(y)$, defined in (11), that existed before the variation V. The second point of view, invoking the
search for a receiving vector 'parallel' to that of the transducer, implies that the receiver can use the correct normalizing function $Q(y)$, i.e. that one which exists after the variation has become effective. It had been explained that this
$Q(y)$ must be determined by synchronous (x = y) tuning of transducer and receiver, and must then be stored in the receiver.
Repeating this procedure after a variation V of the link losses amounts essentially to a recalibration of the receiver,
requiring access to or cooperation by the transducer.

In practice, therefore, the first-mentioned point of view
is of greater interest. The error $\tilde{\delta}x$ is to be determined that
exists when no recalibration is performed, i.e. when the 'old'
$Q(y)$ is used for evaluation. It will be shown that by proper
choice or design of the filters F_1 and F_2 this error δy can be
made negligibly small.

Simultaneously, this discussion will give an estimate of
the systematic error δy that exists in a multi-channel sensor
(M >> 1) in which the detector signal $S(x,y)$ is evaluated directly rather than the normalized signal $s(x,y)$ defined in (8).
Such a simplified evaluation, ignoring completely the influence of $Q(y)$, may be justified if the particular application
rules out the possibility of determining $Q(y)$ by a recalibration procedure. In that case, too, the design of the filters
must be chosen so that the error $\tilde{\delta}x$ is sufficiently small.

The influence of a loss variation $V(\nu)$ on the sensor output
can be determined under fairly general conditions if the vari-
ation is small, $|V(\nu) - 1| << 1$. With $V(\nu)$ being effective, the
previous detector signal $S(x,y)$ is modified to a new signal

$\tilde{S}(x,y)$. Here and below, ~ over any symbol indicates its relation to $V(\upsilon)$. In the evaluation procedures, instead of $s(x,y)$ the modified function $\tilde{s}(x,y) = \tilde{S}(x,y)/Q(y)$ is evaluated, and the position $\tilde{y}$ of its maximum in the y scale is determined. Using

$$\tilde{s}(x,y) = s(x,y) \cdot [1 + \tilde{\delta}S/S] \qquad (40)$$

$$\tilde{\delta}S \equiv \tilde{S} - S = <wF_1(x)|(V - 1)F_2(y)w> \qquad (41)$$

and equations (8) and (10) with the 'old' position $\hat{y} = x$, the condition $d\tilde{s}/dy = 0$ yields directly the desired error

$$\tilde{\delta}x \equiv \tilde{y} - \hat{y} \approx \frac{b^2}{2} \frac{d}{dy} \left(\frac{\tilde{\delta}S(x,y)}{S(x,y)} \right) . \qquad (42)$$

This result shows that a loss variation with a flat spectrum, $V(\upsilon) = $ const, does not produce an error $\tilde{\delta}x$, because in that case $\tilde{\delta}S$ is proportional to S, so that the derivative in (42) vanishes. Moreover, for a spectrum that is not flat, (42) shows that the error $\tilde{\delta}x$ is proportional to the square b^2 of the channel width in the x scale. This fact indicates that by subdivision of the full-scale range X into a sufficiently large number of channels, $M = X/b$, the error resulting from a loss variation can be made negligibly small, as $\tilde{\delta}x \sim M^{-2}$.

From a practical viewpoint, the most important loss variation is probably that caused by a variation $\tilde{\delta}L$ of the fiber length. Denoting the spectral attenuation of the fiber by $\alpha(\upsilon)$, the transmission of a fiber of length L is $\exp(-\alpha L)$. The loss variation associated with $\tilde{\delta}L$ is $V(\upsilon) \approx 1 - \alpha(\upsilon)\tilde{\delta}L$, and the signal modification (41) is

$$\tilde{\delta}S(x,y) \approx - \tilde{\delta}L \int \alpha(\upsilon) F_1(x,\upsilon) F_2(y,\upsilon) W(\upsilon) d\upsilon . \qquad (43)$$

5.1 Two-Channel Sensors

For sensors of type (i) with two discrete receiving channels A and B, evaluation is by channel interpolation. Although it is possible to find $\tilde{\delta}S$ from (41) as outlined, its derivation is simpler from the general interpolation parameter η defined in (17) - (19). Its modification $\tilde{\delta}\eta$ gives directly the approximate relative variation of the output of the two-channel sensor, i.e.

$$\frac{\tilde{\delta}x}{X} \simeq \tilde{\delta}\eta = \frac{1}{\gamma} \frac{r}{(1 + r)^2} \left[\frac{\tilde{\delta}S(x,A)}{S(x,A)} - \frac{\tilde{\delta}S(x,B)}{S(x,B)} \right] \qquad (44)$$

In the special case where the channels are represented by two narrow spectral bands centered at wavenumbers ν_A and ν_B, and assuming a smoothly varying attenuation spectrum $\alpha(\nu)$, the integral in (43) may be approximated by

$$\tilde{\delta}S(x,y) \simeq S(x,y) \: [1 - \alpha(\nu_y) \: \tilde{\delta}L] \qquad (45)$$

with $y = A,B$. Near mid-range, $r \simeq 1$, and the relative error $\tilde{\delta}x/X$ turns out from (19)

$$\tilde{\delta}x/X \simeq (\alpha_1/4\gamma) \: (\nu_B - \nu_A) \: \tilde{\delta}L \quad . \qquad (46)$$

Here the attenuation spectrum of the fiber has been assumed to be smooth enough to be approximated within the source bandwidth by the first terms of a series expansion

$$\alpha(\nu) = \alpha_o + \alpha_1(\nu - \nu_o) + \dots \qquad (47)$$

about some wavenumber ν_o intermediate between ν_A and ν_B.

The result (46) shows qualitatively what also could be expected intuitively for this sensor (i) with two channels at separate wavenumbers ν_A and ν_B. The relative error $\tilde{\delta}x/X$ increases in proportion to the variation $\tilde{\delta}L$ of the fiber length, to the spectral separation $(\nu_B - \nu_A)$ of the two channels, and to the slope α_1 of the attenuation spectrum (47). It is independent of the spectrally flat attenuation term α_o. For a reduction of the effect of a loss variation $V(\nu)$ it is advantageous, therefore, to choose ν_A and ν_B close to each other. Moreover, to reduce the detrimental influence of the crosstalk factor γ in (46), the spectral bands at ν_A and ν_B should be narrow and non-overlapping. It is recognized, then, that a general price to be paid for link-independent spectral encoding lies, to a substantial part, in the design and sophistication of the spectral filters that define the two channels.

To illustrate (46) by a practical example, a sensor is considered that has two channels at the wavelengths $\lambda_A = 1/\nu_A = $ 930 nm and $\lambda_B = 1/\nu_B = 820$ nm, available e.g. from two LEDs. Over this spectral range the attenuation of a fiber may increase or decrease (depending upon the loss mechanism) by as much as 4 dB/km, hence $|\alpha_1(\nu_B - \nu_A)| \lesssim 1$ km^{-1}. For a tolerable variation of $|\delta x/X| \lesssim 0.01$, the maximum permissible length variation is found from (46) to be $|\delta L| \lesssim 40$ m if there is negligible crosstalk, and shorter by the factor γ in the presence of crosstalk.

Much better independence from variations of the link losses can be achieved if the two channels A and B are represented by two finely periodic and carefully interleaved filter spectra. Such an encoding can be realized at moderate expense by a birefringent Fabry-Perot filter whose interference orders differ by an odd multiple of 1/2 for the two orthogonal polarizations [8]. With a filter thickness d_o and mean refractive index n_o, the periodicity of the filter spectrum is $\Delta\nu = 1/2n_od_o$. For a coarse estimate of $\delta x/X$, the periodic spectra of the filters are approximated here somewhat arbitrarily by $F_1(A,\nu) \cdot F_2(A,\nu) \simeq \cos^2(\pi\nu/\Delta\nu)$ for the first channel and, interleaved, $F_1(B,\nu) \cdot F_2(B,\nu) \simeq \sin^2(\pi\nu/\Delta\nu)$ for the second channel. Evaluation of (44) near mid-range ($r \simeq 1$) yields the relative error due to a length variation δL in the form

$$\frac{\delta x}{X} \simeq -\frac{\delta L}{2\gamma}\frac{1}{P}\int_0^\infty \alpha(\nu)\ W(\nu)\ \cos(2\pi\nu/\Delta\nu)d\nu \quad . \qquad (48)$$

In deriving this expression, the approximation $\int W(\nu) \cos^2 (\pi\nu/\Delta\nu)d\nu \simeq P/2$ has been made. Obviously, the integral in (48) is a Fourier-Cosine-integral, expressing the amplitude of the 'ripple' in the attenuation spectrum $\alpha(\nu)W(\nu)$ at that critical periodicity $\Delta\nu$ that matches the period of the encoding filters. This is not surprising, because such a ripple or isolated sharp features in the spectrum $\alpha(\nu)W(\nu)$ could affect both channels in different ways. However, when the period $\Delta\nu$ is chosen sufficiently small (e.g. $\Delta\nu \simeq 20$ cm^{-1}, corresponding to 0.5 mm path difference in the sensor [8]), the ripple amplitude becomes

generally very small because $\alpha(\nu)$ and $W(\nu)$ are typically smooth
spectra. The only conceivable mechanism by which a disturbing
ripple could be produced is an interference (e.g. in a window)
with a path difference of $1/\Delta\nu$, matching that of the encoding
filters. Such a path difference must be avoided in the design
of the system. A more detailed analysis shows that similar errors $\tilde{\delta}x$ may also be caused, to a smaller extent, by path differences that are integer multiples of $1/\Delta\nu$.

The mentioned arguments and some experimental evidence indicate that it should be possible in practice with a system like
that of Ref. [8] to achieve a residual error $\tilde{\delta}x/X < 0.01$ for
variations of the fiber length of the order $\tilde{\delta}L = 0.1 - 1$ km.
Such an improved immunity, then, is recognized to result from
the use of more sophisticated filters.

5.2 Multi-Channel Sensors

For sensors of type (ii), fiber-coupled spectrometers, the
error caused by a variation of link losses may be estimated
from (42) by introducing the mean channel width $b = X/M$ and
observing that $d/dy = (d\nu/dy)d/d\nu$. For the tuning rate an average value $d\nu/dy \approx B/X$ is used, where B is the optical bandwidth
that corresponds to the full-scale range X. With (45), then,
except possibly for a sign

$$\frac{\tilde{\delta}}{X} \approx \frac{B}{2M^2} \tilde{\delta}L \cdot \frac{d\alpha(\nu)}{d\nu} \qquad . \qquad (49)$$

This equation expresses quantitatively what had been stated -
so far - only qualitatively, i.e. the immunity to variations
of fiber length in sensors employing fiber-coupled spectrometers. Like in two-channel sensors, the cause of the error (49)
is the slope $d\alpha/d\nu$ of the attenuation spectrum of the fiber. As
there is typically not much choice possible concerning the value of that slope, the desired immunity must be designed into
the sensor by using a sufficiently high resolution, i.e. a
large number M of channels. There is a limit to M, on the other
hand, set by the power budget, as had been explained.

Considering for illustration again the wavelength interval from 820 - 930 nm and a variation of the fiber attenuation by 4 dB/km over this interval, and the sensor with M = 100 and $RQ_{SN} = 10^3$ from the example of the previous section, equation (49) shows that the error $\tilde{\delta}x/X$ remains below the resolved $\delta x = 10^{-3} X$ if the fiber length is not changed by more than 20 km (disregarding the associated effects on the power budget). Obviously, the large number (M = 100) of channels renders this system fairly immune to variations of fiber length in all cases of practical interest.

Finally, the other type of multi-channel sensors is considered, i.e. type (iii), fiber-coupled interferometers. Concerning the immunity against variations of the link losses, this encoding scheme (iii) is unique among all other schemes discussed here because, by its very priciple, it generally permits the highest possible immunity. The reason is that for these sensors, encoding with $F_1(x,\nu) = \cos^2(\pi\nu x)$ and decoding with $F_2(y,\nu) = \cos^2(\pi\nu y)$, the normalizing function Q(y) is essentially constant, dQ/dy = 0, regardless of the shape of the spectrum W(ν), provided only that y is not too small. As a consequence, the two arguments merge that had been forwarded at the beginning of this Section. In the limit Q(y) = const, there is no need to distinguish between an 'old' and a 'new' function Q(y). Thus, variations of the link loss may well change the signal/noise ratio and the resolution RQ_{SN}, but they do not affect the calibration. In fact, as had been pointed out earlier, Q(y) = const. means that the measurand can be decoded directly from the position (x = $\hat{y}$) of the maximum of the non-normalized signal S(x,y). This general situation does not change in the presence of a loss variation V(ν). Hence, the error $\tilde{\delta}x$ vanishes in this limit.

A more detailed analysis must determine the conditions under which Q(y) is constant and remains so after a loss variation V(ν). From the definition (11), from (8), (10), and with $F_1F_2 = \cos^4(\pi\nu y)$, it follows that

$$Q^2(y) = \frac{3}{8} P + \frac{1}{2} \int W(\nu)\cos(2\pi\nu y)d\nu + \frac{1}{8} \int W(\nu)\cos(4\pi\nu y)d\nu. \quad (50)$$

Here, the two integrals depend on y. Apart from a normalizing factor, they represent the Fourier components of the power spectrum $W(\nu)$ at the periodicities $\Delta\nu = 1/y$ and $\Delta\nu = 1/2y$. Any sharp spectral features or a 'ripple' in $W(\nu)$ may contribute to such Fourier components. However, if y exceeds a suitably high value y_{min}, these periodicities become so fine that for most smooth spectra of interest (halogen lamp, LED) the integrals in (50) vanish. More precisely, these integrals may be viewed as representing the spectral coherency or 'visibility' function [22] of the spectrum $W(\nu)$. When y is much larger than the coherence length l_c of the spectrum $W(\nu)$, the visibility drops to zero, and Q becomes independent of y.

These considerations show the significance of using a source with a smooth, broad spectrum, i.e. with a short coherence length. The shorter l_c is, the less does Q(y) depend on y, and the smaller is the error $(\hat{y}-x)$ of evaluation. In this sense, a halogen lamp or a light-emitting diode in connection with a broad-band detector are ideal light-sources for this type (iii) of spectral encoding whereas a laser is unsuitable.

Moreover, these considerations also show the role of variations $V(\nu)$ of the link losses. In the presence of such variations, the product $V(\nu)W(\nu)$ appears in (50) instead of $W(\nu)$. If the spectrum $V(\nu)$ is of comparable or better 'smoothness' than $W(\nu)$, the loss variation does not cause a significant error δx in the evaluation. On the other hand, if $V(\nu)$ has sharper spectral features than $W(\nu)$, for example a narrow absorption line or an interference 'ripple', it may produce an error δx near such values of the measurand that correspond to the relevant Fourier components of $V(\nu)W(\nu)$. In particular, an interference ripple resulting from a path difference x_r (e.g. in a window) would produce a y-dependency of Q(y) near the arguments $y = x_r$ and $y = x_r/2$, thus causing evaluation errors $\tilde{\delta}x$ there.

Their magnitude can be estimated by considering more generally the error δy that may incur by evaluating the measurand

x from the position y_S of the maximum of the non-normalized
signal $S(x,y) = s(x,y)Q(y)$, rather than from the maximum ($\hat{y}=x$)
of the properly normalized signal $s(x,y)$. With (8), (10), (11),
and $b \simeq \bar{\lambda}/3$ (for a fringe-identifying system with $\bar{\lambda} = 1/\bar{\nu} =$
mean wavelength, see Sec. 3.2), and using Q^2 rather than Q,
this evaluation error can be expressed as

$$\delta y \equiv y_S - x \simeq \frac{\bar{\lambda}^2}{36} \frac{1}{Q^2} \frac{d(Q^2)}{dy} \qquad (51)$$

For a constant Q this error vanishes. Else, the error depends
in general on the actual value of the measurand. It may vary
strongly throughout the range X, peaking near those values y
that are associated with the major Fourier components (50) of
$W(\nu)$.

Before illustrating this type of error by an example, it
should be noted that any real optical spectrum $W(\nu)$ has, of
course, strong Fourier components at very low spectral ripple
frequencies whose periodicity is comparable to the total optical bandwidth B. These components do not produce an error of
the type jsut discussed, however, if the range of variation of
y is suitably restricted, $|y| > y_{min}$, for example by a mechan-
ical stop. The limit y_{min} must be chosen to be at least several coherence lengths of the source. A typical light-emitting
diode ($\bar{\lambda} = 830$ nm), e.g., has a smooth, near-Gaussian spectrum
with a FWHM bandwidth of $B \simeq 700$ cm^{-1}, corresponding to a coherence length $l_c \simeq 6$ μm. Consequently $Q(y)$ becomes independent of y here if $y_{min} \simeq 20$ μm is chosen.

An evaluation error may result, on the other hand, if that
same spectrum is modulated with some features of fine spectral
periodicity, e.g. a ripple of a small relative amplitude a_i and
periodicity ν_i, as it may result from interferences in a window
or in a fiber connector. For that case (50) yields a y-depen-
dence of Q^2 near the points $y_1 = 1/\nu_i$ and $y_2 = 1/2\nu_i$, extending over a width of the order of l_c about either point. The
maximum slope there is $Q^{-2}d(Q^2)/dy \simeq a_i\bar{\nu}/3$, and from (51) the
maximum evaluation error $(y_S-x)_{max} \simeq a_i\lambda/108$ is found. Whether

this error is significant or not depends on the magnitude of a_i and on the accuracy of the system.

This example also illustrates the possible influence of a loss variation $V(\nu)$. If the spectrum $V(\nu)$ has sharp spectral feature of the type mentioned, $Q(y)$ may vary near certain points y_i within the range X, and the resulting error $\tilde{\delta}x$ would be given by (51) as discussed.

6. CONCLUSIONS

The procedures of spectral encoding and decoding have been discusses which may be used in multimode fiber-optic sensors to achieve independence of the sensor output signal from variations of the link losses and to permit interchangeability of transduces without recalibration. Spectral encoding/decoding can be based on the digital scheme of channel-selection or on the analog scheme of channel-interpolation (or equivalently: ratio-evaluation). Many sensors employ both schemes simultaneously. It may not be possible then to make a sharp distinction between both schemes, i.e. to give a unique definition of the channel widths. Concerning the resolution, however, a precise quantitative distinction is not necessary, as the discussion of the power budget has shown that the crucial expression for resolution is $RQ_{SN} = (M-1)R_1Q_{SN}$. Here, the number (M-1) of signal-channels and the single-channel resolution R_1 appear as factors of a product, together with the confidence margin Q_{SN}. For different choices of the channel width, these factors may change, but their product remains invariant.

The possibilities of spectral encoding are of particular relevance for fiber-optic instrumentation in industrial plants. There, measurands exist typically in singel-channel analog form e.g. position, angle, temperature, force, pressure, ... They must be determined with a resolution of $R = 10^2 \ldots 10^4$ or better (corresponding to 6 - 12 bits) and with time constants in the range of $10^{-3} \ldots 10^1$ s. Thus, the information rate represented by a measurand is of the order of $R=10^0\ldots10^5$ bits/s. Although conversion of an analog measurand into a proportional optical intensity is usually very simple (e.g. by shutter-type modulators), the transmission of such a single-channel analog intensity-modulated signal over a fiber-link is not possible with the required resolution. On the other hand, the fiber has an extremely large information-carrying capacity C. Considering only binary digital modulation, which is completely immune to practical variations of link losses, a 1 km long multi-mode step-index fiber permits transmission at up to $\sim 10^7$ bits/s and can accomodate - in principle - of the order of $\sim 10^7$ parallel spectral channels (see Fig. 11). This corresponds to a combined information-carrying capacity of C $\sim 10^{14}$ bits/s. This is some 9 orders of magnitude above the rate coming from a typical sensor. Consequently, the fiber-

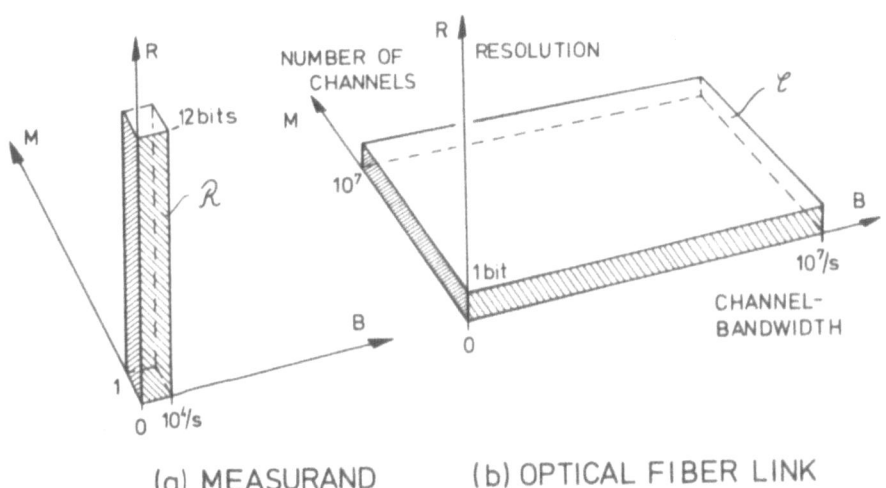

(a) MEASURAND   (b) OPTICAL FIBER LINK

FIGURE 11. Schematic representation of the information rate R originating from a typical industrial sensor (a), and of the information-carrying capacity C of a binary-digital fiber-optic link (b). Both R and C are given by the volumes of the blocks indicated. An ideal transducer must convert the single-channel high-resolution measurand to a multi-channel low-resolution optical signal, so as to accomodate R in C.

optic transmission of the value of a measurand requires spectral encoding. The transducer in the sensor must convert the single-channel high-resolution analog measurand into a multi-channel low intensity-resolution optical signal (ideally into

a binary digital signal as implied in Fig. 11), which may then
be transmitted reliably over the fiber link. In the terminology used above, this conversion is denoted by the 'channel-selection' procedure, whereas the remaining analog intensity-mo-
dulation is represented by the 'channel-interpolation' procedure.

The various encoding schemes discussed differ in the degree
to which they provide this conversion, but also in the amount
of hardware that they require for this conversion. Two-channel
sensors employ the least conversion, relying completely on
channel-interpolation, but they can operate internally with
simple intensity modulation. Their immunity to variations of
the link losses depends on the sophistication of their spectral filters, requiring transmission characteristics with
steep slopes and high peak transmission (i.e. low crosstalk
and large efficiency T_F). Multi-channel sensors offer more of
the mentioned analog-to-digital conversion and are, therefore,
superior by principle in their immunity to variations of the
link losses. Generally, their spectral filters may be simpler,
permitting e.g. more crosstalk, but their modulators (which
may be merged with the filter) must be more sophisticated so
that they can spread the analog input signal among the various
output channels.

From the viewpoint of immune signal transmission in industrial systems, then, sensor schemes are to be preferred that
rely as much as possible on encoding by channel-selection,
which necessarily means that they use a larger number of channels. Such sensors should also permit the best interchangeability of transducers, because their operation is affected least
by variations and tolerances in the signals of individual
channels. In practice, of course, many other factors, such as
cost, size, and engineering aspects are important in the design and industrial applicability of fiber-optic sensors.

REFERENCES

1. Proceedings of the 1st International Conference on Optical Fibre Sensors, OFS'83, London; IEE Conference, Publication No. 221 (1983).
2. Proceedings of the 2nd International Conference on Optical Fiber Sensors, OFS'84, Stuttgart; VDE-Verlag (1984).
3. Third International Conference on Optical Fiber Sensors, OFS'85, San Diego, Ca.; Technical Digest, Optical Society of America (1985).
4. Brenci M., Conforti R., Falciaci R., Mignani A.G., and Scheggi A.M.: Thermochromic transducer optical fiber temperature sensor". In Ref. 2, pp. 155-160 (1984).
5. Markle D.R., McGuire D.A., Goldstein S.R., Patterson R.E., and Watson R.M.: In "Advances in Bioengineering", D.C. Viano (Ed.), Am. Soc. Mech. Eng., New York (1981).
6. Guiliani J.F., Wohltjen H., and Jarvis N.L.: Reversible optical wave-guide sensor for ammonia vapors. Opt. Lett. 8, pp. 54-56 (1983).
7. Scheggi A.M., Brenci M., Conforti G., Falciaci R., and Preti G.P.: Optical fiber thermometer for medical use. In Ref. 1, pp. 13-16 (1983).
8. Dabkiewicz Ph and Ulrich R.: Fiber-optic angular sensor with interleaved channel spectra. Opt. Lett. 11, pp. 543-545 (1986).
9. Hutley M.C.: Wavelength encoded optical fibre sensors. In Ref. 2, pp. 111-116 (1984).
10. Bosselmann Th. and Ulrich R.: High-accuracy position-sensing with fiber-coupled white-light interferometers. In Ref. 2, pp. 361-364 (1984).
11. Beheim G: Remote displacement measurement using a passive interferometer with a fiber-optic link. Appl. Opt. 24, pp. 2335-2340 (1985).
12. Al-Chalabi S.A., Culshaw B., and Davies D.E.N.: Partially coherent sources in interferometer sensors. In Ref. 1, pp. 132-135 (1983).
13. Bosselmann Th.: Spektral-kodierte Positionsübertragung mittels fasergekoppelter Weißlichtinterferometer. PhD-Thesis, TU Hamburg-Harburg (1985).
14. Bornes T.H.: Photodiode array Fourier transform spectrometer with improved dynamic range. Appl. Opt. 24, pp. 3702-3706 (1985).
15. Cielo P.G.: Fiber optic hydrophone: Improved strain configuration and environmental noise protection. Appl. Opt. 18, pp. 2933-2937 (1979).
16. Quick W.H., James K.A., and Coker C.E.: Fiber optic sensing techniques. In Ref. 1, pp. 6-9 (1983)
17. Jones B.E. and Spooncer R.C.: Photoelastic pressure sensor with optical fiber links using wavelength characterization. In Ref. 1, pp. 173-177 (1983).
18. Schwartz M.: Information transmission, modulation, and noise. New York: McGraw Hill, 1980.
19. Zurmühl R.: Praktische Mathematik. Berlin: Springer, (1957)
20. Abramovitz M. and Stegun I.A. (Eds.): Handbook of Mathematical Functions. Washington: NBS Series 55, (1964).

21. Decker Jr. J.A.: Hadamard-Transform Spectroscopy. In Spectrometric Techniques. Vanasse G.A. (Ed.), New York: Academic Press, (1977).
22. Born M. and Wolf E.: Principles of Optics. Oxford: Pergamon Press, (1965).

FIBRE OPTIC INTENSITY MODULATED SENSORS.

R S MEDLOCK

BROWN BOVERI KENT PLC.

1. INTRODUCTION
    Appendix 1 gives examples of the types of modulating signals used for
measurement purposes and most of these have been applied to modulate light
in fibre optic sensor systems. However radiant signals embrace a wider
field than that of optics so there are a few examples of certain forms of
radiant energy being capable of modulating optical systems eg the
Cherenkov effect.   In order to examine modulating techniques in a logical
way it is usual to classify them as five separate effects resulting in
change of intensity, wavelength, phase, polarization or time.   This
classification can be arbitrary depending on how one frames the rules. For
example, an interferometric sensor can be regarded as a device measuring
either a time delay or a phase lag.   Again a photoelastic force sensor
modulates polarised light but the measurement can be in the form of an
intensity signal.   In spite of the ambivalence of the classification system
it is generally the simplest and the preferred method of analysing
modulating techniques.   The characteristics of optical fibres are important parameters in the modulating process.   A variety of characteristics
and materials can be found in the various types of fibre, - monomode,
multimode, birefringent and coated.

Monomode fibres are used where the modulation requires the maintenance of
light coherence as in the case of time or phase modulation generated in
interferometers.   Multimode fibres are useful for intensity and wavelength modulation.   Birefringent fibres are used with polarising
techniques.   Coated fibres offer opportunities for compound modulation
which means that the coating can provide a primary sensory effect which
in turn modulates an optical signal in the fibre by virtue of strain or
birefringence.   In addition there are miscellaneous fibres offering
specific characteristics, examples being liquid core and scintillation
types and no doubt as sensor technology advances there will be further
introductions.   Modulation techniques can be applied intrinsically ie
some optical property of the fibre can be exploited for measurement
purposes over a length of fibre which can then become a distributed
sensor.   Alternatively the fibre can function simply as a conductor of
light to an extrinsic sensor which modulates the incoming light and
returns it through the same or separate fibre to a means of detection.

2. INTENSITY MODULATION
    This was employed in the earliest optical sensor developments.   It
offers the virtues of simplicity, reliability and low cost to sensors
employing one of its many configurations.   The analogue applications of
this technique have one main drawback;   for good accuracy and stability
it needs some form of referencing to avoid errors arising from source

intensity variations, variable losses in fibres and connectors, and sensitivity changes in detectors. Some clever techniques have been developed to provide referencing but this adds to the complexity and cost Figs. 1-10 illustrate the principles of some members of this family of sensors which range from simple on/off switches through analogue modulating types to digital output devices.

2.1 Fibre displacement

Many sensors of the intensity type are modulated by a small linear or angular displacement. Sensors involving relative displacement of fibres are shown in figs. 1a-d.

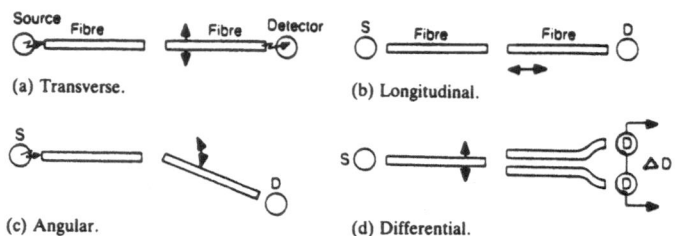

Fig 1   Fibre displacement sensors

2.2 Shutter modulation

Shutter types (figs. 2a-f) have fixed fibres with collimating lenses and variable aperture devices operated by a primary displacement sensor such as a diaphragm, bourdon tube, thermal expansion element or a turbine type of flowmeter. The shutter need not be made of solid material. Liquids in tubes can be used in a similar manner but if the liquid is transparent its function is more complex as its refractive index in relation to that of the fibre, determines the light coupling between the input and output fibres. Fig 2d shows a special type of shutter modulator which provides Moiré fringes. Such a system can measure displacement by the method of counting fringes. This avoids the referencing problem but introduces ambiguity in measurement.

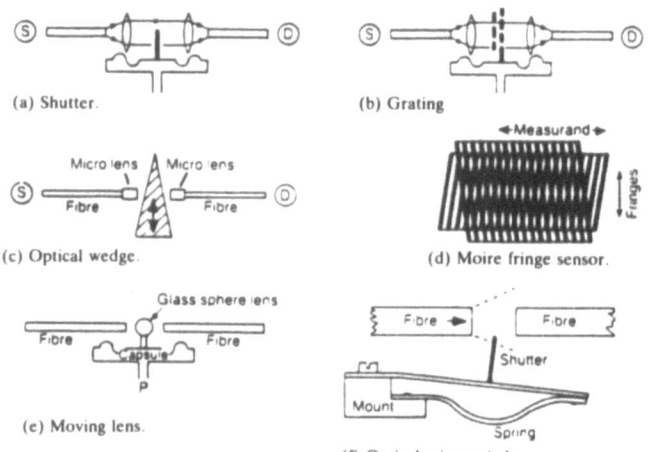

Fig 2   Shutter modulated sensors

2.3 Reflective

One of the earliest fibre optic sensors was the Fotonic*(1) or reflective sensor which can take many forms - multicore, twincore or single fibre with coupler. (figs. 3a-c) The characteristic of these sensors is typically that shown in fig.3d. Fig 3e shows a particular application of reflective modulation in the sensing of the float position of a variable area flowmeter. Light reflection has also been used for level measurement. The principle is shown in fig.3f. As the level changes, light angled on to the surface of the liquid is reflected and moves laterally. This movement can be measured with a multi-element charge coupled device.

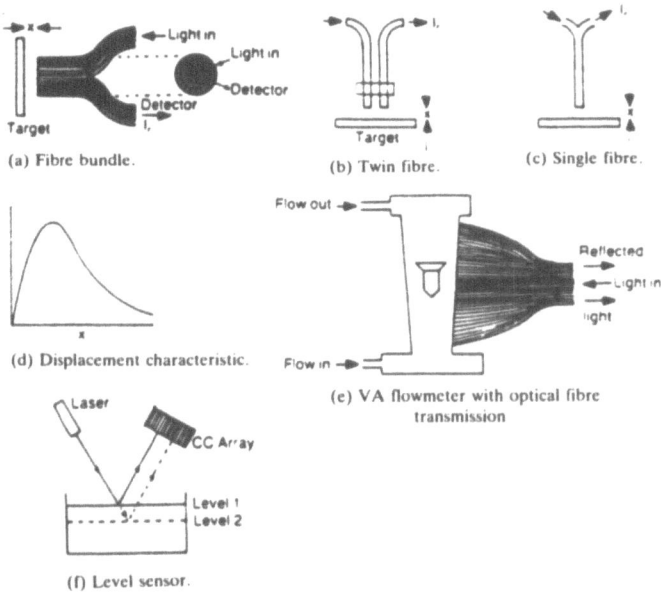

(a) Fibre bundle.

(b) Twin fibre.

(c) Single fibre.

(d) Displacement characteristic.

(e) VA flowmeter with optical fibre transmission

(f) Level sensor.

Fig 3   Reflective sensors

2.4 Fibre Loss

A fourth type of intensity modulation relies on losses from the core or cladding. Such devices vary the degree of fibre bending which results in the loss of the higher modes (ie the light rays which are internally reflected near the critical angle). A typical example is the microbend sensor (2) which is illustrated in fig 4a. It can be shown the loss is a maximum for a given fibre and a small displacement when the bending is applied periodically with a bend pitch 'x' where

$$x = C \pi r n (NA)^{-1}$$

C is a core factor = 1.42 for step index fibres (2 for graded index), r is the core radius and n is the core index and refraction. This type of sensor is applied to a commercial vortex shedding flowmeter to measure the frequency of the shedding by detecting the small pressure change generated by each vortex (3). Fig 4b shows another type of vortex detector consisting simply of a fibre stretched across a pipe. (4)

134

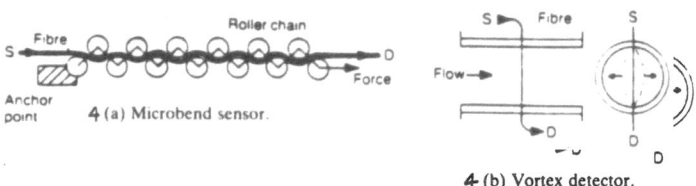

4 (a) Microbend sensor.

4 (b) Vortex detector.

Fig 4 illustrates other means of modulating the loss from a fibre. The losses can be from a normal multimode fibre (as with the microbend sensor) or from its unclad core (fig.4c) or from its cladding. Fibres have been made in which the indices of the core and cladding are nearly equal but have a differential temperature coefficient. If for example the temperature rises, the indices could approach equality with increasing loss from the core.

Conversely ordinary plastic coated silica fibre can act as a low temperature detector because the coefficient of refractive index with temperature is greater for the silicone cladding than it is for a silica and at a temperature below -25 C the cladding increases its index to a value above that of the core with a resulting extraneous loss of light (5) (6).

For measuring temperatures above ambient the role of the core and the cladding can be reversed and this can be achieved by having a liquid-in-glass light guide in which the liquid has a higher index than glass at ambient conditions but decreases fairly rapidly with temperature. There are not many liquids with the required high index but hexachloro 1,3 - butadiene is one that has been used. Another liquid is 1,4 - difluoro octachlorobutane (7). Influences other than temperature can similarly alter the index of the cladding. This is the principle of an oil in water detector which depends on the diffusion of oil into the cladding to change the index and allow the escape of light from the core (8). Oil in water can also be detected by using a fibre core from which the cladding has been removed. By selecting a core with an index similar to that of the oil (but higher than water) the core loss increases as oil dispersions settle on it. (fig 4d) (9).

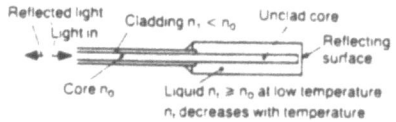

4 (c) Sensitive small range thermometer.

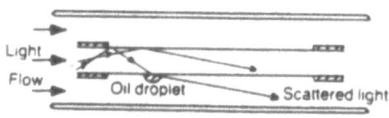

4 (d) Oil in water detector.

A similar principle operates in the case of "cross talk" sensors. These are made by removing short lengths of adjacent cladding from two parallel fibres in contact with each other and bridging the two cores with a transparent material (preferably liquid) which has a high temperature/ index coefficient (10). This device has three applications (i) as a temperature sensor (ii) as an instrument for measuring the refractive

index of a fluid similar to that of the core (iii) As a level sensor in
which the "cross talk" increases as the liquid (which must have an index
of more than 1.40) rises up the sensing gap. This device is suitable as
a brake fluid level sensor. (fig.4e) If the fluid has a low index it
would be necessary to operate the device indirectly by enclosing the
sensor in a small flexible closed end tube filled with a liquid having the
required optimum index.

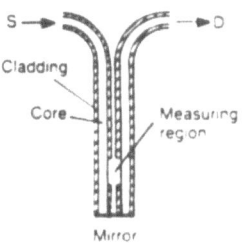

4 (e) 'Cross talk' sensor.

A simple level detector can be made from a loop of fibre as shown in
fig.4f (11). When a liquid starts to submerge the loop the higher light
modes are lost from the fibre into the cladding and the liquid. A
larger attenuation can be achieved by having two fibres (an input and an
output) cemented to a 90 degree glass prism. (fig.4g). In air, the light
ray is reversed through an angle of 180 degrees by internal reflection
within the prism. In liquid, the light escapes from the prism. In
another version (fig.4b) the principle is the same but a single fibre is
used with its end ground to form a 90 degree prism (12). Light is fed
into one arm of a Y coupler and the return light is measured as it
emerges from the second arm.

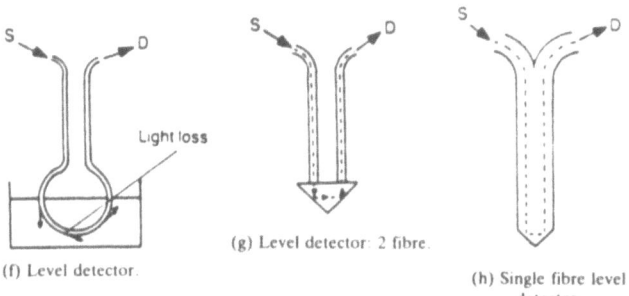

(f) Level detector.

(g) Level detector: 2 fibre.

(h) Single fibre level
detector.

Fig 4 Fibre loss sensors

2.5 Evanescent Field
The next class of intensity modulators involves the phenomenon of
frustrated total internal reflection (FTIR) or attenuated total

reflection (ATR). FTIR, (fig.5a), involves the extraction of energy from the evanescent wave existing near the surface of a prism when light undergoes internal reflection at the prism/air interface. (13). The energy of the evanescent wave decreases exponentially from the prism's surface but for practical purposes can be considered to extend about one micron. A light absorbing surface brought within this distance, extracts the energy and the reflected beam in the prism undergoes attenuation. Such a system can provide sensitive modulation for displacements of 0 to 1 micron. ATR (fig.5b), operates on a different principle. The modulation is brought about by altering the distance of a reflecting surface such as silver from the prism face. (14) When p polarised light in the prism is undergoing total internal reflection at a critical angle of incidence, reflectivity can vary from zero at a gap width of about 1 micron to about 96% for the zero gap.

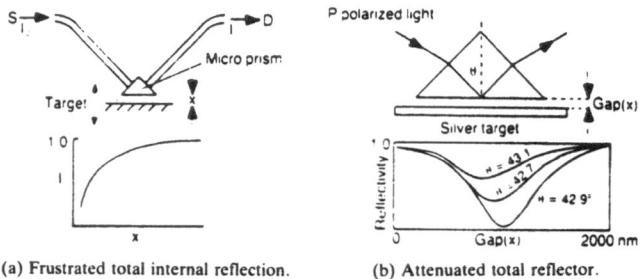

(a) Frustrated total internal reflection.   (b) Attenuated total reflector.

Fig 5   Evanescent field sensors

2.6 Absorption

Intensity modulation by colour change absorption has been used for some types of sensors for both physical and chemical parameters. One of the early developments measured the absorption of an equilibrium mixture of N_2O_4 and NO_2 sealed in a capsule at the end of a fibre bundle (fig.6) Absorption of the light in the gas is temperature dependent. This device was intended for temperature measurement of high voltage transformer windings. (15)

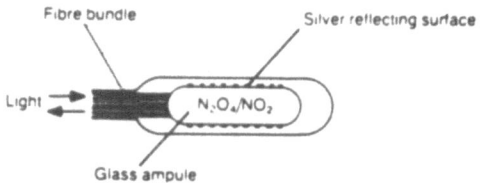

Fig 6   Absorption sensor

2.7 Light Scattering

Turbidity and light scatter are two more parameters which lend themselves to fibre optic measurement by intensity modulation. The presence of oil in water can be measured by the method shown in fig.7a which depends on the measurement of light loss or light scatter from oil droplets in the sample. (16) Another application of light scattering operates in an inverse mode in which a special fibre incorporates reflecting, fluorescing or scattering particles from which light emission is propagated in both directions along the fibre to a detector. This technique can provide a distributed sensor for measuring light intensity. (fig.7b) Light emission in a fibre can also be provided by scintillation and by Cherenkov radiation.

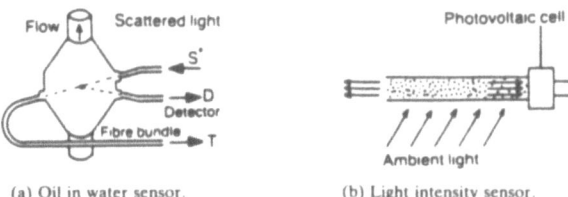

(a) Oil in water sensor.   (b) Light intensity sensor.

Fig 7   Light scattering sensors

2.8 Digital encoding

When measurement is required in digital form free from drift and intensity errors a linear or angular digital encoder can be used. In its simplest form 8 bit encoding will require 8 fibres each being modulated at two levels of light intensity representing 0 and 1 digits. A single transmission fibre system has been devised using eight spectral bands each of which illuminate a separate track on a Gray encoded digitiser. At the receiving end the eight spectral bands are demultiplexed by a diffraction grating and measured by a detector array (fig.8a,b) (17)

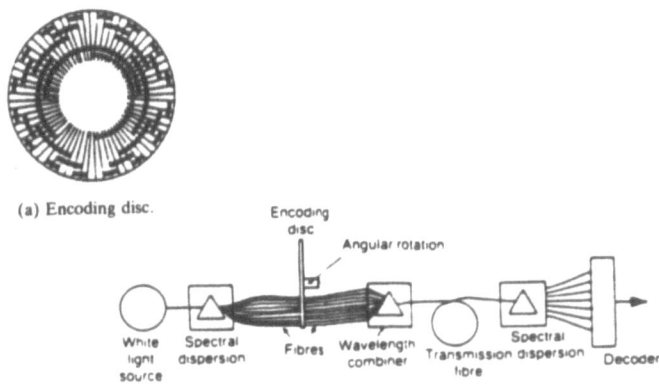

(a) Encoding disc.

(b) Transmission of 8-digit signal on one fibre.

Fig 8   Digital encoding sensor

138

2.9 Refractive Index

Fig.9 shows a refractive index sensor. Light emerging from a fibre into air produces a cone of light whose half angle is θ. If the emerging light then passes into a medium having an index greater than 1.0 the cone angle is diminished and the received flux in the other fibre increases. The ratio of received flux via the liquid to that via the air is a linear function of the refractive index of the medium.(18) The device illustrated in fig.4f can be used for refractive index measurement as well as for liquid level. It was designed for measuring acid level in batteries. (19)

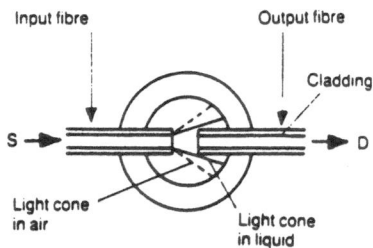

Fig 9   Density sensor

2.10 Electro-optic

Fig.10 shows an integrated optic switch. Coherent light entering a wave guide produced from a titanium diffusion in lithium niobate can be switched electrically by applying a voltage across the electrodes or by surface acoustic wave diffraction.

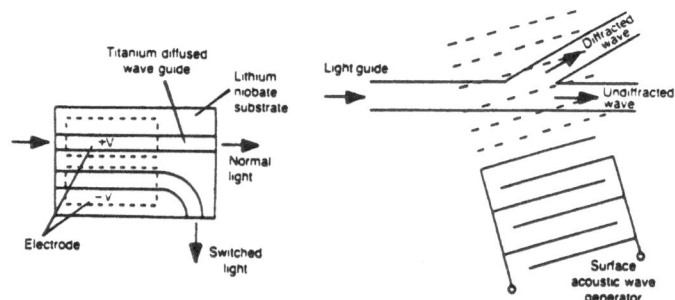

Fig 10   Two methods of switching direction of light in waveguides

3.0 REFERENCING TECHNIQUES

Intensity modulated sensors can suffer from serious errors unless techniques are employed which compensate for variations in light source intensity, losses in optical fibres and connectors, and in variations in detector sensitivities. Methods to overcome these potential sources of error include:-

• Use of digital devices such as the example given in section 2.8

• The two wavelength system. This provides a light source split into
two wave lengths which travel in the same outgoing and return fibres
to and from the sensor. Within the sensor housing they are separated
into two separate paths, one passing through as an unmodulated
reference and the other being subject to intensity modulation. After
modulation the two wavelengths are recombined and conveyed by a single
fibre to the receiving equipment where the two wavelengths are again
separated and measured individually, and ratioed. The efficiency of
such a system of referencing can be affected by alterations in the
spectral intensity of the source, chromatic loss variations in the
fibre, constancy of the filters, and chromatic sensitivity in the
detectors. For efficient referencing the two wavelengths need to be
as close together as possible provided "cross talk" does not occur.
For this reason holographic gratings have been used.

A particular application of the two wavelength referencing technique
has been described for the grating type modulator (fig.2b)[20] The bars
of the grating have been constructed as a narrow band notch filter
which reflects a specific waveband and allows a second waveband to
pass unmodulated.

• In a second method[21] Culshaw used a single wavelength in a balanced
bridge system. The system is shown in fig.11. Two similar LED's
are pulsed at different frequencies eg 1KHz and 10KHz and transmit
light energy through two fibres to the sensor housing which contains
a fibre optic "Wheatstone's Bridge". Each of the two output fibres
carry both frequencies which after filtering provide four signals,
A_1 A_{10}, B_1, B_{10} and the degree of modulation M regardless of losses
in the system can be shown to be equal to

$$M = \frac{A_1}{A_{10}} \times \frac{B_{10}}{B_1}$$

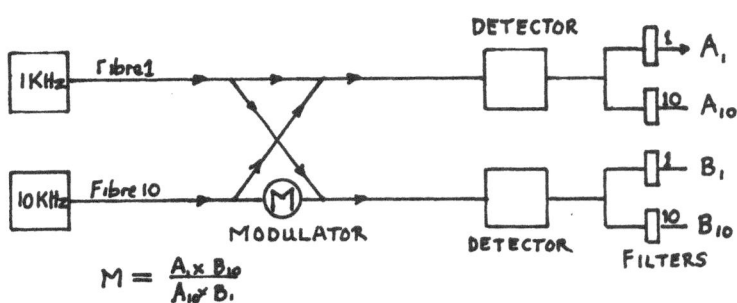

$$M = \frac{A_1 \times B_{10}}{A_{10} \times B_1}$$

FIG 11 SINGLE WAVELENGTH REFERENCING

• A third system is illustrated in fig.12 and is the "Optimitter"
available from Monicell Ltd. Light from the end of an incoming
fibre is passed through a graded index lens to provide a
reproduceable conical emission. The outgoing reference fibre is
similarly fitted with a lens and is mounted in line facing the
incoming fibre. A semi-reflecting mirror is interposed between the
two lenses and its position represents the value of the measured

variable. Light energy transmitted to the reference fibre is
unaffected by the position of the semi-mirror but the light
reflected back into the incoming fibre is a function of mirror
position. The measured variable can be computed from a ratio
function of the two signals. This method of referencing cannot
compensate for variable connector or other losses in the outgoing
and incoming fibres.

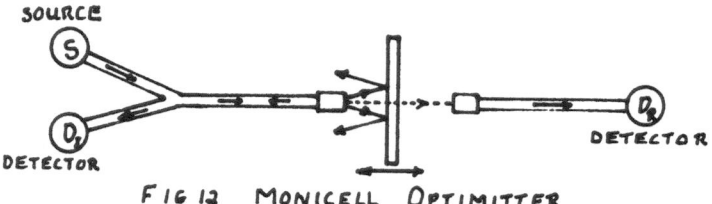

FIG 12   MONICELL OPTIMITTER

R E F E R E N C E S

1.   Mechanical Technology Inc. Latham, New York.
2.   Fields, J N et al. 1980 J Acoust. Soc. Am 67 p816
3.   Bailey Meter Company, USA Tech.Literature
4.   Lyle, J H and Pitt, C W, 1981 Electron. Letts 17 p244
5.   Pilkington Cryogenic Monitoring System: Tech literature.
6.   Murphy, R J and Turner, D M. 1981 US Patent Application 2,062,877
7.   British Patent 1,390,426
8.   Ishikawajima, Harima Heavy Industries (1983) Technical literature.
9.   Pitt, G B et al. Trans I Mar E 1980 92 Pater 8 pp66-75
10. Ramakrishnan, S and Kersten, R Th
     1984   OFS Conf Proc pp105-110 Stuttgart.
11. Sharm, M and Brooks, R E 1980 SPIE 224 pp46-52
12. Geake, J E 1954 J Sci. Instrum. 31 pp260-261
13. Fromm, I 1980 Europatent 0,025,565.
14. Sincerbox, G T and Gordon, S G 1981 Laser Focus. Nov. pp 55-58
15. Montgomery, J D and Dixon, F W 1981 ISA/IMC Prof. Conf.
     "Promecom 81" London, pp81-90
16. Snell, D and Pitt, G D 1983 BHRA Proc Int. Conf. Opt Tech in
     Process Control pp27-41
17. Dakin, J P 1984 Plessey Technical Journal "Systems Technology"
     May No. 38
18. Bell, D et al. West Glamorgan Institute of Higher Education.
19. Spencer, K et al. 1983 IEE Proc. 1st Int.Conf. on Opt.Fibre
     Sensors, pp96-99
20. Spooner R C, "Fibre Optics in Physical and Chemical Sensors"
     IMC Conf., Harrogate, Nov. 1985.
21. Culshaw B, J Phys. E 16 No.10 pp978-986 1983.

APPENDIX 1

EXAMPLES OF MODULATING SIGNALS USED FOR MEASUREMENT PURPOSES

- Physico/mechanical: force, weight, displacement (linear or angular), pressure, differential pressure, vibration sonic energy, velocity, acceleration, flow level, hardness, thickness, viscosity, density, opacity, turbidity, time, moisture content, particle concentration.

- Electrical: voltage, current, frequency, phase angle, inductance, capacitance, resistance.

- Magnetic: field strength, hysteresis, coercivity, induction

- Thermal: temperature, thermal conductivity, heat flow, emissivity.

- Radiation: radiofrequency, infrared, visible, ultra violet, X rays, nuclear.

- Chemical: concentration, composition (gas, liquid, solid), biological activity.

DISTRIBUTED OPTICAL-FIBRE SENSORS

A.J. ROGERS
DEPARTMENT OF ELECTRONIC AND ELECTRICAL ENGINEERING, KING'S COLLEGE
LONDON, STRAND, LONDON WC2R 2LS

1. INTRODUCTION

Optical fibres possess many advantages for measurement sensing. Amongst these are their insulating properties, the intrinsic safety of a passive medium, the immunity from electrical interference, their low weight, and the ease of installation afforded by their flexible geometry.

In addition to these, optical fibres have the advantage of providing what is essentially a one-dimensional measurement medium, allowing either line-integrations or line-differentiations to be performed over any chosen path. The attractions of these features are manifold. The line-integrating property provides the means for attaining large sensitivities, via a long and easily-tailorable path of optical inter-action with the measurand: the optical-fibre gyroscope[1] is a good example of this. In other cases the line-integral of the influencing field is the actual quantity required: the measurement of electric current via the loop-integration of its surrounding magnetic field[2] and of voltage via line-integration of electric field[2] are good examples of this.

However, more interesting from many points of view is the ability to obtain line-differential information from the optical fibre, for this provides the means whereby the spatial distribution of a measurand may be determined over any chosen path. Consequently, both the spatial and the temporal behaviour of the field may be determined simultaneously, yielding, in many cases, a higher level of understanding of the behaviour of the system under observation. Potential applications of distributed optical-fibre sensing are very numerous. For example, the distribution of strain in large, critical structures such as bridges, drains, pressure vessels and aircraft may be monitored continuously. Temperature distributions in equipments such as power transformers, electrical generators, boilers and high voltage cables may be kept under close observation, and this information will be available for the computation of heat flows. Electric and magnetic field anomalies may very quickly be identified and located in electrical networks. In addition to these industrial uses it is clear that spatial information will be of great value in research investigations and experimental rigs, especially since the level of sophistication in the signal interpretation can be much higher in these cases.

Finally, it is interesting to note that a long section (perhaps several kilometres) of fibre, when providing spatially resolved amplitude and phase information about an externally impinging wave-like field, is capable of acting as a large-aperture antenna, thus providing a highly directional polar diagram. A pair of orthogonally-crossed straight fibres, 1km long, could, for example, for e/m waves at 1GHz, provide a reception polar diagram with a half-width of less than 0.02°, in two dimensions.

This paper will describe some of the physical principles available for implementing distributed optical-fibre sensing, and illustrate these by reference to some of the exploratory devices and systems which have been reported to date.

2. PRINCIPLES

We require knowledge of the value of the measurand as a function of position along the length of the optical fibre. Inevitably this implies that each derived measurement value must be identified with a specific section of fibre whose position is known. In order to retain one of the primary advantages of the optical fibre as a sensing medium, that of a passive, intrinsically-safe element, we shall preclude the possibility of the measurement being made in conjunction with some form of active transmitter which identifies, via coding, its own fibre section. Thus we are constrained to identify position from one(or perhaps both) of the fibre ends.

A distinction must be made between two types of distributed sensor. In the first of these it is necessary to make measurements only at discrete, predetermined points (or along specific limited lengths) on the fibre. The measurement system then takes on the character of a series-distributed array of discrete transducers. This we call a quasi-distributed system; in the literature it sometimes has been called a multiplexed system. In the second type the measurement may be made continuously as a function of position at any point along the fibre; this clearly is a much more powerful and flexible arrangement. This we call a fully-distributed or, more often, simply a distributed sensor.

There are several ways by which the required positional information may be acquired at the fibre ends.

The one which springs most readily to mind is the use of a narrow pulse of light launched in at one end, and subsequent measurement of the differential delay for light returning from the different positions.

An example of this type of arrangement for a quasi-distributed system is shown in Fig. 1a [3]. Here a small fraction of the propagating light pulse is split off by means of a coupler at each measurement point, and back reflected (after modulation by the measurand) to return to the launch end, where it is time resolved for positional identification. Fig. 1b shows a similar arrangement which uses transmission couplers. For a continuously-distributed system it is convenient to use the Rayleigh back-scatter from small inhomogeneities in the fibre structure to effect what is essentially a 'radar' (lidar) arrangement, allowing continuous positional resolution. The fraction of the propagating power which is backscattered along the fibre is $\sim 10^{-5}$ per metre. The Optical Time Domain Reflectometry (OTDR) [4] technique, now well established as a diagnostic tool for optical communications systems, is an example of this, and is illustrated in Fig. 2. The time resolution of the backscattered power level determines the spatial distribution of the optical attenuation in the fibre, and thus allows ready identification of fibre breaks, bad splices, abnormally lossy sections of fibre, etc.

The spatial resolution for quasi-distributed systems clearly is pre-determined by the size of the transducing element, which may either itself be a section of the fibre (intrinsic) or not (extrinsic). The propagating pulse and the detection system need have no greater temporal capabilities than those necessary to discriminate amongst the various transducers.

In the case of the fully-distributed system the spatial resolution is determined by whichever is the larger of the pulse width and the detector response time. To fix ideas it should be noted that light, in silica, travels ~ 0.2m in 1 ns, and thus, for resolution of this order, sub-nanosecond pulse and detection techniques will be required.

Other methods of positional identification are possible. For example, the above pulse-propagation-time method can be operated in the frequency domain, by using swept-frequency techniques to synthesise (essentially) a pulse. Yet another method is illustrated in Fig. 3 [5]. Here we see a quasi-distributed arrangement where the identification is made via differential selectivities for various optical frequencies, and this technique could, in principle, be applied also to fully-distributed systems if suitably fabricated fibres were available.

In order to illustrate the above ideas more fully we now turn to some particular systems which have been explored experimentally for certain measurands.

3. DISTRIBUTED TEMPERATURE MEASUREMENT

The measurand which has been given most attention from the point of view of distributed measurement is temperature. This is for two reasons: firstly, there is an urgent continuing need to measure spatial distributions of temperature in such applications as oil-filled power transformers, chemical containers, aerofoils,etc., primarily to monitor for dangerous 'hotspots'. There is also a continuing requirement for distributed fire-alarms in large buildings, for these allow rapid location of the source of a fire outbreak. Secondly, in temperature measurement the required time responses are often relatively slow (several minutes is not uncommon) as a consequence of the normally large thermal capacities of the systems being monitored. This eases detection problems.

3.1 Differential-Absorption Distributed Thermometry (DADT)

One of the first distributed temperature measurement systems to be investigated was the quasi-distributed arrangement, illustrated in Fig. 4, and known as Differential-Absorption Distributed Thermometry [6]. Pulses of light from a dye laser are launched into a multimode fibre, and these are alternately at wavelengths of 605 nm and 625 nm. The fibre is used to connect in series a number of thin (~0.25mm) ruby glass plates, positioned at convenient measurement points. The arrangement makes use of the temperature-dependent absorption edge of ruby glass; the position of the edge rises in wavelength by about 20 nm as the temperature rises from 20°C to 200°C. Light at the 605 nm wavelength lies on the absorption edge and thus suffers a temperature-dependent absorption as it passes through the plates. The 625 nm wavelength lies in the long-wave transmissive region, and light at this wavelength can thus act as a reference level for normalization against the several other effects which can affect the fibre attentuation (e.g. bend loss, joint coupling loss, etc). The individual ruby glass plates are identified by time resolution of the Rayleigh backscatter signals at the two wavelengths. For a chosen plate the light backscattered at 605 nm from a point on the near side of the plate is compared with that from a point on the far side, and this ratio is then normalised by constructing a similar ratio with 625 nm light, and dividing one by the other. The electronics allow ready identification of the individual plates, and of convenient backscatter points. The processor averages over as many pulses as is allowed by the required measurement time response. The time response of the thin ruby glass plates is quite low, less than one second, but this still allows many pulses to be averaged for typical pulsed-laser repetition rates of tens of kilohertz. A typical normalised calibration curve for one of the plates is shown in Fig. 5.

A primary disadvantage of this arrangement is that the absorption at each plate attenuates the signal, and this limits the number of measurement points to about ten, for a practical system with an acceptable dynamic range of temperature measurement.

3.2 Scatter-dependent temperature measurement

The Rayleigh-backscatter coefficient in a fibre depends upon the structure of the core material and, in the case of a liquid core, the structure (and thus also the coefficient) becomes strongly temperature dependent. This fact has been used to measure the distribution of temperature in a liquid-cored fibre [7], via the arrangement shown in Fig. 6. A light pulse of 10 ns duration is launched into a section of multimode fibre with a numerical aperture which is low compared with a co-joined section of liquid-cored measurement fibre. In this way an attempt is made to limit propagation in the measurement fibre to the low order modes, in order to reduce the effect of the temperature on the numerical aperture, within the backscatter process; this results from the temperature dependence of the refractive index, and it opposes the dependence of the backscatter coefficient. Some reported measurements for this system are shown in Fig. 7 and correspond to a scatter co-efficient of 0.42% per deg. C, linear up to the value of 80°C. The maximum temperature of operation of this fibre is limited by the fact that at 160°C the refractive indices of fibre core and cladding become equal, and guiding action is then lost. Higher-temperature operation may be obtained with solid-cored fibres, and some measurements have been made for these [7] which, however, show much lower temperature sensitivities. Upper temperature limits are now set by the thermal behaviour of the coatings, but should allow operation up to several hundred degrees centigrade.

The disadvantages of this technique are: the impracticability, for most operational environments, of liquid-cored fibre; and the low sensitivity of the solid-cored fibre. Furthermore, the dependence of the indication on mode structure is a disadvantage when considering the many types of operational environment which can give rise to mode disturbances.

3.3 Distributed temperature sensing using doped fibre

A distributed temperature sensor which combines the ideas involved in the preceding two paragraphs has recently been reported [8]. It uses fibre, doped with trivalent Neodymium ions (Nd^{3+}), which exhibits a temperature-dependent absorption spectrum. The reported results used a dopant concentration of 5 ppm, this being the value calculated to optimise performance over 200 m of fibre, using OTDR. Clearly, there will be a trade-off, in the choice of dopant level, between sensitivity per unit length on the one hand, and the distance over which sensible measurements can be made on the other. This trade-off is the result of the reliance on an attenuation measurement. Measurements at 904 nm optical wavelength showed the dependence of attenuation on temperature to be linear over the range -50°C to 100°C, with a temperature coefficient of ~ 0.18%.°C^{-1}. The reported performance was that of a 2°C accuracy over the above temperature range, with a spatial resolution of 15m over a fibre length of 140m.

The main difficulty at present appears to be a non-uniformity of dopant distribution throughout the fibre, which gives rise to a variable (with position) temperature sensitivity. This problem should yield to improved fabrication techniques, however.

3.4 Distributed Anti-Stokes Raman Thermometry (DART)

Distributed Anti-Stokes Raman Thermometry differs in several important respects from the preceding three methods for distributed temperature measurement. Firstly, it is the first method to use an optically non-linear effect, the Raman effect, rather than a linear one. Secondly, it operates independently of the fibre material, allowing the important advantage of allowing all installed optical-fibre systems (e.g. communications systems) to become distributed temperature sensors. Thirdly, it measures absolute temperature and thus (in principle at least) requires no calibration.

The method utilises the spontaneous Raman effect. This is the effect whereby light propagating in a medium is modulated by the molecular vibrations and rotations within it. The effect must be treated at the quantum level for proper quantification, and is described in terms of an incident photon being absorbed by a molecule, which is then raised to a virtual excited state. The molecule may re-emit a photon of different energy (and thus different wavelength) either greater or smaller then the original. In each case the energy difference will be equal to one of the vibrational/rotational energies of the molecule. However, only if the molecule is already in an excited state at the time of photon incidence can it emit a photon of greater energy, by decaying from the excited to the ground state. This is called anti-Stokes Raman radiation and its level will be temperature dependent, since (for thermal equilibrium) the number of vibrationally/rotationally excited molecules will depend directly on the absolute temperature. For the emitted photons of smaller energy the decay is (normally) to an excited state from an original ground state, and these are called Stokes photons. The majority of photons will, of course, be re-emitted at the same energy, and will include the Rayleigh scatter. The Raman spectrum for silica is shown in Fig. 8. The banded nature of the spectrum results from the spread of bond energies characteristic of an amorphous solid. The experimental arrangement which has been devised to make use of this effect is shown in Fig. 9. A high power optical pulse is launched into a multimode optical fibre at frequency ν_i, say, from the Argon laser. At a chosen value of ν (from the Raman spectrum) Stokes and Anti-Stokes backscattered radiation levels at frequencies:

$$\nu_s = \nu_i - \nu$$
$$\nu_a = \nu_i + \nu$$

are measured with a monochromator and time resolved. The ratio of anti-Stokes to Stokes measured power levels is given by:

$$R(T) = (\nu_a/\nu_s)^4 \exp(-h\nu/kT)$$

thus providing T, the absolute temperature, as the only unknown. The factor $(\nu_a/\nu_s)^4$ is a consequence of the Rayleigh scattering).

The experimental system used 15 ns pulses at 514.5 nm wavelength and 5W peak power. The separation interval was chosen so that the Raman wavelengths lay at ±10 nm from the 514.5 nm line, according to convenient peaks in the Raman spectrum.

Results for standard multimode fibre are shown in Fig. 10. Sections of the fibre were set at various temperatures, and these were resolved to the extent of 5K in temperature and 5m in fibre length.

The primary difficulty with this method lies in the very low levels of Raman backscatter, which are down on Rayleigh backscatter by a further factor ~ 10^3. This necessitates long integration times for adequate signal to noise ratio, and the above results required a 100s integration time with a 40KHz laser pulse repetition rate.

For field use a pulsed Ar laser is impractical. Recently a distributed Raman system using an 850 nm semiconductor laser source has been reported [10]. This was capable of measuring temperature to an accuracy of 1K over 1 km, with a spatial resolution of 7.5 m.

As a footnote to this section it may also be noted that it is possible to use the Brillouin effect in fibres [11], in an entirely analogous way to the Raman effect, for distributed temperature measurement. The Brillouin effect has the added advantage that the Stokes and anti-Stokes lines result from the scattering by acoustic waves propagating in the silica, and thus are much closer in frequency to the exciting line, the frequency interval being only of the order of 25 GHz. This means that electronic, rather than optical, discrimination can be used in the detection, and the detector design is much easier.

4. DISTRIBUTED STRAIN AND FORCE MEASUREMENT

After temperature measurement, the measurands which have received most attention for distributed sensing are strain and force. Clearly these may be taken together since it is the effect of strain on the optical fibre which will be sensed, and, correspondingly, any external force or pressure field can be caused to stress the fibre and thus to strain it.

The reasons for the interest in this particular area of measurment are evidently to do with the concern which always arises about the mechanical integrity of large, stressed structures: continuous strain distribution monitoring must, therefore, have its attractions.

4.1 Frequency-Modulation Mach-Zehnder Interferometry

The first method we shall describe in this section is, as in the preceding one, a quasi-distributed arrangement. It is shown in Fig. 11. Each of the several sensors arranged in series along the fibre is a Mach-Zehnder interferometer, that is to say, a two-arm optical-phase 'bridge'. One of the two arms is protected from the external stress field whilst the other is exposed to it. The effect of the resulting strain on the exposed arm is to cause a refractive index change, via the elasto-optic effect. The result of this is a phase displacement of the light propagating within it relative to that in the other arm. When the light propagations from the two arms are re-combined, the resulting optical-wave amplitude will depend upon this phase difference.

Identification of the individual sensors is effected by assigning to each one a characteristic path-length difference between its two arms [12]. This is then interrogated by addressing the system with a CW laser source which is amplitude modulated at a frequency which is itself swept linearly and repetitively (for example, a semi-conductor laser may readily be swept up to tens of GHz). This leads to a frequency difference between the two arms of any particular sensor which is characteristic of the sensor, and the non-linear action of the detection photodiode converts this difference into a beat frequency which identifies that sensor. Any strain experienced by that sensor will be detected as a change in its central beat frequency.

Mach-Zehnder sensors are very sensitive, allowing strains of order 1 in 10^8 to be measured, even in electrically noisy environments.

A serious difficulty arises, however, from the necessarily repetitive nature of the sweep ramp. This sawtooth waveform generates an infinite number of harmonics of the ramp frequency and, for more than about three sensors, the spectrum becomes uncomfortably cluttered, and difficult to interpret.

4.2 Frequency-Modulated Continuous Wave (FMCW) Distributed Sensor

A device which works on similar principles to the preceding one, this time in a fully-distributed mode, is the frequency-modulated distributed sensor. A bi-modal fibre is used. The two allowable propagation modes may be either the two polarization eigenmodes of, say, a high-birefringence fibre (fibre with a in-built, large linear polarization retardance), or two separate propagation modes of what is, essentially, a two-moded 'multimode' fibre. The two modes must, in any event, possess different group velocities. Again, the fibre is addressed with a frequency-ramped amplitude-modulated CW optical signal which is launched into just one of the two modes. If there are no external perturbations on the fibre the light will emerge from the fibre also in that mode. But any perturbation which couples light to the other mode will result in the coupled light, on emergence, being delayed in phase (due to the differing group velocities) by an amount which depends on the distance of the perturbation from the exit end. This phase difference will lead to a modulation frequency difference (due to the frequency ramping) which again will be detected as a beat frequency by the detector photodiode. Further, the magnitude of this beat frequency will indicate the size of the perturbation. The most common type of perturbation in practice will be the geometrical distortion caused by bends and pressure-induced strains.

The above idea has been demonstrated both with birefringent fibre and with two-propagation-moded fibre [13]. It has been shown capable of locating the position of an external pressure point to within 10 m in a fibre length of 300 m, using a semi-conductor laser ramped up to 40 GHz at a 200 Hz rate.

A major problem with this device is that it is difficult to obtain fibres and fibre coatings which provide a reliable and consistent relationship between external force and mode coupling coefficient, and the measurement of the magnitude of the perturbation is thus subject to error. Improvements in fibre fabrication and fibre coating techniques could well alleviate this, however.

4.3 Polarization - Optical Time Domain Reflectometry (POTDR).

The first fully-distributed optical-fibre measurement system to be investigated was polarimetric in form. The idea which was used was an extension of OTDR (see section 2). Whereas OTDR measures the distribution of the fibres opticial attenuation, Polarization-Optical Time Domain Reflectometry (POTDR) [14] measures the distribution of its optical polarization properties, again by time-resolving the Rayleigh backscatter from a propagating pulse. POTDR, of necessity, uses monomode fibre, since only with this type of fibre can one assign to the propagating light a single, identifiable polarization state at each point along the fibre.

The experimental arrangement is illustrated in Fig.12. The polarization state of the backscattered light is determined continuously as a function of time (by measurement of the Stokes parameters [14]) allowing the required distribution of the polarization properties to be determined. It follows that the distribution of any external measurand field which can modify these properties (e.g. force/strain plus electric field, magnetic field, etc.) can now also be measured. Fig.13 shows the results obtained with POTDR when measuring the distribution of bend strain for a fibre wound on a drum of diameter of 185 mm [15].

The primary disadvantages of POTDR are concerned with the interpretation of the returning polarization information. The Rayleigh backscatter level is, of course, low, and the requirement is to measure and process the returning signal in the specified measurement response time, which is often less than 1 ms. The optical and electronic demands which this makes are considerable. Furthermore, not all the required information is available in backscatter, as is illustrated in Fig.14. Any (lossless) optical polarization element is equivalent to a retarder-rotator combination [16] and the effect of a go-and-return passage through the element, provided that the element is reciprocal (i.e. its polarization properties are independent of direction), is to cancel the equivalent rotation. Thus, of the three pieces of information needed to specify fully the polarization properties of the fibre (retardation of the retarder, orientation of axes of retarder, rotation) only two are available, and thus for full definition of the system, a priori knowledge of the polarization properties is required. This requirement is best met by the development of fibres with special, controlled polarization properties.

For non-reciprocal optical elements this feature can be used to advantage, as will be discussed in Section 5.1.

4.4 Non-Linear Pulse-Wave Interaction

A primary difficulty which recurs in all time-resolved backscatter methods is the low level of Rayleigh-backscattered signal. This problem can be overcome by using again the Raman effect, but in an entirely different way from that in Section 3.4, which was for temperature measurement.

Consider the arrangement shown in Fig.15. A pulse of light, with large peak power level, propagates in a monomode optical fibre. As it propagates it excites molecules of the medium into the virtual Raman states. Propagating in the opposite direction in the fibre is a continuous optical wave at a convenient Stokes wavelength. This depopulates its corresponding excited States by stimulated emission. However, the effectiveness with which it does this depends upon the relative polarization states of the pump (exciting) and probe (depopulating) beams. Thus the distribution of the perturbations in the fibre's polarization properties may be sensed, via the temporal behaviour of the emerging probe light: this light will have received a position-dependent gain from its interaction with the pulse (whose instantaneous position is known).

Fig.16 shows the experimental arrangement which was used to explore the above ideas [17]. A 9.5 ns, 5W peak power pulse (pump) was launched into a monomode fibre, and its Raman action was probed by means of a He-Ne laser at 632.8 nm wavelength. The pump-pulse source was a Nd-YAG- pumped dye which provided the pulsed power at 617 nm wavelength, so chosen as to allow the 632.8 nm line to lie on a peak of the Stokes spectrum of silica (see Fig.8). The gain received by the probe will be at a maximum when the two polarization states coincide, and at a minimum when they are orthogonal, varying co-sinusoidally between.

Fig.17 shows the measured effect on the distributed Raman gain resulting from application of controlled stress over ~ 1 metre of fibre. It is clear that its effect is to alter the entire gain distribution. This is to be expected, since a change in the polarization state of the light at one point in the fibre must change the states at all other points, in a determinable but rather complex way. The disadvantage, then, lies in the sophistication required for the detection and processing of the signal in order to extract the required (stress) distribution. Nevertheless, the advantage of a large forward-scattered (as opposed to low back-scattered) signal is an important one.

The general idea of a non-linear pulse/wave interaction may be more readily applicable with other types of non-linear effect and with specially fabricated fibres. It is clear also that such systems must operate with semi-conductor laser sources if they are to become practical for use in the field, and thus high peak power semi-conductor lasers will be required.

5. OTHER FIELD MEASUREMENTS

Although the main concentration has been on temperature and strain measurements, some other measurands have received attention from distributed measurement explorations.

5.1 Magnetic Field

When discussing POTDR in Section 4.3 the point was made that information regarding the equivalent rotator in a reciprocal retarder-rotator pair is not available when using this technique. If the fibre were a pure reciprocal rotator then no information on the polarization properties would be available (launch and exit states would always be the same). The action of a longitudinal magnetic field, however, is to produce a non-reciprocal rotator, via the Faraday magneto-optic effect. This effect has been studied extensively for purposes of optical-fibre current measurement [2] where the line-integral of the magnetic field is obtained. But it is also possible, using POTDR, to measure the spatial distribution of the magnetic field, for the Faraday rotation occurs in the same direction regardless of the direction of propagation of the light, and is thus not cancelled during the return passage through the fibre.

An experimental arrangement which utilised low linear birefringence 'spun' single-mode fibre [18] was able to locate the position of a test solenoid to within 0.4 m, over a length of 15 m [19]. The source used was a mode-locked, cavity dumped argon laser providing 200 ps pulses of 20 nJ energy at 514.5 nm. By averaging over only ten pulses (~1ms) a sensitivity of 10^{-3} Tesla was achieved.

A fully-engineered system would be capable of locating faults in electrical conductors and cables in a variety of applications. As a corollary to this section it may be noted that POTDR will be sensitive to any non-reciprocity present in an optical fibre, and will always indicate it via the presence of double-passage rotation. This can be useful for diagnostics in systems such as optical-fibre gyroscopes, where even very low levels of spurious non-reciprocity may severely limit performance, and where it, therefore, needs to be located and diagnosed [20].

5.2 Electric Field

An external electric field also modifies the polarization properties of a single-mode fibre, via the electro-optical Kerr effect. Its action is to induce a linear retardance (birefringence) in the fibre, proportional to the square of that component of field transverse to the fibre axis. Electric current may be measured by line-integrating the magnetic field around a fibre loop, and the prospect of similarly line-integrating the electric field along a fibre stretched between two points, in order to measure their voltage difference, is an attractive one, especially since this measurement would be independent of electric field distribution. The distribution is highly variable in operational high-voltage environments and is a common source of difficulty in voltage measurement.

The optical voltage measurement problem is much more serious than current, however, for it demands a linear, longitudinal electro-optic effect in the fibre, and such effects exist only in crystalline media. They are emphatically absent from amorphous silica.

However, a proposal for a solution to this problem is shown in Fig.18. The spiralling fibre geometry allows the electric field, lying parallel with the axis of the spiral, to act transversely to the fibre axis. POTDR may now be used to measure the distribution of the electric field along the fibre axis, and this may then be line-integrated electronically to derive the voltage between fibre launch and exit points. This arrangement has not yet been tested experimentally. Alone, it is unlikely to be cost effective; but combined, in the same fibre, with optical-fibre current measurement, it could constitute an economically attractive device for use in power system measurement.

5.3 Quasi-Distributed Directional Hydrophone Array.

The optical-fibre hydrophone [21] is a very sensitive (Mach-Zehnder) detector of acoustic waves, and has been the subject of investigation for sub-oceanic use. The suggestion has been made [3] for an arrangement consisting of a series of hydrophones in a quasi-distributed array, similar to the configuration in Fig.1a. Such an array would be highly directional and capable of accurate submarine location, for example.

Presently, the optical components available do not appear to have the performance necessary for the very demanding sub-oceanic environment, but rapid advances are currently being made in the area of component quality.

6 DISCUSSION AND CONCLUSIONS

It is clear that the one-dimensional nature of the optical fibre as a measurement medium offers important possibilities for its use in distributed measurement. Distributed optical-fibre measurement combines all the accepted advantages of optical-fibre sensors with a simultaneous knowledge of both the spatial and temporal distribution of the measurand. The implications for the monitoring and diagnostics of extended structures in civil and aeronautical engineering, in industry and in research are thus considerable.

Although many ideas have been explored in the laboratory the subject is still in its infancy, and there is much more work to be done before viable, engineered systems become available for field use. However, the pioneering work of the past few years has demonstrated their basic technical feasibility.

Many other technical possibilities remain to be explored. Notable amongst these are the fabrication of a range of special fibres, each optimised for a particular measurement function: fabrication techniques need to be brought to bear on fibre materials, dopants, geometries, polarization control and coatings, in order to produce the system performances necessary for economic attractiveness. Then there is a range of non-linear optical effects yet to be explored in regard to their usefulness in the distributed measurement context: Stark effect for electric field measurement, Zeeman for magnetic field, and two-photon absorption for strain field, are examples of these. Finally, there is the technique, already being explored for point chemical sensors, of utilising the evanescent field in the fibre cladding for sensing purposes. This could be done either by sensitising the properties of the cladding material, via etching or ion implantation techniques, or by using special fibre geometries which allow direct access to the evanescent field (e.g. 'D' fibre).

In many types of distributed sensor, the signal processing necessary to extract the measurand distribution is complex. This presents problems which would be eased by performing some of the reduction processing at the optical level, using integrated-optical processing packages. These packages are already being developed, primarily for optical communications purposes. They undoubtedly also will have a significant impact on the design of distributed optical-fibre measurement systems.

The use of distributed optical-fibre sensors as highly directional arrays again is one with many attractions, and is one which has not yet been explored to any extent experimentally. Work in this field could well prove to be of considerable value in several diverse application areas.

To conclude then: the possibilities for passive, interference-free, insulating, easily-installed, distributed measurement sensors are attractive. The applications prospects are good, the technical problems are interesting and challenging; but the economic questions are complex and as yet unresolved.

Much has yet to be done, but the potential rewards provide ready justification for the effort which is to come.

REFERENCES

1. Vali, V. and Shorthill, R.W: 'Fibre Ring Interferometer', Appl. Opt. Vol 15, No 5, pp 1099-1100, 1976.
2. Rogers, A.J: 'Optical Measurement of Current and Voltage on Power Systems', IEE Journal of Electric Power Applications, Vol 2, No 4, pp 120-124, 1979.
3. Nelson, A.R., McMahon, D.H. and Gravel, R.L: 'Passive Multiplexing System for Fibre-Optical Sensors', Appl. Opt, Vol 19, No 17, pp 2917-2920, 1980.
4. Barnoski, M.K. and Jensen, S.M: 'Fibre Waveguides: a Novel Technique for Investigating Attenuation Characteristics', Appl. Opt. Vol 15, No 9 pp 2112-2115, 1976.
5. Jones, B.E: 'Simple Optical Sensors for the Process Industries using Incoherent Light' IMC Symposium on 'Optical Sensors and Optical Techniques in Instrumentation, London, Nov. 1981.
6. Theocharous, E: 'Differential Absorption Distributed Thermometer' Proceedings of the 1st International Conference on Optical Fibre Sensors (OFS '83),London, IEE Conference Publication No. 221, pp 10-12, 1983.

7. Hartog, A.H: 'A Distributed Temperature Sensor Based on Liquid Core Fibres' J. Lightwave Techn., LT1, No 3, pp 498-589, 1983.
8. Farries, M.C. et al: 'Distributed Temperature Sensor using Nd^{3+}-Doped Optical Fibre', Elect. Lett., Vol 22, No 8, pp 418-419, April 1986.
9. Dakin, J.P., Pratt, D.J., Bibby, G.W. and Ross, J.N: 'Distributed Anti-Stokes Ratio Thermometry', Proceedings of the Third International Conference on Optical-Fibre Sensors (OFS '85), San Diego, USA, Postdeadline Paper, Feb. 1985.
10. Hartog, A.H., Leach, A.P., Gold, M.P: 'Distributed Temperature Sensing in solid-core Fibres', Elect. Lett. Vol 21, No 23, pp 1061-1062, Nov. 1985.
11. Ippen, E.P. and Stolen, R.H: 'Stimulated Brillouin Scattering in Optical Fibres', Appl. Phys. Lett., Vol 21, No 11, pp 539-541, 1972
12. Giles, I.P., Uttam, D., Culshaw, B. and Davies, D.E.N: 'Coherent Optical-Fibre Sensors and Modulated Laser Sources', Elect. Lett, Vol 19, No 1, pp 14-15, Jan 1983.
13. Younquist, R.C. and Franks, R: (Private Communication)
14. Rogers, A.J. 'Polarization-Optical Time Domain Reflectometry : a New Technique for the Measurement of Field Distributions' Appl. Opt. Vol 20, No. 6, pp 1060-1074, 1981.
15. Ross, J.N. 'Birefringence Measurement in Optical Fibres by POTDR', Appl. Opt. Vol 21, No 19, pp 3489-3495, 1982.
16. Clark Jones, R: 'A New Calculus for the Treatment of Optical Systems: Pt II' Journal of the Optical Society of America, Vol 31, pp 493-499, 1941.
17. Farries, M.C. and Rogers, A.J: 'Distributed Sensing Using Stimulated Raman Interaction in a Monomode Optical Fibre', Proceedings of the 2nd International Conference on Optical-Fibre Sensors, Stuttgart, (OFS '84) Paper 4.5, pp 121-132, 1984.
18. Barlow, A.J., Ramskov-Hansen, J.J. and Payne, D.N: 'Anisotropy in Spun Single-Mode Fibres', Elect. Lett. Vol 18, No 5, pp 200-202, 1982.
19. Ross, J.N: 'Measurement of Magnetic Field by POTDR', Elect. Lett, Vol17 No 17, pp 596-597, 1981.
20. Rogers, A.J: 'Polarization Properties of Monomode Optical Fibres : use of POTDR to Determine Spatial Distributions' Proc. of the International Conference on Fibre-Optical Rotation Sensors and Related Technologies, MIT, Cambridge, Mass. Nov. 1981, Springer-Verlag, pp 208-214, 1982.
21. Giallorenzi, T. et al: 'Optical-Fibre Sensor Technology' IEEE, J. Quantum Electronics, Vol 18, pp 626-641, 1982.

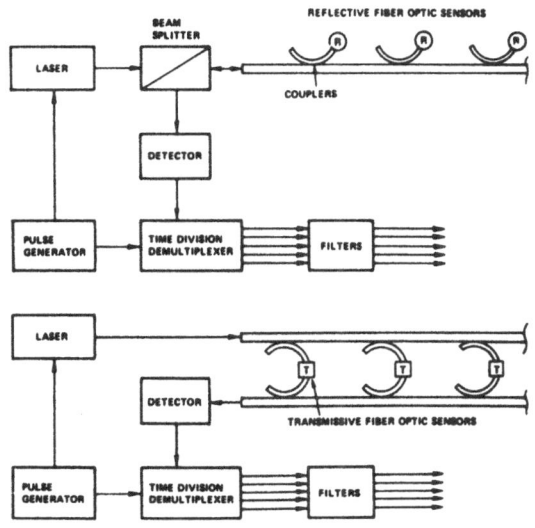

Fig. 1 Quasi-distributed Systems with Temporal Resolution

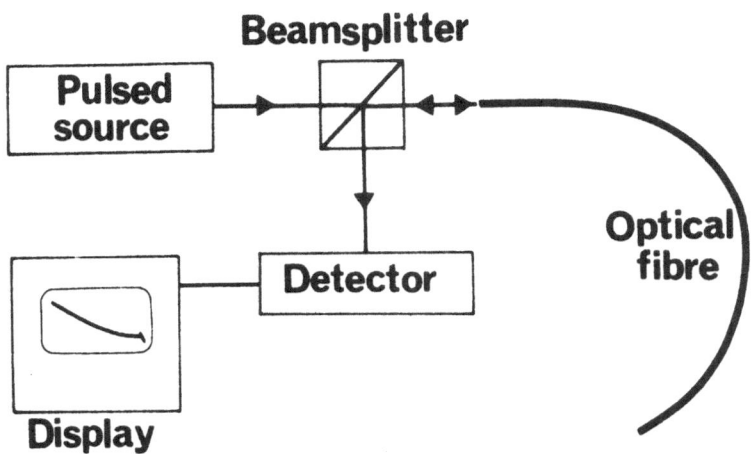

Fig. 2 Optical Time-Domain Reflectometry

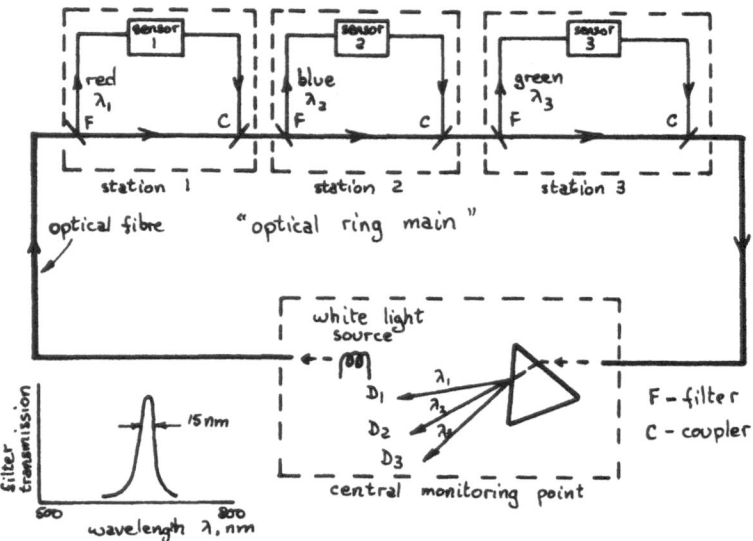

Fig. 3   Quasi-distributed System with Wavelength Resolution

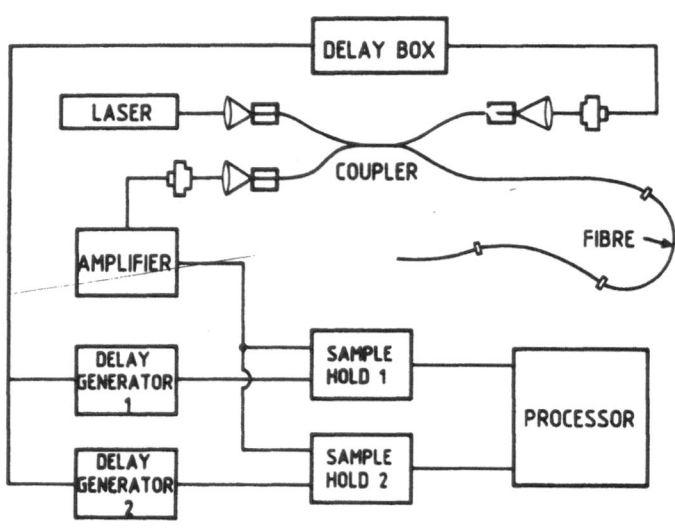

Fig. 4   Differential-Absorption Distributed Thermometry (DADT)

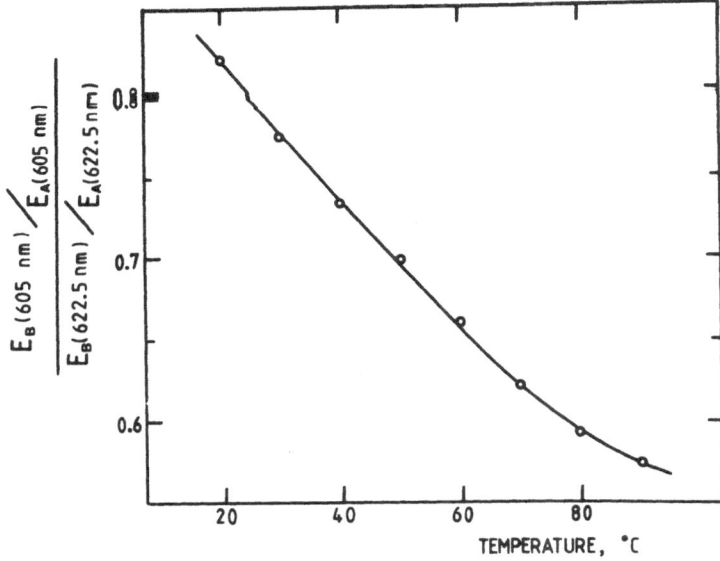

Fig. 5   DADT : Dependence on Temperature

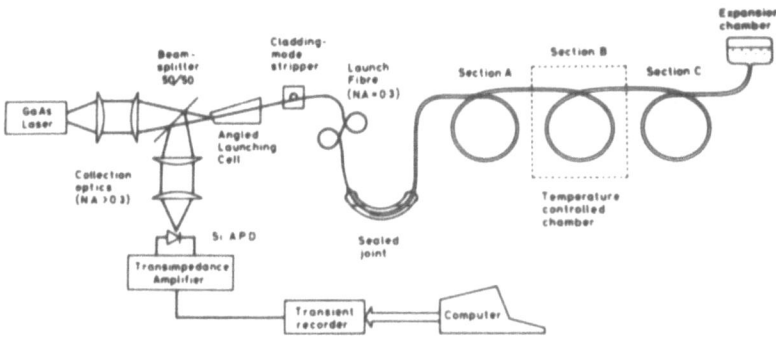

Fig. 6   Scatter-Dependent Temperature Measurement (SDTM)

158

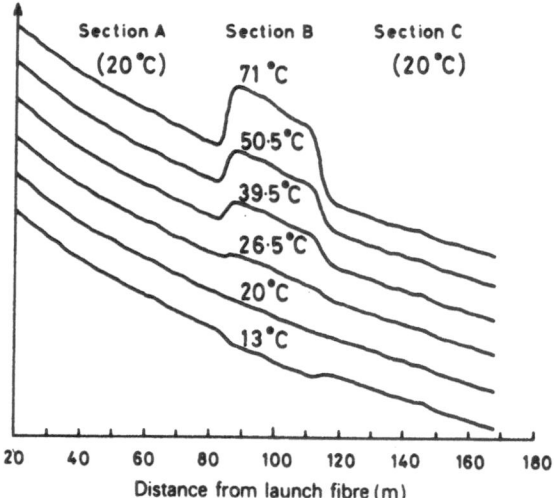

Fig. 7   SDTM with Liquid-Cored Fibre

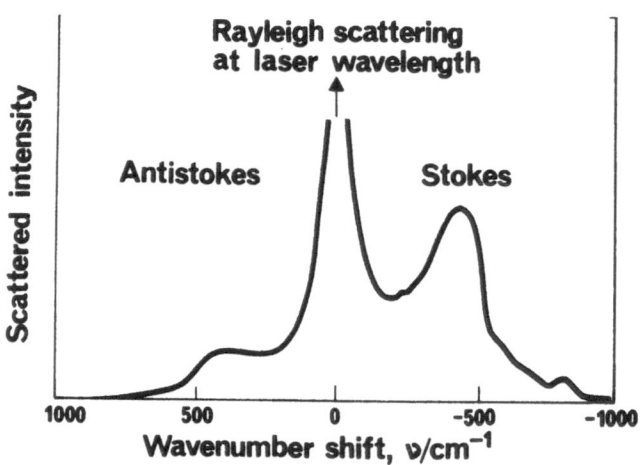

Fig. 8   Raman Spectrum for Silica

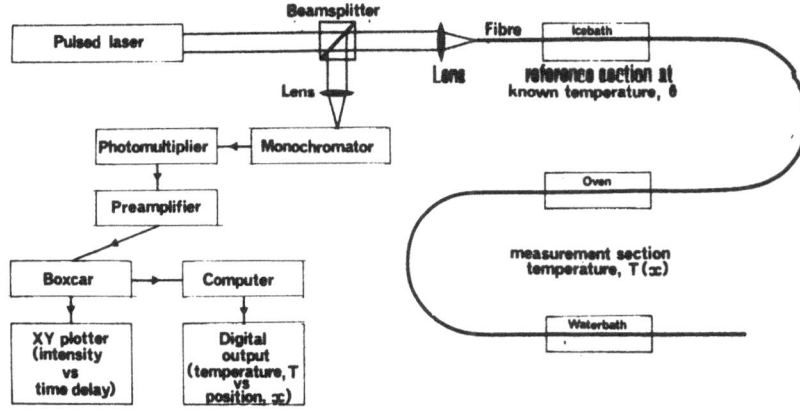

Fig. 9 Distributed Anti-Stokes Raman Thermometry (DART)

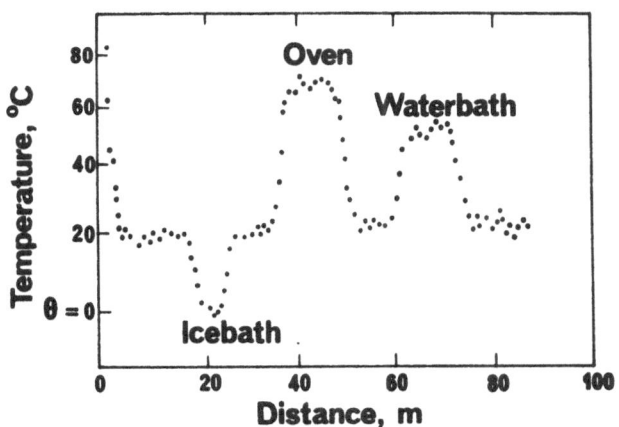

Fig. 10 A Temperature Distribution Measured with DART

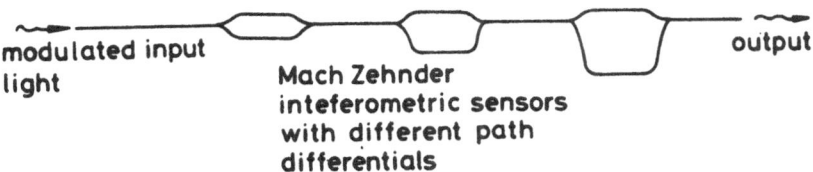

Fig. 11 Distributed Mach-Zehnder Transducers

160

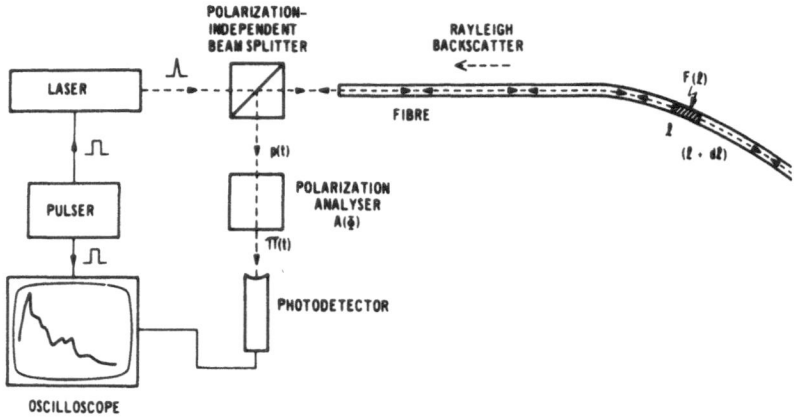

Fig. 12   Polarization-Optical Time Domain Reflectometry (POTDR)

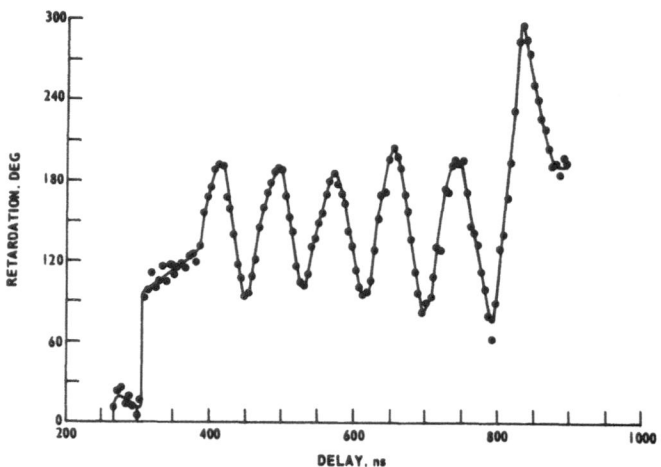

Fig. 13   Distributed Bend-retardation Measurement using POTDR

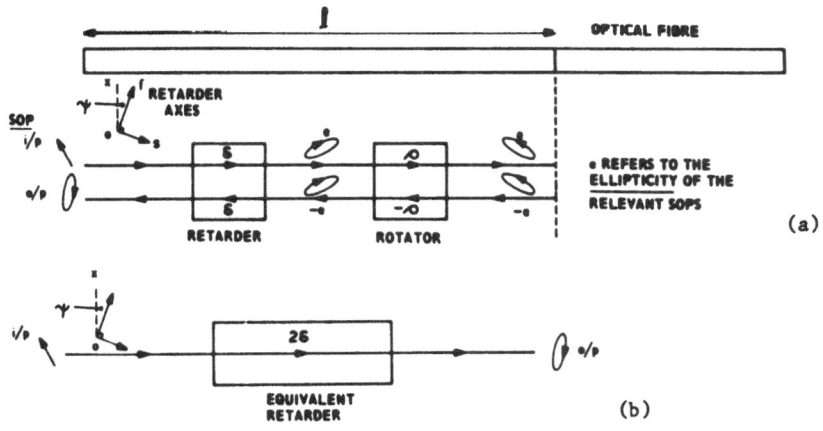

(a)

(b)

Fig. 14   Equivalent Retarder-Rotator for Reciprocal Optical Element

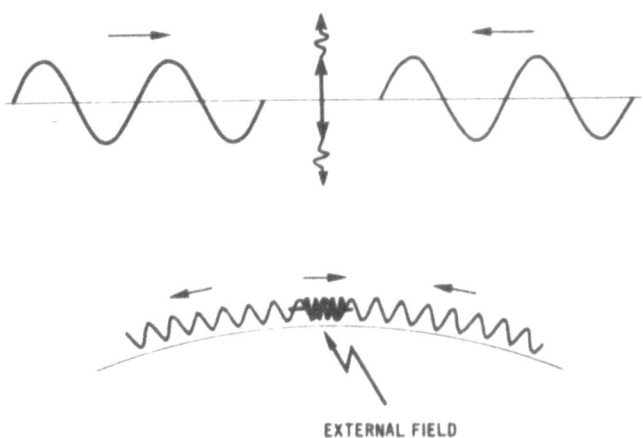

Fig. 15   Non-Linear Pulse/Wave Interaction

162

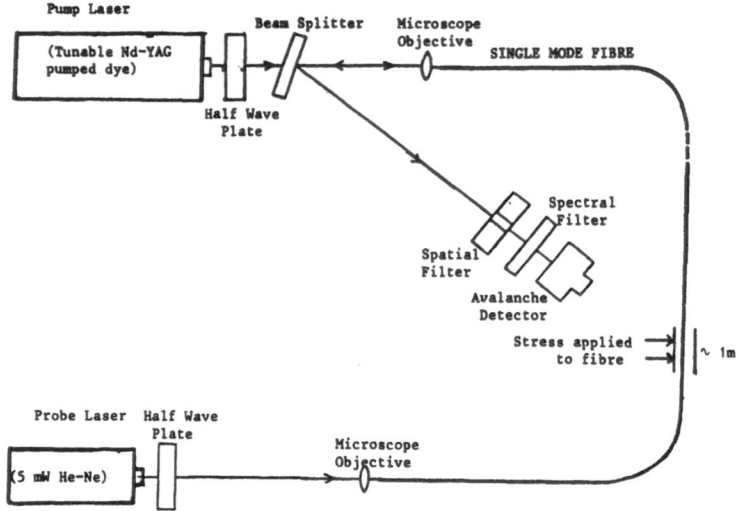

Fig. 16   Arrangement for Pulse/Wave Raman Interaction Dsitributed
Measurement

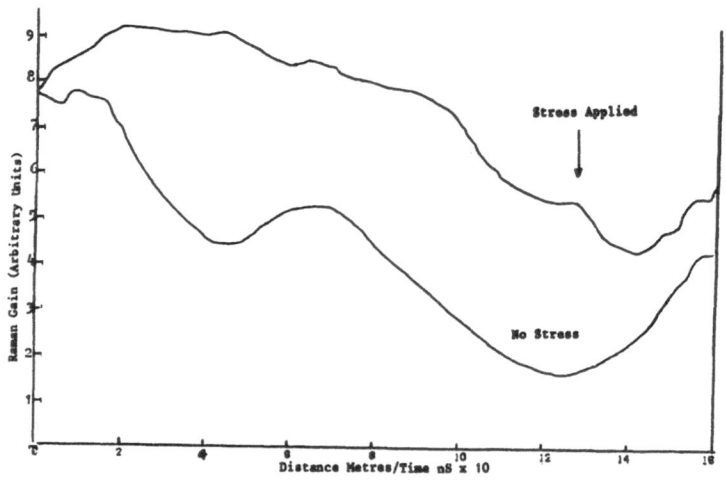

Fig. 17   Local Stress Effect in Distributed Raman System

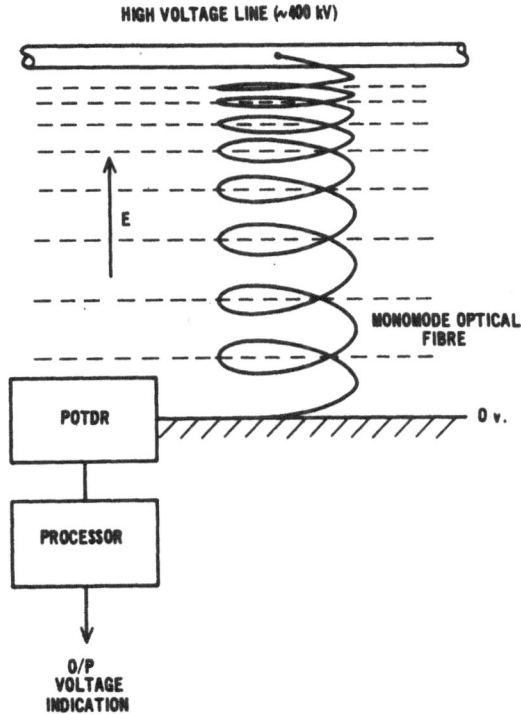

Fig. 18 Voltage Measurement using POTDR

DISTRIBUTED AND MULTIPLEXED FIBRE OPTIC SENSOR SYSTEMS

Brian Culshaw
Department of Electronic & Electrical Engineering
University of Strathclyde
204 George Street
Royal College Building
Glasgow G1 1XW

ABSTRACT

This paper presents a detailed theoretical assessment of the numerous forms of multiplexed optical fibre transducer systems and derives criteria whereby their potential performance may be compared. This leads to estimates for the maximum number of sensor elements which may be addressed and the resolution available at each of these elements. The paper then examines a selection of the available data on multiplexed systems and compares the achieved performance with that which is apparently possible.

INTRODUCTION

The prospect of a completely optical multiple sensor optical fibre network has been recognised for some years. The potential advantages of such a network arise largely in the fact that the entire system is non-electrical. Furthermore, a multiplexed system permits a more economic use of the tremendous data carrying capacity of an optical network. Most sensors barely scratch the surface of this data carrying ability and, even though all multiplexed networks must, by definition, lose some optical power, there is still a great deal left to form a realisable network.[1,2,3]

Whilst there have been numerous experimental reports of trial systems, very little has been published on networks and fundamental system analysis and there are few, if any, projections available to indicate the potential performance limitations of multiplexed and distributed fibre optic sensors. Perhaps, this is due to the multiplicity of possible networks which may be configured. However, a more careful examination shows that it is quite feasible to extract realistic themes and to classify multiplexing systems. This paper presents one possible approach to such classification based upon some fundamental observations on the power budget for a general multiplexed system. This leads to the hypothesis of three types of multiplex systems – namely fully distributed sensors, single source multiplexed discrete sensors and multiple source multiplexed discrete sensors. The paper presents a simple but useful theoretical approach to analysing such systems, draws some conclusions concerning potential performance and presents some brief case studies based upon published experimental data.

SOME FUNDAMENTAL OBSERVATIONS

A generalised distributed optical fibre system is shown in Figure 1. It consists of an optical source (or sources) and an optical detector (or detectors) coupled to a signal processing function and usually to a source modulation function which is linked into the signal processing. The system has two simple functions.

- To measure the optical signal returned from the network and
- To identify the component of this signal which originates from each sensor or sensing element in the network.

It is, therefore, – at least in principle – simply a matter of comparing the various available systems using the ease, both technically and economically, with which they are capable of performing the above two functions as a criterion.

Before venturing further into the analysis it is important to appreciate that a general multiplexed system consisting of N sensors (or resolvable sensing intervals) each operating with a bandwidth B and with a number of resolvable intervals S will carry a total information content C of [4]:

$C = N.B \log_2(1 + S)$ bits/sec

In a real system S will be of the order of a fraction (typically 0.2) of the per-sensor signal to noise ratio. We can, therefore, define a system bandwidth B_s and a total system signal to noise ratio SNR :

$$B_s = N.B$$

$$SNR = 5.S$$

This division reflects what is probably the natural means whereby the sensors would be multiplexed onto the system: that is in effect using some form of frequency multiplexing. However, it would also be conceivable to stack the sensors in amplitude slots within the same total bandwidth . In which case, the required signal to noise ratio is now 5.S.N whilst the bandwidth remains at B. It is a classic result from communication theory that the latter arrangement will be less economical in power than the former. If we assume shot noise limited detection, then it is simple to show that the first division requires a total of :

$25.S^2.N.B$ photons/second

at the receiver, whilst the latter requires :

$25.S^2.N^2.B$ photons/second

For an analogue sensor with a typical industrial specification requiring 0.1% resolution, S will be 10^3 so that the required number of photons at the receiver is of the order of :

$2.5 \times 10^7.N.B$ photons/second

This typically corresponds to approximately 10pW per sensor element per Hertz of bandwidth. However, we should recognise that at these low power levels it is unlikely that shot noise limited detection can be achieved. It is readily shown [5] that shot noise limited detection requires :

$$R_L \gtrsim \frac{0.1}{\text{Optical Power}}$$

It is, therefore, impractical to anticipate shot noise limited detection for optical powers of less than or about 1 microwatt.

It is important to interpret the concept of shot noise limited detection correctly in the context of analogue systems. The shot noise limit corresponds, in effect, to the statement that the detector is acting as a photon counter. However, in an analogue system the achievable resolution is dictated by the Poisson statistics operating on the rate of arrival of photons at the detector. Consequently shot noise limited detection in the conventional sense of the word is not required. A more careful examination of the problem indicates that the receiver noise may exceed the shot noise by a factor $M^{\frac{1}{2}}$ where M is the number of photons arriving on average in the observation interval of the detector. Our 10 picoWatt signal is, therefore shot noise limited with the load resistance of the order of $10M\Omega$ For a detector in which the load resistance is reduced to say 100 $k\Omega$ and the required power per sensor is increased to 100 pico Watts.

The power required when a number of sensors are "stacked" depends on a rather complex combination of the coding system used to multiplex the various sensors and the dominant noise forces. However, if we assume that the system is optical shot noise limited in the manner described then the required optical power will go as N^2 if the sensors remain stacked within the same bandwidth and as N if the sensors occupy N times as much bandwidth (where N is here the number of sensors in the network).

The second requirement on the multiplexing system is to identify the individual sensors or sensing points in the multiplexing network. This implies that each sensor must have its own individual addressing label. The detailed requirements depend upon the system type (Figure 2). In the distributed sensor (Figure 2A), the location of the sensing point is part of the information to be extracted from the observation system. In this case the classical radar approach [6] determines the necessary source bandwidth required to achieve the appropriate range resolution. In contrast the multiplexed point sensor system (Figure 1) involves only stationary sensors whose position remains constant with time and, therefore, may be programmed into the detection electronics at the installation phase. This may involve either a calibration step with all the sensors in a known position, or a prior knowledge of the installation process in which all delay lengths are calibrated. Finally in the multi-source sensor (Figure 2b) each sensor is located by virtue of the wavelength appropriate to that particular element. This concept has the advantage that no distance measuring is necessary in the decoding process against which must be weighed the additional complexity of the optical system. Polarisation coding may also be used in multi-source systems.

The most important observations which one can make on the distributed and multiplexed systems are that the total information carried by the system is determined entirely by the number of sensors on the network and the required resolutions for each sensor. This, in turn determines the necessary receiver bandwidth. However, if range resolution is also required the necessary receiver bandwidth may be considerably increased - with consequent noise penalties - though again, in principle, by suitable signal processing, this excess detector bandwidth may be reduced. In the remainder of this paper these fundamentals are utilised to discuss the relative merits of a number of multiplexor architectures.

DISTRIBUTED SENSORS AND REFLECTOMETERS

The principal source of loss in most optical fibres is the Rayleigh backscatter. The magnitude of the scattered radiation depends upon the effective cross section of microscopic inhomogenieties in the guiding region the fibre. Some of the scattered radiation is collected by the fibre and guided back towards the source. The collected fraction depends upon the numerical aperture of the fibre and, therefore, the total radiation returned towards the source may be modulated by altering either the scattering cross-section or the numerical aperture [7]. In addition to the backscattered radiation which originates throughout the fibre length there are optical power reflections from discontinuities e.g. at the end of the fibre or at a connector. Backscattered radiation is consequently used as a diagnostic tool in testing fibre optic installations [8].

The scope for using this phenomenon as an environmental sensor depends primarily on the efficacy with which the measurand of interest may be made to interact with the scattering characteristics of the fibre. As examples liquid core fibres [9] are found to have scattering cross sections which are strongly temperature dependent, whilst careful measurement of the polarisation properties of the backscatter [10] can give information about certain types of mechanical stress that may be applied to a monomode fibre. There are also backscatter systems which are, in fact, quasi-point sensors, where, for instance, microbend loss may be introduced using local mechanical strain [11].

In analysing such systems the starting point is to determine a typical number of photons arriving at the detector. For a fibre with a scattering loss characterised as α_s Nepers per metre and negligible other loss, the power scattered at a length 1 from the source from a total length δl is given by :

$$P_s = P_0\,\alpha_s\,\delta l\,e^{-\alpha_s l}$$

and from this fraction F given by :

$$F = (NA)^2/4$$

will be collected. The total backscattered power arriving at the receiver is :

$$P_{sr} = FP_0\alpha_s\int_0^L e^{-2\alpha_s l}\,dl = \frac{F_0 P}{2}\{1 - e^{-2\alpha_s L}\}$$

The total backscattered power is obviously a maximum as the total attenuation tends to infinity. However, in practice, a total attenuation of 3dB would be a reasonable design criteria.

It is convenient to normalise discussion of optical power to an average launch power of 1 milliWatt. For a fibre with a numerical aperture of 0.1 and 3dB total loss it is then straightforward to show that the total power reflected to the receiver is of the order of 625 nW which corresponds to approximately 4×10^{12} photons/second. In order to achieve a 1 metre resolution, the receiver must be capable of resolving 10 nanosecond intervals (since it is the round trip time which is important). In each such interval of the order of 4×10^4 photons arrive at the detector so that the maximum possible resolution in each 10 nanosecond interval is 0.5%. However, it is likely that thermal detector noise will modify this to a resolution of 10% or worse. Illuminating the fibre with 10,000 pulses per second will improve this by a factor of 100 – assuming that all the integration processes can be achieved. Thus in principle it seems that about 0.1% resolution should be attainable in the bandwidth of 1Hz for a system involving 1,000 resolvable intervals along the sensor fibre but assuming perfect signal integration. This does, of course, require elegant and advanced electronics – which is intrinsically expensive. Additionally the assumed resolution of 0.1% is implicitly assuming a 100% potential modulation depth. This is impractical since this would correspond to cutting off the light altogether. Assuming a 10% modulation depth implies that 1% resolution over a usable dynamic range is achievable. This gives a usable – but not terribly impressive – potential sensor system which must have very special applications in order to justify its cost and complexity.

Typical results for a liquid core distributed temperature sensor are shown in Figure 3 whilst in Figure 4 a subtle temperature sensor which measures the ratio of the Stokes and anti-Stokes Raman backscatter is shown.[12] The backscattered Raman signal is very small (approximately 10^{-3} of the Rayleigh backscatter). However, the method has the advantage of being absolute since the measured ratio is :

$$R(T) = \left\{ \frac{\lambda_s}{\lambda_a} \right\}^4 e^{-h\nu/kT}$$

which after correction for differential scattering (the 4th power terms) permits temperature on any fibre regardless of core construction or modal properties. A typical spectrum is shown in Figure 5.

Fully distributed sensors are simple to instal but require special fibres and advanced signal processing to optimise their use. Additionally the range of possible measurands is restricted and it is only temperature which inherently affects the core properties of a fibre. Any attempt to measure other properties must incoporate suitable temperature compensation and this further complicates the signal processing. There are advantages in these distributed sensors primarily concerned with their small size, relatively rapid response time (though limited by the integration via the electronics) and their truly distributed nature. However, it is important to realise that only measurands which affect the guiding properties of the core either by interacting with the core itself or with the cladding material may be used in such systems. All fibres are temperature sensitive in this context – though some more than others – and it should also be appreciated that fibres with special cladding, which may be fabricated from biologically active material for example, are in effect invariably local point sensors which do not require coupling into the fibre link. Such sensors do, as we shall see, have important advantages in terms of their overall loss budget and may make a useful contribution to future system applications.

MULTIPLEXED POINT SENSORS

A discussion of the properties of multiplexed point sensor systems may usefully start with a consideration of the basic power budget. If we set a resolution of 0.1% in 1Hz bandwidth per sensor as the target specification, we find that typically a received power of 100pW per sensor is required. For 1mW of available launch power into the system we have, therefore, an absolute maximum of 10^7 resolvable sensors available in the system. However, this observation should be tempered by a consideration of the onset of optical carrier shot noise. It is likely that at 1mW, the receiver will be shot noise limited so that the optical carrier:noise ratio will limit the performance. In a 1Hz banwidth this is 78dB. To achieve the nominal 0.1% resolution requires of the order 37dB SNR in the bandwidth of interest. Therefore, the maximum number of sensors at one milliwatt received power is of the order 15,000. This figure is derived on the basis that the system multiplexes into a given bandwidth. So, depending on the multiplexing scheme, an absolute maximum ranging from 10^7 (frequency multiplexing) to 10^4 (amplitude multiplexing) may be expected. This, of course, neglects two important practical considerations - the loss per sensor and the efficiency of the coding/decoding network.

In any point sensor network there is an inevitable loss involved in attaching the sensor element into the network. Suppose a total of N sensors are to be attached to the network and each sensor has a loss of xdB then, again, assuming 100 pW per sensor and a launch power of 1mW we find that the absolute maximum sensor count is given by:

$$\frac{10^7}{N} < 10^{Nx/10}$$

The importance of the per-sensor insertion loss may then be judged by inserting x = 0.1dB into the above relationship observing that the maximum value of N is then of the order of 500 whilst for x = 1dB the maximum for N is of the order of 50.

This, of course, has simply covered the effect of coupling a sensor into the network. An additional requirement of this class of sensor network is that the different sensors must be spaced at different time intervals from the receiver. This implies either directly or indirectly that a directional coupler element must be used. If the coupling ratio of this element is a fraction k then since each sensors signal must go through the element twice, there is (see Figure 6) an effective loss of $10\log_{10}(k^2)$ for each sensor. If the couplers all have the same ratio (see below), then on average each sensor signal will go through half of the couplers on both outward and return journeys so that an average additional loss of $10N\log_{10}(1-k)^{-1}$ is incurred. We then obtain for a total maximum per-sensor loss of 70dB corresponding to 1mW launch power and 100pW per sensor, that the maximum value of N is given by :

$$\left(10N\log_{10}(1-k)^{-1} + Nx/10\right) + 10\log_{10}N < 70$$

If we take as an example a 20dB coupler corresponding to k = 0.01 and a
0.5dB loss per sensor we find a maximum value then of approximately 90.
These very simple order of magnitude estimates of system performance lead
to an important conclusion which is that for any practical discrete sensor
multiplexed system the maximum number of resolvable analogue sensors will
be of the order of a few tens. This number may only be increased if the
required resolution is eased or if the optical power is increased.
However, the essential logarithmic relationship does imply that even
relaxing these requirements will make relatively little difference to the
total permissable sensor count. Real improvements are only available by
reducing the per-sensor loss to the minimum.

It is intuitively obvious that the optimum network design involves the same
power arriving at the detector per sensor. This may be argued simply from
the observation that if this is not the case then some sensors must have
too much power arriving at the detector. If we examine the generalised
coupler network shown in Figure 7 with the coupling ratios shown then we
see that the power return from nth sensor is given by :

$$P_n = P_0 k_n^2 \prod_{p=1}^{n-1} (1-k_p)^2$$

From this it is immediately deduced that the optimum value for k_n is :

$$k_n = (N-n+1)^{-1}$$

This estimate, of course, has neglected the effect of losses within the
fibre and of splicing/connector losses. These will both tend to reduce the
values of the coupling ratios for the sensors nearer the source and the
detector. It is also interesting that even the perfect topology in which
the ideal coupling ratios can be achieved and which is otherwise totally
loss free, has an intrinsic loss per sensor of N^2 in contrast to the N used
in the earlier simplistic arguments. Therefore, even our perfect loss free
network would be incapable of supporting more than approximately 3,000
sensors. These optimum networks will inevitably be very difficult to
realise in practice. In the first instance, because of the requirement for
carefully controlled coupling ratios for each sensor. It is, therefore,
useful to consider the case where the coupling ratios are identical for
each element and where the system performance is dictated by the sensor
with the maximum loss. If we take as an example k = .01 then the maximum
loss will be at the (N-1)th sensor and will be :

$$(1-k)^{2(N-2)}$$

The general relationship is then :

$$20(N-2)\log_{10}(1-k) + 2k < 70$$

Taking the case of our 1% couplers this result in a maximum sensor count of approximately 17 transducers. It is important to realise that in a network of this nature the first sensor for which the total loss is only 40dB's and the last for which is the total loss is 70dB's will impose considerable strain on the signal processing network especially with regard to dynamic range unless attenuation is introduced into the sensors which are closer to the transmit/receive electronics. Consequently networks using constant coupling ratios for each sensing elements are very inefficient in their use of optical power. Some compromise may be reached by utilising a range of couplers with the lowest value of k nearest the source and receiver.

The basic power budget of all the multiplexed fibre optic point sensors follows very similar lines to those described above. It is in the overall system architecture that the major operational distinctions become apparent. In particular this applies to source modulation and coding formats of which there is a considerable variety. In effect the source modulation scheme provides a means whereby various weighted sums of all the sensor outputs may be obtained. If we have N sensors each of bandwidth B then simple sampling considerations indicate that at least 2NB measurements must be made on the system per second. Therefore, in principle, the modulation applied to the system source need only have a total bandwidth of 2NB. In this statement it is assumed that the sensors are sensibly stationery so that the weightings in the sum for each value of the modulation function remain constant for all time. If the sensors move, then a different approach is required resulting in increased bandwidth in the modulation function. It is also interesting to note that coefficients in the weighted sum need not be known prior to the installation of the network. It is always possible to incorporate an initialisation procedure which will enable the system to measure the weightings.

Source modulation schemes may be categorised according to whether they are totally incoherent, exploit coherence in the modulation or exploit coherence in the optical wave form. The former two may be implemented with any type of optical source, whereas the latter requires a laser whose coherence length is specified by the system parameters. Figure 8 shows these classifications schematically. The simplest incoherent system is straightforward pulse modulation (Figure 8a). Here the sensor output appears as a pulse train which is, in general, easily decoded. However, the pulse length must be shorter than the intersensor delay requiring correspondingly high bandwidth electronics (and therefore increased noise levels) and the maximum source duty cycle is less than N:1. Even though the peak power in the pulse may be increased it is often difficult to do so, since many optical sources are peak power limited and sometimes optical fibre non-linearities (especially in single mode sytems) can cause significant spurious signals. Source modulation coding which entails continuous operation is often preferable. By definition this is usually lower in bandwidth and often higher in average power. Figure 8b shows the elements of a chirp modulated system. Here the various sensors appear at different points in frequency space. This point is a permanent characteristic of the sensor and the amplitude of the spectrum centred around this point with a sinc envelope imposed upon harmonics of the chirp frequency is a measure of the output of each individual sensor. The signal processing for these systems may be centred around standard SAW technology but there are implications in the sinc spectrum which imply that for crosstalk to be kept below 1% the sensors should be separated in frequency space by, at least 10 times the chirp repetition rate[13]. Yet another variation which also uses coherent modulation shown in Figure 8c. In this system sinusoidal intensity modulation is applied at, at least, N/2 frequencies. The weighted sums which constitute the in-phase and quadrature components at the detector are a function of the phase delays to the various sensors. This system has the advantage of simple and easily controlled modulation functions and requiring minimal bandwidth electronics compared to the preceeding two. A trial system (Figure 9) has demonstrated the viability of this approach and the entire source, detection and processing electronics may be obtained at a component value of the order of hundreds of dollars[14].

Detailed performance comparisons of these three basic systems is beyond the scope of the current discussion. Criteria for comparison should include complexity of signal processing, the effect of numerical errors in any computational procedures, the straightforward power levels available and optical signal to noise ratios and the system compatibility with various types of sensor. However, it is perhaps worth emphasising that all these systems are compatible with sensors utilising any modulation scheme with the possible exception of interferometers (since the sorce is incoherent). The concepts may also be extended to include balanced systems using wavelength or other forms of dual channel referencing[15,16].

Interferometric sensors require a source whose coherence length exceeds the maximum path difference to be experienced in the network. Interferometric sensors may be coded either through varying the path difference in each of the interferometers constituting each sensor, or by varying the time delay between source, the various interferometers and the detector. In the latter system the interferometers are effectively intensity modulators and the systems illustrated in Figure 8 are directly applicable. Two systems specifically suitable for the multiplexing of interferometers are shown in Figure 10. The optical frequency modulated chirp system (figure 10a) is analogous to the system in figure 8b. The output signal from each sensor is itself a baseband frequency as opposed to an intensity modulation of the base band frequency in the former system. The effective baseband frequency is measured as a ratio of appropriate harmonics of the chirp repetition rate and requires stringent control on the source modulation waveform to be fully effective. Good FMCW optical waveforms may be generated using a helium neon laser and an external modulator. However, a more convenient technique involves modulating the current of a single mode semiconductor laser though the latter is far more difficult to control[17,18,19]. An alternative approach is the optical correlation system shown in Figure 10b. Here a short coherence length source - typically a light emitting diode - is used to energise the interferometer network. The detector is then a reference adjustable interferometer and the setting of this interferometer is modified to seek the maximum interference signal corresponding to operating on the zero fringe with an incoherent source, provided the path differences on each interferometer the source coherence length then each interferometer may be individually identified. An air path demonstration[20] illustrated the principle and a more detailed discussion on sensor limits and similar systems criteria is given in reference 21.

Optically coherent multiplexing is apparently much more difficult to implement with any accuracy than the simple systems using coherent modulation functions. This primarily stems from the requirement for extremely stable optical source waveforms. In principle these could be supplied using an injection locked or cavity controlled semiconductor laser along with an external modulator. Direct modulation of semiconductor lasers has been shown to be very difficult to achieve with sufficient precision and external modulation of gas lasers – which is simple to implement in practice – is operationally undesirable. The optical correlation system requires a stable and accurately calibrated reference interferometer and it is still far from clear how this may be effectively implemented in practice.

Interferometric sensors and interferometric sensor multiplexors may also be realised using dual moded fibre, either involving stress or form induced birefringence or taking a single mode fibre for, say 1.3 micron use and operating at 850nm. The advantage of these systems is that couplers are not required and there are no splices so that much of the loss budget is eliminated. The snag is that spatial resolution along the fibre may only be achieved in reflection so that the comments on backscatter in the earlier sections of this paper must be brought to bear.

MULTIPLE SOURCE MULTIPLEXED SYSTEMS

Multiple source sensor systems almost invariably involve wavelength coding – as shown in Figure 11. The principal and most obvious contrast between this and the systems previously described is that here it is quite feasible to multiplex several transducers at one point on the optical fibre link. In addition to wavelength coding one may also use orthogonal modulation techniques involving, for instance, linear and circular polarisation [22] or quadrature modulation of a phase coherent carrier.

In contrast with the distributed point sensors, multiple source systems are limited in their performance only by the quality of the components used to distinguish one source from the others throughout the network. These characteristics are usually a function of the available technology though all do have some form of fundamental physical limitation. For instance spectrometers are restricted in their resolution by, in effect, the F number of the input optics [23].

The power budget for the straightforward multiple source multiplexed system is made very simple since only one source feeds each sensor. These loss mechanisms are, therefore, simply the total loss of the fibre link involved and the loss per splitting element including the effects of connectors/splices. If we take the latter as 3dB in total (which may be slightly optimistic) and again assume a 100 pico Watt necessary thermal noise limited received power per sensor element, we then find that with 1mW of launched power a total loss of 70dB is permissable. However, this implies a maximum of 23 sensors on the network unless a "tree" configuration is used where each splitter, in effect, feeds a number of sensors in parallel. Therefore, on power budget criteria we find a similar overall potential sensor count to a realistic single source multiplexed system. However, it should be emphasised that substantial improvements are possible by reducing the insertion loss per device.

Crosstalk in distributed point sensor systems is a function of the stability and accuracy of the interrogating electronics. In multiple source systems crosstalk is a function of the ability of the wavelength (or other optical parameter) multiplexing devices at each sensor head. Each splitter (see Figure 12) really sends the majority of the light of the appropriate wavelength into the sensor, whilst allowing the majority of the unwanted light to pass by. Small fractions of each section of the incident spectrum pass through the wrong path. Crosstalk occurs when light from sensor n is passed through all the other sensors and is modulated by changes in the other sensors. The effect of the chain of sensors is additive, so that if a fraction f of the input light in the unwanted part of the spectrum is passed through each sensor, then the total possible crosstalk signal is of the order $(N-1)f$. For example, a system of 20 sensors requiring a resolution of 0.1% will require a minimum crosstalk of -43dB to be totally sure that the crosstalk level will be acceptable. However, since it is unlikely that all the N-1 sensors which are involved in generating the crosstalk will go from fully on to fully off, then considerable relaxation of this nominal specification is possible. A crosstalk level of perhaps -35dB will probably be adequate.

The concepts of wavelength multiplexed sensor systems have been known for at least 5 years. During that time a few trial systems have been built but there is minimal published data on their performance. In comparing the theoretically achievable performances of these networks with single source systems, it is apparent that both rely upon low insertion loss sensor heads and for a given insertion loss at the sensor head, both will give comparable performance in terms of the maximum achievable number of sensors. In terms of the overall system, the multiple source concept requires more sources and detectors but is significantly simpler in terms of electronic signal processing.

DISCUSSION

It is useful to divide these systems into the two categories in the title of the paper - distributed sensors and multiplexed sensors.

Distributed sensors are usually variations on the theme of optical time domain reflectometry. Their features may be generalised as :

- Spatial resolution is entirely dictated by pulse rise times and so high speed (and, therefore, noisy) electronics is required. The minimum bandwidth will exceed 100MHz, thought the sensing bandwidth will be much lower.

- The total backscattered signal would be of the order of 30dB below the incident signal. This total signal is distributed along the sensor length.

- Distributed sensors are limited to measurands which modulate the fibre core. This usually means temperature though strain (via microbending or via variable coupling induced in polarisation maintaining fibre) is another possible measurand.

- Pulse repetition rate is limited by the total round trip sensor transit time and duty cycles of 0.1% or less are typical.

- Peak input power levels may be limited in certain fibre types (especially single mode) due to induced non-linearities.

In contrast multiplexed systems may be summarised as having the following major features:

- Multiplexed point sensors are compatible with all fibre optic sensor types and all fibre sensor modulation mechanisms.

- The power budget for all multiplexed systems is limited by the per sensor insertion loss <u>independently</u> of detailed sensor system architecture.

- All systems require "special" components for optimum operation. This includes special ratios on directional couplers and/or finely tuned wavelength multiplexing elements. Some compromises are inevitable in the use of "standard" components.

- Most multiplexed systems are conceived as multimode systems. In many cases single mode may offer potentially improved techical performance.

- All systems would benefit from a standard sensor response which may, for example, be a range of fractional transmittances or reflectances. A near equal optical power distribution between all the elements in the system will clearly optimise crosstalk and signal processing procedures in general.

The single source systems which have been proposed and, in some cases, experimentally investigated differ from multiple source systems largely in the mechanisms for crosstalk between sensors. It has become apparent in the preceeding discussion that the power constraints are broadly similar in determining the maximum number of sensors allowed in each system. In a single source system one principal practical limit on achievable crosstalk stems from the temporal stability of the modulation function applied to the optical waveform. Changes in this result in apparent shifts in the positions of the sensors in the network so that the output from a given real sensor is derived from a weighted sum of sensors in the positions in which the networks thinks the sensors are located. This in turn gives rise to crosstalk, the detailed nature of which depends upon the interrogation system used. Single source systems also involve, sooner or later some digital signal processing and the effects of analogue to digital conversion and rounding errors in computation are often significant. With careful design of the signal processing algorithms, this may be minimised, though it may imply the use of 16 (or even 32) bit arithmetic. Great care is always necessary in the derivation of the constants in the setting up of the system. In a multiple source system there is a standing crosstalk term due to imperfection in the wavelength (or other optical parameter) splitting devices in the network. However, there is an additional term which may arise from drift in the source parameter (usually wavelength) with time, temperature and bias conditions. This obviously becomes more serious the more sources involved in the network and may be readily quantitified given the detailed parameters of a particular system. Multiple source systems do, however, offer the advantage that no digital processing is required unless, of course, some curve fitting is neede d for linearisation.

There is the scope for hybrid systems of which the most obvious is a time and wavelength multiplexed point sensor network. However, given the constraints on insertion loss per sensor it appears that, at present, it is unlikely that such hybrid networks will offer any significant advantages over, what may be termed "single theme" networks. In the longer term if the insertion loss per sensor can be minimised then there could well be enhanced capacity available.

Distributed and multiplexed fibre optic sensor networks have been demonstrated and can achieve practically useful performance. The distributed system is limited by the use of high speed electronics and the requirement for a measurand interaction which modifies the guiding properties of light within the fibre. Multiplexed systems appear to be constrained by the achievable levels of insertion loss per sensor and in the forseable future it seems unlikely that more than a few tens of sensors could be linked within a given network regardless of the details of its architecture. The broad bandwidth analogue electronics required in the distributed network is no longer necessary, though considerable sophistication is often required in the analogue and digital processing within a multiplexed system.

ACKNOWLEDGEMENTS

The author would like to thank members of the Optoelectronics Group at the University of Strathclyde for their contributions to this work - especially Deepak Uttam and Janus Mlodzianowski and would also like to acknowledge numerous discussions with members of the fibre optic sensor community in formulating the treatment of the problems presented here. In particular Alan Rogers, John Dakin, Peter McGeehin and Den Davies must have special mention.

REFERENCES

1. A R Nelson, D G McMahon and R L Gravel: "Passive multiplexing system for fibre-optic sensors", Applied Optics, 19, pp2917-20, 1980
2. B Culshaw: "Optical fibre and signal processing" Peter Perignus, Stevenage 1984
3. A J Rogers: "Intrinsic and extrinsic distributed optical fibre sensors", Proc. SPIE 566 (Fibre Optic and Laser Sensors III), San Diego 1985.
4. M. Schwartz: "Information transmission modulation and noise", McGraw, Hill, New York , 1980
5. A Yariv: "Introduction to optical electronics", Holt, Reinhart & Winston, New York 1976
6. S Al-Chalabi, B Culshaw, D E N Davies, I P Giles and D Uttam: "Multiplexed optical fibre interferometers; analysis based on radar systems", Proc IEEE- Part J 132, 2, p150 (April 1985)
7. M Kerker: "The scattering of light and other electromagnetic radiation", Academic Press, New York, 1969
8. D Marcuse: "Principles of optical fibre measurements", Acadmic Press, 1981
9. A G Hartog: "A Distributed temperature sensor based on liquid core fibres", Journal of Lightwave Technology, LT1, No.3, pp498-509, 1983.
10. A J Rogers: "Polarization optical time domain reflectometry - a technique to measure field distributions", Applied Optics 20, 6 p1060-74, 1981
11. C K Asawa, M K Bronowski and S K Yao: "Microbending of optical fibres for remote force measurements", US Patent No.4,463,254 July 31st, 1984
12. J P Dakin, D J Pratt, G W Bibby and J N Ross: "Distributed optical fibre Raman temperature sensor using a semiconductor light souce and detector"., Electronics Letters, 21, p569, 1985
13. A R Nelson, D H McMahon and H Van de Vaart: "Multiplexing system for fibre optic sensors using pulse compression techniques", Electr. Lett. 17, pp263-264, 1981
14. J Mlodzianowski, D Uttam and B Culshaw: "A multiplexed system for analogue point sensors", IEE Colloquium on Distributed Optical Fibre Sensors, 12 May, 1986, London
15. I P Giles, B Culshaw and J Foley, "A Balancing techique for optical fibre intensity modulated transducers", Proc. 2nd Intnl. Conf. on Optical Fibre Sensors, p117, Stuttgart, September 1984 (VDE Verlag, Berlin)
16. B E Jones and R C Spooncer: "An optical fibre pressure sensor using a holyographic shutter modulator with a two-wavelength intensity referencing", Proc. 2nd Intnl. Conf. on Optical Fibre Sensors, September 1984, Stuttgart, Proc. from VDE Verlag Berlin
17. I P Giles, D Uttam, B Culshaw and D E N Davies: "Coherent optical fibre sensors with modulated laser sources", Electronics Letters 19, No.1 p14, 1983
18. D Uttam, I P Giles, M S Nre and B Culshaw: "The principles of remote interferometric optical fibre strain measurement", Proc BHRA Conf. optical Tech. in Process Control, The Hague, June 1983 (BHRA, Cranfield UK)

19. D Uttam and B Culshaw: "Semiconductor lasers in advanced interferometric optical fibre sensors", Volume No.132, No.2, p184, June 1985
20. S A Al-Chalabi, B Culshaw, D E N Davies: "Partially coherent sources in interferometric sensors", Proc. 1st Intnl. Conf. in Optical Fibre Sensors, London 1983, p132 (IEE Conference Publication No.221)
21. J L Brooks, R H Wentworth, R C Youngqist, M Tur, B Y Kim and H J Shaw: "Coherent multiplexing of fibre interferometric sensors", IEEE Journal of Lightwave Technology, LT-3, No.5, pp1062-1072, 1985
22. A J Rogers: "Optical fibres in comunications and measrurement", CEGB Research No.16, pp3-17, August 1984
23. J J Francis: "The design of optical spectrometer", (Chapman & Hall, 1969)

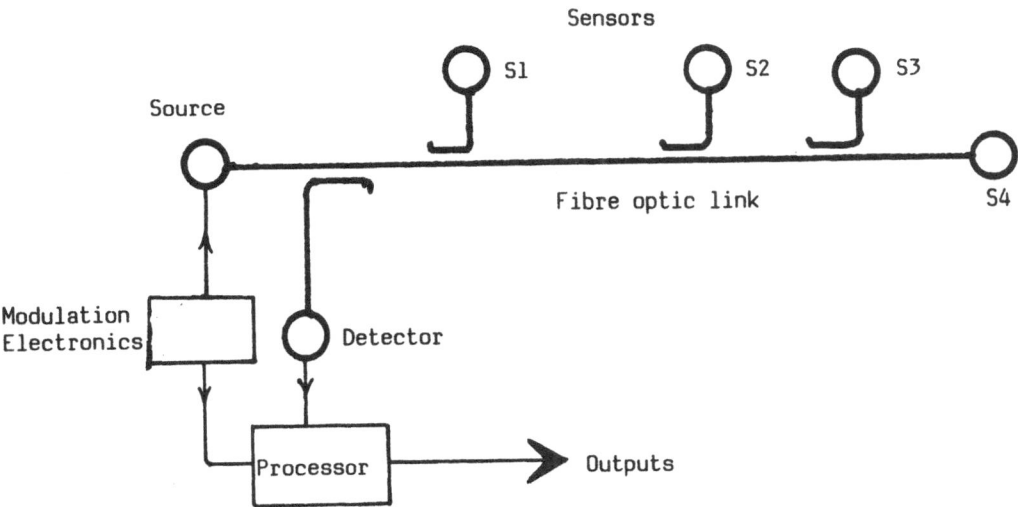

Figure 1:   Generalized multiplexed sensor system

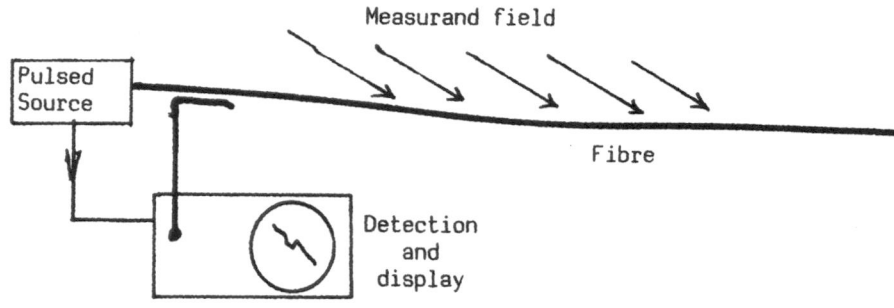

Figure 2a: Distributed sensors

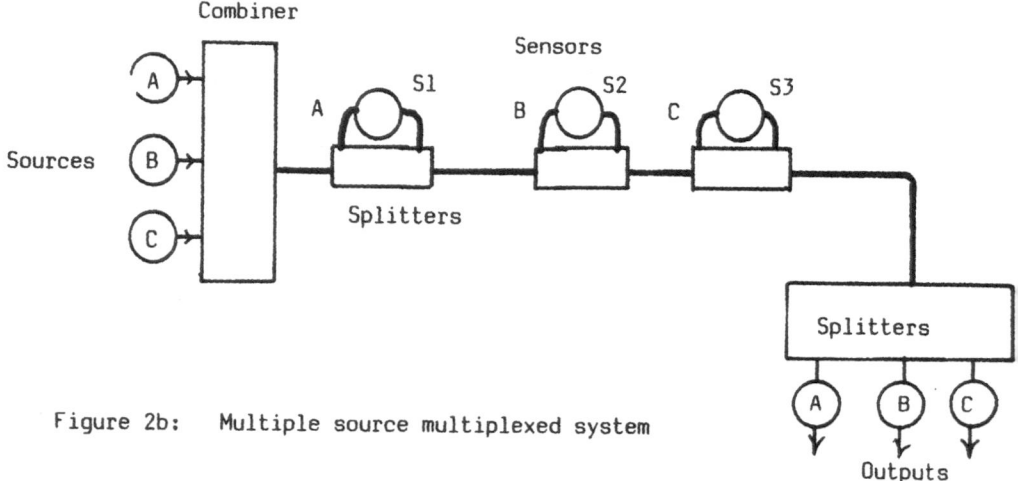

Figure 2b: Multiple source multiplexed system

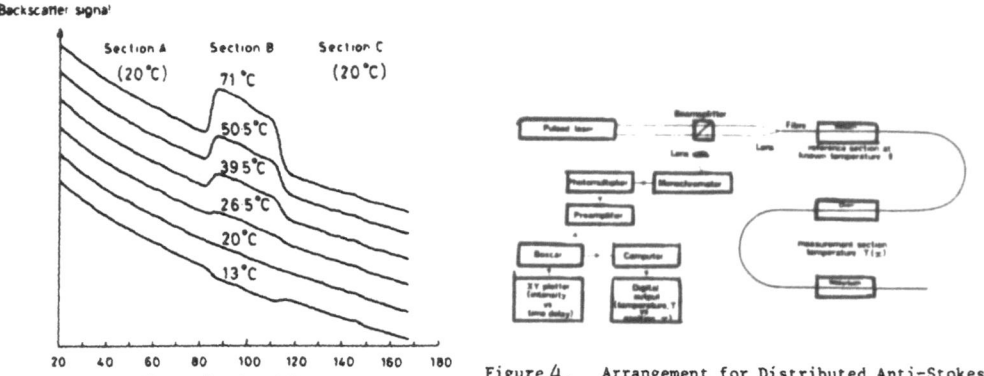

Figure 3. SDTM Results with Liquid-Cored Fibre.[9]

Figure 4. Arrangement for Distributed Anti-Stokes Ratio Thermometry (DART).[12]

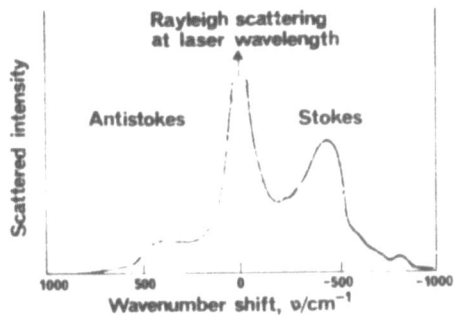

Figure 5. Raman Spectrum for Silica Fibre.

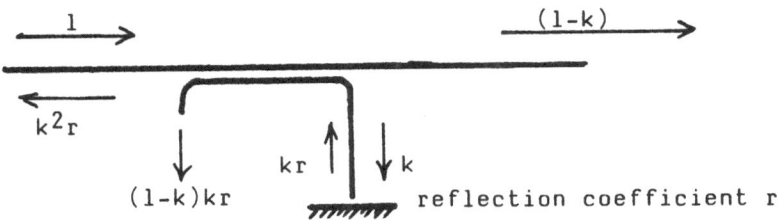

Figure 6   Power budget through single directional
           coupler, power coupling k.

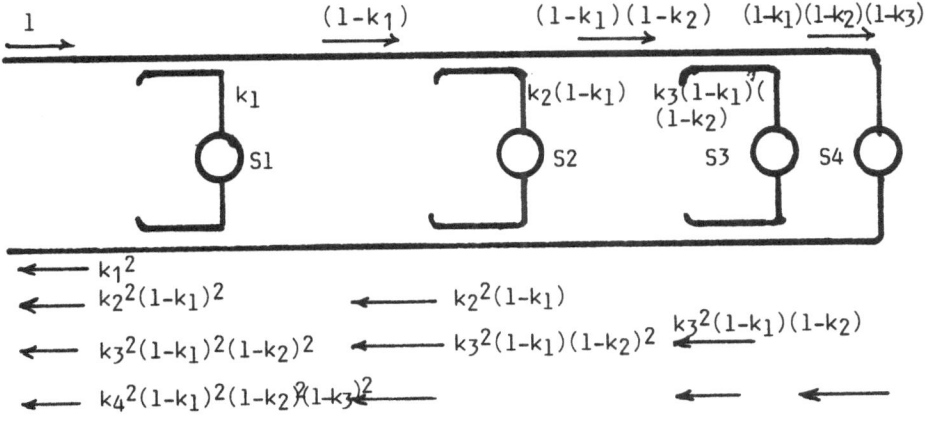

Figure 7   Network power budget - transmission through S1 - S4 taken
           as unity.

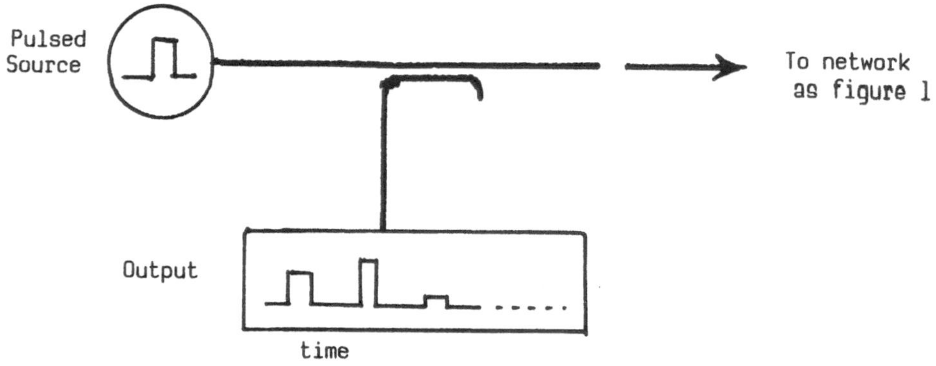

182

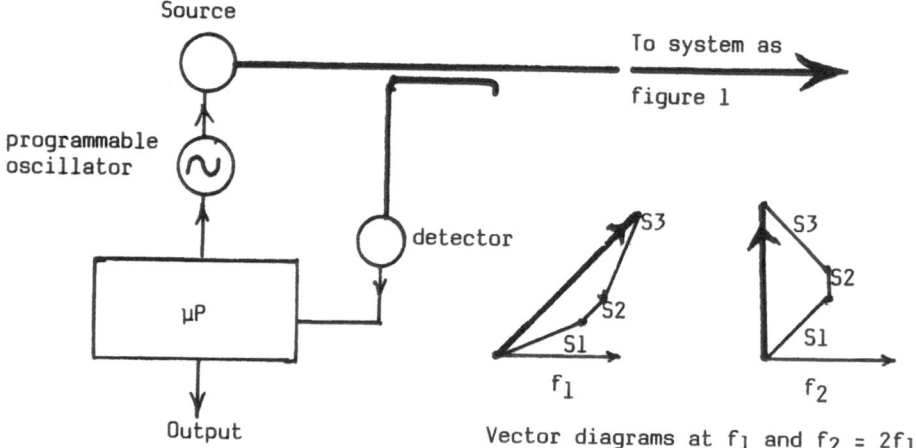

Figure 8a   Pulse modulated system

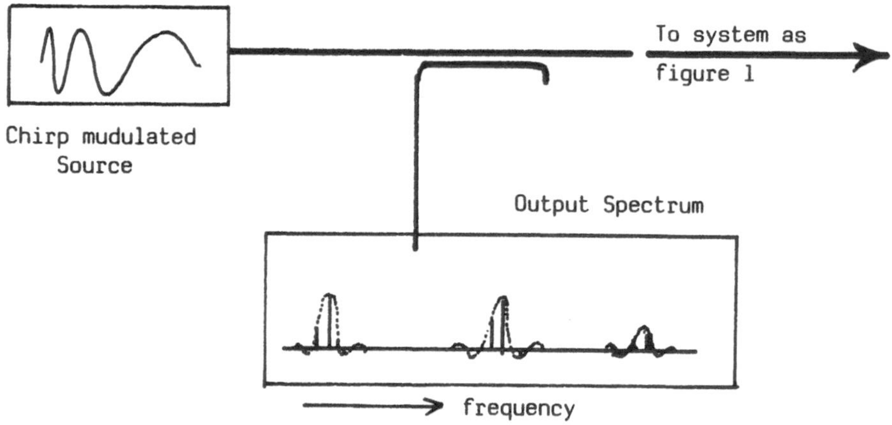

Figure 8b: Chirp modulated sensor system

Vector diagrams at f_1 and $f_2 = 2f_1$

Figure 8c: Switched frequency multiplexing

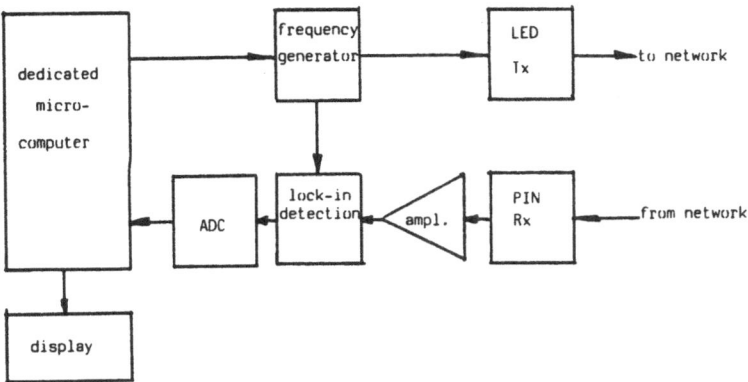

Figure 9. Block diagram of overall Multiplexing System

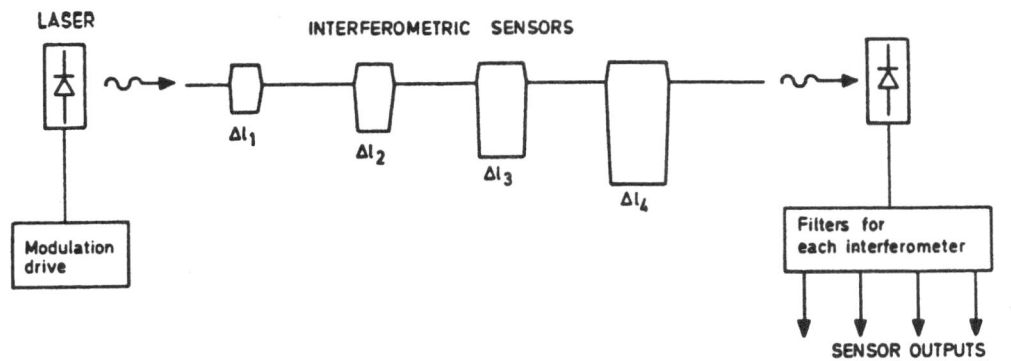

Fig.10a Sensor highway based on FMCW

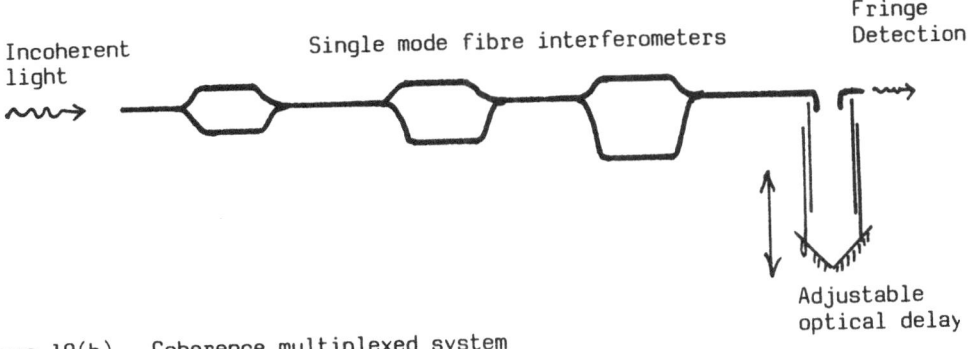

Figure 10(b)   Coherence multiplexed system

184

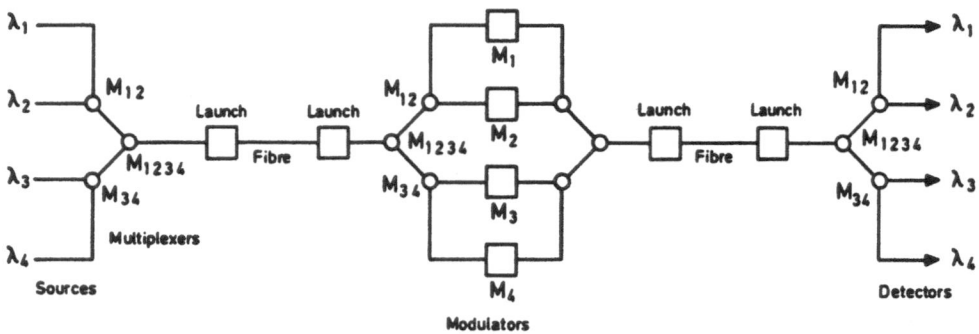

Figure 11. **ANALYSIS OF WAVELENGTH MULTIPLEXER**

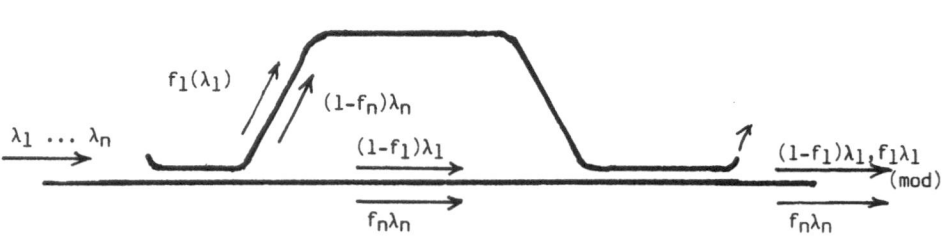

Figure 12

FIBER OPTIC TEMPERATURE SENSORS

William H. Glenn
United Technologies Research Center
Silver Lane
E. Hartford, CT 06108
U.S.A.

1. INTRODUCTION

Temperature measurement is essential for industrial process control, aircraft and engine monitoring, seismic and medical instrumentation. environmental sensing and a wide range of other areas. Fiber optic sensors offer some unique advantages over conventional sensors, particularly in environments that are thermally or electromagnetically harsh. When properly configured, they can provide extremely precise measurements and can have very rapid response.

There are many physical effects that can be used as the basis for a temperature sensor; any fiber optic sensor generally displays some temperature sensitivity. Sensors may be categorized as incoherent or coherent. In the first class only variations of the light intensity are involved in the sensing process while in the second, variations in the optical phase are involved even though the final output may be an intensity variation. In this sense polarimetric sensors are coherent since it is the relative phase between the two orthogonally polarized components of the light signal that determines the state of polarization. A number of fiber optic temperature sensors of both classes are reviewed here. In the case of coherent sensors, the measurement of temperature is usually complicated by the sensitivity of the device to strain. Some techniques to resolve this problem are discussed.

The effect of temperature on sensors designed to measure other parameters must always be considered. The effect of gross variations in the sensor's environment can usually be evaluated in a straightforward way. Even in a constant temperature environment, however, there are statistical fluctuations in the optical properties of a fiber due to the fact that it is at a finite temperature. This can set a lower limit to the minimum detectable optical phase shift. A simplified analysis of this effect is presented in another paper by the present author appearing in this volume (1).

2. INCOHERENT SENSORS

Temperature sensors have been demonstrated that are based on the temperature dependent variation of the transmission, fluorescence or emission of a sensing element. The sensing element can be a separate component or it can be a section of the fiber itself. Several examples are discussed below.

2.1 Transmission

Glassy or crystalline media that have been doped with trivalent rare earth oxides are very attractive for optical temperature sensors. The rare

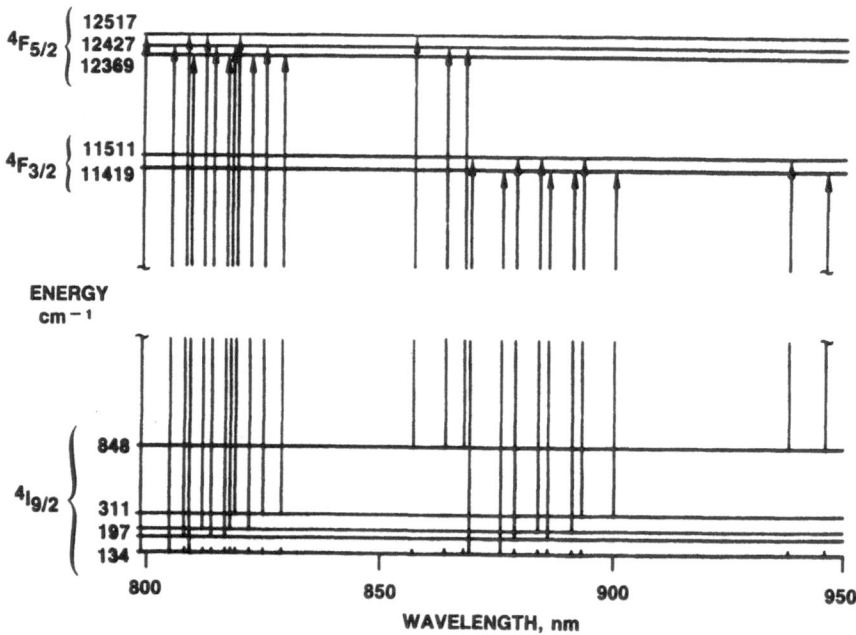

FIGURE 1. Energy levels and optical transitions in Nd:YAG.

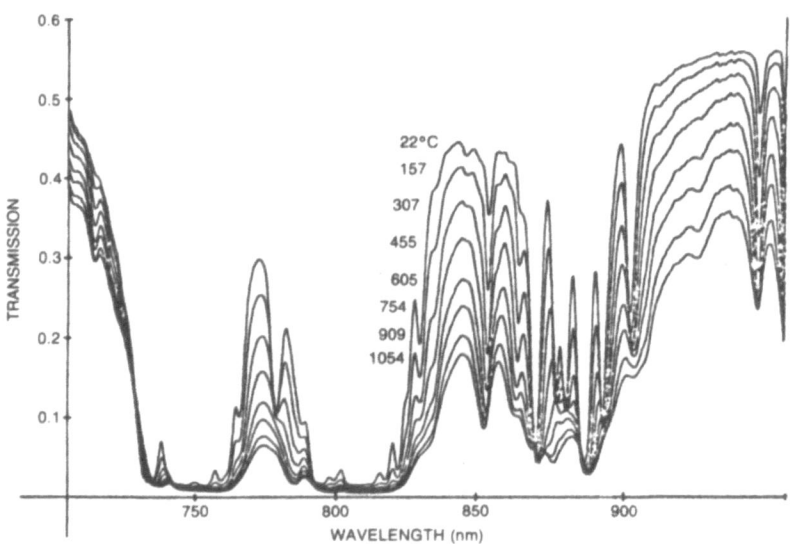

FIGURE 2. Absorption spectrum of Nd:YAG.

earths have numerous absorption lines in the visible and near infrared regions of the spectrum and the relative strength of these lines varies strongly with temperature. The reason for this variation may be understood from Fig. 1 which shows a portion of the energy level diagram of Nd in YAG. The visible and near infrared absorptions are due to transitions from the low lying manifold of energy levels to higher lying states. At absolute zero, all the ions would lie in the ground state and transitions could only take place from this level. As the temperature increases the number of ions in the ground state decreases and the higher lying levels become populated in accordance with Boltzmann's principle. As a result, the absorption strength for transitions originating in the ground state decreases and the absorption strength for transitions originating in the higher levels increases. This leads to a temperature dependent absorption spectrum as illustrated in Fig. 2.

The absorption spectra of rare earth ions in glass hosts are similar to the spectra in crystalline hosts although the lines are broadened considerably. Fiber optic temperature sensors have been demonstrated in which the sensing element is a section of fiber whose core was doped with Nd or Eu. (2,3).

Two prototype rare earth sensor systems were evaluated. One used a fiber with a core of borosilicate glass doped with 15 weight percent Eu_2O_3 and the other a fiber with a core of an aluminosilicate glass doped with 5 weight percent Nd_2O_3. The borosilicate glass had a softening temperature of $450^{\circ}C$ while the alumonosilicate glass could be operated to $800^{\circ}C$. The Nd sensor system will be described since it had the wider temperature range.

The Nd doped sensing fiber was drawn from a preform made with a polished rod of the 5 percent Nd doped glass placed inside a fused silica tube which served as the cladding. This preform was then drawn down to the desired fiber diameter in several stages. A final diameter of 150 μm was chosen to match the diameter of Corning type 1505 fiber that was used to transmit the optical probing signal to the sensing fiber and to return the signal from the fiber to the detector. The Corning fibers were fused to the sensing fiber whose length was typically 2-4 cm. The optimum length of sensing fiber is one that provides a transmission of 1/e (37%) at the midpoint of the temperature range to be sensed. Higher absorption gives a greater fractional change of transmitted power with temperature but leads to a lower signal strength and a poorer signal-to-noise ratio. With a heavier doping the sensor could be made shorter or a lighter doping could be used for a longer sensor. The sensing fiber was formed into a loop to fit in a small probe housing with the input and output fibers side by side. The transmission spectrum of a 3.7 cm sensor is shown in Fig. 3.

The simplest approach to a sensor system using the rare earth fiber would be one that measures the transmission of the fiber at a temperature sensitive wavelength. This approach, however, is sensitive to incidental losses such as those that could be encountered in connectors or splices between the sensor and the detector. To minimize this problem a two wavelength approach was taken. The transmission of the fiber was measured at each wavelength and the ratio of the two transmissions was used as the temperature dependent parameter. In this approach any incidental losses that are independent of wavelength are cancelled out. Ideally one would like to choose one wavelength where the transmission increases with temperature and the other where the transmission decreases. The two wavelengths chosen for the sensor were 840 nm and 860 nm. The transmission at 840 nm always decreases with increasing temperature whereas the

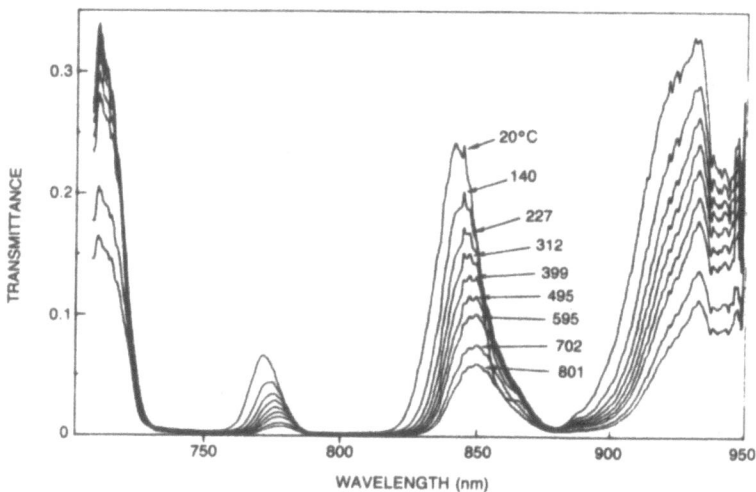

FIGURE 3. Absorption spectrum of Nd doped fiber.

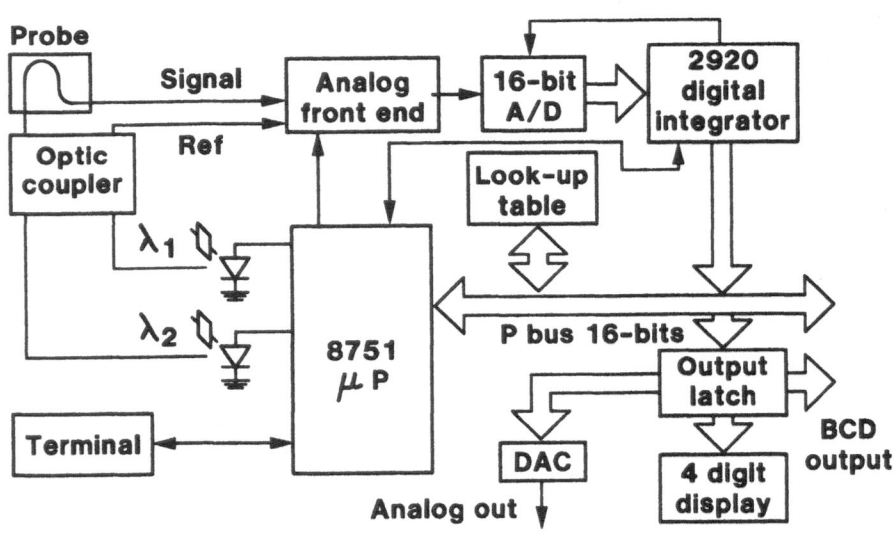

FIGURE 4. Rare earth temperature sensors.

transmission at 860 nm increases up to 500°C and then decreases slightly.
With these two wavelengths there is good sensitivity below 500°C and
reduced sensitivity at higher temperatures.

Two LEDs were used as the light sources. The emission from each diode
was collimated with a graded index microlens, wavelength filtered with a
ten nm bandwidth filter, and then each was focussed onto one of the input
ports of a 2x2 fiber optic coupler with another microlens. One of the
output ports carried the composite signal to the sensor and the other port
went to a reference detector to measure the input signal levels. The LEDs
were pulsed on and off sequentially so that at the detectors the wavelength
being probed could be determined. A schematic of the entire sensor system
is shown in Fig. 4. The multiplexer sorted out which detector was being
read for conversion to digital format. Six parameters were sensed for a
temperature measurement, the input and output power at each of the two
wavelengths and the dark current from each detector when both LEDs were
off. A digital integrator integrates the signal over the pulse duration.
The microprocessor subtracted off the dark currents and calculated the
ratio of the transmission at the two wavelengths. This ratio was then
converted to temperature by a calibration curve stored in a lookup table.
This rather complex electronic system was designed to achieve high speed
operation and is certainly not necessary for most applications.

The response of the sensor system is shown in Fig. 5. The quantity
plotted is the ratio of the transmissions at the two wavelengths. The
response increases monotonically with temperature and shows the expected
reduction in sensitivity at the higher temperatures.

Some problems that were encountered with this sensor were errors due
to drift in the spectra of the LEDs and errors due to inadequate signal
levels. The use of thermally stabilized laser diodes could improve the
performance. Another problem was that the neodymium spectrum did not allow
the choice of a very good reference wavelength. In the case of europium,
this latter problem does not exist. Europium has a few well isolated
absorption lines and large regions that are completely transparent. The
absorption of europium is much less than that of neodymium so that a longer
sensor (10-20 cm) is required.

Fiber optic temperature sensors based on the temperature dependent
transmission of semiconductors have been reported by Kyuma et. al (4). The
bandgap of semiconductors and consequently the position of the absorption
edge is temperature dependent. If a sample is probed with a wavelength
lying on the slope of the absorption edge, the transmitted power will vary
with temperature. In the work cited, platelets of GaAs and CdTe were used
as the sensing elements. Thin platelets of the semiconductors were placed
between two fibers and the transmission was probed. The source used was
an AlGaAs LED operating at 880 nm with a spectral width of 150 nm. The
transmission of the samples showed an approximately exponential decrease
with increasing temperature as the absorption edge moved to longer
wavelengths. A sensor system was reported using a two wavelength time
division multiplexing scheme virtually identical to the one described
above. An InGaAsP LED at 1.27 μm was used as a a reference signal. The
change in the transmission ratio for the two wavelengths was less than 1 dB
for a connector loss of 20 dB. Accuracies of ± 1°C over the range
from -10°C to 300°C were reported.

2.2 Fluorescence

The fluorescent emission from phosphors and fluorescent media is
temperature dependent. Fluorescence arises from the radiative decay of

190

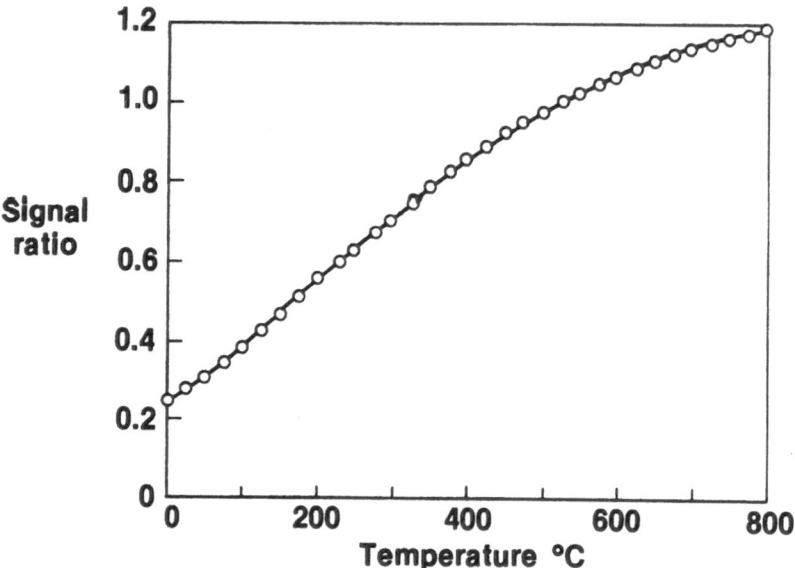

FIGURE 5. Neodymium temperature sensor response.

excited electronic energy levels. This process works in competition with
the nonradiative decay processes due to phonon excitation. These processes
become stronger with increasing temperature, i.e., the decay becomes faster
which also decreases the total fluorescence yield. Temperature
measurements may be made either by monitoring the relative strengths of
different emission lines or by monitoring the decay time of the
fluorescence. A fiber sensor based on the fluorescence of Nd:glass has
recently been reported (5). A small glass sample was coupled to the end of
a fiber and was excited by the 805 nm, square wave modulated emission from
a miniature LED. The fluorescence decay rate decreased linearly over the
temperature range from 0 - 100°C. This type of sensor is attractive
because of the small, low power excitation source. Other sensors based on
the fluorescence of rare earth phosphors are available commerically (6).

2.3 Emission

Optical pyrometry allows the measurement of temperature by relating
the measurement of the optical power radiated by an object in selected
regions of the optical spectrum to the temperature of the object. It is an
old and well thought out area of technology. Pyrometric measurements may
be implemented using fiber optics, and commercial equipment is available to
perform these measurements.

Another type of fiber optic pyrometer suitable for gas temperature
measurements has recently become available (7). This device uses a
sapphire rod-fiber that has a metallic coating on a small portion of its
tip. This forms a small cylindrical black body cavity. The emitted light
is carried by the sapphire guide to a detection system. The quoted
operating range of this sensor is 500 - 2000°C.

On the research level, however, one issue should be clarified because it is one that causes much confusion. Should one image the object onto the fiber? What happens if the image is behind the fiber? These questions are irrelevant. A fiber and any assoicated lens system has a numerical aperture, which may depend on the position in the output aperture. A thermal source emits a certain number of Watts/(square cm-steradian.). The maximum collected power by a fiber optic and lens system is equal to the integral of the emission over the numerical and physical aperture. Put in another way, the maximum power that may be collected is achieved by putting the fiber-lens system in contact with the object to be sensed. Alternately, if light is sent down the fiber, and if the spot size falls within the object to be sensed, then the fiber optic-lens system will receive the maximum power that can be collected.

Pyrometry of an isolated hot object is relatively straightforward. In many cases of practical interest the situation is complicated by the presence of extraneous reflected radiation. As an example, one might want to sense the temperature of the combustion chamber liner in a furnace. This temperature is high but the temperature of the burning gasses in the combustor is higher. Most of the radiation received by an optical pyrometer would consist of relfected radiation from the burning gasses rather than from the metal surface of interest. A possible solution to this problem lies in the technique of active pyrometry.

Active pyrometry relies on the well established principles of synchronous detection. If the temperature of an object can be modulated, as could be done with a pulsed laser beam, then the thermal emission would also be modulated at the same frequency as the excitation. The reflected radiation would be unmodulated and could thus be discriminated against. If a measurement is made at two different frequency bands and a ratiometric approach is used to determine the temperature, then the ratio is independent of the absolute value of the temperature rise caused by the modulation.

The basic concept is shown in Fig. 6 A modulated radiation source, here a CO_2 laser, causes the temperature of the object to be sensed to fluctuate slightly. The thermal emission is

$$P = A \ \varepsilon(\nu) \ \frac{\nu^3}{e^{h\nu/kT}-1}$$

In the visible and near IR, and for moderate temperatures, this is well approximated by

$$P = A \ \varepsilon(\nu) \ \nu^3 \ e^{-h\nu/kT}$$

If the temperature is modulated by an amount ΔT, then

$$\Delta P = A \ \varepsilon(\nu) \ \frac{h\nu^4}{kT^2} \ e^{-h\nu/kT} \ \Delta T$$

This may be done at two separate wavelengths (frequencies ν_1, ν_2). The ratio of the two modulated power components is

$$R = \frac{\Delta P(\nu_1)}{\Delta P(\nu_2)} = \frac{\varepsilon(\nu_1)}{\varepsilon(\nu_2)} \left(\frac{\nu_1}{\nu_2}\right)^4 e^{-h(\nu_1 - \nu_2)/kT}$$

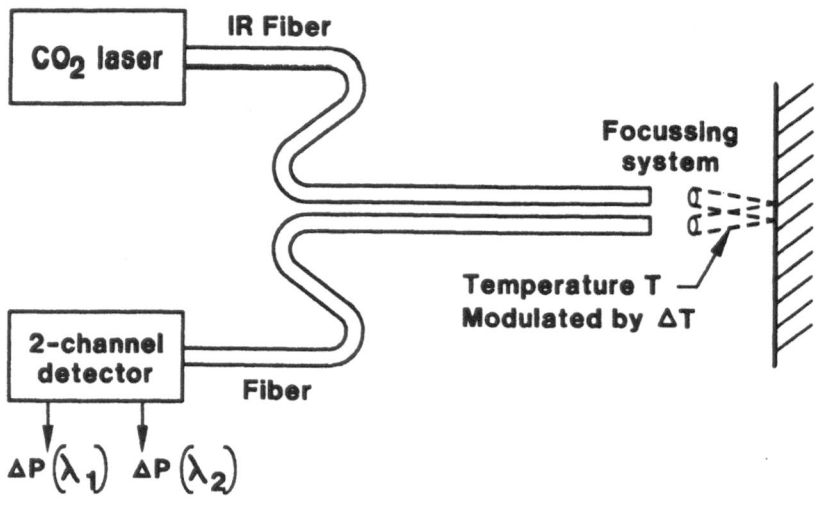

FIGURE 6. Active Optical Pyrometer Concept.

This ratio is seen to be independent of ΔT and also of any geometrical factors involved in the light collection efficiency. A very simple verification of this concept has been carried out at the author's laboratory. A CO_2 laser modulated the temperature of an electrically heated metal sample. Two spectral regions were monitored, one corresponding to the full response of a silicon detector and another, with the same detector blocked by a 100% reflectivity Nd laser mirror. The ratio of the two powers showed a monotonic increase of about 30% over the temperature range 800-1000°C. This approach should be suitable for measurements in the presence of a high levels of background radiation. It is more suitable for static rather than dynamic measurements since it uses time integration to discriminate against the background.

3. COHERENT SENSORS

Coherent optical temperature sensors all depend in some way on the change in phase of a signal as a result of a temperature change. The fractional change in path length in a fiber $\partial n_r/\partial T = 10^{-5}/°C$. The phase shift $(2\pi/\lambda)\Delta(nl)$ is approximately 1 rad/°C-cm for a wavelength of 632.8 nm, the He-Ne laser wavelength. In a coherent optical system it is possible to detect optical phase shifts of 10^{-6} radians or less so that extremely high temperature sensitivity can be realized. In most cases, this sensitivity is too high. It leads to problems with lead effects and ambiguities and means must be found to minimize these problems.

3.1 Interferometers

A standard form for a coherent fiber optic sensor consists of a Mach-Zehnder interferometer. One arm of the interferometer is the sensing fiber and the other is a reference arm. The reference arm usually contains a modulator or frequency shifter so that the relative optical phase shift between the two arms can be converted to an electrical phase shift in the detected signal. This type of sensor has been discussed in great detail by many authors (see for example (8)).

To be useful as a practical, localized temperature sensor the problems associated with lead effects and ambiguities. A potentially attractive approach to the lead and localization problem lies in the technique of coherence multiplexing that has been discussed by Brooks, Kim and Shaw (9). This can be implemented in many ways; a simple embodiment will be described here. The sensing element consists of a remotely located, unequal path length, fiber optic Mach Zehnder interferometer (MZI). It is illuminated by a source of short coherence length. The excitation may equally well be thought of as an extremely short time duration pulse (τ = lcoh/c). In this view, the output from the MZI consists of two pulses, separated in time. These do not interfere with one another. At the detector, the signal is passed through another identical unequal path length interferometer. At the output of this interferometer, interference can now occur. The signal that passed through the short arm of the sensing MZI and the long arm of the reference MZI combines with the signal that passed through the long arm of the sensing MZI and the short arm of the reference MZI. There are extraneous signals, i.e., from the long-long and the short-short paths, but these do not present a problem if a modulation or frequency shifter is placed in one arm of the reference MZI. This has the effect of removing the desired signal from baseband and converting it to an electrical phase measurement. It also allows the elimination of the extraneous signals.

Interferometreic sensors of this type will generally have a problem with ambiguities since they measure a phase rather than an absolute path length change and there is a $2n\pi$ uncertainty. This may be resolved by using two (or more) wavelengths

$$\phi_1 = 2\pi \frac{(\Delta n l)}{\lambda_1}$$

$$\phi_2 = 2\pi \frac{(\Delta n l)}{\lambda_2}$$

$$\phi_1 - \phi_2 = 2\pi \left(\frac{1}{\lambda_1} - \frac{1}{\lambda_2} \right)(\Delta n l) = 2\pi \frac{(\Delta n l)}{\lambda_s}$$

$$\lambda_s = \frac{\lambda_1 \lambda_2}{\lambda_2 - \lambda_1}$$

The use of two wavelengths allows the creation of a "synthetic" wavelength λ_s which can be much longer than the optical wavelengths λ_1 and λ_2. This serves as a coarse scale and the individual λ s as a vernier.

3.2 Twin Core Sensor

The twin core sensor is an interferometric sensor that has received considerable attention at the author's laboratory (10), (11). The basic concept is illustrated in Fig. 7. The sensing element is a fiber having two closely spaced single mode cores in a common cladding. Because of the proximity of the cores, the evanescent field from each core overlaps the other, leading to a mutual coupling. If light is injected into one of the cores, it will gradually leak ito the other core until finally all is in the unexcited core. The process then reverses leading to a periodic intechange of light between the cores as the light propagates down the fiber.

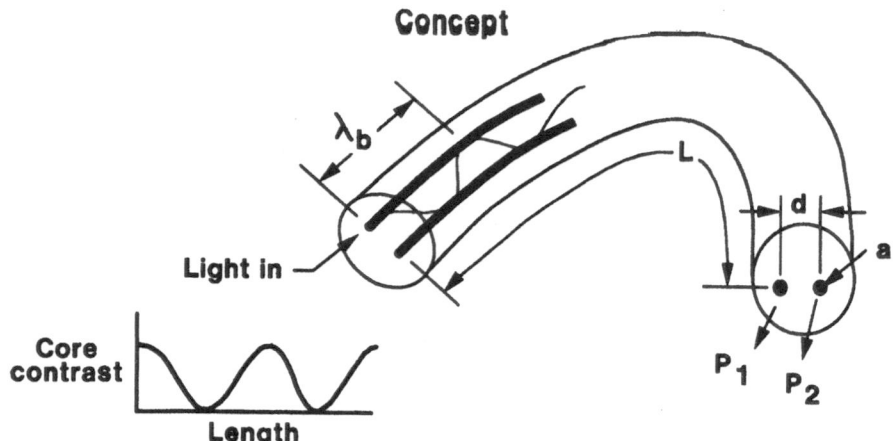

External disturbance changes L and beat length λ_b

Core contrast $(P_1 - P_2) / (P_1 + P_2)$ a function of $\phi = \pi (L/\lambda_b)$

FIGURE 7. Twin core sensor.

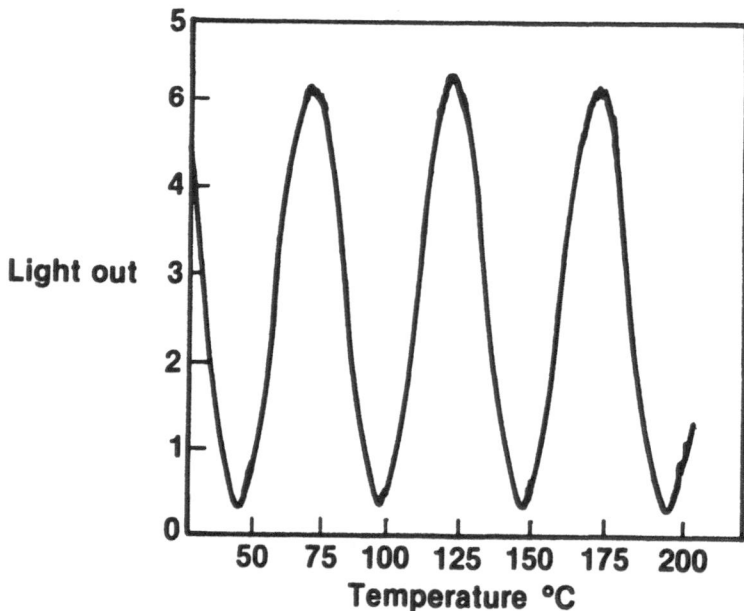

FIGURE 8. Twin core sensor temperature response.

The beat length $_b$ depends on the geometry and material parameters of the fiber, and can range from infinity down to a few tenths of a millimeter. At the output of the fiber, a certain distribution of light is observed. This is characterized by the contrast ratio

$$R = \frac{P_1 = P_2}{P_1 + P_2}$$

This is independent of the absolute power level and varies sinusoidally down the length of the fiber. As the temperature of the fiber is raised, the fiber expands and the material parameters change slightly. The net effect is to move to a different point on the sinusoidal R vs 1 curve. Typical repsonse of a twin core fiber to temperature changes is shown in Fig. 8. If the fiber is subjected to an axial strain, similar effects occur. This potential complication can be turned to an advantage. When properly configured, the twin core fiber can be used to sense both quantities independently. This will be discussed later.

A more rigorous view of the behavior of the twin core fiber is presented in Fig. 9. Here the propagating modes of the whole structure are considered. The lowest order modes are the symmetric and the antisymmetric modes shown. In the former, the electric fields in the two cores are in phase while the latter they are 180° out of phase. The curves show the effective index for the two modes as a function of the V parameters. Also shown is the index for a single core of the same size. For low V values there is a significant index and hence velocity of propagation difference.

A condition in which only one core is illuminated consists of a linear superposition of the symmetric and the antisymmetric modes. The fields add in one core and cancel in the other. As the modes propagate down the fiber, they slip out of phase. The points at which the phase difference is 180° are those where all the light is in the originally unexcited core. The twin core fiber is a differential interferometer with an effective wavelength λ_b.

The sensitivity to both strain and temperature is a potentially troublesome issue that must be addressed. Actually, many of the envisioned applications of this sensor involved its use as a strain sensor rather than as a temperature sensor and ways were sought to minimize the temperature sensitivity. A detailed analysis of the dependence of $\partial\varphi/\partial T$ on the geometrical and material parameters was carried out. A result of such an analysis is shown in Fig. 10. For a proper choice of parameters it is possible to make $\partial\varphi/\partial T = 0$. This is a potential solution if the sensor is to be used only for strain measurements. A more general approach is shown in Fig. 11. For small changes, the phase of the contrast function depends linearly on the temperature change and the strain. The coefficients of preportionality depend on the material and geometrical properties, the temperature and the wavelength. If the phase of the contrast function is measured at two wavelengths, a set of two equations in T and ϵ are obtained. If the fiber is designed so that the matrix of coeficients is more singular, then the relation may be inverted to give T and ϵ separately.

One of the major problems in applying such a sensor is the localization problem, i.e., the elimination of lead-length problems. Suitable input and output couplers have been fabricated for the twin core fiber. The input coupler consists of a single core, single mode fiber that is fused to the twin core fiber in such a way as to couple light into only one of the two cores. The output coupler was formed from two multimode

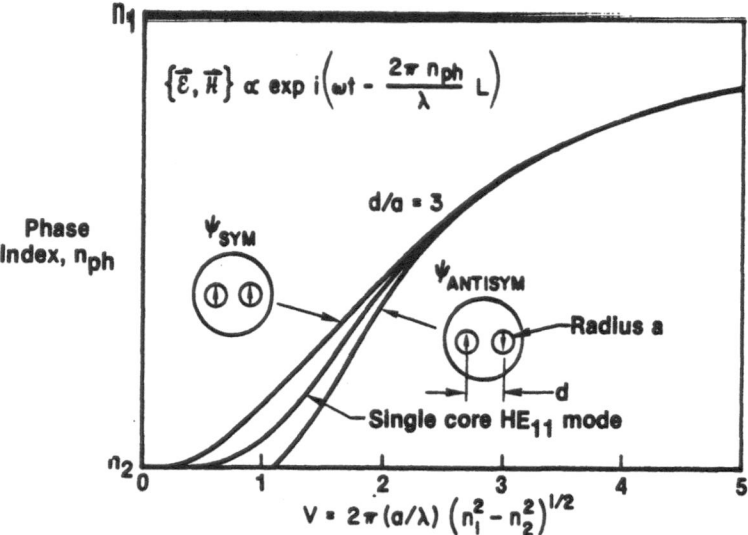

FIGURE 9. Modes of a twin core fiber.

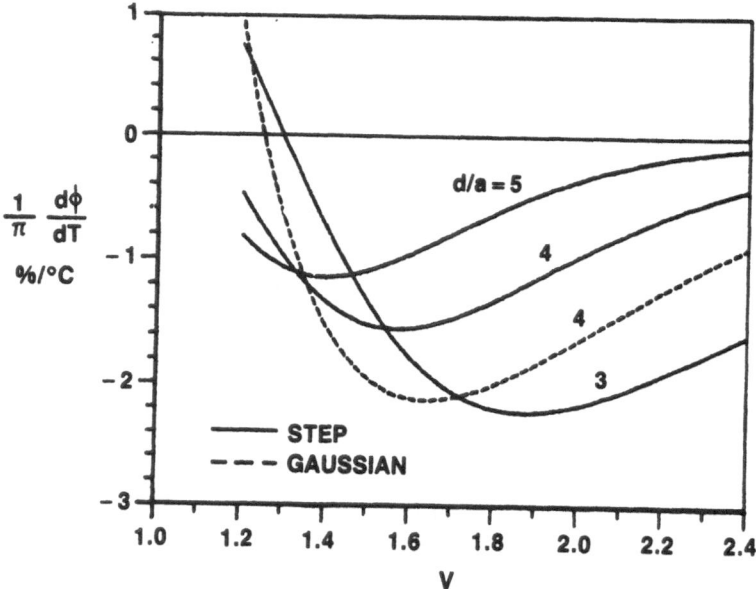

FIGURE 10. Temperature sensitivity of twin core sensor.

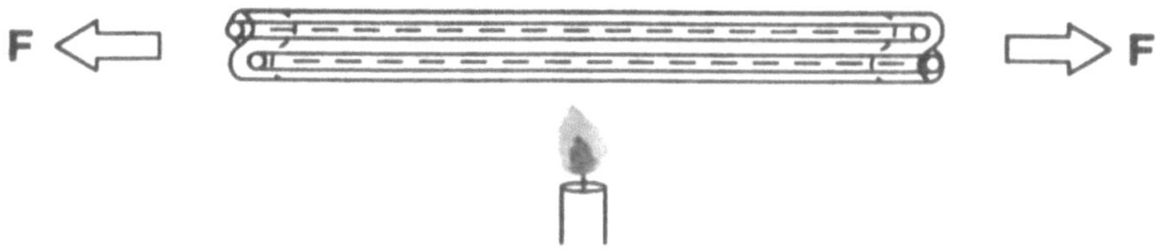

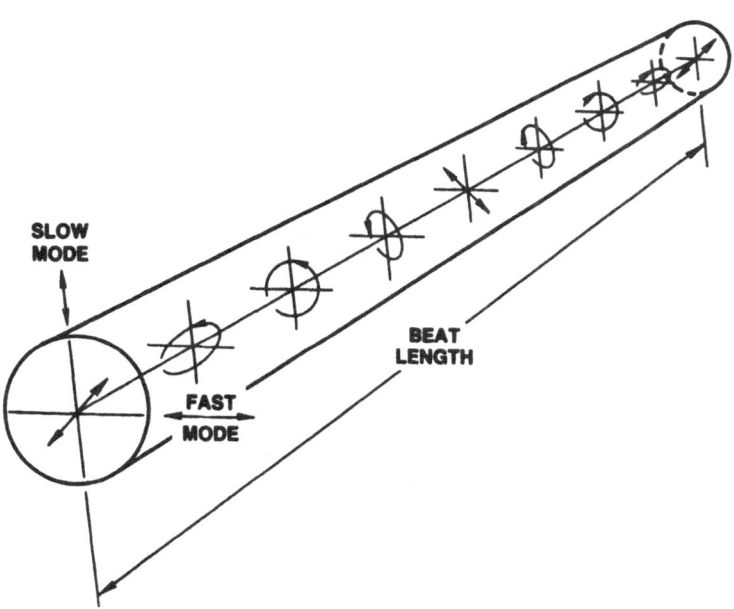

$$\delta \phi_1 = A\,(T_1\,\lambda_1\,)\,T + B\,(T, \lambda_1\,)\epsilon$$
$$\delta \phi_2 = A\,(T, \lambda_2)\,T + B\,(T, \lambda_2)\epsilon$$

FIGURE 11. Dual function fiber optic sensor.

SLOW
MODE

FAST
MODE

BEAT
LENGTH

FIGURE 12. Polarimetric sensor concept.

fibers which were fused together over a short length. The fused section was then drawn down to the same OD as the twin core fiber (basically forming a short length of multimode twin core fiber). This was then fused to the output end of the twin core sensor.

The twin core sensor, while still experimental, offers the promise of simultaneous temperature and strain measurements. If may be mounted on the surface of a component or, in the case of an object made of composite materials, it may be embedded in the structure during fabrication.

3.3 Polarimetric Sensors

Temperature sensors may be implemented using polarization preserving fiber. If light is injected into a length of polarization preserving fiber with polarization at 45° with respect to the principal axis of the fiber, the state of polarization will vary periodically along the fiber length as illustrated in Fig. 12. The polarization preserving fiber is a differential interferometer and in this respect is similar to the twin core fiber. The twin core fiber relies on the temperature dependence of the phase shift between the symmetric and the antisymmetric modes. The polarametric sensor relies on the temperature dependence of the phase shift between the two states of polarization

$$\phi = \frac{2\pi}{\lambda} (n_f - n_s) \, 1$$

Fiber optic temperature sensing using polarization preserving fiber has been reported by Eickhoff (12) and by Corke et al.(13). In the latter work a 24 cm length of polarization maintaining fiber with a silvered end was coupled to a 90 meter length of polarization maintaining fiber as the lead. The axis of the sensing fiber were at 45° with respect to the lead fiber. One polarization mode of the lead fiber was excited, resulting in equal excitation of the modes of the sensing fiber. After a double pass through the sensing fiber, the light was resolved by the lead fiber into its two modes of polarization. With this arrangement only phase changes between the two modes in the sensing fiber affect the output. This experiment demonstrated a high sensitivity and the insensitivity of the measurement to effects occuring in the lead fiber.

4. CONCLUSIONS

A number of fiber optic temperture sensors, both coherent and incoherent, have been received including a dual function sensor capable of measuring both strain and temperature. Many viable techniques for temperature sensing have been demonstrated in the laboratory. Several of these techniques have carried to the point where there are now commercially available fiber optic temperature sensing instruments. These remain a variety of needs including multifunction sensors, multiplexed sensors to provide a set of point temperature measurements, distributed sensors to provide a contuum of measurements, and survivable sensors with wide bandwidth that can operate in high temperatures, with wide bandwidths in contaminating environments. These are some of the areas of current research.

REFERENCES

1. W. H. Glenn: Thermodynamic Limitations to the Measurement of Phase Shifts in Optical Fibers, this volume.

2. E. Snitzer, W. W. Morey and W. H. Glenn: Fiber Optic Rare Earth Temperature Sensors, Presented at the International Conference on Fiber Optic Sensors, London, England, 26-28 April, 1983.
3. W. W. Morey, W. H. Glenn and E. Snitzer: Fiber Optic Temperture Sensor, Proceedings of the 29th International Instrumentation Symposium, Albuquerque, New Mexico, 2-6 May, 1983.
4. K. Kyuma, S. Tai, T. Sawada and M. Nunoshita: Fiber-Optic Instrument for Temperature Measurement, IEEE J. Quantum Electronics QE 18 pp. 676-679, April, 1982.
5. R. V. Grattan and A. W. Palneri: Simple Inexpensive Neodymium Rod Fiber Optic Temperature Sensor, Paper THFF5, Third International Conference on Optical Fiber Sensors, San Diego, California 13-14 February, 1985.
6. "Fluoroptic Temperature Sensor": Luxtron Co. Mountain View, California 94043.
7. Accufiber Co.; Vacouver, Washington 98661
8. T. G. Giallorenzi, J. a. Bucaro, A. Dandridge, G. H. Sigel, Jr., J. H. Cole, S. C. Rashleigh and R. G. Priest: Optical Fiber Sensor Technology, IEEE J. Quantum Electronics QE 18, pp. 626-665, April 1982.
9. J. L. Brooks, B. Y. Kim and H. J. Shaw: Coherence Multiplexing of Fiber-Optic Interferometric Sensors, paper Th BB4, Third International Conference on Optical Fiber Sensors, San Diego, California, 13-14 February 1985.
10. G. Meltz, J. R. Dunphy, W. W. Morey and E. Snitzer: Crosstalk Fiber-Optic Temperature Sensor, Appl. Opt. 22 pp. 464-477, 1 February 1983.
11. G. Meltz and J. R. Dunphy :Twin Core Fiber-Optic Strain and Temperture Sensor, Paper THFF6, The Int. OFC.

12. W. Eickhoff, Temperature Sensing by Mode-Mode Interference in Birefringent Optical Fibers, Opt. Lett. 6 pp. 204-206, April 1981.

13. M. Corke, A. D. Kersey, K. Liu and D. A. Jackson: Remote Temperature Sensing Using Polarization - Processing Fibers, Electron Lett. 20, pp. 67-69, 19 January 1984.

GUIDED-WAVE CHEMICAL SENSORS

A.L. HARMER
BATTELLE
Geneva Research Centres
7, Route de Drize
1227 Carouge/Geneva
Switzerland

1. INTRODUCTION
Chemical sensing with optical fibres is one of the more interesting areas of optical fibre sensors, and as classical chemical instrumentation frequently employs spectrometry there is a natural extension to using optical fibres as light guides in spectrometers. A number of people have reviewed work in optical fibre chemical sensors (1, 2, 3, 4) and optical fibre biomedical sensors (5, 6). Optical fibre sensors have been applied to measurement of a large number of parameters:

Refractive Index (including liquid-level)
Absorption and Colorimetry
Fluorimetry
Reflectivity
Scattering, Nephelometry and Turbidity
Biomedical

Chemical sensors can be divided into four categories:
i   The fibre acts as a light conductor only
ii The fibre allows interaction of the light in the fibre with the surrounding medium
iii The fibre is enclosed in a chemical jacket, which monitors a selective chemical reaction
iv The fibre has an external jacket which interacts physically (e.g. by absorption or diffusion into the jacket) with a chemical species. This is an indirect means of sensing chemical changes.

In this article, the term optical fibre implies any light guide and includes integrated optic waveguides and light conductors of special geometries as seen in the examples given.

2. SPECTROMETERS
Special spectrometers need to be developed for fibre optics; with appropriate optics to match the fibre numerical aperture and the small core size, to miniaturise the spectrometer for mounting and integration with the processing electronics, and to allow multiple connection of separate sensing heads (fibres) with a centralised instrument. Conventional spectrometers have been modified

to allow coupling with several fibres, or a star coupler can be used to divide the light into multiple channels for monitoring several locations (8), or to monitor a single location from different positions (9).

Mechanical switching techniques are being developed to connect a single fibre from the spectrometer with an array of fibres connected to different probes (10).

A number of fibre optic spectrometers or wavelength selection devices have been developed for wavelength multiplexing of communication data channels on a single fibre. One such spectrometer is a small block of BK 7 glass (20 x 20 x 98mm) with a spherical mirror at one end and a grating at the other (11). The fibres are attached to a window in the middle of the grating. This is designed for an NA of 0.3. Losses of 1.8 dB to 2.2 dB are reported for a 9 channel multiplexer with a graded-index 50/125 fibre as input and 9 step-index 80/125 fibres as output. The crosstalk between neighbouring channels (40 nm separation) is around - 30 dB but decreases to - 50 dB with fewer channels.

A miniaturised spectrometer with CCD array, for mounting on a printed circuit card inside the electronics, has been developed for a film thickness monitor (13, 14). Light from the input fibre is focussed by a 40 mm achromate lens onto a blazed reflection grating with 300 lines/mm. A 50 element CCD array gives a spectral resolution of 10nm. The spectrometer is 10 mm thick and 80 mm long.

Spectrometers made by integrated optics have also been suggested (15).

3. ABSORPTION MEASUREMENTS
3.1. Gas Monitors
Important developments have been made to use fibre optic transmission for gas detection, particularly for explosive gases. An important limitation is the fibre transmission spectrum, which for silica fibres extends only to about 1.8 μm, whereas most of the standard "finger-print" absorptions used to identify simple molecules (CO, CH_4, C_2H_6, N_2O, etc) are their vibrational spectra occuring in the mid IR region (2 μm to 6 μm). Thus overtone spectroscopy is used, measuring higher harmonics of these fundemental absorptions (16, 17, 18), the absorptions occurring between 1.2 and 1.7 μm. Absorption coefficients are low: for the CH_4 harmonic the absorption coefficient is $9.3.10^{-6}$/ppm.m. The lower explosive limit for CH_4 is 5% and the higher explosive limit 15%. Detections of 0.8% (or 400 ppm) of the lower explosive limit has been achieved with fibre optic cavities (19).

To increase the sensitivity a long cavity length is used, sometimes with multiple passes, achieved by reflection from a corner cube or concave mirrors (20). Mechanical design of a stable cavity is critical for multiple reflections and to refocus the light back into the small core of a fibre.

The optical measurement system generally includes an IR LED, interference filters and dual photodetectors (20), and a differential absorption technique for signal and reference channels (19). Further developments need to be made to provide a stable high-resolution optical detection system at low cost.

Remote Measurements have been made at 5 Km distance (21) with a resolution of 5.3% of the lower explosive limit of CH_4. Fields trials have been conducted over a six-month period on a North Sea oil rig with excellent results (22).

3.2. Absorption and Reflectance Measurements
Remote absorption and reflectance measurements have been performed in many applications and commercial equipment is becoming available using fibre optic probes. Examples of such applications are :

(i)   Measurement of alcohols content in gasoline by overtone spectroscopy in the region of 1550 nm (23) Linearity of 1% was obtained for concentrations from 0-10% alcohol content.

(ii)   Photometry of biological tissues (24).

(iii)   Thin film monitor during deposition (25).

4. FLUORESCENCE
Fluorence techniques with fibre optics have been developed for remote monitoring of different chemical species such as $UO_2{}^{++}$, Actinides, Sulphates, inorganic chloride, H_2S, etc. (26,27,28,29).
Applications are in nuclear power stations, pollution monitoring, detection of trace impurities in ground and surface water. Fibre optic probes offer advantages of localised measurement in remote locations (the optical instrumentation can be 1km away), with high sensitivity and high coupling efficenciency of the exciting and fluorescent light.

Fluorescence is excited by laser (e.g. Argon) and the back-scattered fluorescent light picked up by the fibre and analysed by a spectrometer and photon counting techniques. Where a single fibre is used to carry both the exciting and fluorescent light, the natural fluorescence in the fibre can limit the lower detection limit (3.10^{-6}) for Rhodamine B dye detected by a 600 µm PCS fibre) (29). This can be overcome using two seperate fibres.

Different probe geometries have been used: a simple fibre end (31), a ball-shaped lens (28) a semi-permeable membrane (28). For uranium ion detection sensitivies of better than 10^{-14} molar can be achieved with enhancement by phosphate coprecipitation (26).

Other applications of fibre-optic fluorimetry include the

detection of stomach cancers by UV excitation with an endoscope (32), the detection of coal tar contamination on the sknin with a sensitivity of a few ng/cm^2 (32), in mapping the distribution of algae (plankton) in lakes (33), and in detecting herbicides in plant leaves. Water cooled fibres have been used to measure the Na and OH concentrations inside gas flames (34).

4.1 Raman Spectroscopy

Raman spectroscopy has also been performed on gas flames, measuring N$_2$ in a CH$_4$ air flame at 2100° K (35). A 60 μm diameter, 20 m long fibre was used. Raman spectroscopy has also been performed with fibre optic bundles (36). Light from an argon laser was used, and conducted by a glass bundle to a cavity of two spherical mirrors where multiple passes enhances the excitation. The Raman light was picked up at right angles through a horn shaped fibre bundle with a rectangular aperture 0.8 by 9.7mm and analyzed by a double spectrometer. Basic problems in improving sensitivity are :

(i)   To have efficient coupling in and out of fibres. These do not match the f number of the monochrometer.
(ii)   To minimize unwanted fluorescence and raman light generated in the transmitting fibre.
(iii)   To improve the excitation intensity in the sample volume where scattered light is collected.

5. OXIMETRY

In-vivo oximetry has been developed using fibre optic spectrometry (37,38). In general, reflectance of haemoglobin is measured at two wavelengths; one wavelength around 650 nm which is the peak reflectance of oxyhaemoglobin, and the second wavelength around 850 nm which is the isobestic point where satured and unsaturated haemoglobin have the same absorption (39). The concentration of oxyhaemoglobin is

% HbO$_2$ = A - B (R_1/R_2)

where A, B are constants depending on probe geometry and R_1, R_2 are the reflectance at the first and second wave length. Filters (39) and LEDs (40) at these wavelengths have been used. Typical performance (41) is:

Accuracy: less than 2% error of vitro spectrometric measurement
Range: 60 - 90% HbO$_2$
Linearity: 2.7%
Interference: 0.7% change for 0.1 pH change.

Such devices have been combined with fibreoptic pressure sensors for intravascular monitoring, and for determination of cardiac output by measuring dye dilution of injected indocyanine green dye (which absorbs strongly at 805 nm near the isobestic wavelength) into the bloodstream (42,43).

6. SCATTERING

Light scattering is an important measurement in many applications, for the determination of impurities in liquids and for measurement of particle size. One sensor which has been commercialised is an oil-in-water monitor for measuring the oil pollution in effluent water. For marine use the specifications for ballast dumping are 1000 ppm ± 10 ppm. For other areas e.g. electricity generating stations, oil content is measured as 0-2 ppm ± 0.1 ppm (44).

The marine oil-in-water monitor consists of measurement of scattered light from a laser diode at 850 nm connected to a fibre optic lead placed in the water flow (45). Straight-through and sideways scattered light is collected.; the angular dependence allows distinction between large and small particles. The device shows some temperature sensitivity : for light crude oils the change is less than 0.2% /°C, for heavy crudes there is a larger change. The assembly includes an automatic window cleaning system to counter dirt accumulation.

High pressure cells, up to 250 bars pressure, for turbidity measurement, have been built with up to 50 cm optical path length (46). The construction requires accurate alignement of the optics and high pressure seals. These have been used for differential measurement of particle contamination in cooling water. Alternatively, dual fibre optic probes measure back scattered light through 180° (47).

Measurement of surface scattering from opaque surfaces (48) has been made with a fibre optic reflectance probe, for determination of demineralization of teeth enamel, milk-fat content and the whiteness of the eyeball. A reference fibre-optic probe on a white $BaSO_4$ calibrated surface provides a differential measurement.

7. REFRACTIVE INDEX AND LIQUID-LEVEL

Measurement of refractive index of a liquid can be made by measuring the critical angle 0 of the angle of refraction at the fibre liquid interface given by Snell's law

$$n_1 \ . \ \sin \theta_1 = n_2 . \sin \theta_2$$

where n_1, θ_1 are the refractive index and angle of incidence of the fibre respectively and n_2, θ_2 are for the liquid. The measurement is usually made by measuring the total transmitted light down a fibre. As the refractive index n_2 of the liquid approaches that of the fibre so less light is transmitted, and greater resolution is obtained.

A variety of probe geometries are possible:
(i)   Prisms: (generally 45° prism), and other flat surfaces are attached to the fibre end. This requires machining of the fibre and mounting, and attenuation losses in air are usually high, typically 10 dB or more, due to alignment of the fibre ends on the prism (49, 50, 51).

(ii)   A declad fibre, in which the optical cladding is
        removed and replaced by the liquid is equivalent
        to a variable numerical aperture measurement (52).
        This technique has been used in a liquid chroma-
        tograph in which a He-Ne laser beam is injected
        through aspheric optics down a small, 1 mm diame-
        ter glass rod. The rod is built into a chroma-
        tography cell of $7\mu L$ volume, and the sensitivity
        is 1.10^{-5} refractive index units (53).

(iii)   A fibre can be bent in a single continuos curve,
        either a U bend or a spiral of multiple turns.
        The fibre may also have the cladding removed (49,
        54, 55). An improvement is to use a series of
        bends in alternate directions (56).

   Problems associated with refractive index measurements
are the change in refractive index with temperature, and
contamination of the probe with dirty liquids. Automatic
temperature compensation can be achieved by using the
variation of the light output from an LED with temperature,
or by having two probes of different material with
different refractive index temperature sensitivities (56).

8. pH Sensing

A number of pH sensors have been developed based on a
hydrogen ion-permeable membrane which encloses a colorimetric pH indicator. An optical fibre reads the colour of the
indicator at two wavelengths, one wavelength with maximum
intensity change for a small pH change and the second
wavelength for normalisation which is insensitive to pH
changes.

Goldstein and Peterson (57) developed a pH sensor based on
phenol red which is covalently bound onto polyacrylamide
microspheres of 5 - 10 μm diameter. These are packed into a
cellulose dialysis tube, of 0.3 mm internal diameter, which
is permeable to hydrogen ions. Polystyrene microspheres of
1 μm diameter are included in the packing to scatter and
reflect back the light coming from a Y-guide probe. Phenol
red shows a maximum intensity change at 560 nm and a second
wavelength at 600 nm is pH insensitive. The performance
achieved is

        Physiological range: 7.0 to 7.4 pH
           Accuracy: ± 0-01 pH
        Temperation coefficient: 0.017 pH/°C
        Variation to ionic strength: ± 0.01 pH for 11% ionic
                                     change.

Practical studies to develop the sensor for clinical use
and to field test the device have been persued. (58, 59,
60).

This sensor has been used as a foetal scalp monitor with
the sensor contained in a spiral stainless steel microtube

which could be screwed into the skull.

A more recent development employs bromothymol blue as an indicator absorbed on a styrene-divinylbenzene copolymer which is attached to a plastic fibre optic bundle (61). The probe tip was covered with a PTFE membrane which acts as a permeable membrane and ligh reflector.

These pH sensors require a correction for the difference between pH and pK of the dye. However a practical technique has been evaluated, using a double fluorescence indicator system with indicators of different valence values (63), in which both pH and ionic strength can be measured simultaneously.

Measurement of pH can also be performed by fluorescence. For fluoresceinamine, covalently coupled onto cellulose the fluorescence increases in its acidic state. However concentration quenching occurs due to the proximity of the immobilized fluorescent molecules. (62).

9. FLUORESCENT QUENCHING SENSOR: O2 SENSOR

An interesting technique of measuring O2 concentration illustrates the principle of fluorescence quenching. A fluorescent dye is absorbed on a polymeric substrate membrane which is oxygen permeable. Absorption of oxygen through the membrane reduces the fluorescent efficiency of the dye. Two dyes have been examined.

In the first perylene dibutyrate is absorbed on porous polypropylene (64). Excitation is at 468 nm and emission at 514 nm. Two filters are used to measure the emitted and scattered exciting light. The fluorescent efficiency drops from 40% in air to 16% in 1 atmosphere of oxygen. The accuracy is 1 torr partial oxygen pressure.

A second sensor is based on pyrenebutyric acid with excitation at 342 nm and emission at 395 nm (65).

10. COMPETITION BINDING: GLUCOSE SENSOR

A sensor for measuring glucose concentration operates on reversible competition binding, similar to the immunoassay principle (66). The sensor consists of a hollow dialysis tube of 0.3 mm internal diameter, sealed at one end and attached to a single optical fibre at the other. A carbohydrate receptor Concanavalin A is immobilised on the inner surface of the tube.

A high molecular weight carbohydrate, in this case fluorescein labelled dextran, is imprisoned in the tube and competes with glucose, which diffuses in, for binding to the Con A. The level of fluorescence from the fluorescein excited by light from the fibre is a direct measure of glucose concentration. The response is linear from 50 to 400 mg % glucose and response time is 7 minutes at 24°C. Drift over 15 days operation was 15% for unknown reasons.

11. IMMUNOLOGICAL ASSAY

Chemical and biomedical constituents in the body are measured by two main techniques : biochemical analysis and immunoassay. Biochemistry employs well established techniques, such as titration and chemical reactions in liquids, to determine the concentration of sodium and chloride ions, cholesterol, glucose, iron, etc with concentrations normally in the range 10^{-3} to 10^{-6} molar. For many other important biochemical species e.g. steroids and hormones, drugs and metabolites, bacterial and viral antigens, concentrations are from 10^{-6} molar to as small as 10^{-15} molar, and immunoassay techniques are employed.

Immunoassay measures a specific reaction (6), such as an antibody (Ab) reaction with an antigen (Ag)

$$Ab + Ag \leftrightarrows Ab \ Ag$$

A four step technique is used : a known quantity of a labelled antigen is added to the unknown quantity of antigen in the sample. The label is a radioisotope, an enzyme, a fluorophor or luminescent marker. The mixture of antigens is reacted with a known quantity of antibody attached to a suitable substrate such as plastic beads, glass beads, pegs or paper tags. The next step is to mix and incubate the reactants to complete the chemical binding of antigen to antibody. This is a followed by a seperation step to remove the antibody substrates which can be done by means of centrifuge, filtration, magnetic particle separation (for magnetic substrates), electrophoresis, solid-phase separation or chromatography. The labelled antigen may then be estimated from the amount attached to the substrates, or from antigen left in solution, and by substraction the unknown antigen concentration is calculated.

This technique is slow (taking several hours) due to the separation step, gives only the total concentration of antigen once the reaction is complete, and uses expensive materials such as radioisotopes.

11.1 EVANESCENT WAVE SPECTROSCOPY

The principle of using an optical technique for immunoassay is based on internal reflection spectroscopy or evanescent wave spectroscopy. The evanescent wave of light bound within a waveguide penetrates into the surrounding liquid by a fraction of the wavelength and can measure reactions taking place on or near the surface of the guide within the order of a wavelength of light. The intensity of evanescent wave may be increased by plasmon resonance.

A guided mode (represented by a ray of light, will remain within the waveguide if the angle of incidence at the waveguide-liquid interface is greater that the critical angle θ

$$\theta = \sin^{-1} (n_2 / n_1)$$

where n_1, n_2 are the refractive indices of the waveguide and liquid respectively. The electric field E of the evanescent wave, associated with the bound mode, penetrates a given distance z outside the guide :(67)

$$E = E_0 \cdot \exp (-z / d_p)$$

where E_0 is the electric field at the waveguide-liquid interface and d_p is the penetration depth, or the distance at which E_0 falls by $1/e$ of its value

$$d_p = \frac{\lambda}{2 \pi n \ (\sin^2 \theta - \sin^2 \theta_c)^{1/2}}$$

For typical values (68), a quartz substrate, n, = 1.46, immersed in water n_2 = 1.33, for λ = 300 nm and θ = 75° (10° larger than θ_c = 65.6°) the penetration depth is 102 nm.

The size of an 1gG molecule (immunoglobulin) is approximately 10 nm by 6 nm, which is therefore smaller than the penetration depth.

11.2 SURFACE REACTION MEASUREMENT
An antibody is immobilised on the surface of the waveguide and the light measures the antigen-antibody reaction.

Several methods can be employed; measuring the light scattering directly from the antigen or the total absorption, or the fluorescence if a labelled antigen is used. In this case the fluorescence is also collected and trapped within the waveguide itself. An enhancement effect (as high as 50 times the sideways emitted light) is due to preferential coupling of fluorescent light with the same electric field intensity as the evanescent exciting wave, and this has been verified experimentally and theoritically (69, 70).

Measurements with evanescent wave spectroscopy have been made using both glass slides as a waveguide and optical fibres (a PCS fibre without the cladding).

For FITC-labelled 1gG the sensitivity achieved is around 10 nM concentration for scattering changes and 5 nM concentration for fluorescence (68).

The technique eliminates the lengthy seperation and washing stages, and a dynamic rate monitoring can be used making measurments possible within 5 - 10 minutes. This reduces the time, cost and complexity of the measurement compared to conventional techniques.

A large number of different biochemical systems have been studied by optical measurement on continuous surfaces, using ellipsometry, attenuated total reflection, interference techniques and total internal reflection fluorescence. (Ref. 6 lists 45 different experiments). It is therefore likely that evanescent wave spectroscopy will become widely applied in the future.

11.3 SURFACE PLASMON RESONANCE

Surface plasmon resonance occurs at a dielectric - metal layer interface where light which is total reflected within the dielectric induces a collective oscillation in the free electron plasma at the metal boundary. The conditions for this to occur are that the momentum of the photons should match the surface plasmons on the opposite surface of the metal film (71). This occurs at some critical incident angle of light. The resulting effect is to produce a large change in the reflection coefficient at this resonance angle.A typical set-up employs a silver film of 50nm thickness evaporated onto a glass plate or prism. The sample is placed on the silver layer. The intensity of the electric field of the evanescent wave within the sample layer is enhanced by two orders of magnitude compared to a simple dielectric interface. Changing the sample properties shifts the resonance angle, providing a highly sensitive means of monitoring the surface reaction. An angular resolution of a typical measurement set-up is better than 5.10^{-2} deg. (72).

Gas sensing has been achieved by coating a 43 nm layer of silicone-glycol copolymer onto a silver-coated (56 nm) glass substrate (73). The anaesthetic gas halothane swells the film, a swelling of 0.1% results in a shift of 10^{-2} deg. The resulting sensitivity is 3.10^{-5} deg/ppm and the response is highly linear to 0.5% volume with a time response of a few milliseconds.

Immunoassay reactions have been studied. A single monolayer 5 nm thick of IgG produces a shift of 0.6°. (71) Injection of antihuman globulin (a-IgG) which binds onto the IgG results in a further shift of 0.9°. From the initial time derivative it is possible to determine the concentration of a-IgG, to less than 2 µg/ml.

Film layer thickness of antibody human serum albumin (anti-HSA) bonded to HSA have been calculated and measured by surface plasmon resonance (74). The absorption of HSA on a silver film shifts the resonance peak by 1.1° corresponding to a film thickness of 6 nm (HSA molecular dimensions are 6 x 6 x 2 nm). Reaction with anti-HSA shifts the peak by 3.2° corresponding to 22 nm or several (up to four) mono-layers.

12. CHEMICAL SENSING BY PHYSICAL MEASUREMENT
A number of chemical sensors have been suggested that use physical interaction with the fibre as a means of sensing. It should be remembered that fibre interferometers have a very high sensitivity and can be used to measure displacement, strain, pressure, etc.

(i) A hydrogen sensor has been examined which uses a fibre coated with Palladium (75). The fibre is in a Mach-Zehnder interferometer and absorption of hydrogen introduces a phase shift ammounting to 1 fringe for 0.2% hydrogen.

Phase changes of 10^{-6} fringes can be measured on sensitive equipment.

Temperature effects are equivalent to 2 ppm $H_2/°C$.

(ii) Photacoustic spectroscopy is a spectroscopic method where the light absorbed is measured by the heat produced and corresponding pressure changes in a closed cell. This cell can use optical fibres to excite the absorption (76); or the measurement of the pressure waves can be made with a fibre-optic Mach-Zehder interferometer. (77).

13. CONCLUSIONS
Fibre optic technology offers many exciting possibilities for future chemical sensing.

The present state of use of the technology may be summarised as having the following feature:

- Miniaturised spectrometers with non-moving scan (CCD readout)
- Multiple heads and centralised processing electronics
- Remote measurements, up to 5 km distance
- Special cells: such as high pressure cells, multiple light paths in the same cell
- Special fibre probes: such as lens-ended fibres for remote fluorescence
- Probe packaging, in hypodermic needles for medical measurements.

Problems posed with present sensors are:

- Fibre coupling problems
- Optical cell design for reproducibility
- Long term stability
- Limited dynamic range
- Problems in sensor design involving mass transfer stage for the reagent and analyte in different phases, and non-reversible reactions.

212

Future developments in fibre optics should offer the following possibilities:

- All fibre probe cell adapted to specific applications
- I.O. Chip spectrometer
- Tunable wavelength solid-state sources
- Wide spectral range transmission fibres (both near
  U.V. and mid-I.R.)
- Miniaturised sample hangling (e.g. microfluidics) for
  mass transfer, mixing, etc.

The problems foreseen with further miniaturisation and
small probe size are probe interchangeability, accuracy and
calibration of small sample volumes, and contamination.

This paper has shown that fibre optic technology is already
being applied to a wide range of chemical and biomedical
sensors. It is certain that applications will continue to
grow in the future.

REFERENCES

Reviews
1. C. Nylander: Chemical and Biological Sensors. J. Phys E.
   Sci Instrum 18, 736 - 750 (1985).
2. T. Hirschfeld: J.B. Callis and B.R. Kowalski, Chemical
   Sensing in Process Analysis Science, 226, 312 - 318
   (1984).
3. W.R Seitz: Chemical Sensors based on Fiber Optics
   Analytical Chemistry, 56, 16A - 33A (1984).
4. I. Chabay: Optical Waveguides Analytical Chemistry, 54,
   1071A - 1080A (1982).
5. J.I. Peterson and C.G. Vurek: Fiber Optic Sensors for
   biomedical applications. Science 224, 123 - 127 (1984).
6. J.F. Place: R.M. Sutherland and C. Dähne:
   Opto-electronic Immunoassay at continuous surfaces.
   Biosensors 1 (1985).
7. G. Boisde, C. Linger and J.J. Perez: Developments
   recents de la spectrophotometrie par fibres optiques
   pour le contrôle "in situ" de l'uranium VI en solution.
   Fifth ann. symp. on Safeguards and Nuclear Mat. Manag.,
   France, p. 203 208 (1983).
8. S. Klainer, T. Hirschfeld, H. Bowman, F. Milanovich D.
   Perry and D. Johnson: A monitor for detecting
   nuclear-waste leakage in a subsurface repository.
   Science Division Annual Report, Lawrence Berkeley Lab.
   LBL-11981 (1981).
9. Trott G.R., Furtak T.F. Rev. Sci. Instrum. 51, 1493
   (1980).
10. T. Hirschfeld:"Remote analysis by fluorescence, Raman and
    absorption measurements over optical fibres". European
    Conf. on industrial line spectrographic analysis, Rouen,
    France. June 1985. p.1 - 1 (1985).
11. J.P. Laude, J. Flamand, J.C. Gautherin, D. Lepere, P.
    Gacoin, F. Bos and J. Lerner: "Stimax, a grating
    multiplexer for monomode or multimode fibres".

European Conf. Optical Communications, Geneva, p 417 - 420 (1983).

12. J.P. Laude, J.C. Gautherin, D. Lepere, J. Flamand, P. Gacoin and F. Bos: Multiplexeurs multiples a fonction optique partagée. Opto, 19, 29 - 31 (1984).

13. H.E. Korth: Film Thickness monitoring with a fiber optic realtime spectrometer.
Second Int. Conf. Opt. Fibre Sensors, Stuttgart, 219 - 222 (1984).

14. H.E. Korth: A computer integrated spectrophotometer for film thickness monitoring. Journal de Physique 44, C10 p 101 - 104 (1983).

15. P.K. Tien and R.J. Capik: A thin film spectrograph for guided-waves. Topical Meeting on Integr. and Guided-wave Optics, Nevada, p TuB3-1 (1980).

16. H. Inaba, T. Kobayashi, M. Hirama and M. Hamza: Optical-fibre Network System for Airpollution monitoring over a wide area by optical absorption method. Elec. Lett. 15 749 - 751

17. A. Hordrik, A. Berg, D. Thingbo: A fiber optic gas detection system. Ninth Europ. Conf. Opt. Commun, Geeneva, 317 - 320 (1983).

18. K. Chan, H. Ito and H. Inaba : An optical fiber based gas sensor for remote absorption measurement of low-level CH4 gas in the near-infrared region. J. Lightwave Tech. LT-2, 234-237 (1984).

19. K. Chan, H.Ito and H. Inaba : Remote sensing system for near-infrared differential absorption of CH4 gas using low-loss optical fibre link. Appl. Ont. 23, 3415-3420 (1984).

20. S. Stueflotten, T. Christensen, S. Iversen, J.O. Hellvik, K. Almas, T. Wien, A. Graav: An infrared fibre optic gas detection system. Europ. Conf. Industrial line spectrographic analysis, Rouen, p. 3.1-3.5 (1985).

21. K. Chan, T. Furuya, H. Ito and H. Inaba : Full optical remote measurement of CH4 gas in the near-infrared using a 5 Km long low-low optical fibre link. Opt. and Quantum Elec. 17, 153-155 (1985).

22. S. Stueflotten : Fibre optics in Offshore Systems. Ericsson Review, F-61, 24-27 (1984).

23. L.A. Hilliard : Application of single optical fibres to remote Absorption measurements. Anal. Proc. 22, 210-211 (1985).

24. S.Ji, G. Luthen, M. Kessler : Some quantitative aspects of micro light guide photometry of biological tissues. Ilphen aan den Rijn, Netherlands (Sijthoff and Noordhoff) p. 237-268 (1979).

25. H.M. Runciman, W.B. Allan and J.M. Ballantine : A thin film monitor using fibre optics. J.Sci. Instrum. 43, 812-815 (1966).

26. S. Klainer, T. Hirschfeld, H. Bowman, F. Milanovich, D. Perry and D. Johnson : A monitor for detecting nuclear-waste leakage in a subsurface repository. Report : Lawrence Berkeley Laboratory LBL-11981, UC-70 (1981).

27. F.P. Milanovic, T. Hirschfeld : Process, product, and waste stream monitoring with fibre optics. Advances in Instrumentation. Proceeding of ISA Int. Conf. 38, Pt. 1, 407-418 (1983).

28. F.P. Milanovich, t. Hirschfeld : Remote fibre fluorimetry. Intech, March 1984, 33-36 (1984).

29. J.P. Dakin and A.J. King : Limitations of a single optical fibre fluorimeter system due to background fluorescence. First Int. Conf. Opt. Fibre Sensors, IEE Vol. 221, 195-199 (1983).

30. R.E. Grojean and J.A. Sousa : Bifurcated fiber luminometer. Rev. Sci. Instrum. 51, 377-378 (1980).

31. G.C. Huth, A.D. Profio and D.R. Doiron : Early lung cancer detection with laser fiberoptic bronchoscope. Laser and Electro-optic 2, 35 (1978).

32. T. Vo Dinh and R.B. Gammage : Fibre optic monitor of skin contamination. Chemistry and Industry, Sept. 707 (1980).

33. T. Lund : A fibre optics fluorimeter for algae detection and mapping. First Int. conf. Opt. Fibre Sensors, IEE. Vol 221, 190-194 (1983).

34. G. Kychakoff, M.A. Kimball-Line, R.K. Hanson : Fiber optic absorption/fluorescence probes for combustion measurements. Appl. Opt. 22, 1426-1427 (1983).

35. A.C. Eckbeth : Remote detection of CARS employing fibre optic guides. Appl.Opt. 18, 3215-3216 (1979).

36. R.E. Brenner and R.K. Chang : Utilization of optical fibres in remote inelastic light scattering probes in Fibres Optics, Advances in Research and Developmennt' ed. B.Bendow and S.S. Mitra, Plenum Press New York. p. 625-640 (1979).

37. C.C. Johnson : Biomed. Sci. Instrum. 10, 45 (1974).

38. M.H.J. Landsmann : Fiberoptic reflection photometry Verenigde Reproductive Bedrijve, Groningen, Netherlands (1975).

39. B.G. Gamble, P.G. Hugenholtz, R.G. Monroe, M. Polanyi and A.S. Nadas : The use of fiberoptics in clinical cardiac catheterization. I. Intracardiac Oximetry Circution, 31, 328-343 (1965).

40. C.C. Johnson, R.D. Palm, D.C. Stewart and W.E. Martin : A solid state fiberoptic oximeter. J. Assoc. Advancement of Med. Instrum. 5, 77-83 (1971).

41. E.A. Woodroff and S. Koorajian : In vitro evaluation of in vivo fibreoptic oximeter. Med. Instrum. 7, 287-292 (1973).

42. M.L. Polanyi : Dye Curves. Ed. D.A. Bloomfield (University Park Press, Baltimore), 267-284 (1974).

43. M.L.J. Landsman, N. Knop, G.A. Mook, W.G. Zijlstra : Pfluegers Arch. 379, 59 (1979).

44. P. Extance, G.D. Pitt : Intelligent Turbidity Monitoring. Int. Conf. on Opt. Tech. in Flow Monitoring and Control, Hague (BHRA) 43-54 (1983).

45. D. Snel, G.D. Pitt : Oil Content Monitoring. Int. Conf. on Optc. tech. in Flow Monitoring and Control, Hague (BHRA) 27-42 (1983).

46. J.J. Perez : Perspectives des Mesures Spectrométriques pour l'analyse de ligne industrielle et de contrôle des eaux. Eur. Conf. Industrial Line Spectrographic Analysis, Rouen, 2.1-2.12 (1985).

47. L. Papa, E. Piano and C. Pontiggia : Turbidity Monitoring by Fiber Optics Instrumentation. Appl. Opt. 22, 375-376 (1983).

48. P.C.F. Borsboom, J.J. Ten Bosch : Fiber optic scattering monitor for application on bulk biological tissue, paper and plastic. S.P.I.E. Vol. 369, 417-421 (1983).

49. N.S. Kapany and D.A. Pontarelli : Photo refractometer. Extension of Sensitivity and Range. Appl. Opt. $\underline{2}$, 425 (1963).

50. N. Abuaf, O.C. Jones and G.A. Zinner : Optical Probe for local void fraction and interface velocity measurements. Rev. Sci. Instr. $\underline{49}$, 1090-1094 (1978).

51. M.A. Vince, H. Breed, G. Krycuk and R.T. Lahey : Optical probe for high-temperature local void fraction determination. Appl. Opt. $\underline{21}$, 886-892 (1982)

52. N.S. Kapany and J.N. Pike : Fibre Optics. A Photorefractometer. J. Opt. Soc. Am. $\underline{47}$, 1109-1116 (1957).

53. D.J. David, D. Shaw, H. Tucker and F.C. Unterleitner. Rev. Sci. Instr. $\underline{47}$, 989 (1976).

54. E. Karrer and R.S. Orr : J. Opt. Soc. Am. $\underline{36}$, 42-46 (1946).

55. T. Takeo and H. Hathori : Jap. P. Appl. Phys. $\underline{21}$, 1509-1512 (1982).

56. A.L. Harmer : Optical Fibre Refractometer using attenuation of cladding modes. First Int. Conf. on Opt. Fibre Sensors, London, I.E.E. Vol 221, p. 104-108 (1983).

57. J.I. Peterson, S.R. Goldstein, R.V. Fitzgerald Anal. Chem. $\underline{52}$, 864 (1980).

58. D.R. Markle, D.A. Mc Guire, S.R. Goldstein, R.E. Patterson, R.M. Watson in 1981 Advances in Bioengineering, Ed. D.C. Viano (Am. Soc. of Mech. Eng., New York, p 123 (1981).

59. G.A. Tait, R.B Young, G.J. Wilson, D.J. Steward, D.C. Mac Gregor. Am. J. Physiol. Heart Circ. Physiol. $\underline{12}$, H 1027 (1982).

60. R.M. Watson, D.R. Markle, Y.M.Ro, S.R. Goldstein, D.A. Mc Guive, J.L. Peterson, R.E. Patterson. Am. J. Heart Circ. Physiol. $\underline{15}$, H232 (1984).

61. G.F Kirkbrigth, R. Narayanaswamy and N.A. Welti Fibre-optic pH Probe based on the use of an immobilised colorimetric indicator. Analyst. $\underline{109}$, 1025 - 1028 (1984).

62. L.A. Saari and W.R. Seitz. pH sensor based on immobilized fluoresceinamine Anal. Chem $\underline{54}$, 821 - 823, 1982.

63. N. Optiz and D.W. Lubbers. New fluorescence photometrical techniques for simultaneous and continuous measurements of ionic strength and hydrogen ion activities. Sensors and Actuators $\underline{4}$, 473 - 479 (1983).

64. J.I. Peterson, R.V. Fitzgerald and D.K. Buckhold. Fibre Optic probe for in vivo measurement of oxygen partial pressure. Anal. Chem. 56, 62 - 67, 1984.

65. D.W. Lubbers and N.Opitz. Optical Fluorescence Sensors for continuous measurement of chemical concentrations in biological systems. Sensors and Actuator $\underline{4}$, 641 - 654, 1983.

66. S. Mansouri and J.S. Schultz. A miniature optical glucose sensor based on affinity binding. Bio/Technology, 885 - 889, Oct. 1984.

67.R.M. Sutherland, C. Dähne, J.F. Place and A.S. Ringrose.
Optical detection of antibody - antigen reactions at a
glass liquid interface. Clin. Chem. 30/9, 1533 - 1538
(1984).

68.R.M. Sutherland, C. Dähne, J.F.Place and A.S. Ringrose.
Immunoassays at a quartz-liquid interface: Theory,
Instrumentation and Preliminary Application to the
Fluorescent Immunoassay of Human Immunoglobulin G.
J. Immunological Methods 74, 253 - 265. (1984)

69.Lee E.H, Benner R.E., Fenn J.B. and Chang R.K.
Angular distribution of fluorescence from liquids and
monodispersed spheres by evanescent wave excitation.
Appl. Opt. 18, 862 - 870. (1979)

70.Carniglia C.K., Mandel L. and Drexhage H.
Absorption and emission of evanescent photons.
J. Opt. Soc. Am. 62, 479 - 486. (1972)

71.B.Liedberg, C. Nylander and I. Lundstrom. Surface Plasmon Reasonance for Gas detection and biosensing.
Sensors and Actuators, 4, 299 - 304, 1983.

72.I.Pockrand, J.D. Swalen, J.G. Gordon and M.R. Philpott.
Surface Plasmon Spectroscopy of Organic Mondlayer Assem-
blies.
Surface Science 74, 237 - 244, 1977.

73.C.Nylander, B.Liedberg and T.Lind.
Gas Detection by means of surface plasmon resonance.
Sensors and Actuators 3, 79 - 88, 1982.

74.M.T. Flanagan, R.H. Pantell.
Surface Plasmon Resonance and Immunosensors. Electronics
Letters 20, 968 - 970, 1984.

75.M.A. Butler. Optical fiber hydrogen Sensor. Appl. Phys.
Lett. 45, 1007 - 1009 (1984).

76.D.H. McQueen. A simplified open photoacoustic cell and
its applications.
J. Phys. E.:Sci. Instrum. 16, 738 - 739, 1983.

77.D.H. Leslie, G.L. Trusty, A. Dandridge and T.G. Giallorenzi.
Fibre-Optic Spectrophone. Elec. Lett. 17, 581, 1981.

Fiber LDA System

Takashi Nakayama
Mitsubishi Electric Corp., Central Research Laboratory

1. INTRODUCTION

Velocity is one of the most interesting physical quantities to detect for the control of industrial manufacturing processes as well as for the determination of vector components of a moving object in the physical science. Among the variety of methods to measure the velocity, it is desirable to detect velocity without disturbing the motion of the moving object. An optical beam aiming at a moving object is supposed to be the best fit for this kind of demand, provided that the optical power is low enough not to disturb the state of the moving object.

Other outstanding advantages of the optical beam methods are : (i) remote sensing capability, that is, to apply for the measurement in adverse environment, (ii) precise velocity measurement in the very short period of time, that is, to apply for the real time measurement for monitoring and control of the moving object, (iii) the moving object can be solid state or liguid state, that is, to apply for the discrete object as well as for the fluid flow.

Most of these advantages of optical method are shared with the microwave method. But because of the wavelength difference, optical beam method can detect the velocity of much smaller object. Another advantage of optical beam method is that it is much easier to guide the beam, which will be discussed in detail, later.

We start with the physical basis of Laser Doppler Velocimeter, where we briefly review the doppler effect and the significant advantage of having laser as a light source (Chapter 1). In Chapter 2, we review the Laser Doppler Velocimeter System, which is the preceding system of Fiber Optic Laser Doppler Velocimeter. Optical fiber is introduced in the system in Chapter 3 which is the main subject of this lecture. In Chapter 4, application examples of Fiber LDV are described. Further research efforts on Fiber LDV Sytem are discussed in the last Chapter 5.

2. OPTICAL CONFIGURATIONS AND ELECTRICAL SIGNAL PROCESSING OF LDV SYSTEM

2.1. Optical Configurations of LDV System

When a focused laser beam hits a moving object, the scattered light comes out of the moving object, the scattering pattern depends on the incident laserwave length and on the size of the object. The scattered light carries the information on the velocity of the moving object, that is, the frequency of the scattered light shifts toward higher side if the moving object approaches to the detector of the scattered light. This is the Doppler effect. In Fig.1, $\bar{a}i$=unit vector in the direction of the incident beam, $\bar{a}r$=unit vector in the direction of scattered beam which is not scattered by the moving object. Doppler frequency shift ω_0 is given by

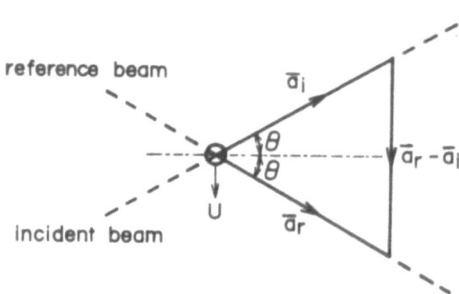

Fig.1 Incident and scattered beams

$$\omega_0 = (\overline{k}_r - \overline{k}_i).\overline{U} \qquad (1)$$

where k_r is the wave vector of the scattered beam and k_i is the wave vector of the incident beam, $\overline{U}$ is the velocity of the moving object.

$$\overline{k}_i = (\frac{2\pi}{\lambda_i})\,\overline{a}_i \;,\; \overline{k}_r = (\frac{2\pi}{\lambda_r})\overline{a}_r \qquad (2)$$

Since the frequency shift is very small compared to the frequency of the laser beam, $|\overline{k}_r| \approx |\overline{k}_i|$. This approximation leads to ;

$$\omega_0 = |\overline{k}_i|\,(\overline{a}_r - \overline{a}_i).\overline{U} = \frac{4\pi U}{\lambda_i}\sin\theta \qquad (3)$$

With the fixed angle of θ and laser wave length λ_i, the velocity of the moving object U can be calculated from the measured value of ω_0. Fig.2 is called the foward-scatter reference beam heterodyne system, because foward-scatter beam mixed with the reference beam is detected by the photo-diode. If we detect the backward-scatter light with the reference beam, then it is called the backward-scatter reference beam heterodyne system. The reason why we use the heterodyne system is that the frequency shift is so small compared with the frequency of laser itself, which is very difficult to detect the frequency shift directly. The problem for the application of the heterodyne system is that exact optical alignment of the

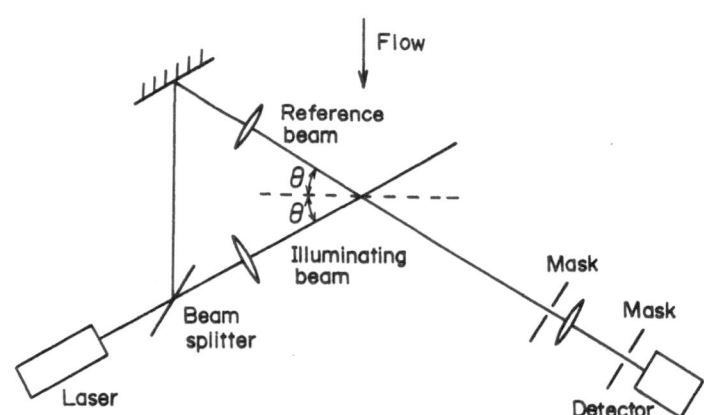

Fig.2 Foward-scatter reference beam heterodyne system

optical components on the optical bench is necessary in order to get satisfactory signal-to-noise ratio. Also, it is very difficult to get system reliability. Particularly, vibration effect is significant. Another optical configuration is the foward-scatter differential beam heterodyne method. This is illustrated in Fig.3. Two scattered lights ① and ② of which Doppler frequency shift are opposite, are combined (heterodyne) at the surface of the detector. The light intensity at the detector I is :

$$I \propto (E_2 + E_2)(E_2^* + E_2^*) = A_1^2 + A_2^2 + 2A_1 A_2 \cos\{(\overline{k}_2 - \overline{k}_1).\overline{U}t + \phi\}$$

Therefore, the Doppler frequencey shift is :

$$\omega_0 = (\overline{k}_2 - \overline{k}_1).\overline{U} = \frac{4\pi U}{\lambda_i} \sin\theta \qquad (4)$$

which is exactly the same expression as in eq.(3)
Again, we can make use of the back-scatter light for this system. The problem for the application of this system is obviously the same as the reference heterodyne type.

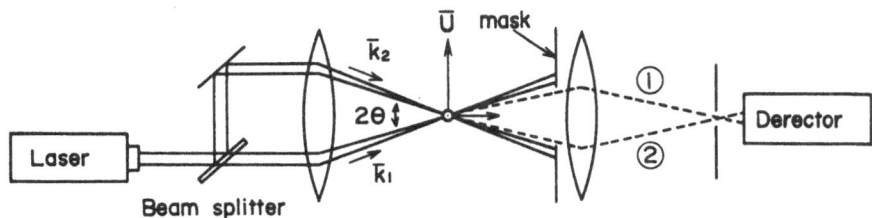

Fig.3 foward-scatter differential heterodyne system

2.2. Electrical Signal Processing

Here we consider the scattering is the Mie type, that is, the size of the moving objects are dominantly larger than the wavelength of the laser. The output of the detector is shown in Fig.4. Fig.5 is to show the output signal of one moving object passing through the measuring volume. In Fig4, high frequency component is the Doppler signal, that is, its frequency is

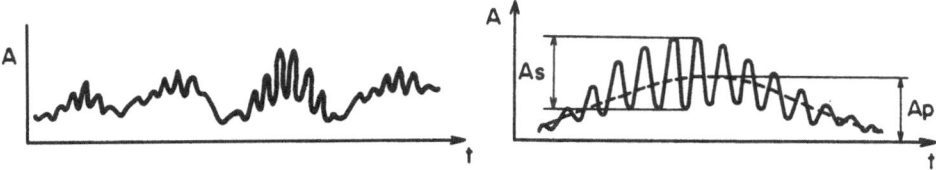

Fig.4 Output of the detector

Fig.5 Output of the detector for a
       particle

the Doppler shift frequency. The low frequency component is the rate of passing objects through the measuring volume. The figure of merit of the Doppler signal is As/Ap in Fig5. Therefore the signal processing starts with filtering out the low frequency component.

We can classify three signal processing methods followed by the high pass filter. The first method is the frequency anlysis. This is to get the Doppler frequency distribution. Very often, the spectrum profile for each frequency scan is integrated in the certain period of time. The most probable frequency is considered to represent the average velocity of the moving objects. The second method is simple in principle, to count the peaks of high frequency component, that is, frequency counter. Again, the statistical method is introduced to define the average Doppler shift frequency. The last one is called the frequency tracking method, which is most suitable for the continuous densly distributed objects. The block diagram of the frequency tracker is shown in Fig.6. The Doppler shift frequency f_0 is mixed with frequency f_s from V.C.O. (Voltage Controlled Oscillator). The difference (f_s-f_0) is led to frequency discriminator through I.F. filter. Output voltage from the frequency discriminator in a certain time interval is integrated. The output voltage from the integra-

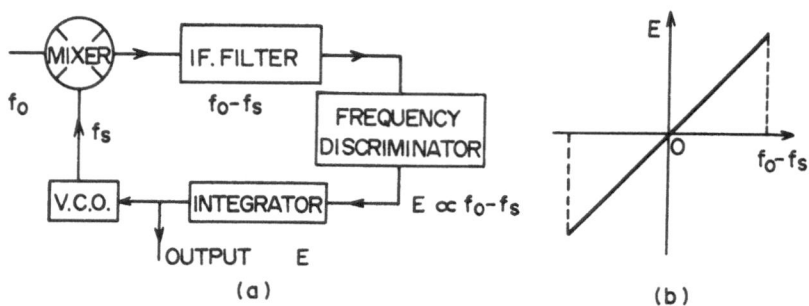

Fig.6(a) Block diagram of
frequency tracker

Fig.6(b) Output voltage vs
Doppler shift

tor is proportional to f_s-f_o. When this voltage is applied to V.C.O, V.O.C is so acted as to send frequency $f_s=f_o$. Therefore, E represents the voltage proportional to Doppler shift frequency f_o.

3. OPTICAL FIBER LDV [4)]

As we see from chapter 2 that the optical configurations without optical fiber are rather difficult for the stable operation and the cost of the system is not low enough for wider use in the industry as well as in the scientific research.

In order to eliminate the problem of optical precise alignment, bulky heavy optical bench, the laser beams are guided by the optical fiber. Optical path by the optical fiber guide gives us the design flexibility for the optical configuration. Due to the fact that the transmission loss of silica fiber is reduced to 0.1 1dB/km, we do not have any problem as far as the optical signal loss is concerned. Also transmission characteristics are not degraded by severe environment. Disadvantges of optical fiber are (1) the light intensity in the optical fiber is reduced by the difficulty of coupling of fiber to the light source (2) multimode fiber which has larger optical aperture, gives disturbance on the phase of optical signal. If we replace the multimode fiber with single mode fiber or rather with the polarization maintaining fiber, this problem can be solved.

Another problem which blocks the wider use of the optical configurations mentioned in chapter 2 is that, the gas laser is used as a light source. Gas laser such as He-Ne laser has a very high coherency, but it is bulky, and its power consumption is rather high. To cope with the obstacle, a semiconductor laser diode is going to replace the gas laser. LDV system with optical fiber and a laser diode has been in the developmental stage.

3.1. Reference Beam Heterodyne System [4),5),6)]

When we make use of optical fiber, laser beam out of the optical fiber has to be collimated on the moving object. Practical simple probe was built for this purpose. Graded index rod micro lens was attached at the end of the fiber which guides the laser beam aiming at the moving object. Fig.7 shows the improvement of the precision of Doppler shift frequency by the collimating effect of micro-lens. Fig.8 shows basic diagram of the

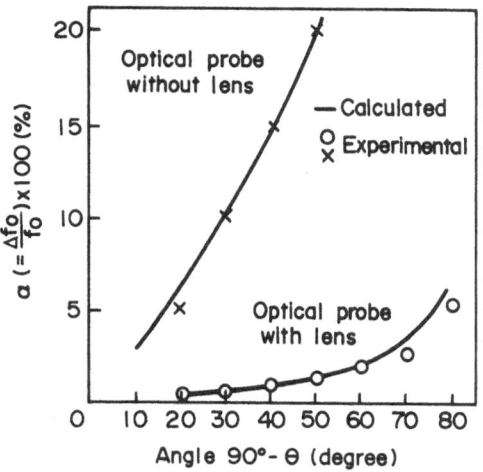

Fig.7 Collimating effect of the micro-lens

fiber optic reference beam heterodyne system. Here the reference beam is the reflected beam at the interface of lens with the probing fiber. The light source used is gas laser (He-Ne). In order to sense the direction of frequency shift, f_b was introduced by ultrasonic diffraction modulator, which is shown in Fig.9. Fig.10 shows the experimental results of velocity measurement from the Doppler shift frequency. The upper frequency is limitted by the bandwidth of the receiver. Fig.11 shows the Fiber optic LDV with a laser diode. Here, we have two additional components, that is, the external mirror and the Faraday isolator.

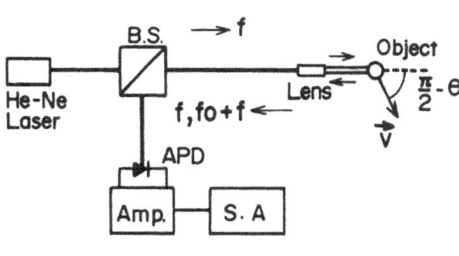

Fig.8 Fiber optic reference beam heterodyne

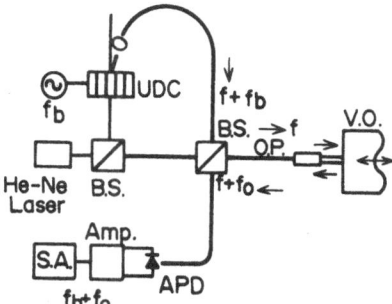

Fig.9 Fiberoptic r.b. heterodyne with ultrasonic modulator

Since the coherency of a semiconductor laser is not so good as a gas laser, the effective cavity length of the LD was extended by the the external mirror. The line width of LD with external cavity is 300 kHz while that of LD itself is 5 MHz. Since the coherency of LD improved significantly, we can make use of LD as a coherent light source. The role of Faraday isolator is to avoid the unstable modes of LD by the reflected light. As we see from the block-diagram of the LD-OFLDV system, we see the problem of coherency of LD and Faraday isolator.

To eliminate the Faraday isolator, the $\lambda/4$ plate at the end of microlens was deviced in the optical configuration. Fig.12 shows the system and the device. The linearly polarized incident beam out of polarization beam splitter (PBS) is launched into the polarization maintaining single mode fiber. Since the Doppler shifted reflected beam passes through $\lambda/4$ plate twice, the polarization rotates 90°, so that PBS blocks this reflected beam to go back to the laser diode. The reference beam is reflected by the

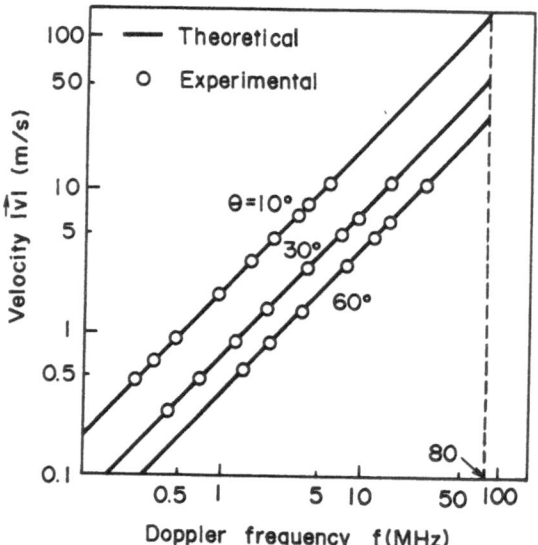

Fig.10 Experimental results of velocity

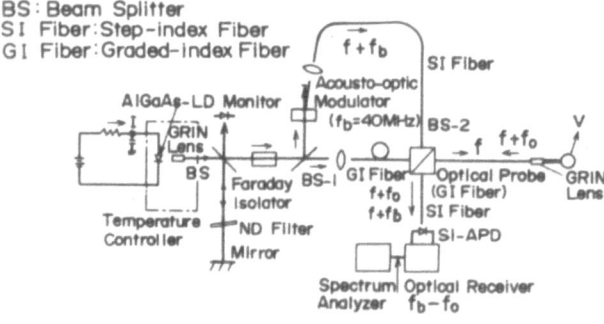

BS: Beam Splitter
SI Fiber: Step-index Fiber
GI Fiber: Graded-index Fiber

Fig.11 Fiber-optic LDV with a laser diode (1)

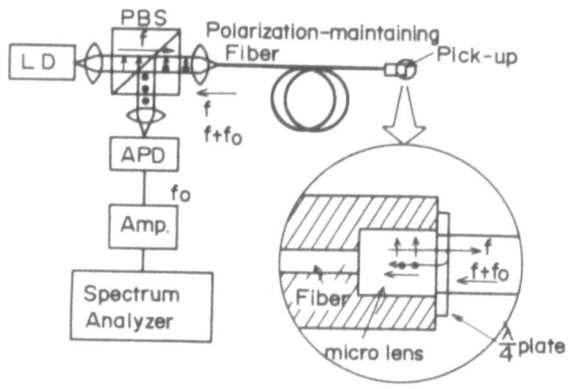

Fig.12 Fiber-optic LDV with LD (2)

outside surface of the $\lambda/4$ plate thus again the reflected beam polarization rotates 90° which is blocked by PBS. Therefore no reflected beam disturbs the LD mode, and the heterodyne detection occurs at the surface of the receiving device (APD).

3.2 Differential Heterodyne System

As we mentioned in 3.1, we make use of micro-lens to focus the laser beam out of the fiber or into the fiber for the differential heterodyne system. The optical configuration of the system with the probe structure is shown in Fig.13. As we see from Fig.13, differential frequency of scattered light from two direction :
$$f + f_0 - (f - f_0) = 2f_0$$
is picked up by the detector (APD). Therefore, twice of the Doppler shift frequency is measured. Relative locations of the two illuminating optical paths and the pick-up optical path are fixed as are shown in Fig.14. The significance of this structure is that the probe structure of this method gives the better precision of measuring the Doppler shift frequency than that of the reference method. From eq.(3) in Chapter 2, the angleθ between pick up optical path and the direction of the moving object it has to be precisely fixed to get the precise Doppler shift frequency, which is the difficult problem in practice. In case of

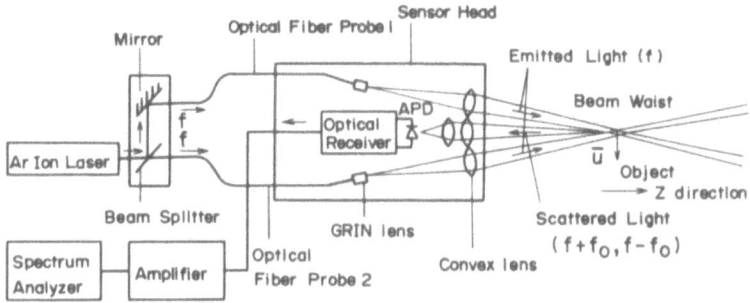

Fig.13   Fiber-optic differential heterodyne system

reference beam heterodyne (Fig.8).

$$\omega_0 = \frac{4\pi U}{\lambda_i} \sin \theta$$

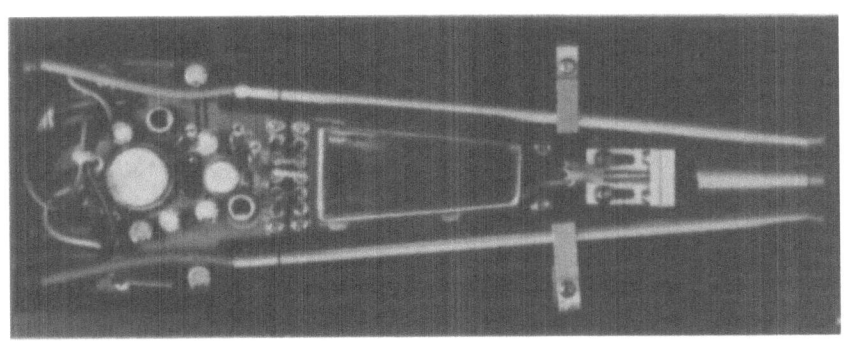

Fig.14   Probe structure of differential heterodyne system

If we have the error $\Delta\theta$, then the error in ω_0 ($\Delta\omega_0$) is

$$\Delta\omega_0 \simeq \frac{4\pi U}{\lambda_i} \cos\theta . \Delta\theta \tag{5}$$

In case of differential heterodyne with the probe shown in Fig.14,

$$\omega_0 = \frac{4\pi U}{\lambda_i} \sin (\frac{\pi}{2} - \theta) = \frac{4\pi U}{\lambda_i} \cos \theta$$

With the error in θ ($\Delta\theta$), $\Delta\omega_0 \simeq \frac{4\pi U}{\lambda_i} \sin\theta . \Delta\theta$ (6)

In practice, θ is a small angle, therefore $\Delta\omega_0$ in eq.(5) is much larger than that of eq.(6).

4.   APPLICATIONS OF OPTICAL FIBER LDV [7).8)]

We limit ourselves to the particular fields of application. One is for the manufacturing process applications and the other is for the medical method. From eq.(3) in Chapter 2, the angle between pick up optical path and the direction of the moving object it has to be precisely fixed to get system is shown in Fig.15. This type is the optical configuration of differencial heterodyne system. The optical probe emitts two illuminating laser beams. Optical instrumentation, in this case, the gas laser is mounted in the upper part of this unit and the electrical signal processing unit is in the lower part. The velocity ranges are (1) 1 m/min ~ 20 m/min, (2) 10 m/min ~ 200 m/min, (3) 100 m/min ~ 2,000 m/min and the velocity (m/min) is digitally displayed. In Fig.15, the application examplers for the manufacturing process in the industry are listed up.

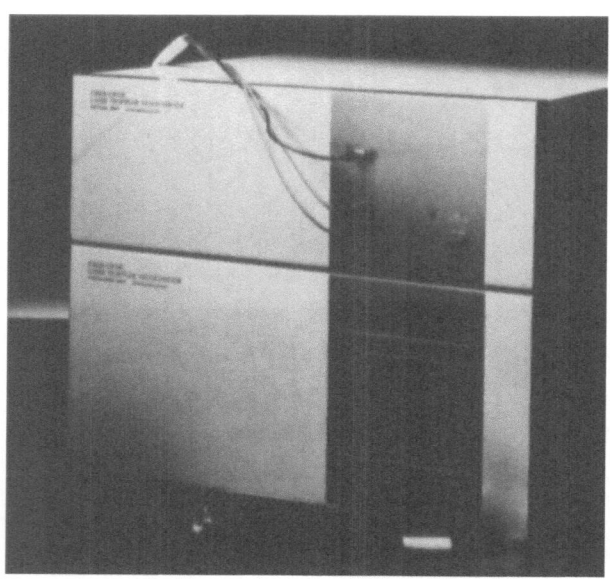

Fig.15 Commercially available OFLDV

For the metal industry, OFLDV are used to control of continuous roll milling or to control the uniformity of the coating material of the metal sheet. For the paper and film industry, the system is used to control the rolling speed. Fig.17 indicates the measurment of hot steel pipe in the metal industry. In all cases, non contact method of OFLDV system is the great advantage.

Another field application is for the medical electronics. The blood stream velocity measurement in the blood circulation system is one of the most valuable information for medical science. The velocity measurement by OFLVD system has the following advantages; high special resolution

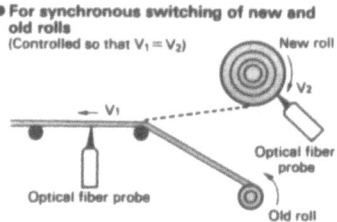

- For velocity control of continuous rolling
- For velocity measurements before and after rolling

- For length measurement and synchronous cutting of extruding lines for steel plate and nonferrous materials

- For synchronous switching of new and old rolls
(Controlled so that $V_1 = V_2$)

Fig. 16 Application examples in the manufactering processes

Fig.17 The measurement of hot
steel pipe

(~100 m); high temporal resolution
(~8 msec); excellent accessibility
to objects. The system is shown in
Fig.18. This system is the
reference beam heterodyne system,
almost as same as the system shown
in Fig.11. The probe was specially
designed for the velocity measurement of blood as shown in Fig.19.
In order to fix the angle between
optical fiber and the direction of
blood stream, special mount was
designed. Heparin (500 units/kg)
was administered intravenously to
prevent coagulation. The experimental result is shown in Fig.20.
The experiment was performed for
the blood stream of the artery of
an adult mongrel dog.

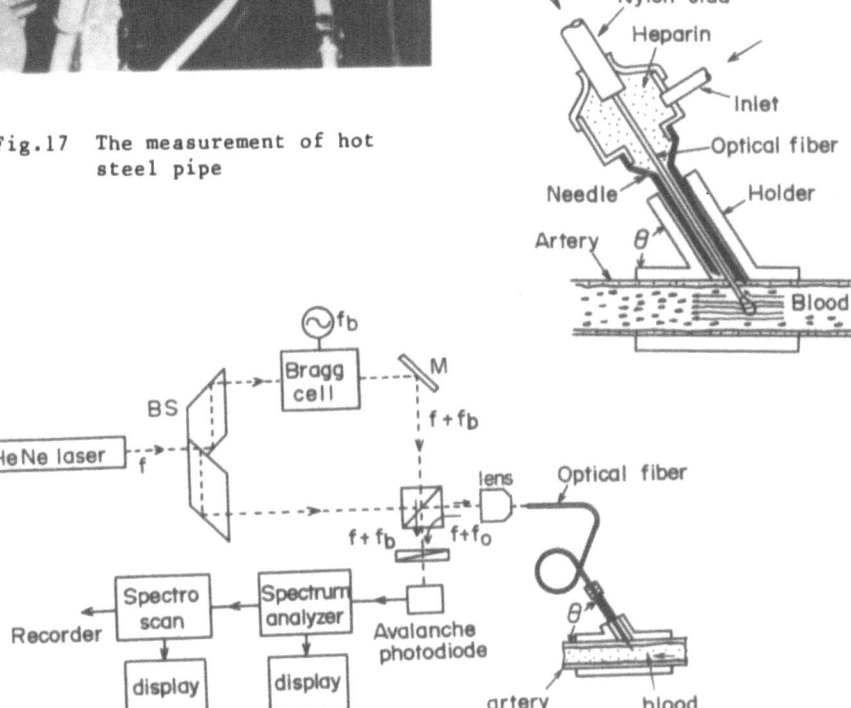

Fig.18 Reference beam heterodyne system
for the blood velocity

Fig.19 Special probe structure

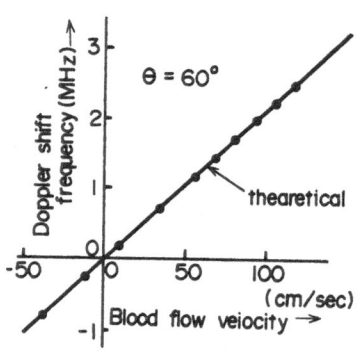

Fig.20 Experimental result
of blood stream

5. FURTHER IMPROVEMENTS OF OPTICAL FIBER LDV SYSTEM

First of all, the number of optical components has to be reduced to improve the reliability of the system. This means the integration of optical components such as beam splitter, polarizer, acousto-optic modulator. In order to enhance the SN ratio, single-mode polarization maintaining fiber should be used throughout the system. Laser power should be raised to increase the detected signal level for measuring the flow rate of gas. The problem of replacing the gas laser with the semiconductor laser is its coherency. In Fig.11, the coherency of semiconductor laser was improved by the external mirror. Another approach is to device the structure of LD. Quantum well structure with distributed feed-back may be one of the best structure for the highly coherent source. Electronic feed-back loop to the laser diode by wide band amplifier to reduce the FM noise of spontaneous emission can be realized by the opto-electronic integrated circuit on a semiconductor chip.

Reference
(1) T.S.Durrani and C.A.Greated "Laser Systems in Flow Measurement" Plenum Press, New York (1977)
(2) F.Durst, A.Melling and J.H. Whitelaw "Principles and Practice of Laser-Doppler Anemometry" Academic Press, London (1979)
(3) Association of the Measurements of Flow,ed. "The Basics and the Applications of LDV" Nikkan-Kogyo Shinbunsha, Tokyo (1978)(in Japanese)
(4) K.Kyuma et al, "Laser Doppler velocimeter with a novel optical fiber probe" Applied Optics, 20 , pp.2424~27, (1981)
(5) K.Kyuma et al, "Fiber-optic laser Doppler velocimeter using an external-cavity semiconductor" A.P.L. 45, pp.1005~1006, (1984)
(6) H.Nishihara et al, "Use of a laser diode and an optical fiber for a compact laser-Doppler velocimeter" Optics Letters, 9, pp.65~ 67 (1984)
(7) Laser Doppler velocimeter, Catalogue of Mitsubishi Electric Corp., (1985)
(8) H.Nishihara et al., "Optical-fiber Doppler velocimeter for high-resolution measurement of pulsatile blood flows" Appl. Optics, 21, pp.1785 ~1790 (1982)

POLARIZATION PHENOMENA IN OPTICAL FIBERS

T. OKOSHI

DEPARTMENT OF ELECTRONIC ENGINEERING, UNIVERSITY OF TOKYO,
7-3-1 HONGO, BUNKYO-KU, TOKYO 113, JAPAN

1. INTRODUCTION

1.1 Degeneration of polarization modes in a single-mode optical fiber

In ordinary axially-symmetrical single-mode fibers, two mutually-independent orthogonal HE_{11} modes can be propagated. In the framework of Cartesian coordinates, these are the HE_{11x} mode which has the principal electric field component in the x-direction, and the HE_{11y} mode which has principally the y-component of the electric field. We hereafter omit the suffix 11 for simplicity and call these HE_x and HE_y modes. On the other hand, any linearly-polarized HE_{11} mode can be expressed as a combination of two circularly-polarized, clockwise and counter-clockwise rotating HE_{11} modes. Hence, we may equivalently say that in an axially symmetrical fiber, two mutually-independent, circularly-polarized, clockwise and counter-clockwise rotating modes can be propagated. We hereafter call these clockwise and counter-clockwise HE_{11} modes the HE^+ mode and HE^- mode.

If the fiber is completely round, all of the HE_x, HE_y, HE^+ and HE^- modes are degenerated; i.e., no difference exists among their propagation constants. Moreover, if the fiber is completely straight, no coupling exists between these modes. This means that if one of these modes is launched at the input, the same mode will appear at the exit.

1.2 Polarization states in an axially nonsymmetrical fiber

However, an actual optical fiber is neither completely axially-symmetrical nor completely straight. As the result the polarization state of the propagated wave is subject to unstable fluctuation when ambient condition changes. In a multimode fiber, such instability usually causes little trouble except for its possible effect on modal noise. However, in a single-mode fiber, the following problems arise (1)(2).

(1) In an optical fiber communication or optical fiber sensing system, the received signal level fluctuates (usually, the sensitivity is degraded) when the receiver is sensitive to the polarization. This situation takes place when an optical IC (integrated circuit) is used in the receiver, or when the heterodyne-type or homodyne-type optical detectors are used, in which a matching of the polarization state is required between the received signal and the local oscillator (3).

(2) In measurements using a single-mode fiber, such as

magnetooptic current sensing or laser gyroscope, the
polarization instability deteriorates the measurement
accuracy or the bit-error rate.
(3) Even when a fiber is designed to be axially symmetrical,
slight elliptical deformation exists. This residual
ellipticity separates the propagation constants of two
orthogonal HE_{11} modes which otherwise degenerate with each
other, and causes the so-called polarization mode
dispersion in the group delay. (This effect, however,
causes a trouble usually only in high-bit-rate, long-
distance optical fiber communications.)

When only the polarization-mode dispersion (the group-delay
difference: item (3)) is the problem, a direct solution is to
fablicate the fiber as completely circular as possible.
Various efforts toward this target have been reported (4)(5)
(6)(7).

However, in many fiber optic measurement or communications
systems, the use of a round fiber does not solve the problem.
In many cases the solution is given by the use of a
polarization-maintaining fiber over the entire length of the
fiber channel, or by a polarization-state control scheme (8).
In addition, the so-called polarization-diversity (9) can be
used effectively in heterodyne/homodyne-type communications;
however, so far not in measurement schemes.

The purpose of this paper is to review and compare various
polarization control techniques such as the polarization
maintaining fibers and the polarization-state control schemes.
In the following we start with the consideration of the
general "state-of-polarization" (often abbreviated as SOP for
short) of light in a single-mode optical fiber.

2. MATHEMATICAL EXPRESSIONS OF GENERAL STATE-OF-POLARIZATION
2.1 Completely polarized, partially polarized, and unpolarized lights

We consider first a completely monochromatic (coherent)
light. If we express its electric field by a vector
$\underline{E}(x,y,z,t)$ and observe its motion at a fixed point (x,y,z),
the end point of the vector traces along a straight line, or a
circle, or an ellipse, i.e., generally speaking, an ellipse.
Such a light is called a completely polarized light, and the
above three cases are referred to as linear polarization,
circular polarization, and elliptical polarization,
respectively.

On the other hand, an "unpolarized" light also exists. A
typical example is a thermal radiation from an isotropic
surface of a heated body. In such a case the end point of the
vector moves entirely randomly. A light having properties
between these two extremities, i.e., the completely polarized
and unpolarized lights, is called a partially polarized light.
In this case the movement of the vector end is neither
completely regular nor completely random.

Any existing light is more or less partially polarized, and
can be expressed as a sum of completely polarized and
unpolarized lights (10).

2.2 Expression of partially coherent light (coherency matrix)

We consider a light which is propagated in z-direction, having an average angular frequency ω and certain amount of frequency fluctuations (in other words, phase fluctuation) around it. The electric field vectors of such a light in the x and y directions at a fixed point can be expressed generally as

$$E_x(t) = a_x(t) \cos[\omega t + \dot{\phi}_x(t)]$$

$$= a_x(t) \, \mathrm{Re}[\exp\{j\omega t + j\phi_x(t)\}] \qquad (1)$$

$$E_y(t) = a_y(t) \cos[\omega t + \phi_y(t)]$$

$$= a_y(t) \, \mathrm{Re}[\exp(j\omega t + j\phi_y(t)\}] , \qquad (2)$$

where $a_i(t)$ and $\phi_i(t)$ (i=x, y) denote amplitude and phase noises in the respective directions, respectively. These quantities are constant in completely polarized lights.

Next, we consider the intensity of such a partially polarized light. For the discussion in later subsections, we consider a case when a phase delay ε is given to the y-component of the electric field, and compute the intensity to be observed after the light passes through a linear polarizer tilted from the x-direction by θ. Then the electric field in the θ-direction can be expressed as

$$E_\theta(t;\varepsilon) = E_x \cos\theta + E_y \, e^{-j\varepsilon} \sin\theta . \qquad (3)$$

Therefore, the intensity is given as

$$I(\theta,\varepsilon) = E_\theta(t;\varepsilon) \cdot E_\theta(t;\varepsilon)^*$$

$$= J_{xx}\cos^2\theta + J_{yy}\sin^2\theta + J_{xy}e^{j\varepsilon}\cos\theta \sin\theta$$

$$+ J_{yx}e^{-j\varepsilon}\sin\theta \cos\theta , \qquad (4)$$

where J_{xx}, J_{yy}, J_{xy}, J_{yx} are components of the following matrix:

$$J = \begin{bmatrix} \langle E_x E_x^* \rangle & \langle E_x E_y^* \rangle \\ \langle E_y E_x^* \rangle & \langle E_y E_y^* \rangle \end{bmatrix} = \begin{bmatrix} \langle a_x^2 \rangle & \langle a_x a_y e^{j(\phi_x-\phi_y)} \rangle \\ \langle a_x a_y e^{-j(\phi_x-\phi_y)} \rangle & \langle a_y^2 \rangle \end{bmatrix} . \qquad (5)$$

Matrix J is called the coherency matrix. The diagonal components J_{xx} and J_{yy} are real quantities expressing light intensities in the respective directions. The sum of these, i.e., the trace of the matrix:

$$\mathrm{Tr}[J] = J_{xx} + J_{yy} = \langle E_x E_x^* \rangle + \langle E_y E_y^* \rangle \qquad (6)$$

expresses the total intensity. The off-diagonal components $J_{xy}=\langle E_x E_y^* \rangle$ and $J_{yx}=\langle E_y E_x^* \rangle$ are always complex-conjugate with each other.

The statements in the above paragraph convinces us that the number of degree of freedom of an arbitrarily polarized light is four including total intensity, and the number is three if we disregard the total intensity and only think of the state-of-polarization. It is known that all the four components of J can be expressed as simple functions of six values of I (E9.(4) measured for various combinations of θ and ε (10).

2.3 Decomposition of a partialy polarized light to completely polarized and unpolarized lights

Any partially polarized light can be regarded as a sum of completely polarized and unpolarized lights.

We note first that for a completely unpolarized light $I(\theta,\varepsilon)$ is irrelevant to θ and ε;

$$I(\theta, \varepsilon) = const. \tag{7}$$

Hence, if we write $J_{xx}+J_{yy}=I_0$, the coherency matrix

$$J = \frac{1}{2} I_c \begin{bmatrix} 1 & 0 \\ 0 & 1 \end{bmatrix}. \tag{8}$$

On the other hand, for a completely polarized light, the relations to be satisfied are $J_{xy}=J_{yx}^*$, and

$$J_{xy}J_{yx} = \langle a_x a_y e^{j(\phi_x-\phi_y)}\rangle \langle a_x a_y e^{-j(\phi_x-\phi_y)}\rangle \tag{9}$$

$$= \langle a_x^2\rangle \langle a_y^2\rangle = J_{xx}J_{yy}.$$

In other word, for a completely polarized light,

$$det[J] = J_{xx}J_{yy} - J_{xy}J_{yx} = 0 \tag{10}$$

We should also note that J_{xx} and J_{yy} are both real.

With the above conditions in mind, we show that any J matrix can be expressed uniquely as

$$J = J^{(1)} + J^{(2)} \tag{11}$$

where $J^{(1)}$ and $J^{(2)}$ denote the completely polarized and unpolarized components, hence

$$J^{(1)} = \begin{bmatrix} B & D \\ D^* & C \end{bmatrix}, \quad J^{(2)} = \begin{bmatrix} A & 0 \\ 0 & A \end{bmatrix} \tag{12} \tag{13}$$

where $A \geq 0$, $B \geq 0$, $C \geq 0$, and $BC-DD^*=0$. We can obtain immediately from Eqs.(11)-(13)

$$A + B = J_{xx}, \quad D = J_{xy}$$

$$D^* = J_{yx}, \quad A + C = J_{yy} \tag{14}$$

where the J components denote those of the given partially polarized light. From Eq.(14) and the condition $BC-DD^*=0$, we obtain

$$(J_{xx} - A)(J_{yy} - A) - J_{xy}J_{yx} = 0. \tag{15}$$

Hence, if we take the root with negative sign,

$$A = \frac{1}{2}(J_{xx} + J_{yy}) - \frac{1}{2}\sqrt{(J_{xx} + J_{yy})^2 - 4|J|} \tag{16}$$

and consequently,

$$B = \frac{1}{2}(J_{xx} - J_{yy}) + \frac{1}{2}\sqrt{(J_{xx} + J_{yy})^2 - 4|J|} \tag{17}$$

$$C = \frac{1}{2}(J_{yy} - J_{xx}) + \frac{1}{2}\sqrt{(J_{xx} + J_{yy})^2 - 4|J|} \tag{18}$$

$$D = J_{xy}. \tag{19}$$

If we take the root with positive sign in computing A(Eq.(16)), we find that both B and C must be negative. Hence A, B, C and D are determined uniquely, and the foregoing statement has been proved.

2.4 Degree of polarization

The total intensity of a partially polarized light is given as

$$I_{total} = Tr[J] = J_{xx} + J_{yy} \tag{20}$$

(see Eq.(6)), whereas the intensity of its completely polarized component is expressed, from Eq.(12), as

$$I_{pol.} = Tr[J^{(1)}] = B + C$$
$$= \sqrt{(J_{xx} + J_{yy})^2 - 4|J|}. \tag{21}$$

The ratio of I_{pol} and I_{total} is called the degree of polarization, and is expressed generally, from the above two equations, as

$$P = \frac{I_{pol.}}{I_{total}} = \sqrt{1 - \frac{4|J|}{(J_{xx} + J_{yy})^2}}. \tag{22}$$

The degree of polarization P has the following features.
(1) For a completely polarized light, J=0 and hence P=1.
(2) For an unpolarized light, P=0.
(3) For any (partially polarized) light, $0 \leq P \leq 1$ because $|J| \leq J_{xx}J_{yy} \leq (J_{xx}+J_{yy})^2/4$ always holds.

3. GRAPHICAL EXPRESSIONS OF COMPLETELY POLARIZED STATE

If we strict our discussion to the completely polarized light, there are two methods for graphical expression of the state of polarization (SOP).

3.1 Poincaré sphere

Poincaré proposed in 1892 a spherical graph for expressing general SOPs, which is now well known as the Poincaré sphere (11). The SOP of any completely polarized light, i.e., an arbitrary elliptical polarization, can be represented by a spot on the Poincaré sphere (12)(13) shown in Fig.1, where
(1) The longitude represents twice the tilt angle α of the elliptical polarization, and

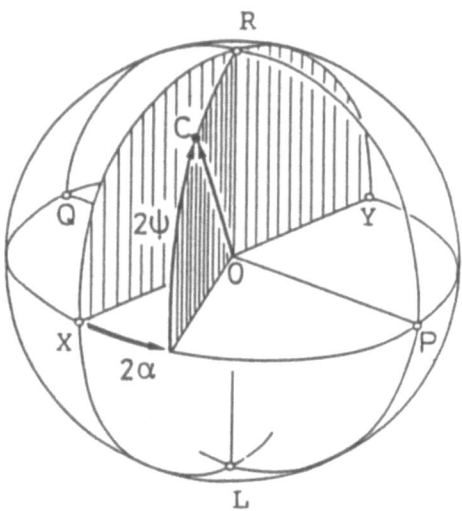

Fig.1 The Poincaré sphere (after Ulrich (12)).

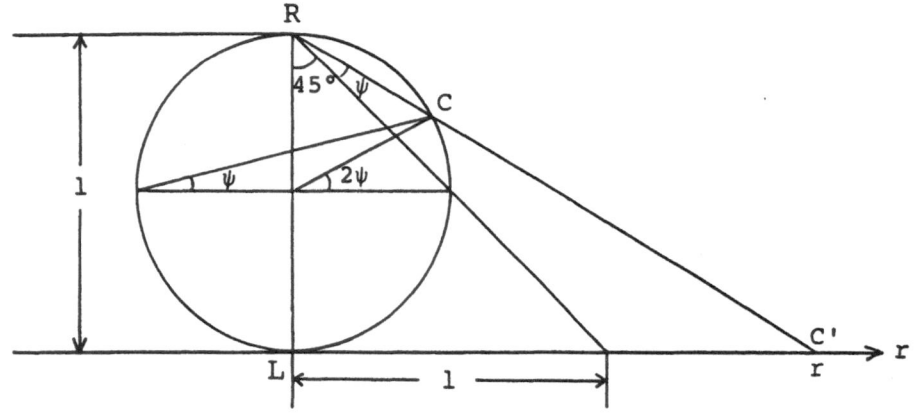

Fig.2 Cross-section of the Poincaré sphere and its stereoscopic projection onto a plane touching the sphere at its South Pole.

(2) The angle ψ in Fig.1, i.e., one half of the latitude,
   corresponds to the ellipticity (12).
Some characteristic spots on the sphere: R(North Pole),
L(South Pole), and X, Y, P, Q on the equator, correspond to
   R: Clockwise circular polarization,
   L: Counterclockwise circular polarization,
   X: Horizontal linear polarization,
   Y: Vertical linear polarization,
   P: Linear polarization with +45° tilt angle,
   Q: Linear polarization with -45° tilt angle,
respectively.
   The Poincaré sphere has widely been used to express the SOP
of a light. However, it has a drawback in that it can not be
drawn or printed on a paper because it is curved.

3.2 Planar charts for expressing an SOP of completely
    polarized light
   So far, basically two methods of expressing an SOP on a
planar chart have been proposed. In 1954, Jerrard wrote a
tutorial paper which is often cited at present (13). In this
paper Jerrard explained the Poincaré sphere as the retro-
stereoscopic projection from the $(\dot{a}_y/\dot{a}_x)$ complex plane. This
plane can be used to express any SOP as discussed by Takenaka
(14). The second chart proposed by Okoshi is obtained as a
stereoscopic projection of the Poincaré sphere from a
fictitious light source at the North Pole R onto a plane
touching the sphere at the south Pole L; see Fig.2 (15).
   Figure 3 shows this second chart drawn by a computer. Some
key parameter values are also shown to facilitate the usage.
   This figure indicates that four groups of characteristic
curves exist: equi-R circles, equi-ϕ circles, the concetric
equi-ψ circles, and radial straight lines giving equi-α
contours. The implications of these parameters are described
in the literature (15).

4. FLUCTUATION OF STATE-OF-POLARIZATION (SOP) IN SINGLE-MODE
   FIBERS

4.1 Measured SOP fluctuation in single-mode fibers
   Until early 1980s, it was believed by most of the
specialists that the SOP of light at the exit end of a
single-mode fiber would fluctuate very drastically if the
fiber was an ordinary round one, in other words, if it was not
a polarization-maintaining type. Such a common feeling
probably originated from that at that time thinly coated
fibers (before cabling) were usually used in laboratory
experiments.
   In the fall of 1982, British Telecom researchers reported
that the SOP fluctuation was not so drastic when the fiber was
cabled and enstalled (16). This report was significant
historically because it seems to have switched the main stream
of the fiber polarization research from the polarization-
maintaining fiber to more flexible countermeasures against the
SOP fluctuation, including various SOP-correction devices.
(We should note, however, that when the ambient condition of a
fiber, the temperature for example, is disturbed drastically,
the SOP also fluctuates appreciably.)

Fig.3 The proposed chart for expressing a state-of-polarization
drawn by a computer (after Okoshi (15)).

4.2 Degradation of degree of polarization in single-mode fibers

Within the authors' knowledge, nobody has reported the measurement of the degradation in the degree of polarization P in a single-mode fiber as a function of its length. If such a measurement is performed, the result will depend critically upon the linewidth of the light source as well as the characteristics of the fiber in a very complex manner. However, such a measurement would be worthwhile to perform at least to evaluate the validity of the theory.

5. STATE-OF-POLARIZATION CONTROL DEVICES
5.1 Classification of SOP-control schemes
The SOP-control devices have been developed mainly to match the SOPs of signal and local oscillator lights in heterodyne or homodyne receivers (8). They convert an arbitrary, often fluctuating SOP of an incident signal light to a prescribed (in most cased linear) polarization equal to that of the local oscillator light, or convert a linear polarization of the local oscillator light to an SOP equal to the fluctuating signal SOP. In the following we consider the former case.

So far six SOP-control schemes have been proposed and experimented with. The SOP controlling elements used in these schemes are, in a chronological order, electromagnetic fiber squeezers (17), electrooptic crystals (18), rotatable fiber coils (19), rotatable phase plates (20), Faraday rotators (21), and rotatable fiber cranks (22). These schemes can be classified into three groups according to their SOP-conversion principles called in Ref.(8): Type I (fiber squeezers and electrooptic crystals), Type II (rotatable fiber coils, phase plates and fiber cranks), and Type III (Faraday rotators). In addition, many variations and combinations are possible between these basic three types.

The six polarization-state control schemes mentioned above are tabulated in Table 1.
5.2 Features of various SOP-state control schemes
Various schemes are compared in the following with respect to four technical requirements: (1) insertion loss, (2) endlessness in control, (3) temporal response, and (4) presence or absence of mechanical fatigue.
(1) Insertion Loss
The requirement for low insertion loss can practically be satisfied only by all-fiber-type devices. This is because an polarization-state controller is most probably used in a single-mode fiber circuit, in which the loss will increase appreciably (at least several dB, or more) if a non-fiber element is inserted. An all-fiber-type device can be spliced with much lower insersion loss (≈ 0.2dB). Among the six devices described above, four devices can satisfy this condition (see Table 1).
(2) Endlessness in Control
This is an important requirement because the nature of the polarization-state fluctuation in actual fiber channels is not known, and hence "resetting" might become necessary in those attempts in which the control range is limited. Among the six

Table 1. Features of six polarization-state control schemes proposed
so far (after Okoshi (8)).

Polarization-state control schemes	Insertion loss	Endlessness in control	Temporal response	Mechanical fatigue
Fiber squeezers (Ulrich)	Low	No	Medium	Present
Electrooptic crystals (Kubota et al.)	High	No	Fast	Absent
Rotatable fiber coils (Lefevre)	Low	No	Slow	Present
Phase plates (Imai et al.)	Medium	Yes	Slow	Small
Faraday rotators (Okoshi, Cheng, et al.)	Low	No	Fast	Absent
Rotatable fiber cranks (Okoshi, Fukaya, et al.)	Low	Yes	Slow	Present

Table 2. Classification of various polarization-maintaining
optical fibers.

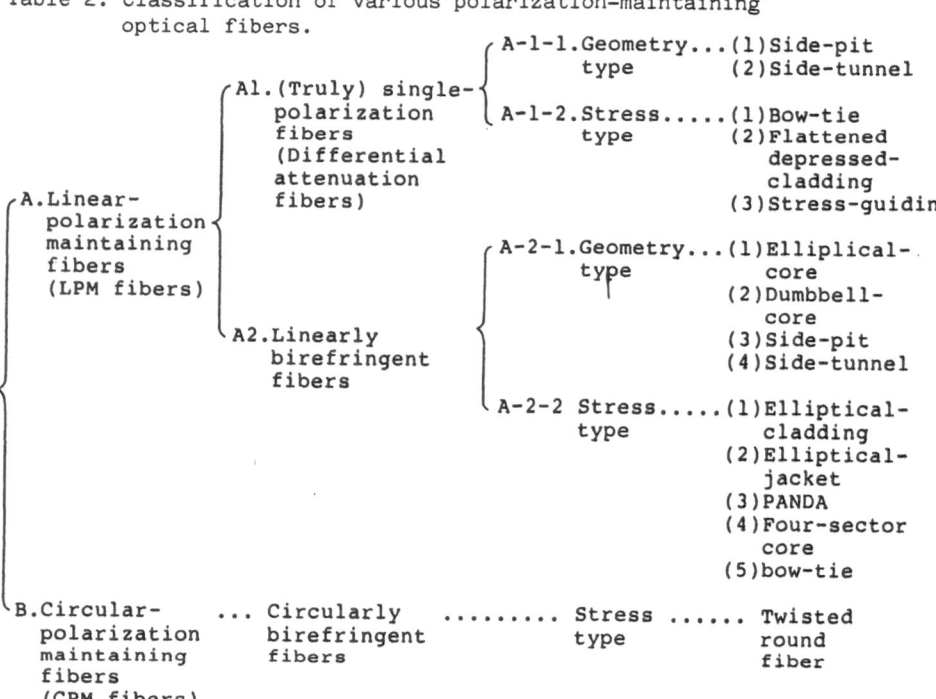

devices listed in Table 1, only phase-plate and rotatable
fiber-crank schemes can satisfy this requirement.
(3) Temporal Response
  All the mechnical schemes have poor temporal response as
compared with all electronic ones, i.e., the electro-optic and
Faraday devices.
(4) Presence or Absence of Mechanical Fatigue
  All the mechanical schemes have more or less the possibility
of mechanical fatigue.

6. POLARIZATION-MAINTAINING OPTICAL FIBERS
  In many cases, a more essential and complete solution to the
SOP fluctuation is the use of the polarization-maintaining or
single-polarization fibers (1)(2), which is the subject of
this section.
6.1 Classification of polarization-maintaining optical fibers
  A comment should first be given as to the presently existing
confusion in terminology. Various terms such as
polarization-maintaining fibers, polarization-holding fibers,
single-polarization fibers or SPSM (single-polarization
single-mode) fibers are now used in this area, in a somewhat
confused manner, without widely-accepted definitions. In the
following, the present author uses tentatively "single-
polarization fiber" to denote the so-called "truly" single
polarization fiber, and use "polarization-maintaining fiber"
to denote generically the single-polarization fiber and
"birefringent" fiber.
  According to the above terminology, various polarization-
maintaining fiber schemes are classified as shown in Table 2
(23). Short comments on each type follow:
A. Linear-Polarization Maintaining Fibers (LPM Fibers)
  Fibers belonging to this category are designed so that only
one of the two linearly polarized modes (HE_x or HE_y modes) can
be propagated. These fibers can further be classified as A-1
and A-2:
A-1. Truly Single-Polarization Fibers (Differential
     Attenuation Fibers)
  The fiber is designed so that the transmission losses for
the HE_x and HE_y modes are largely different; hopefully so that
one propagation mode is cut off in a specific frequency (i.e.,
wavelength) range. Such a scheme was first called the
"absolutely single-polarization fiber" (2), and later also
called the "truely single-polarization fiber," or simply the
"single-polarization fiber," or the "differential attenuation
fiber" in contrast to the "birefringent fiber." This scheme
can further be classified as
    A-1-1. Geometry-Induced Single-Polarization Fibers,
           examples of which are side-pit and side-tunnel
           fibers, and
    A-1-2. Stress-Induced Single-Polarization Fibers, example
           of which is the differential-attenuation-type bow-
           tie fiber.
A-2. Linearly Birefringent Fibers
  In this case the fiber is designed so that the propagation

constants of HE$_x$ and HE$_y$ modes are different. As has been predicted by theory, when the propagation-constant difference increases, the two polarization modes are "decoupled," and the launched one will remain travelling with a very little power transfer to the other over an appreciable length of the fiber. This scheme can further be classified as

A-2-1. Geometry-Induced Birefringent Fibers, in which the propagation-constant difference is produced by an axially nonsymmetrical refractive index distribution. The elliptical and dumbbell core fibers belong to this category. However, we should note that the stress-induced birefringence (see the next item) also exists in such fibers, and also that the truly single-polrization fibers such as side-pit and side-tunnel fibers can be used also as birefringent fibers at frequencies outside the truly single-polarization region.

A-2-2. Stress-Induced Linearly Birefringent Fibers, in which the propagation-constant difference is produced by an axially nonsymmetrical internal stress. The elliptical cladding fiber, elliptical jacket fiber, PANDA fiber, four-sector-core fiber, and birefringence-type bow-tie fiber are included in this category.

B. Circular-Polarization Maintaining Fibers (CPM Fibers)

A round (axially symmetrical) fiber is twisted to produce a difference between the propagation constants of the clockwise and counter-clockwise circularly polarized HE$_{11}$ modes. Thus, these two circular polarization modes, i.e., the HE$^+$ and HE$^-$ modes, are decoupled. These fibers, therefore, may also be called the "circularly birefringent" fibers. (We should note that a "truly single circular-polarization fiber" has not ben invented.)

6.2 History of research and development of polarization-maintaining fibers

As early as the mid 1970s, some of the optical fiber specialists who originally came from microwave research area believed that axially nonsymmetrical single-mode fibers would become common in future, like rectangular waveguides at microwave frequencies. Vigorous research toward such fibers, however, started in late 1970s. Desciption of the history of the research and development of polarization-maintaining fibers will offer the best way to understand various ideas behind the polarization-maintaining structures. The short history can conveniently be divided into the following four periods:

1976-1977

In this first period, the modal birefringene caused by an unintentional elliptical deformation of the core was the principal subject of the research. Intentional polarization-maintaining schemes were not investigated.

1978-1979

Various kinds of linear-polarization maintaining (LPM) fiber schemes, such as elliptical core fibers, dumbbell core fibers, stress-induced (elliptical cladding) fibers were proposed, mainly by the investigators of Bell Laboratories. The side-pit fiber was proposed in Japan; this was the first proposal of the concept of the truly single-polarization (differential attenuation) fiber.

1980-1981
The center of research seems to have once moved from the United States to Japan and Europe in this period. Researchers of CNET (France) proposed an entirely new scheme, the circular-polarization maintaining (CPM) fiber, in the fall of 1980 (24). Various novel LPM fibers, such as novel type of elliptical cladding fiber, elliptical jacket fiber, side-pit fiber (25)(26), PANDA fiber (27), and four-sector-core fiber were proposed and/or experimented with in Japan.

1982-1985
The development of various fibers continued. On the other hand, British Telecom people reported in the fall of 1982 that the polarization-state fluctuation was not so drastic when the fiber was cabled and enstalled (16). This report led to a new discussion on the polarization-maintaining capability that would be required in practical systems.

Three new proposals for the polarization-maintaining fiber structure appeared in 1982. The first is the side-tunnel fiber (28) in which the refractive-index pits in side-pit fiber is replaced by vacant tunnels. The second is bow-tie fiber (29)(30) which is a version of the PANDA fiber but the stress-producing part is shaped like a bow-tie to maximize the birefringence. Thirdly, the possibility of realizing a truly single-polarization fiber taking advantage of the stress-induced birefringence was discussed by Eickhoff (31).

In 1983, experiments of stress-induced truly single-polarization (differential attenuation) fibers were reported; these were the measurement of the truely single linear-polarization characteristics of "bent" bow-tie fibers (32)(33), and that of the "flattened" depressed-cladding fiber by Simpson et al. (34)(35). The PANDA fiber was improved further (36)(37). Snyder and Rühl proposed a new concept of the anisotropic differential attenuation fiber (38).

Two other significant achievements were reported in 1983 by Birch et al. (39) and Okamoto et al. (40). The first paper demonstrated experimentally that a truly single-polarization transmission was possible regardless the wavelength. The second paper showed theoretically that a low-dispersion (several ps/km.nm) was obtainable over a wide wavelength range (1.5-1.7μm) with bow-tie or PANDA structure.

7. SUMMARY AND CONCLUSION
The basic properties of the polarization state and its fluctuation, the mathematical and graphical representations, actual fluctuation of state-of-polarization in single-mode fibers, various polarization-state control schemes, and finally the classification and characteristics of various polarization-maintaining fibers have been described.

240

REFERENCES

1. Kaminov LP: Polarization in Optical Fibers. IEEE Jour. of Quantum Electron., Vol.QE-17, No.1, pp.15-17, Jan. 1981.
2. Okoshi T: Single-Polarization Single-Mode Optical Fibers. IEEE Jour. of Quantum electron, Vol.QE-17, No.6, pp.879-884, June 1981.
3. Okoshi T: Heterodyne and Coherent Optical Fiber Communications: Recent Progress (Invited). IEEE Trans. on Microwave Theory Tech., Vol.MTT-30, No.8, pp.1138-1149, August 1982.
4. Schneider H, Harms H, Rapp A, and Aulich H: Low-Birefringence Single-Mode Fibers; Preparation and Polarization Characteristics. Appl. Opt., Vol.17, No.19, pp.3035-37, Oct. 1, 1978.
5. Normann SR, Payne DN, Adams MJ and Smith AM: Fabrication of Single-Mode Fibers Exhibiting Extremely Low Polarization Birefringence. Electron. Lett., Vol.15, No.11, pp.309-311 May 24, 1979.
6. Barlow AJ, Payne DN, Hadley MR, and Mansfield RJ: Production of Single-Mode Fibers with Negligible Intrinsic Birefringence and Polarization Mode Dispersion. European Conf. on Opt. Commun. (ECOC'81), Paper No.2-3, September 8-11, 1981, Copenhagen.
7. Barlow AJ, Ramskov-Hansen JJ, and Payne DN: Birefringence and Polarization-Mode Dispersion in Spun Single-Mode Fibers. Appl. Opt., Vol.20, No.17, pp.2962-2968, Sept. 1, 1981.
8. Okoshi T: Polarization-State Control Schemes for Heterodyne or Homodyne Optical Fiber Communications (invited). IEEE/OSA J. Lightwave Tech., Vol.LT-3, No.6, pp.1232-1237, Dec. 1985.
9. Okoshi T, Ryu S and Kikuchi K: Polarization-Diversity Receiver for Heterodyne/Coherent Optical Fiber Communications. Proc. of IOOC'83, Paper No.30C3-2, June 27-30, 1983, Tokyo.
10. Born M and Wolf E: Principles of Optics. Pergamon Press, 1975 (Fifth Edition).
11. Poincare H: Theorie Mathematique de la Lumiere. Gauthiers-Villars, Paris, 1982, Vol.2, Chapter 12.
12. Ulrich R and Simon A: Plarization Optics of Twisted Single-Mode Fibers. Applied Optics, Vol.18, No.13, pp.2241-2251, July 1, 1979.
13. Jerrard HG: Transmission of Light through Birefringent and Optically Active Media: the Poincare Sphere. J. Opt. Soc. Am., Vol.44, No.8, pp.634-640, Aug. 1954.
14. Takenaka H: A Unified Formalism for Polarization Optics by Using Group Theory. Japan J. Appl. Phys., Part I --- Vol.12, NO.2, pp.226-231, Feb. 1973, Part II --- Vol.12, No.11, PP1729-1731, Nov. 1973.
15. Okoshi T: A Planar Chart Equivalent to Poincare Sphere for Expressing State-of-Polarization of Light. to be published in IEEE/OSA Jour. of Lightwave Tech.

16. Hodgkinson TG, et al.: Experimental 1.5um Coherent Optical Fiber Transmission System. Paper No.AXII-5 (Technical Digest pp.414-418), ECOC'82, Sept. 1982, Cannes, France.
17. Ulrich R: Polarization Stabilization on Single-Mode Fiber. Appl. Phys. Lett., Vol.35, No.11, pp.840-842, December 1979.
18. Kubota M, Oohara T, Furuya K, and Suematsu Y: Electro-Optical Polarisation Control on Single-Mode Optical Fibres. Electron. Lett., Vol.16, No.15, p.573, July 17, 1980.
19. Lefevre HC: Single-Mode Fiber Fractional Wave Devices and Polarisation Controllers. Electron Lett., Vol.16, No.20, pp.778-780, September 25, 1980.
20. Imai T, Nosu K and Yamaguchi H: Optical Polarization Control Utilising an Optical Heterodyne Detection Scheme. Electron. Lett., Vol.21, No.2, pp.52-53, January 17, 1985.
21. Okoshi T, Cheng YH, Kikuchi K: New Polarization-Control Scheme for Optical Heterodyne Receiver Using Two Faraday Rotators. Electron. Lett., Vol.21, No.18, pp.787-788, August 29, 1985.
22. Okoshi T, Fukaya N, and Kikuchi K: New Polarization-State Control Device: Rotatable Fiber Cranks. Electron. Lett., Vol.21, No.20, pp.895-896, September 26, 1985.
23. Okoshi T: Review of Polarization-Maintaining Single-Mode Fiber (Invited). The 4th Int'l Conf. on Integrated Opt. and Optical Fiber Commun. (IOOC'83), Paper No.28A4-1, June 27-30, 1983, Tokyo.
24. Jeunhomme L, and Monerie N: Polarization-Maintaining Single-Mode Fiber Cable Design. Electron. Lett., Vol.16, No.24, pp.921-922, Nov. 20, 1980.
25. Okoshi T and Oyamada K: Single-Polarization Single-Mode OPtical Fiber with Refractive-Index Pits on Both Sides of the Core. Electron. Lett., Vol.16, No.18, pp.712-713, Aug. 28, 1980.
26. Hosaka T, Okamoto K, Sasaki Y, and Edahiro T: Single-Mode Fibers with Asymmetrical Refractive-Index Pits on Both Sides of Core. Electron. Lett., Vol.17, No.15, pp.191-193, March 5, 1981.
27. Hosaka T, Okamoto K, Miya T, Sasaki Y, and Edahiro T: Low-Loss Single-Polarization Fibers with Asymmetrical Strain Birefringence. Electron. Lett., Vol.17, No.15, pp.530-531, July 23, 1981.
28. Okoshi T, Oyamada K, Nishimura M, and Yokota H: Side-Tunnel Fiber: An Approach to Polarization-Maintaining Optical Waveguiding Scheme. Electron. Lett., Vol.18, No.19, pp.824-826, Sept. 16, 1982.
29. Birch RD, Varnham MP, Payne DN,and Tarbox EJ: Fabrication of Polarization-Maintaining Fibers Using Gas-Phase Etching. Electron. Lett., Vol.18, No.24, pp.1036-1038, Nov.25, 1982.
30. Ourmazd A, Birch RD, Varnham MP, Payne DN and Tarbox EL: Enhancement of Birefringence in Polarization-Maintaining Fibers by Thermal Annealing. Electron. Lett., Vol.19, No.4, pp.143-144, Feb. 17, 1983.

31. Eickhoff W: Stress-Induced Single-Polarization Single-Mode Fiber. Optics Letters, Vol.7, No.12, pp.629-631, December 1982.
32. Varnham MP, Payne DN, Birch RD, and Tarbox EJ: Single-Polarization Operation of Highly Birefringent Bow-Tie Optical Fibers. Electron. Lett., Vol.19, No.7, pp.246-247, March 31, 1983.
33. Varnham MP, Payne DN, Birch RD, and Tarbox EJ: Bend Behavior of Polarizing Optical Fibers. Electron. Lett., vol.19, No.17, pp.679-680, Aug. 18, 1983.
34. Simpson JR, Sears FM, and MacChesney JB: Single-Polarization Fiber: Topical Meeting on Opt. Fiber Commun. (OFC'83), Paper No.TuA2, Feb. 28-March 2, 1983, New Orleans.
35. Simpson JR, Stole RH, Sears FM, Pleibel W, MacChesney JB, and Howard RE: A single-Polarization Fiber. IEEE/OSA Jour. of Lightwave Tech., Vol.LT-1, No.2, pp.370-374, June 1983.
36. Shibata N, Sasaki Y, Okamoto K, and Hosaka T: Fabrication of Polarization-Mintaining and Absorption-Reducing Fibers. IEEE/OSA Jour. of Lightwave Tech., Vol.LT-1, No.1, pp.38-43, March 1983.
37. Sasaki Y, Hosaka T, Takada K, and Noda J: 8km-Long Polarization-Maintaining Fiber with Highly Stable Polarization State. Electron. Lett., Vol.19, No.19, pp.792-794, Sept. 1, 1983.
38. Snyder AW and Rühl F: New Single-Mode Single-Polarization Fiber. Electron. Lett., Vol.19, No.5, pp.185-186, March 3, 1983.
39. Birch RD, Varnham MP, Payne DN, and Okamoto K: Fabrication of a Stress-Guiding Optical Fibre. Electron. Lett., Vol.19, No.21, pp.866-867, Oct. 13, 1983.
40. Okamoto K, Varnham MP, and Payne DN: Polarization-Maintaining Optical Fibers with Low Dispersion over a Wide Spectral Range. Applied Optics, Vol.22, No.15, pp.2370-2373, Aug. 1, 1983.

Integrated Optical Sensors

R.Th.Kersten
Schott Glaswerke
Hattenbergstr. 10
D-6500 Mainz
F.R.Germany

1. Introduction

The history of Integrated Optics may be compared with the political situation in Europe during the Middle Ages: complicated and sometimes impossible to follow. The only difference can be seen in the time period: Integrated Optics is only 17 years old.

It started with a paper published in Bell System Technical Journal (December issue) in 1969 [1], where Miller published the dream of a new future of integrated circuits, which will use photons instead of electrons.

Scientists working in the fields of optics, communications and microwaves got euphoric about this new aspect and worldwide various efforts started to realize these new components. At that time the main application of this exciting field was seen in optical communications; but very soon the optical computer was discussed, too. Optical communications took a large step forward in 1970, when Corning Glass Works announced for the first time a low loss fiber (20 dB/km at a wavelength of about 800 nm)[2]. Very soon it became clear, that so-called multimode step-index fibers would not offer the expected properties with respect to bandwidth. Therefore the development in optical communications was concentrated on single-mode fibers; this was a real push for Integrated Optics, because it was already obvious, that Integrated Optic components would offer the most interesting advantages using single-mode technology.

In the middle of the 70s, however, it was claimed that graded-index multimode fibers would provide the same properties as single-mode fibers, but avoid the adjustment problems because of a much larger core diameter. This was the death for Integrated Optics almost; only signal processing people survived in the field and the effort in new developments in Integrated Optics was reduced very much. Fortunately new bandwidth limiting effects were discovered very soon with respect to graded-index fibers and with the end of the 70s Integrated Optics became again a new born research and development area.

Since that several semi-commercial Integrated Optic elements have been realized, investigated and used in prototypes. Beside the use in optical communications the new field of fiber optic sensors gave new accents to the world of Integrated Optics, i.e. the gyro on the chip [3].

The first principle proposals for Integrated Optic components given by Miller in 1969 do not differ very much from todays designs. The question, why these components are still under development and not commercially available is quite easy to answer: The technological processes to produce Integrated Optic components have not yet reached an economical stage.

2. Fundamentals

The light guiding in Integrated Optics is based - as in optical fibers - on the phenomenon of total internal reflection. Therefore the light is confined to an area of higher refractive index with respect to its surroundings. The materials usable for the realization of Integrated Optic circuits should have low loss at the working wavelength.

In principle very different waveguiding structures have been proposed and realized up to now. We have to distinguish between slab waveguides (Fig. 1a) and channel (or strip) waveguides (Fig. 1b). Unfortunately the nomenclature still is very inconsistent: slab waveguides sometimes are called two-dimensional (light is guided with respect to two dimensions, i.e. transversal and longitudinal) and channel waveguides three-dimensional (guiding in all three dimensions).

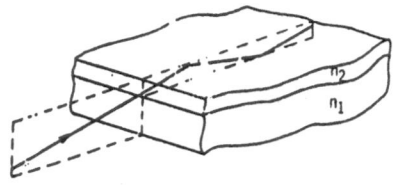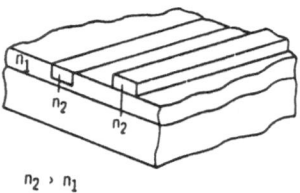

$n_2 > n_1$

Fig.1: Schematics of slab- (a) and strip-waveguide (b)

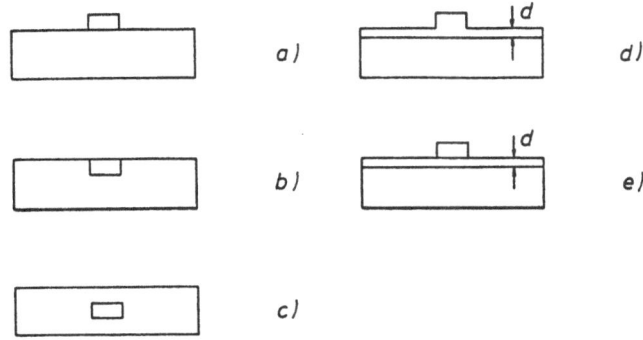

Fig.2: Different types of strip waveguides

While slab waveguides are important for fundamental investigations as well as for optical signal processing, channel waveguides are used in connection with fiber optics.

In the field of channel waveguides various different guiding
structures have been investigated (Fig. 2). The most simple
channel waveguide (Fig. 1b and Fig. 2a) is a high index strip
(mostly of rectangular shape) on top of a low index substrate.
Similar to optical fibers the number of guided modes depends
on the dimensions, the refractive indices and the wavelength.
Because the guide has no circular symmetry (like fibers), we
have to distinguish between lateral and transversal modes. The
larger the dimensions of the guide and the larger the
refractive index difference between guiding and surrouding
material, the higher will be the number of guided modes.
Therefore one disadvantage of the structure shown in Fig. 2a
becomes obvious: despite the base of the channel guide the
whole guide is surrounded by air causing a high index
difference. Therefore to build a single-mode waveguide,
dimensions have to be reduced to very small values (in the
range of some micrometers or even less); this usually results
in great technological problems.

To get around this problem, the guide can be implanted into
the substrate (Fig. 2b); only one side of the guide is adjacent to the air interface. This still is a disadvantage with
respect to guiding loss: the amount of scattered light depends
on the surface roughness as well as on the index difference.
Therefore such guides have to be prepared with high precision
to avoid additional losses: to reduce the additional loss
caused by surface roughness below 1 dB/cm, the guide/surface
roughness has to be below 50 nm!

This again can be avoided by burying the guide quite well

below the substrate surface (Fig. 2c). Now all guide-substrate interfaces have the same, usually small refractive index difference, and the preparation of low-loss waveguides is much easier. However, special technologies only offer the possibility to prepare such buried guides.

Especially in the field of Integrated Optics with semiconductor materials (i.e. III/V-compounds) two other guiding structures are important: Fig. 2d shows the so-called ridge-waveguide: First a slab waveguide is deposited on the substrate; then the slab thickness is reduced outside the guide (e.g. by chemical etching) below the cut-off thickness of the fundamental mode. Therefore the light is confined to the thicker part of the slab structure. Very similar is the strip-loaded channel waveguide (Fig. 2e). Here a slab structure, which is too thin to guide any mode, is loaded with a strip of (usually a different) high-index material, forming together with the slab a channel waveguide.

In the literature there is no standard classification of waveguides. However, to avoid any misunderstanding in the following, we will use the classification as given in Table 1.

Table 1: Classification of Integrated Optic Waveguides

passive:	waveguides with fixed properties, like beam-splitters, curvatures, crossings, etc.
passive and controllable:	waveguides with changeable properties like modulators, switches, etc.
active:	optical active waveguides like lasers, amplifiers, parametric oscillators, etc.

single-mode : multimode

Integrated Optic circuits can be composed of a few fundamental waveguiding structures, which are shown in Fig. 3. In Fig. 3a the most simple element is depicted: a straight guide, which connects other Integrated Optic elements. It has to be mentioned that curved guides are allowed, too, but no corners (like in electrical ICs). Besides, the allowable radius of curvature depends very much on the refractive index difference and usually is very large (with respect to the wavelength of light).

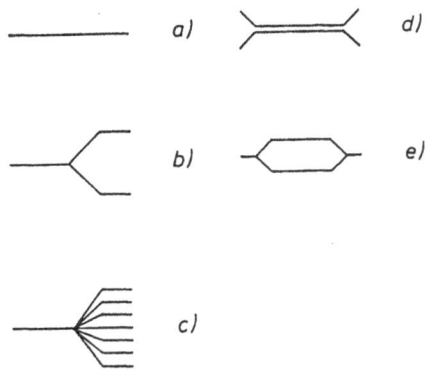

Fig.3: Fundamental waveguide structures

Fig. 3b shows a branch or Y-junction, which is used either to split or to combine the light. A similar, but more complicated structure is shown in Fig. 3c. It has to be stressed, that the elements shown in Figs. 3a-3c can be either single-mode or multimode waveguides and that the structures shown are sketches; realized waveguides usually are more complicated.

In case of multimode waveguides, structures as Fig. 3b, c are very difficult to design, because every mode behaves individually. Therefore the function depends very much on the mode excitation. This already gives a first hint that Integrated Optics mainly is a single-mode technology.

A further element is shown in Fig. 3d. It is the so-called directional coupler, which plays a very important role with respect to modulators and switches. It is a very mode-selective device and can be used as a single-mode element only. If light is coupled to one of the input ports, it is coupled forth and back periodically between the two adjacent guides. It depends on the coupling length (all power is coupled from one guide to the other) and the device length, if the coupler is in the cross-state (all power emerges from the non-excited waveguide) or the parallel state. The theory on directional couplers is very complex, especially because the coupling starts already, if the two guides are approaching each other.

Finally, especially in the field of optical sensors, the so-called Mach-Zehnder-Interferometer has to be mentioned. It is a single-mode structure and is shown in Fig. 3e; it consists mainly of two Y-branches. Light coupled to the interferometer is divided into the two arms and later on combined again. If the path length difference between the two arms is half-wave-

length or an odd multiple of that, the two waves interfere de-
structively at the combining Y-branch. This results in the
generation of the second order mode, which cannot be guided
and therefore radiates into the substrate.

3. Materials

Up to now we have discussed principles of Integrated Optics
without considering any realization. As already mentioned, a
low loss material is needed, which allows for the preparation
of waveguides (i.e. structured areas with increased refractive
index).

In the moment three material systems are important with respect to Integrated Optic technology:

- glass
- dielectric crystal
- semiconductor (III/V-compounds)

In Table 2 a summary is shown about material, technology and
waveguide losses (straight, single-mode channel waveguide).

Whereas for glass the ion-exchange and for $LiNbO_3$ (a crystal
with high electrooptic constant) the Ti-indiffusion are the
most important technologies, for InGaAsP no preference can be
given. It is likely that MOCVD will be the most suitable tech-

Table 2: Integrated Optic Materials

Material	Glass	Crystal [$LiNbO_3$]	Semiconductor [InGaAsP]
Technology	Ion-Exchange Sputtering CVD	Ti-Indiffusion Proton-Exchan.	LPE MOCVD MBE
Loss	0.01 dB/cm	0.1 dB/cm	1 dB/cm

nology, because of the possibility to prepare large substrates
with sufficient homogeneity.

Obviously glass is used for passive waveguides like bifurca-
tions or combiners. Therefore it is a multi- and a single-mode
technology.

Both $LiNbO_3$ and InGaAsP offer the possibility to build passive but controllable waveguides using mainly the electrooptic

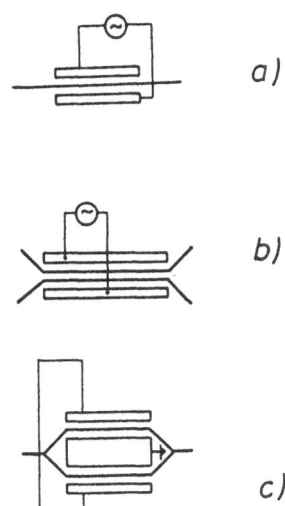

Fig.4: Various passive controllable elements

effect. This can be illustrated by using the fundamental waveguiding structures from Fig.3.

-   If an electrooptic material is used, a simple straight
    waveguide can be converted into a phase modulator by
    putting two electrodes on top of the substrate (Fig.4a)
    along the waveguides. The voltage applied to the elec-
    trodes will change the refractive index in the waveguide
    via the electrooptic effect. This results in a phase
    modulation of the guided wave.

-   A directional coupler can be made a switch or a modu-
    lator by using electrooptics. As shown in Fig.4b two
    electrodes are placed on top of the two coupled wave-
    guides; because the electrooptic effect depends on the
    direction of the electric field with respect to the
    crystal orientation, an applied voltage will lower the
    refractive index in one guide, but increase it in the
    other. In this manner the coupling length can be
    changed, and we can switch between the cross and the
    parallel state.

-   The phase modulator already discussed in Fig.4a can be
    combined with the Mach-Zehnder-Interferometer configu-
    ration (Fig.4c); by applying a voltage, the phase dif-
    ference between the two interferometer arms can be

changed. In this manner either destructive or constructive interference can occur at the combining Y-branch. Such an element will be used as an on/off-switch or an intensity modulator. With respect to the directional coupler it has the advantage of being simpler, but shows the disadvantage of radiating light into the substrate (when in off-state) which may interfere with other Integrated Optic elements on the same substrate.

Finally the use of III/V-compounds gives us the possibility to fabricate active elements, like lasers or light amplifiers. Therefore this technology offers the broadest spectrum for the realization of complex Integrated Optic circuits. On the other hand the technology is very complicated, because each element (laser; guide; detector) to be integrated on a common substrate has to be realized with different material combinations to allow for good performances (high gain; low loss; high absorption).

It is estimated, that semiconductor Integrated Optics will stay in the research for another 10 to 15 years. $LiNbO_3$ has been used for most of the laboratory demonstrations and now the first integrated optic element (a Mach-Zehnder-Interferometer) is offered commercially. Because glass is restricted to passive components only, the interest was quite small. However, new fiber optic concepts for local area networks or distributed sensors call for passive and simple (equivalent to cheap) optical elements (mainly splitters and combiners). It is doubted, that fused fiber couplers can be mass-produced and offered at low price. Therefore glass Integrated Optics has gained importance within the last few years.

4. Losses

Obviously the same loss mechanisms as in optical fibers occur in Integrated Optic circuits, namely

- scattering
- absorption
- radiation

Scattering is caused by two different effects: by volume scattering and by surface scattering. While in fiber optics the volume scattering exceeds the surface scattering effects, it is vice versa in Integrated Optics. The reason for that is, that the fibers are downscaled with respect to dimensions during the fabrication process, i.e. the large preforms are pulled to very small fibers. By that any surface defect is reduced. Integrated Optics, however, are produced at an 1:1 scale. As already mentioned, even a very small roghness at a waveguide/air interface causes large losses. To overcome this problem, buried waveguides are advantageous. Also the use of a graded index distribution diminishs the influence of geometrical distortions.

Because of the short waveguide length in Integrated Optics the absorption losses can usually be neglected.

Radiation losses may occur at waveguide imperfections, i.e if the waveguide geometry changes abruptly (for example caused by a masking error). Also at the boundary of a straight and a bend guide such radiation loss cannot be avoided. However, the most severe effect of radiation losses is due to sharp bends. The minimum radius of curvature which is allowed for negligible radiation loss, depends on the waveguide structure, namely the index difference, the wavelength and the index distribution as well as the mode number. Theory is very complicated and mostly the limits are found empirically.

There is a trade-off between allowable small radius of curvature (which is needed for a high integration density) and the ease of technology, i.e. the waveguide dimensions. As already mentioned, the waveguide dimensions depend on the refractive index difference. If the difference is high, the dimensions for building single-mode waveguides become very small (too small for any fabrication process), the scattering gets high, but the radius of curvature can be made very small. Therefore one has to find a compromise. This also depends on the waveguide material used. While ion-exchange in glass allows refractive index differences up to 0.1, Ti-indiffusion in $LiNbO_3$ only gives 0.05; using III/V-compounds allows for 0.1 or even higher. Therefore in $LiNbO_3$ the radius of curvature usually is in the region of some centimeters, in glass some millimeters and in III/V-compounds some 100 micrometers.

5.Technology

Table 3 gives a summary of the different materials which can be used for Integrated Optics. For the three classes of materials we will use in the following the typical representatives as given in Table 2, i.e. glass, $LiNbO_3$ and InGaAsP.

5.1 Fabrication of glass waveguides

Glass waveguides are mainly fabricated by an ion-exchange process. The principle is shown in Fig.5. The glass with the ion M_1^+ is immersed in a salt with ion M_2^+. If the salt is heated and gets liquid, a thermal ion-exchange can occur: the ions M_1^+ in the glass are exchanged with the ions M_2^+ of the salt melt. By using an appropriate combination of ions (e.g. exchange of Na-ions in the glass with Ag-ions of the melt), the index of refraction will be increased. However, -

252

Table 3: <u>Comparison between Integrated Optic materials</u>

	Glass	Crystal	Semiconductor
Technology	simple	moderate	complex
Loss	very low	low	medium
Integration Density	low	low	medium
Electrooptic and similar Effects	very low	high	high
Optical Amplification	none	small	high
Nonlinear Effect	very low	high	?
Availability	very good	medium	bad

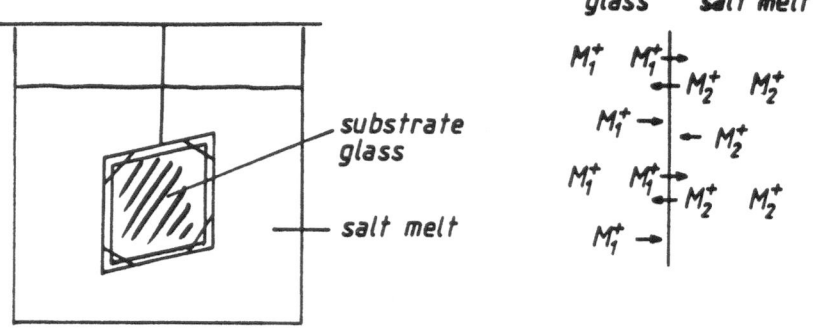

Fig.5: Thermal ion-exchange in glass

and this is a very important advantage of this technology - an index decrease is possible, too. A summary of usable ions is shown in Table 4. Also the possible index differences are given as well as the ionic radii. If two ions with a large difference in the ionic radius are exchanged, high mechanical stress in the glass will occur; this mostly destroys the surface.

For a production it is also important, that the used materials are easy to handle, for example are non-toxic. The fourth

Table 4: $\underline{M^+}$-ions for the salt bath

Ion	Ionic radius / CN* (pm)	Δn_d	LD_{50}^{**} / salt (mg / kg)
Li^+	59/4 76/6	0,02	710/Li_2CO_3
Na^+	99/4 102/6	-0,02-0,002	1955/$NaNO_3$
K^+	138/6	0,009	1894/KNO_3
Rb^+	152/6	0,01	1200/$RbCl$
Cs^+	167/6	0,04	1200/$CsNO_3$
Ag^+	126/6	0,10	2820/Ag_2O
Tl^+	150/6	0,10	25/Tl_2SO_4

*   CN: coordination number of the ion
** LD_{50}: a dose which is lethal to 50 % of the animals tested

column in Table 4 gives the lethal dose for the various salt baths. As can be seen, the silver-ion exchange is the most desirable, while the Tl^+-exchange should be avoided. The ion-exchange shown in Fig.5 usually is very slow, and to fabricate multimode-waveguides takes very long (up to days). Therefore mainly the field-assisted ion-exchange is used, as shown in Fig. 6. The principle is the same, however, front and back of the substrate are electrically separated and the exchange salt is at the front side only. By applying a voltage (a field of about 50V/mm is commonly used), the exchange is accelerated. While during the thermal exchange both sides of the substrate are treated in the same manner, the field assisted exchange causes a higher index at the surface and could give a lower index at the back of the substrate. This again may lead to opposite surface tensions of the substrate; if the substrate is too thin,it will be bent.

The two processes (thermal and field-assisted) also result in different index profiles, as can be seen from Fig.7. The ther-

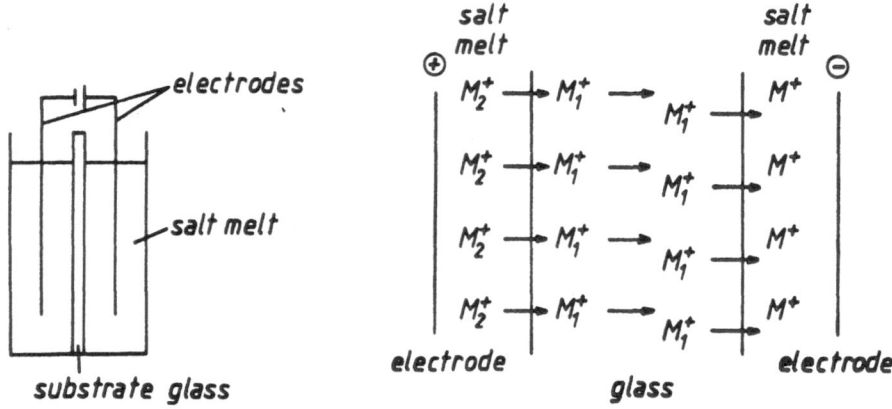

Fig.6: Field assisted ion-exchange in glass

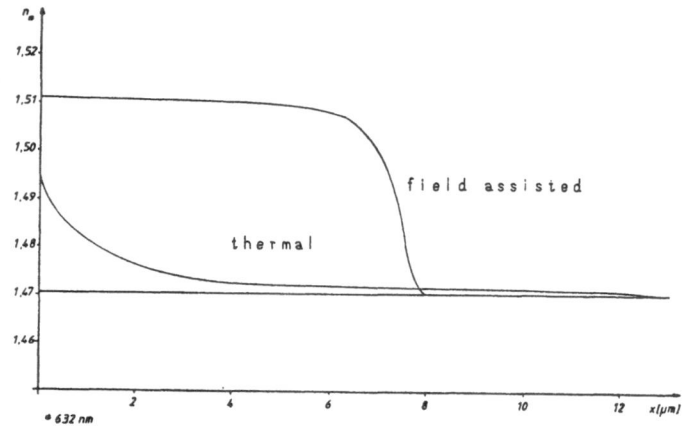

Fig.7: Refractive index profile for ion-exchanged guides

mal exchange shows an exponential decrease of the refractive index, while the field assisted exchange gives a step-like profile. By annealing, this step-like profile can be changed to a gaussian-like profile. Also buried waveguides can be realized very easily by performing a second field assisted exchange.

The reduction in processing time is about one order of magnitude with respect to the thermal exchange. Especially in the case of multimode-waveguides (which should be adopted to multimode fibers, i.e. 50/125 μm graded-index fibers), only

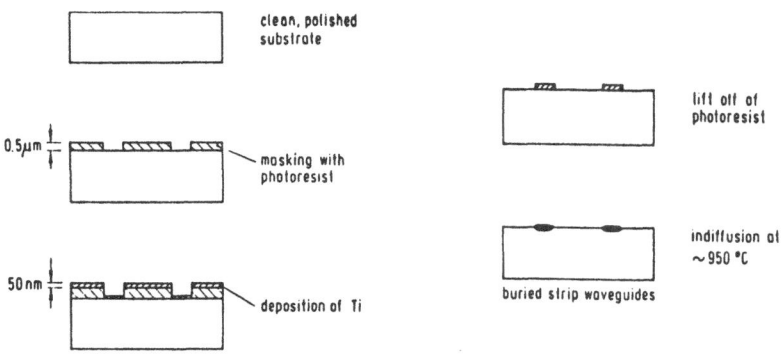

clean, polished substrate

masking with photoresist

0.5 μm

deposition of Ti

50 nm

lift off of photoresist

indiffusion at ~950 °C

buried strip waveguides

Fig.8: Ti-indiffusion into LiNbO$_3$

the field assisted exchange is usable.

5.2 Ti-indiffusion into LiNbO$_3$

The process of indiffusion of metals into LiNbO$_3$ was first investigated by Bell Labs in 1974 [4]. The most suited metal is Titanium. The whole process is shown in Fig.8. Starting with a substrate with polished surface (LiNbO$_3$-substrates today are available with diameters up to 3.5 inches) a mask is put on top by using conventional photolithography. This photoresist-mask is used to put a structured thin (20 to 70 nm) Ti-film on the LiNbO$_3$-surface. This metallic film is indiffused at about 1000°C for 5 to 8 hours. Thus a waveguide is built. The refractive index difference which can be achieved is about 0.05, but usually less. Because LiNbO$_3$ is a two-axis crystal, the waveguides show birefringence.

A still unsolved problem is the so-called outdiffused waveguide: During the high temperature in-diffusion process also Li-ions outdiffuse from the crystal surface forming a higher index layer which acts as a surface slab waveguide. This may cause leakage from any strip waveguide into this slab. It is still not well understood how this layer is built during the diffusion. Using a water saturated atmosphere during the diffusion seems to avoid this problem.

Also other problems have to be stated with LiNbO$_3$:

a. The so called optical damage in LiNbO$_3$ depends on impurities (mainly Fe^{2+}) and does not allow to guide high power density through Ti:LiNbO$_3$ waveguides because the waveguiding is temporarily destroyed by this effect. Experimentally the power limits are some 10µW in case of single-mode waveguides at 633 nm. The effect vanishes at wavelengths well above 1.3µm. As already stated, the effect depends very much on the crystal quality. Today LiNbO$_3$ with Integrated Optic quality is offered, but the problem still exists.

b. The process parameters to form a certain waveguide depend very much on the crystal. Unfortunately the LiNbO$_3$-crystal can be pulled with different ratios of Li:Nb; therefore for the Ti-indiffusion process it is necessary to know the exact Li:Nb ratio of the crystal to be processed.

All these difficulties are not so severe for fundamental experiments, but up to now they did not allow for a high yield mass-production of such devices.

Besides LiNbO$_3$ also LiTaO$_3$ could be used. But because of the low Curie-temperature of this material (around 650°C instead of 1250°C for LiNbO$_3$) the crystal has to be poled after the diffusion process.

5.3 III/V-compounds

In the case of III/V-compounds, namely InGaAsP, several different technologies can be used without any preference. This technologies are already used in the production of optoelectronic circuits like semiconductor lasers or LEDs. Because these technologies have been already described in detail elsewhere [5], they are not discussed in this content.

6. Processing steps

To build Integrated Optic waveguides, it is not sufficient to create an area of high index within a lower index substrate. A lot of different processing steps are necessary to build complex Integrated Optic circuits. This is shown schematically in Fig.9: Besides the fabrication of the waveguide we need an etching process either to interrupt waveguides (groove with mirror) or to avoid interaction between two adjacent waveguides (separating groove). Thin film technology is necessary too, to build antireflecting coatings, mirrors or buffer layers. Buffer layers are needed, if the waveguide is not burried and metallic electrodes have to be put on top of the waveguides. These metallic electrodes would attenuate the guided light; to separate guide and electrode, a dielectric, optically low loss buffer layer is introduced. Finally the edge preparation of the waveguide is not a trivial problem to solve, because the edge between front end and substrate surface has to be very sharp to avoid any destruction of the waveguide end.

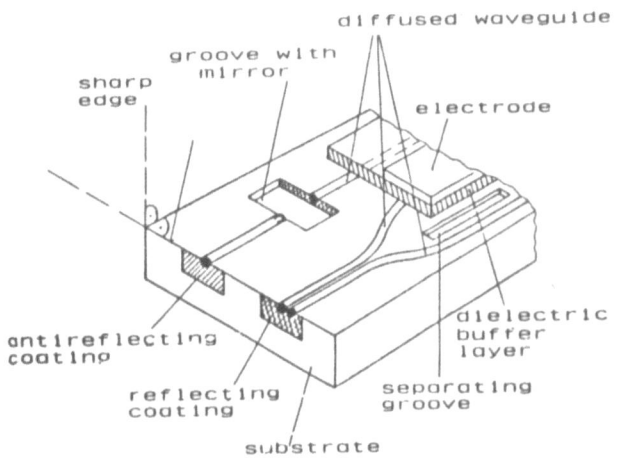

Fig.9: Complicated IO-structure showing various
processing steps

7. Coupling

Until now we have assumed that the light is already travelling
in the Integrated Optic waveguide. In case of III/V-compound
Integrated Optics this assumption may be valid, because light
can be generated and detected in the Integrated Optic circuit
itself. However, in case of glass or $LiNbO_3$-waveguides the
light has to be coupled in and out a tiny little waveguide.

Several schemes have been used. For experiments the prism-
coupler is a very easy to use coupling method (Fig.10a). Because the coupling to a special mode depends on the coupling
angle, this is also an instrument for measuring waveguides. A
second possibility is the grating coupler (Fig.10b). However
the efficiency depends very much on the optimization of the
grating period with respect to the waveguide, i.e. the guided
mode. Therefore it is not very well suited for the investigation of unknown waveguides. Also the fabrication (usually
with interferometry) is very complicated.

The most convenient method is the end-fire coupling using a
microscope objective or - more practical for commercial applications - a fiber (Usually only the microscope-coupling is
called "end-fire coupling", whereas direct fiber access is
called "butt-coupling".). In both cases the endface-polishing
of the substrate is a severe problem. With respect to butt

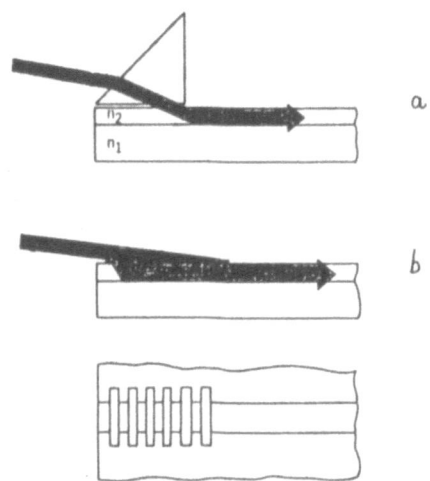

Fig.10: Coupling into slab waveguides by means of a prism-
film-coupler (a) or a grating coupler (b)

coupling of the fiber, other problems have to be faced:

- The fiber alignment with respect to the Integrated Optic
  waveguide is difficult, because the guide is invisible in
  most cases.
- The fiber has to be fixed without disturbing the alignment.
  The glue (usually UV-hardened) has to have a very low
  shrinkage; otherwise also the alignment will be changed
  causing increased coupling losses.
- Finally the mode-field of the Integrated Optic waveguide and
  the fiber should be identical or very similar to have low
  coupling losses.

On the other hand, source-fiber coupling is easy (with respect
to source-Integrated-Optic coupling). Therefore the fiber-pig-
tailed Integrated Optic circuit will be the most practical
configuration for commercial applications.

8. Integrated Optic for/as sensors

Up to now we only discussed the fundamentals of Integrated Optics and their realization. But what are the advantages to use
integrated optics for/as sensor? The main advantages are the
following:

- Miniaturisation
- Reduced adjustments
- New possibilities

This causes

- Cost reduction
- Technical improvement
- New possibilities

In the following we will discuss some examples of Integrated Optic circuits for/as sensors, which partly have already been used in prototypes, but some are just proposals.

8.1 Evanescent field sensor

While evanenscent field sensors play an important role in fiber optic sensors, also configurations have been proposed using Integrated Optics. A typical arrangement with a Mach-Zehnder-Interferometer is shown in Fig.11a [6]: One arm of the interferometer is protected against the surrounding (arm 1), while the other is exposed. Because the optical field is not totally constricted to the waveguide, but some of the field is travelling as an evanescent wave at the guide-surface interface, any change of the surrounding medium will influence the guided mode. In this manner, chemicals can be measured with respect to their refractive index. Though absolute measurements are very difficult, a high relative sensitivity can be reached. An estimate about the sensitivity is shown in Fig.11b, where I_0 is the light intensity coupled to the Mach-Zehnder-Interferometer, I_x the output intensity, which depends on the phase difference between the two arms. The phase difference is given by the difference of the optical lengths of the arms (L_1N_1 and L_2N_2, respectively). It is assumed, that the change of the refractive index of the surface influences linearly the mode propagation (which is not

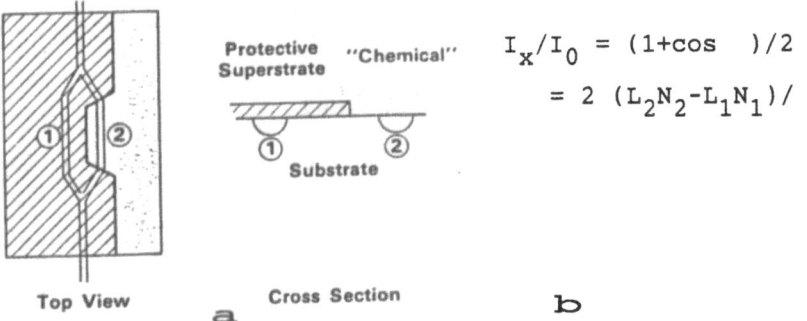

Protective Superstrate "Chemical"

Substrate

Top View   Cross Section

$$I_x/I_0 = (1+\cos\quad)/2$$
$$= 2\ (L_2N_2-L_1N_1)/$$

a   b

Fig.11: Mach-Zehnder-Interferometer as chemical sensor

260

true). Then index-differences much lower than 10^{-5} can be detected.

8.2 Integrated Optics and the optical gyro

One of the main pushes for Integrated Optics came from the development of optical gyros [3]. The optical gyro is based on the Sagnac effect; three different configurations are possible (Fig.12):

- the active gyro, where mostly a HeNe-laser is used in a ring configuration;
- the passive resonator structure, where the resonator can be made of a fiber or an Integrated Optic ring;
- the passive interferometer, which uses a long length of fiber and which usually is called "Fiber Gyro".

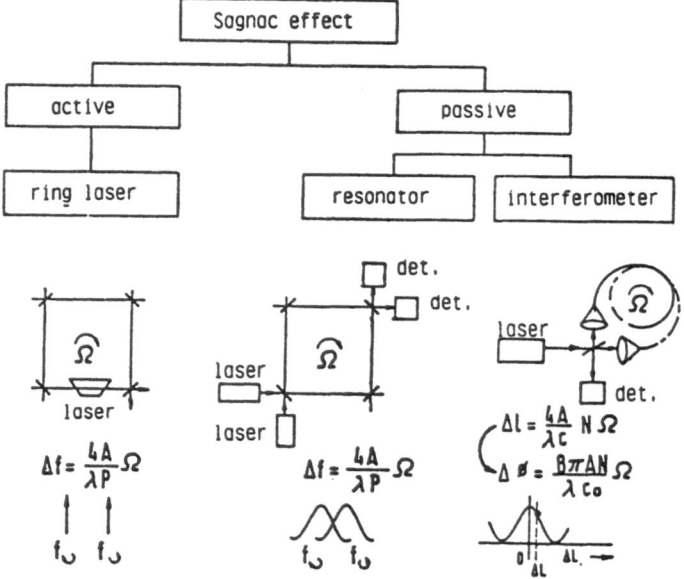

Fig.12: Principles of optical gyros

In principle all three configurations can use Integrated Optics. However, we will restrict the discussion to the use of Integrated Optics in the case of the fiber gyro.

In Fig.13 the schematic arrangement of a fiber gyro is shown. A lot of different optical elements are needed and using bulk optics (Fig.14) leads to a very large and heavy instrument, because any mechanical instability will influence the performance of the sensor. Therefore such arrangements show dimen-

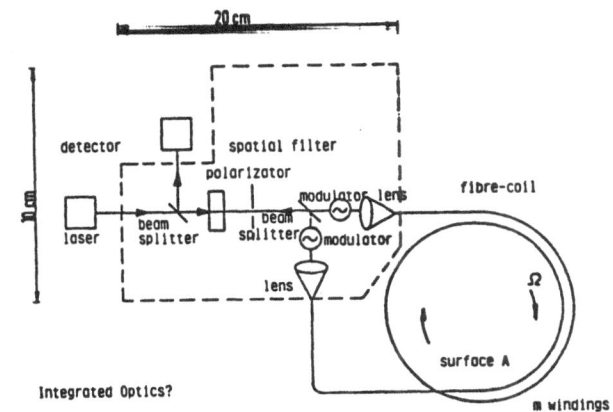

Fig.13: Schematic configuration for a fiber gyro

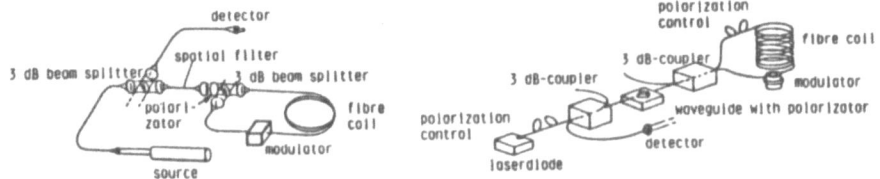

Fig.14: Fiber gyro with
bulk optics

Fig.15: The "All-Fiber-Gyro"
after Stanford

sions in the order of 20cm x 10cm (see Fig.13). A first step
has been made by the so-called "All-Fiber-Gyro" proposed and
realized by people from Stanford University [7]; the arrangement is shown in Fig.15. All elements are made from fiber-optic components. However, there are still mechanical elements
involved, like the polarization controller as well as the piezoelectric modulator. Therefore the next consequent step would
be the use of Integrated Optic circuits; the minimum configuration of such a circuit is shown schematically in Fig.16.
Beam splitters are realized using Y-branches. The polarizer is
simply made by a metallic overlay, which attenuates only the
TM-mode (E-field perpendicular to the substrate surface). The
spatial filter is a simple single-mode waveguide connecting
the two Y-branches. Finally the modulator is made as a phase
modulator; this requires the use of an electrooptic material.

In the first experiments such an Integrated Optic circuit has
been used by different groups; the chip was made with Ti-indiffused LiNbO$_3$-waveguides. However, the performance of the
gyro was surprisingly bad. Later the main problems have been
investigated:

- The degree of polarization achieved with the metallic surface layer was not sufficient.
- The spatial filter (some millimeters of single-mode waveguide) was not efficient enough.
- The reciprocity of the whole element was not good enough because of small deviations caused by masking and technological errors.

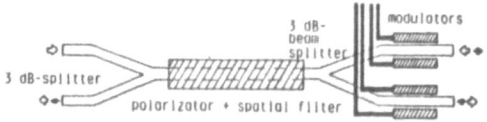

Fig.16: Minimal configuration of an IO-chip for the gyro

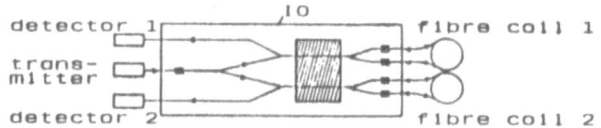

Fig.17: IO-chip for a two-axis fiber-gyro

Therefore the space filter was replaced by some meters of single-mode fiber (if possible, a single-polarization guiding fiber) and the Integrated Optic circuit was cut into two pieces. Most of commercially announced gyro-prototypes are using such a configuration.

A further step is to use a more complex Integrated Optic circuit, to realize a two-axis gyro (Fig.17). In this case, a single source can be used reducing the coupling problems.

8.3 Fabry-Perot sensors

A special class of fiber optic sensors uses the Fabry-Perot effect [8]. With a special evaluation scheme this fiber optic sensor is among the few which are independent from transmission lines to and from the optical sensor. Therefore this concept plays an important role in the field of fiber optic sensors.

With this respect, Integrated Optics can be used in two ways. Very "conventionally", as explained already in case of the gyro, all the optics needed to feed and detect the light to/from the sensor can be integrated avoiding a bulky and

therefore very unreliable arrangement. However, we will concentrate on a Fabry-Perot sensor using Integrated Optics.
Which parameters can be detected using the Fabry-Perot:

- Temperature
- Pressure (equivalent to force, acceleration, acoustics)
- E-Field (i.e.current)
- H-Field

Both temperature and pressure can be measured with glass as
sensitive medium, while for the measuremnt of E- and H-field
an electrooptic material like $LiNbO_3$ must be used. The sensitivity is highest for temperature, lowest for H-field.

A typical arrangement is shown in Fig.18. A Fabry-Perot is
made in $LiNbO_3$ by Ti-indiffusion. The resonator is realized

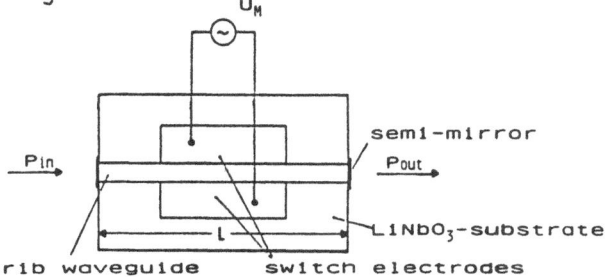

Fig.18: Fabry-Perot-Interferometer as a sensor

by putting semi-reflecting mirrors on the ends of the waveguide. To achieve high reflectance, the mirror should be
placed perpendicular to the waveguide, which is somewhat cum-
bersome to realize. For some special evaluation of the signal,
two electrodes are placed on the substrate forming a phase
modulator. If possible, such a configuration should be avoided, because then electrical connections are needed at the
sensor and the main advantage of an optical sensor (no electrical or metallic connection between sensor and evaluation
equipment) is lost. But the sensor shown in Fig.18 will work
without the electrodes also.

We will not discuss, how such a sensor works. This already has
been described elsewhere [8]. We will concentrate on possibilities offered by Integrated Optics with respect to this type
of sensor.

The first possible configuration is shown in Fig.19. Several
identical Fabry-Perot sensors are placed close to each other.
By that the spatial change of the parameter to be sensed can
be detected. For example, a laser beam running across the substrate surface and incresing the temperature locally will be

detected with respect to intensity <u>and</u> direction.

In most of the fiber optic sensors the dynamic range is insuf-
ficient. Therefore several similar sensors with varying sensitivity are used together. This is not easy to realize because
of packaging problems. In Integrated Optics this may be solved
as shown in Fig.20. The Fabry-Perot sensors are curved and
show different lengths. Because the sensitivity is proportional to the length of the Fabry-Perot sensor, the arrangement
fullfills this task.

Obviously a lot of similar arrangements can be considered.
However, the main advantage is the compactness of the device
as well as the planar technology.

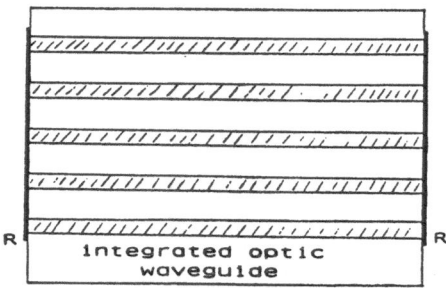

Fig.19: Array of Fabry-Perots for a sensor with spatial
resolution

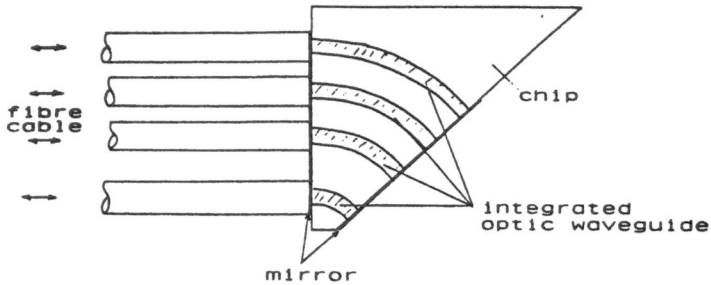

Fig.20: Fabry-Perot sensor array with different resonator
lengths for large dynamic range

9. Problems still to be solved

Though Integrated Optics has seen already a long development time, there are still a lot of (sometimes fundamental) problems to be solved. A simple Integrated Optic arrangement shown in Fig.21 demonstrates the different topics, while one of the most severe restrictions in Integrated Optics is shown in detail in Fig.22: The guides are very small; even if complex structures like a 4x4 matrix are realized, the substrate area needed in lateral direction is then also very small and mainly given by the dimensions of the fibers to be coupled to the guides. But because small radii of curvature are not allowed and the coupling length in directional couplers are in the order of millimeters, the device length goes up to some centimeters. In this special example the 4 x 4 matrix needs a substrate area of 0.2mm x 25mm. In this way a high integration density cannot be realized and the complexity of the Integrated Optic circuit is limited mainly by the device length, or by the maximum substrate length which is available.

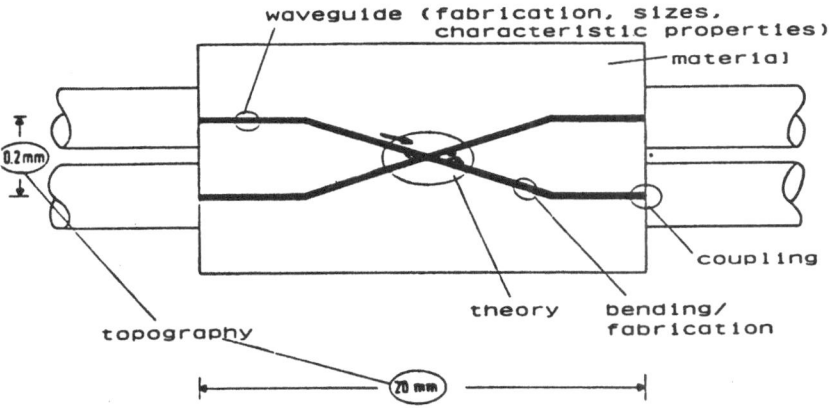

Fig.21: Fundamental Problems of Integrated Optic circuits

10. Conclusion

In principle Integrated Optics offer a very high potential in the area of optical sensors. It can be used either as a sensor (i.e. Fabry-Perot sensor or ring-resonator for the gyro) or for fiber-optic sensors (i.e. fiber gyro). However, there are still a lot of problems to be solved before Integrated Optics gets a commercial product, mainly:

- Coupling with fibers
- Production of suited masks

266

- Reduction of losses
- Appropriate packaging and housing
- ...

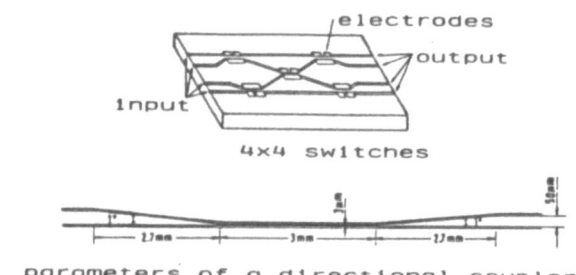

4×4 switches

parameters of a directional coupler

necessary substrate surface (true to scale)

Fig.22: The problem of Integrated Optic topology

11. References

[1] S.E.Miller: Integrated Optics: An Introduction; Bell Syst. Techn. J. 48 (1969) 7, 2059-2069

[2] F.P.Kapron, D.B.Keck, R.D.Maurer: Radiation Loss in Glass Optical Fibers; Appl. Phys. Lett. 17 (1970) 10, 423-425

[3] S.Ezekiel, H.Arditty: Fiber Optic Rotation Sensors; Springer-Series on Optical Sciences, Springer-Verlag

[4] R.V.Schmidt; I.P.Kaminow: Metal Diffused Optical Waveguides in LiNbO$_3$; Appl. Phys. Lett. 25 (1974) 8, 458-460

[5] H.C.Casey, M.B.Panish: Heterostructure Lasers, Part A and B; Academic Press 1978

[6] A.Harmer, Battelle Institute, Geneva: private communication

[7] R.A.Bergh, H.C.Lefevre, H.J.Shaw: All Single-Mode Fiber Optic Gyroscope with Long-Term Stability; Opt. Lett. 6 (1981), 502-504

[8] R.Kist, W.Sohler: Fiber-Optic Spectrum Analyzer; J. Lightw. Tech. LT-1 (1983) 1, 105-110

SOURCES AND DETECTORS FOR FIBER-OPTIC SENSORS

R. KIST

FRAUNHOFER-INSTITUT FÜR PHYSIKALISCHE MESSTECHNIK,
HEIDENHOFSTR. 8, D-7800 FREIBURG, GERMANY (FRG)

1. INTRODUCTION

A fiber-optic sensor (FOS) as shown schematically in Fig. 1 consists essentially of a light source, a fiber link (fiber 1, fiber 2, and connectors C), a detector, and a sensor element. This sensor element might be localized or distributed and is exposed to the light modulating action of the measurand or parameter (P) of interest.

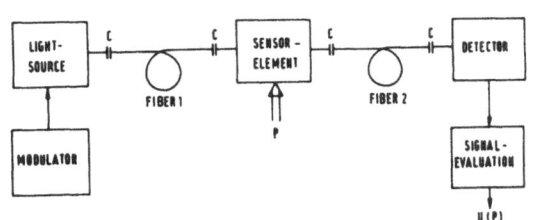

Fig. 1: General scheme of a fiber-optic sensor (C = connector, P = measuring parameter).

The sensor element can be a specially configured or modified part of the fiber link itself (intrinsic FOS) or consist of an external module (extrinsic FOS). A large variety of physical effects can be used to provide modulation of the photon current in the sensing element by action of the external parameter P. This modulation may act upon the amplitude (intensity), phase, polarization (differential phase), or spectral distribution of the light. In general fiber-optic sensors with modulation of intensity or spectral distribution make use of multimode fibers (multimode FOS) whereas monomode fibers are appropriate for phase and polarization modulated sensors (monomode FOS).

In a communications system the components are in general selected such as to provide lowest overall loss, i.e. largest repeater spacing and high signal to noise ratio (low bit error rate) as well as highest bandwidth, i.e. channel capacity and bit rate. Sources therefore have to meet specifications such as high optical output power, large modulation bandwidth, and high coupling efficiency into fibers. Detectors should provide high spectral responsivity (quantum yield), short rise time, and low dark current.

Fiber-optic sensors in contrast to most communications systems represent a large variety of physical features

/1,2,3,4,5/. The many different sensing elements and physical effects involved (bending, squeezing, twisting, stretching, heating, irradiating the fiber, light reflection, refraction, diffraction, Sagnac effect, induced linear and circular birefringence, photoluminescence, black body radiation etc.) ask for careful selection of the sources and detectors specifically suited for a given fiber-optic sensor. This statement may not be of critical importance for many cases that occur in the research laboratory. However, for the design and engineering of prototype FOS that have to meet severe technical and economic standards the appropriate sour- ce and detector selection may become an important issue.

The first aim of the present paper consits in reviewing the essential features and properties of various light sources such as incandescent lamps, gas lasers, light emitting diodes (LED's), superradiant diodes (SRD) and laser diodes (LD) as well as detectors such as photodiodes (PD), PIN-diodes and avalanche photodiodes (APD). The second aim is to discuss a selected set of fiber-optic sensors in view of their specific requirements for appropriate light sources and detectors.

2. CONVENTIONAL SOURCES

Conventional sources suited for operating fiber-optic sensors comprise incandescent lamps with heated filament, i.e. black body radiators, and gas lasers.

2.1 Incandescent lamps

Most fiber-optic sensors use sources of small dimensions and narrow spectral width. However, sensors with modulation of the spectral distribution or with spectral encoding require spectrally broadband sources with in general black body radiation characteristics as shown in Fig. 2. The spectral radiance L   as function of wavelength λ and temperature T is given by Planck's law:

$$L_\lambda = \frac{2\,h\,c^2}{\lambda^5} \left[e^{\,hc/\lambda kT} - 1 \right]^{-1}$$

with c = speed of light, h = Planck's constant, k = Boltz- mann's constant. According to Wien's law L_λ has a maximum at a wavelength λ_{max} = 2898 μm · K/T[K]. As an example Fig. 3 shows the typical normalized spectral distribution of a RCA tungsten lamp operated at 2900 K. Drawbacks of such incandes- cent lamps having optical output powers of typically a few Watts are their low coupling efficiency of the order of a few times 10^{-5}, the relatively short lifetimes (typically 10^3 to a few 10^4 hours), gradual intensity decline, and low frequency response of at most a few 10 Hz.

There are small incandescent lamps of typically 6 mm diameter and 15 mm overall length available /6/ that come in a lensed version (Fig. 4). Their optical output is in the spectral range 400 - 800 nm with powers of about 100 mW. Liftimes are in the range of a few 10^3 to 10^4 hours. These micro-incandescent lamps are well suited for simple intensity modulated sensors and can also be used to excite phosphors and photoluminescent materials.

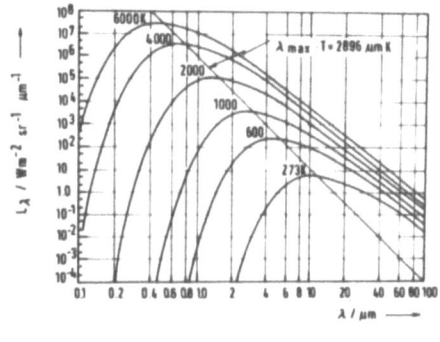

Fig. 2: Black-body spectral radiance as function of wavelength and filament temperature T.

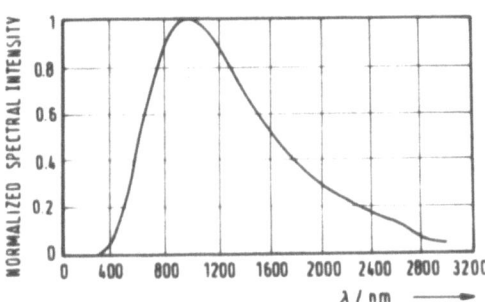

Fig. 3: Normalized spectral distribution of a tungsten lamp at T = 2850 K.

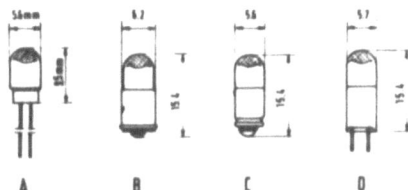

Fig. 4: Various examples of lensed microlamps /6/.

2.2 Gas Lasers

Many laboratory demonstration fiber-optic sensors are operated with standard gas lasers, which provide highly directional monochromatic radiation at power levels of typically a few mW. The most commonly used gas laser is the He-Ne laser radiating at 633 nm with output power of up to 10 mW. The beam has a typical diameter of 0.7 mm with about 1 mr divergence. He-Ne-lasers can have a coherence length of up to several 10^4 m.

For some cases such as Laser Doppler Anemometry (LDA) where high optical power and short wavelength is important the argon ion laser may be appropriate which radiates at 514 nm (main line) and cw output powers of several watts.

3. SEMICONDUCTOR SOURCES AND DETECTORS

Optoelectronic sources and detectors based on suitable semiconductor materials are since many years established in the areas of light control, safety systems, endoscopy for the detection and inspection of holes, structures, and work pieces, the measurement of position, length and angles in tool machines, control of optical instruments, galvanic separation of signal flows and conversion between optical and electrical energy. A large variety of products in the fotoindustry, automobile and computer industry for example make use of the advantageous properties of optoelectronic components such as their small size, low energy consumption and large operation bandwidth. The light emitting diode (LED) and the silicon photodetector are the dominant optoelectronic elements in these fields of application.

With the advent of optical fiber communications a strong need for sources of high optical energy density in the spectral windows of low fiber attenuation (0.83 µm; 1.3 µm and 1.5 µm) triggered worldwide efforts years ago to develop semiconductor lasers or laser diodes (LD). Today a large variety of LED and LD source components are commercially available at low or decreasing cost, and many of them can be used for the development of fiber-optic sensors.

3.1 Materials and Effects

Fig. 5 shows the energy band structure of the semiconductors germanium, silicon, and gallium arsenide in the direction of different crystal axes, indicated by the Miller indices, (111) and (100). The abscissa denotes the momentum $k = 2\pi/\lambda_e = 2\pi \cdot m \cdot v/h$ of electrons and holes moving in the valence and conduction band, respectively. The symbol λ_e denotes the electron de Broglie-wavelength, m and v the electron mass and velocity, respectively, and h is Planck's constant.

As can be seen the lowest energy transition (ΔE) from the bottom of the conduction band to the top of the valence band can occur in Ge and Si only if it is accompanied by momentum transfer (indirect transition), whereas GaAs allows for the highly probable direct bandgap transition without momentum transfer. GaAs is a direct semiconductor, Ge and Si are indirect semiconductors. The band gap energy E_g relates to the wavelength λ of the photon emitted by a band-band recombination as

$$\lambda \text{ [nm]} = 1240/E_g \text{ [eV]}$$

The value of E_g and hence the radiated wavelength can be varied systematically in the $In_{1-x}Ga_xAs_yP_{1-y}$ compound semiconductor system by suitably choosing x and y, which denote the fraction of Ga-atoms replacing In-atoms, and the fraction of As-atoms replacing P-atoms, respectively. Fig. 6 shows in a three-dimensional representation of this quaternary III-V-compound system the dependence of the band gap energy E_g as function of x and y /7/. The dashed curves indicate compounds of equal lattice constant, the solid curves compounds of equal E_g.

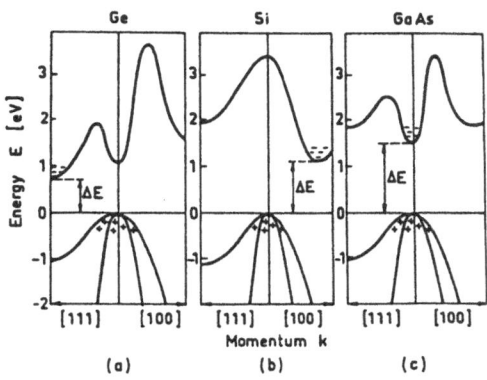

Fig. 5: Energy band struc-ture of a) germanium, b) silicon, and c) gallium arsenide.

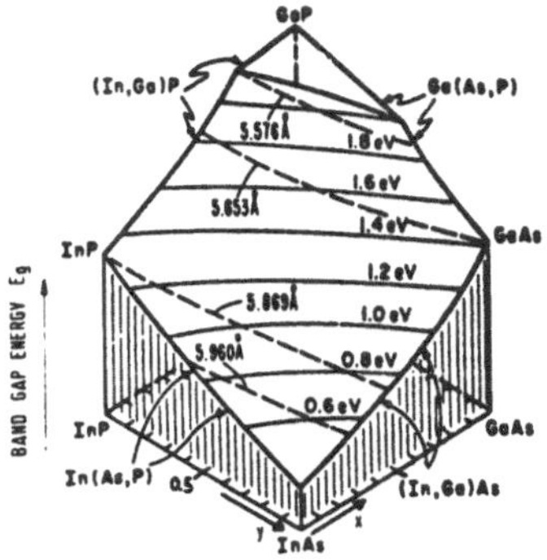

Fig. 6: Band gap energy and lattice constant for the quaternary compound semiconductor system $In_{1-x}Ga_xAs_yP_{1-y}$ /7/.

Fig. 7 shows the band structure of $GaAs_{1-x}P_x$ for various values of the mole fraction x /8/. Pure GaAs (x=o) is a direct semiconductor, whereas for $x \geq 0.4$ the $GaAs_{1-x}P_x$ -compound becomes an indirect semiconductor. Since the band gap energy also varies as function of x the color of the recombination light can be selected by a suitable choice of the mole fraction.

Indirect recombination processes involve the creation of phonons ($\hbar\omega$) that can take over the original momentum of the recombining electron in order to fulfill the requirement for energy and momentum conservation. However, this kind of indirect band gap transition is very inefficient. The transition probability can be enhanced by introducing into the semiconductor material appropriate impurities that act as trapping centers for electrons. An impurity center with a trapped electron captures a hole from the valence band thus forming a bound exciton. Via diffusion of the original momentum of the highly localized trapped electrons efficient direct transitions via annihilation of the excitons become possible. Fig. 8 schematically presents this possibility for the case of nitrogen and zinc impurities in GaP.

In the compounds $In_{1-x}Ga_xAs_yP_{1-y}$ and $Ga_{1-x}Al_xAs$ the refrac-tive index of the material decreases with increasing band gap energy, so that structured systems with layers of different band gap energy and index of refraction can be realized by changing the corresponding mole fraction.

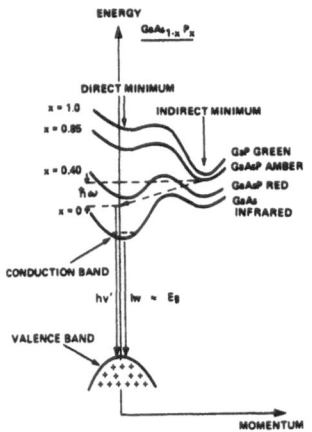

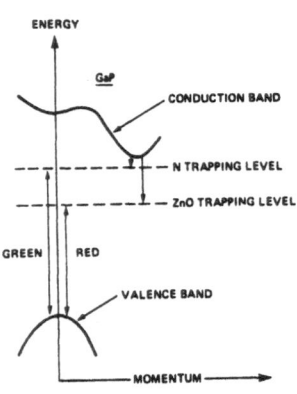

Fig. 7: Direct and indirect band-gap-transitions of GaAs$_{1-x}$P$_x$ /8/.

Fig. 8: Bound exciton recombination in GaP /8/.

Semiconductor p- and n-type materials are produced by diffusing or implanting acceptor atoms (i.e. Ge for GaAs) or donor atoms (i.e. Zn or Te for GaAs) into the pure host crystal, respectively. As indicated in Fig. 9a) acceptor atoms accept an electron from the valance band leaving a hole there, whereas donor atoms donate an electron to the conduction band. Joining a p-type to a n-type material produces a p-n junction according to Fig. 9b). The adjacent free electron and hole populations will move across the junction until an equilibrium potential barrier E_B is established. In order to get a semiconductor light source, charge carriers must be continuously generated and allowed to recombine, a process which is called current injection electroluminescence. This is achieved by applying an external forward bias voltage to the p-n junction.

The current injection is a non-optical pumping process which provides population inversion. As shown in Fig. 10 electrons that have gained sufficient energy by the injection field populate the conduction band up to the Fermi energy E_{FC} leaving a hole population of equal density in the valence band down to the Fermi level E_{FV}. Recombination of electrons at energy E_c' with holes at energy E_v' produces photons of energy $hv = E_c' - E_v'$. This recombination is spontaneous in the case of LED sources. Since each state can be populated by only one electron (Pauli principle) photons with energies larger than the band gap energy ΔE but smaller than the Fermi level difference $E_{FC}-E_{FV}$ cannot be absorbed. Such photons can, however, induce recombinations that produce

stimulated emission. This is the basic process in semiconductor lasers.

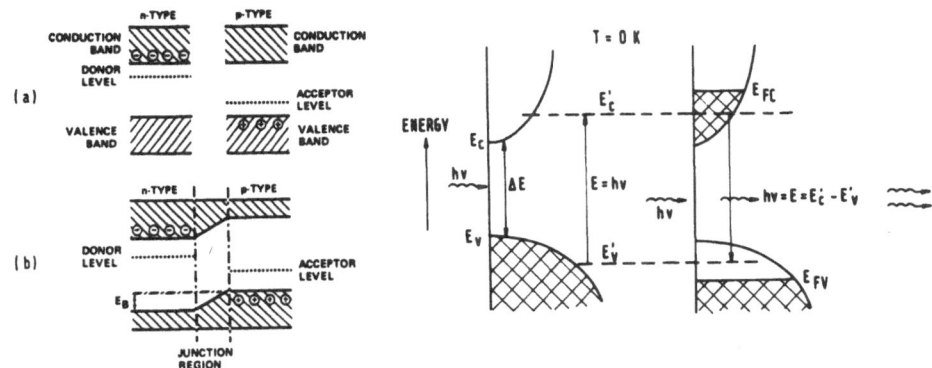

Fig. 9: Energy-band structure of a p-n semicon-
ductor homojunction /8/.

Fig. 10: Population inversion in a semiconductor /9/;
explanations see text.

The simple p-n junction with current injection is called a
homojunction as shown in Fig. 11. Electrons driven by the
forward bias voltage U_F flow accross the junction into the
p-region and recombine in the recombination region within a
few ns. Since the injection current density needed for a
given optical output power and hence the heating rate of the
material increase with increasing recombination region
thickness it is important to reduce this thickness. By
sandwiching a thin semiconductor layer of lower band gap
energy (and higher index of refraction) between two adjacent
layers of higher band gap energy (and lower index of refraction) a double heterostructure (DH) is produced. Recombination occurs predominantly in the thin layer of lower band gap
energy. As already mentioned this can be realized for example
with the compound system $Ga_{1-x}Al_xAs$ (doped with Ge for p-type
and Sn/Te for n-type) by suitably varying the mole fraction x
from layer to layer. The corresponding double heterostructure
is shown in Fig. 12 which also indicates the index step
associated with the active layer, the thickness d of which is
typically a few times 0.1 μm.

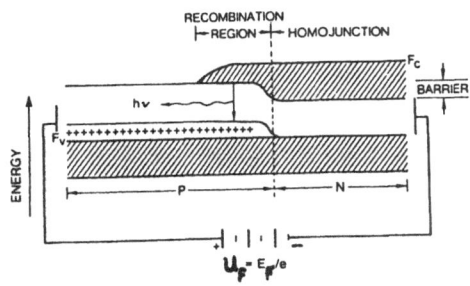

Fig. 11: Energy-band
structure in a forward
-biased p-n homo-
junction /10/.

274

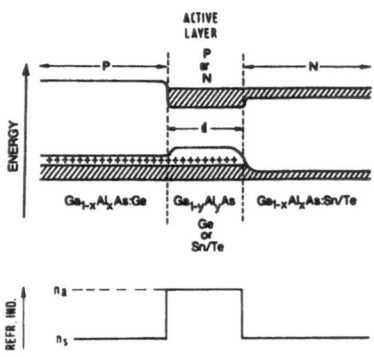

Fig. 12: Energy-band structure in a forward-biased heterostructure /10/ with the refractive index step associated with the active layer.

3.2 Light Emitting Diodes

Light emitting diodes (LED) produce incoherent radiation by spontaneous emission via the electroluminescence process in the semiconductor materials as described in the previous section. This radiation can leave the LED through the surface of the active layer (surface emitter) or trough the active layer edge (edge emitter) as shown in Fig. 13 a) and b). Figs. 14 and 15 show a few examples of the structure of surface emitting LED's in the visible and near infrared. They are produced starting with substrate crystals of n-GaP or n-GaAs onto which the active layer is grown by vapour phase epitaxy (VPE) or liquid phase epitaxy (LPE). In the substrate material which is transparent to the operational wavelength the downward emitted light can be reflected in order to increase the optical yield in lm per Watt electrical input power. The double-hetero(DH)-structured LED in Fig. 15 is of the Burrus-type with the substrate material etched down to the DH structure, which eliminates substrate absorption and enhances the coupling efficiency to a fiber epoxied into the well. This is shown schematically in Fig. 16 a).

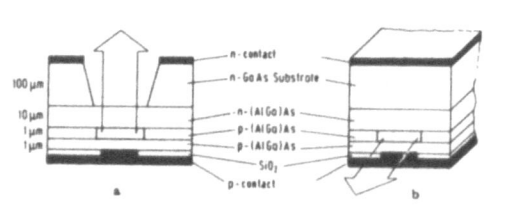

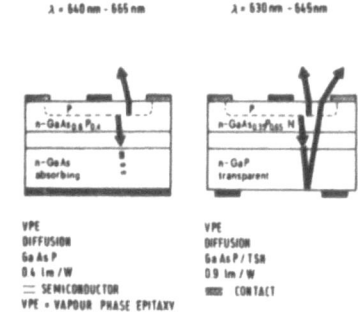

Fig. 13: LED surface (a) and edge (b) emitter /11/.

Fig. 14: Examples for surface emitting LED's radiating in the red /12/; TSN = Transparent Substrate Nitrogen; opt. yield in lm/W.

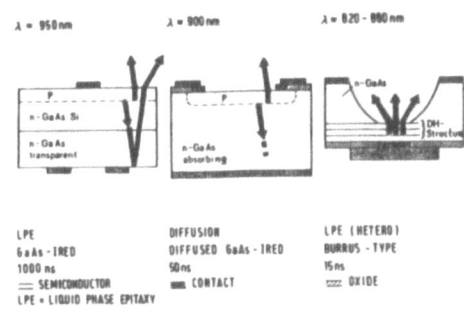

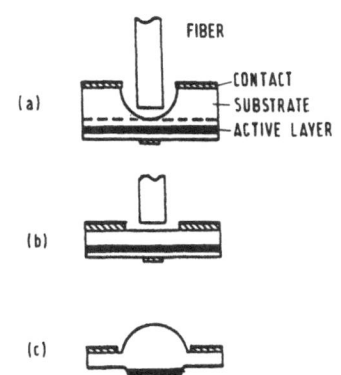

Fig. 15: Examples for surface emitting LED's radiating in the near IR /12/, typical switching times are given in ns.

Fig. 16: Etched well LED to fiber coupling (a), LED to fiber butt coupling (b), domed (lensed) LED (c).

The surface emitting LED has a Lambertian angular distribution of the radiance according to $I(\theta) = I_0 \cdot \cos \theta$, where I_0 is the intensity normal to the active layer, and θ is the angle with respect to the normal direction. Simple butt coupling of a surface emitting LED as shown in Fig. 16 b) provides a total coupling efficiency of less than 0.1 % if the size of the Lambertian source is smaller than or equal to the (multimode) fiber cross section. Since the refractive index of the materials involved is 3.6 (GaAs), 3.5 (GaAlAs) and for IR emitting LED 3.4 (InP), the critical angle for total reflection is about 16°. This means that only a fraction of about 1 % (2 % with reflecting back contact) of the generated light can leave the LED within the critical escape cone. This situation can be improved if the radiating surface is covered by a hemispherical structure (dome or lens) as shematically shown by Fig. 16 c). The escape cone then is increased to 26° in case of a plastic structure which increases the fiber coupling efficiency typically to a few 0.1 %.

The optical power of edge emitting LED's radiated as a beam of typical half-power width of 25° to 30° is generally a few times lower than that of surface emitting LED's, however their coupling efficiency into fibers is higher.

Fig. 17 shows typical optical power versus forward current characteristics of a AlGaAs LED for various temperatures. Fig. 18 shows the normalized emission spectrum for several injection current values. With increasing current the temperature of the active layer increases with the result of decreasing band gap energy and hence increasing peak power wavelength.

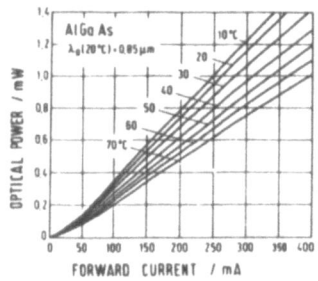

Fig. 17: Optical output power versus forward current of a AlGaAs-LED at various temperatures /11/.

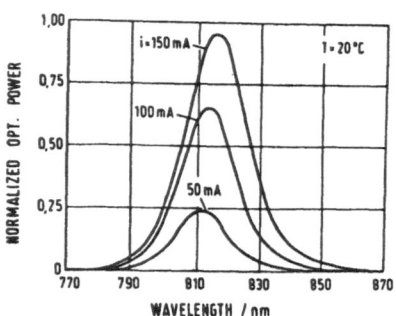

Fig. 18: Normalized AlGaAs-LED emission spectrum for various forward current values /11/.

The table of Fig. 19 summarizes the main features of several LED's emitting in the visible and near IR. λ_0 denotes the center wavelength. The spectral width $\Delta\lambda$ is small for direct band-band recombination materials (mole fractions are not indicated in the table) and much larger for indirect materials. η is the optical yield (optical output power P per electrical input power), ϕ_v the luminosity, and U_F the forward voltage for an injection current value of 20 mA. The rise time τ determines the modulation bandwidth of the LED, it depends on the effective carrier lifetime as well as the junction and parasitic capacitances.

Type	λ_0 [nm]	$\Delta\lambda$ [nm]	η [%]	at 20 mA P [mW] ϕ_v [mlm]	U_F [V]	τ [ns]
IR 950nm GaAs	950	50	12	3 / —	1.3	500
IR 870 nm (GaAl)As	870	80	10	5 / —	1.4	500
Red (GaAl)As	650	20	3	1.1 / 80	1.8	50
Red Ga(AsP)	660	20	0.6	0.18 / 8	1.6	50
Orange Ga(AsP)	625	40	0.6	0.24 / 50	2.0	100
Yellow Ga(AsP)	590	40	0.1	0.04 / 20	2.2	100
Green GaP	565	40	0.2	0.1 / 60	2.4	400
Blue GaN	490	80	0.01	0.01 / 1.4	5	?
Blue SiC	480	50	0.02	0.016 / 1.5	4	?

Fig. 19: Summary of main features of LED's emitting in the visible and near IR /13/.

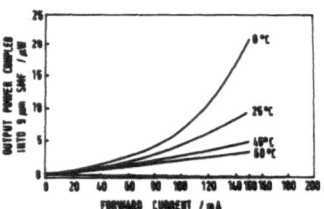

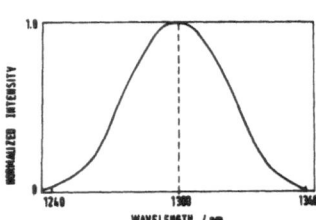

Fig. 20: Normalized spectral intensity and optical power coupled into a single mode fiber for a 1.3 µm LED /14/.

Fig. 20 shows an example of the normalized spectral intensity distribution of a InGaAsP LED peaking at λ_0 = 1300 nm as well as its output power coupled into a 9 μm core diameter single mode fiber as function of the forward current for various LED temperatures.

3.3 Semiconductor Lasers

In section 3.1 it has been outlined that current injection into a p-n junction represents a non-optical pumping process that provides population inversion. Stimulated emission is possible in that situation for photon energies that exceed the band gap energy but are below the difference of the Fermi energies in the conduction and valence band (Fig. 10). The system thus provides optical gain which leads to laser action if the structure is realized within a Fabry- Perot cavity making the active layer a resonant waveguide. The mirrors of this cavity are in general provided by the cleaved end facets of the active layer. Thus a laser diode can be viewed at as an edge emitting LED with mirrored ends.

Fig. 21 a) shows a diode laser as a four port component with an optical output power P at the lasing wavelength λ which both depend on the laser diode current I_{LD} and the laser temperature T. At low values of I_{LD} spontaneous emission dominates and the characteristic is similar to the one of a LED. Onset of stimulated emission is marked by the knee at the threshold current I_{th} above which P increases dramatically as function of I_{LD} (Fig. 21 b)). If the temperature is increased lasing starts at higher I_{LD}-values, the slope of the characteristic being unchanged. The change $\Delta I_{th} / \Delta T$ is typically 0.5 mA/°C, but depends on the laser structure.

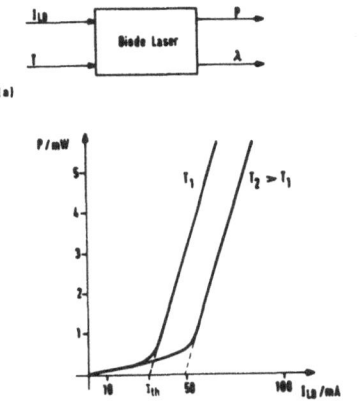

Fig. 21: Laser diode as a four port component (a) and optical power versus laser diode current (b).

The history of semiconductor lasers is marked by the development of many increasingly complex structures in order to realize components with low threshold current, high DC optical power, good confinement of the light guided in the active zone, and high stability of the intensity and wavelength.

Early homodiode and single heterostructure lasers had threshold current densities of the order of 100 kA/cm^2 and 10 kA/cm^2, respectively. They allowed only for pulse operation. Double and multiple heterostructure lasers brought this value below 1 kA/cm^2 allowing for DC operation at room temperature. The active zone was a planar waveguide of thickness of a few 0.1 µm with index guiding of the laser light in the vertical direction. However, the width of the active zone was 100 to 300 µm so that the optical output was laterally multimode, i.e. spectrally broad, and the field pattern not well suited for coupling into a fiber. In addition the onset of lasing was not uniform over the active layer which produced noise due to filamentary lasing.

The breakthrough came with the stripe waveguide lasers which provide lateral confinement of the active region with the advantages of low I_{th}, low operation temperature, single lateral mode operation without filamentary lasing noise, and much enhanced laser to fiber coupling efficiency. There are two classes of stripe waveguide lasers shematically shown in Fig. 22 a) and b), the index guided laser (IGL) and the gain guided laser (GGL), respectively. Both have typical waveguide dimensions of 0.3 µm x 10 µm x 400 µm. The IGL provides lateral guiding by a suitable lateral index profile, whereas in the GGL lateral guiding is achieved by the lateral injection current confinement due to the presence of an isolating (mostly SiO_2) layer with a narrow opening. Fig. 23

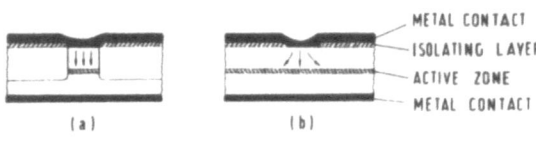

Fig. 22: Schematics of the index guided (a) and gain guided (b) laser.

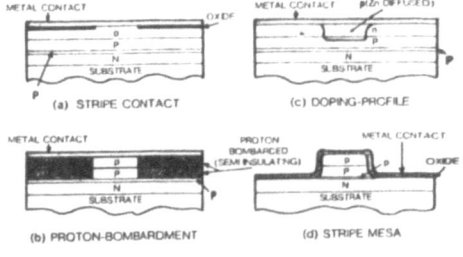

Fig. 23: Examples for various structures of stripe geometry laser diodes /10/.

shows several examples for structures of stripe geometry laser diodes. The oxide stripe structure (a) is the same as Fig. 22 a). The structure of Fig. 23 b) provides lateral current confinement by creating semi-insulating lateral zones of high resistivity by proton bombardement. A complementary method consists in strongly decreasing the resistivity in a stripe region by diffusion of a suitable dopant such as zinc (c). Finally Fig. 23 d) shows a structure in which the stripe is realized as a protruding plateau (mesa) thus providing lateral current confinement. Index guided lasers allow generally for low threshold current (typically 10 to 60 mA) and monomode operation. The astigmatism of the far field is negligible, which allows for easy coupling into fibers. However, their fabrication is more complicated. The gain guided lasers, on the other hand, are easy to fabricate, but have higher threshold current (50 to 120 mA), are multimode and strongly astigmatic. The astigmatism requires optical correction by a cylindrical lens for coupling into a fiber. Due to their shorter coherence length (typically 10 mm as compared to several m for IGL) and lower population inversion rate GGL are less sensitive against optical feedback than IGL. This feedback induces spectral instabilities, noise, and modifications of the laser characteristic.

Yet another GGL structure is shown in Fig. 24. It is the V-groove laser in which current injection is defined by the p (Zn)-diffused V-groove the notch of which provides an electrical contact close to the active layer thus limiting its lateral extent. This laser is suited for high power (up to 60 mW) pulse (100 ns) operation.

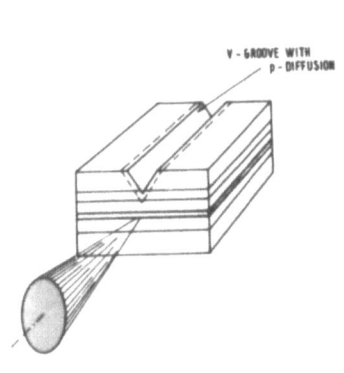

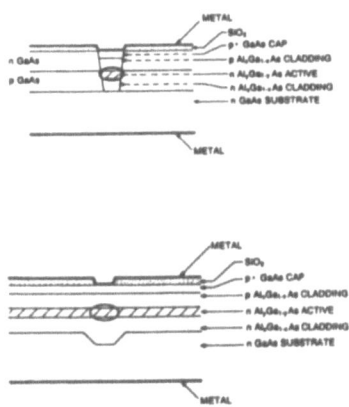

Fig. 24: Principle of a V-groove-structured semiconductor laser.

Fig. 25: Buried-heterostructure (BH) laser (above) and channelled-substrate-planar (CSP) laser (below).

Fig. 25 shows two prominent examples of index guided lasers out of a large variety of realised IGL structures, i.e. the buried-heterostructure (BH, above) and channelled-substrate-planar (CSP, below) laser. In the BH-laser the use of etching and subsequent epitaxy techniques leads to a structure in which the active layer is completely surrounded by lower index (cladding) material /16/. The BH-laser structure provides good transverse mode stabilization, cw output powers of up to 10 mW and threshold current values of typically 20 - 30 mA. In the CSP-laser confinement of the lasing mode is achieved by excess absorption in the substrate outside the active channel which also leads to transverse mode stabilization. Threshold currents are 40 - 90 mA at room temperature, the optical cw output power is up to 15 mW /17/.

Fig. 26 shows the radiation pattern of a laser diode. According to the laws of diffraction the divergence of a coherent light beam is inversely proportional to the source, i.e. the active layer dimensions. Therefore the opening angle of the beam perpendicular to the junction strongly exceeds the one for the parallel beam. A typical example of optical output versus forward current characteristics at various temperatures is shown for a CSP-laser in Fig. 27 /18/. In AlGaAs-CSP lasers the wavelength $\lambda \approx 800$ nm varies as function of the active layer temperature T between adjacent mode hops at a rate of 0.05 nm/K (23 GHz/K) due to the temperature dependence of the optical resonator length. The temperature drift rate averaged over several mode hops which reflects the gain curve drift due to the temperature dependence of the band gap energy is about 0.25 nm/K. Since the active layer temperature changes as function of injection current at a rate of about 67 mK/mA /19/, the laser wavelength can be tuned or modulated indirectly via the injection current according to the curves of Fig. 28. At modulation frequencies below about 1 kHz, the laser temperature starts to follow the current modulation.

Good wavelength stability and much reduced overall thermal wavelength drift (less than 0.1 nm/K) is achieved with the distributed feedback (DFB-)laser which has been fabricated in AlGaAs on a GaAs substrate by the molecular beam epitaxy (MBE) technique /20/. Fig. 29 shows the complex laser structure with the periodically disturbed active zone of periodicity $\Lambda = \lambda_o/n = 256$ nm (n = 3.4 for AlGaAs, λ_o = 870 nm) for second order Bragg reflection (for first order Bragg reflection $\Lambda = \lambda_o/2n$), which provides continuous (distributed) feedback thus eliminating the need for facet mirrors. The periodic structure is fabricated by holographic photolithography and chemical etching. The injection current is laterally confined by an oxide stripe.

Fig. 30 shows the output power versus current characteristics for the AlGaAs DFB laser in pulse operation (threshold current is 160 mA) as well as the stable single longitudinal mode oscillation close to λ_o = 874 nm for injection currents ranging from 170 to 350 mA. The single longitudinal mode oscillation was maintained over a 30 ^{o}C temperature interval without mode hopping, the temperature drift of the wavelength being about 0.07 A/K.

Coupling of laser light into a fiber aims for high coupling efficiency and low optical feedback. This holds for both, fiber-optic communications as well as fiber-optic sensors. Fig. 31 shows various schemes for laser-to fiber coupling: direct butt coupling (a), coupling via a selfoc or grin-lens

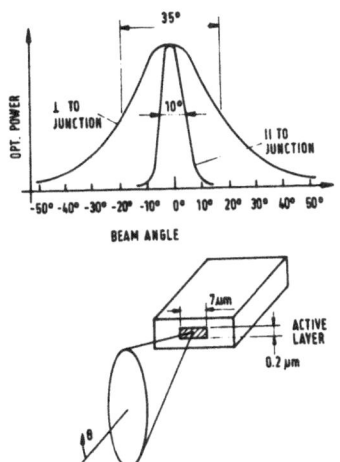

Fig. 26: Radiation pattern of a laser diode.

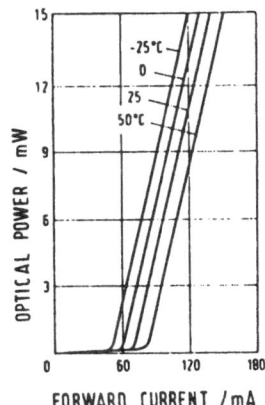

Fig. 27: Characteristics of a CSP-laser for various temperatures.

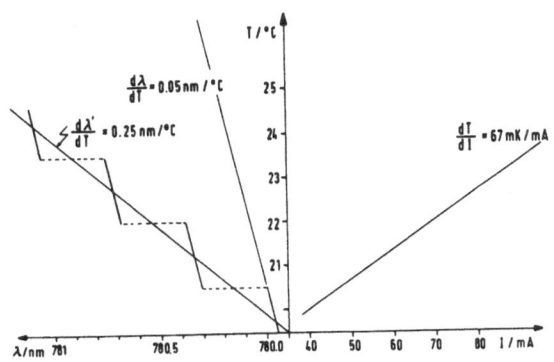

Fig. 28: Variation of the wavelength of a CSP laser as function of temperature and injection current.

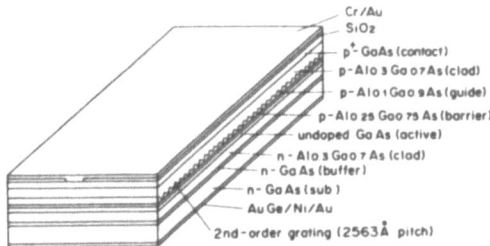

Fig. 29: Structure of a AlGaAs DFB laser /20/.

282

(b), fiber lens coupling for astigmatic laser sources (c), spherical fiber lens coupling (d), and discrete double lens coupling (e). Method (e) allows to insert into the collimated section of the beam an optical isolator (polarizer plus $\lambda/2$-plate) for optical feedback reduction. This is important for example when operating a Michelson or a Fiber-Fabry-Perot Sensor.

Various approaches have been proposed for laser-fiber-coupling with a miniaturized spherical fiber lens. A particulary successful version for coupling a single-frequency laser (λ = 1,3 µm) into a monomode fiber combines a hemispherical lens of radius r = 55 µm fused to a silica cylinder (PCS fiber core) of length l = 960 µm to provide large focal length and hence large distance between lens front and laser facet /21/. Coupling efficiencies of 40 to 70 percent and optical feedback of 10^{-7} of the emitted laser power have been achieved with this method.

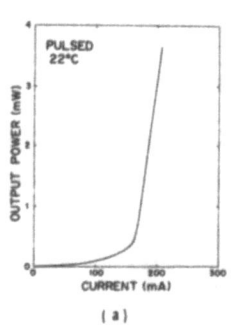 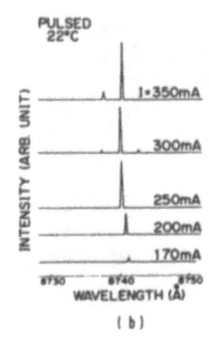

Fig. 30: Characteristic (pulse operation) and single mode oscillation of the AlGaAs DFB laser /20/.

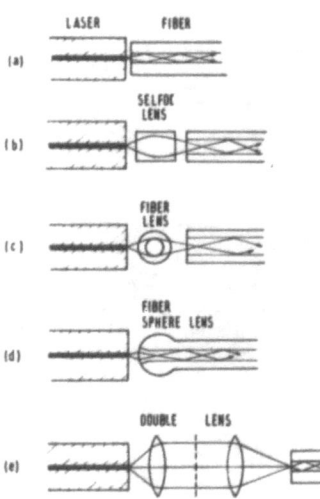

Fig. 31: Various schemes for laser-to-fiber coupling.

3.4 Superradiant Diodes

In edge emitting LED's population inversion can occur at high injection currents leading to stimulated emission. If this stimulated emission allowed only to perform single pass, i.e. if lasing action is avoided, a superradiant diode, SRD (also superluminescent diode, SLD) is formed. Lasing action can be avoided either by AR coating the facets or by inserting an unpumped region near the rear facet as schematically shown in Fig. 32. The spectral distribution as shown in Fig. 33 is about half as broad as for a LED. The angular distribution of the far field (Fig. 34) is narrowed due to the contribution of stimulated emission.

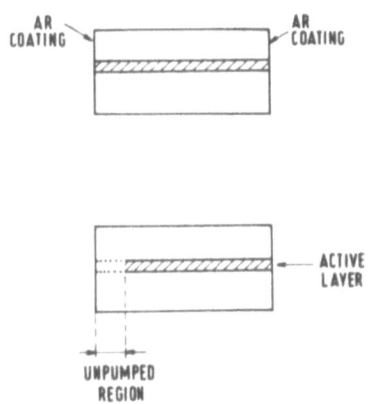

Fig. 32: Schematics of a SRD structure.

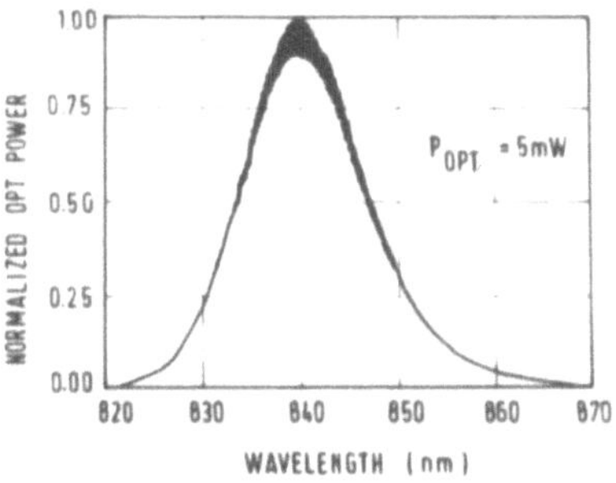

Fig. 33: Spectral distribution of a SRD.

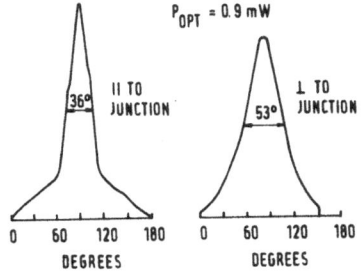

Fig. 34: Far-field angular distribution of a SRD.

Fig. 35 compares the optical power versus current characteristics for a laser diode (LD) and a SRD. A drawback of the SRD is its nonlinear characteristics. It has, however, the advantage of strongly reducing the noise level which in a LD is associated with the laser action around threshold. Short coherence length (less than 50 µm) and coupling efficiencies of 80 % into multimode and typically 20 % into monomode fibers have been achieved with a AlGaAs double-heterostructure SRD /22/.

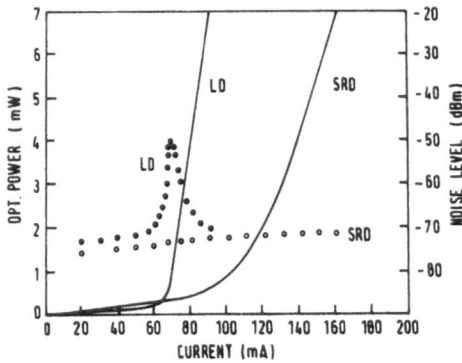

Fig. 35: Comparison between LD- and SRD optical power versus current as well as noise versus current characteristics.

4   DETECTORS

Depending on the light source, the power budget, and the bandwidth (modulation and frequency response) of a fiber-optic sensor an appropriate detector for converting the sensor's optical into an electrical signal must be selected. The basic process for any photodetector is photoabsorption which occurs in semiconductors at optical wavelengths larger than the cutoff wavelength $\lambda_c = hc/E_g$ with the effect of creating free electrons in n-semiconductors, free holes in p-semiconductors, and free electron-hole pairs in intrinsic semiconductors. The photoeffect is characterized by the quantum efficiency η of the semiconductor material, i.e. the probability of electron-hole pair generation per incident photon. This can be converted into the photodetector responsivity $\rho = \eta \cdot e/h \cdot \nu$ given in A/W for photon energy $h\nu$ (e = elementary charge). The photocurrent for a given optical power P is then $i_p = \rho \cdot P$. Fig. 36 shows the responsivity curves of Si- and Ge-photodiodes along with the spectral position of specific III-V compounds and the human eye's responsivity curve. Si is very well suited to detect radiation in the first fiber-optic window at around 800 nm. Its spectral response decreases, however, drastically above 1 µm. In the second and third fiber-optic window Ge is best adapted for optoelectronic energy conversion.

4.1   pn-photodiode

A semiconductor p-n-junction operated in reverse bias condition provides the simplest photodiode the principle of which is shown in Fig. 37. The reverse bias creates a rarefication zone at the p-n-transition in which a large electric field builds up. Photons that are absorbed within the rarefication zone, move fast under the influence of the electric field and provide a large drift current of short response time. Charge carriers produced by photoabsorption outside the rarefication zone move slowly due to diffusion and constitute a small diffusion current of long response time. In the absence of photoabsorption a residual current which is due to thermally generated free charge carriers flows in a reversely biased diode. It is called the dark current.

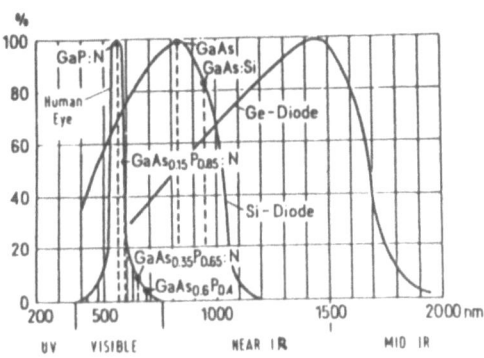

Fig. 36: Spectral response of human eye, of Si- and Ge- as well as several III-V semiconductor detectors.

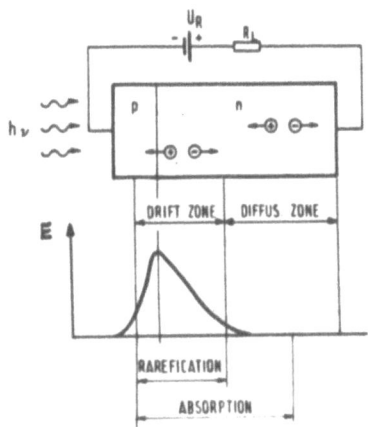

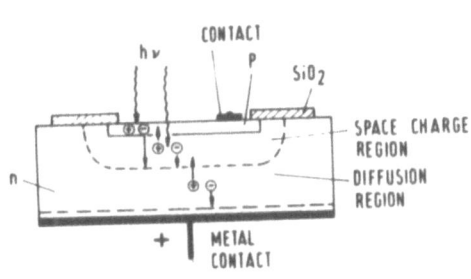

Fig. 37: Principle of a p-n- photodiode.

Fig. 38: Example of a p-n- photodiode structure.

Fig. 38 shows a cross sectional view of a typical p-n-photodiode structure. Since in a simple p-n-photodiode the diffusion zone is much larger than the drift zone, this kind of detector is slow and not well suited for high bandwidth, high sensitivity applications.

4.2   PIN-photodiode

In order to speed up the detector's response the rarefication zone should be as large as the absorption zone. This can be achieved by sandwiching an intrinsic (i) semiconductor layer between the p- and n-zones as shown in Fig. 39. The small amount of free charge carriers in the intrinsic material provide a large electric field across the i-zone so that electrons and holes produced by incident light form a drift current with fast response. Fig. 40 shows the typical structure of a PIN-photodiode.

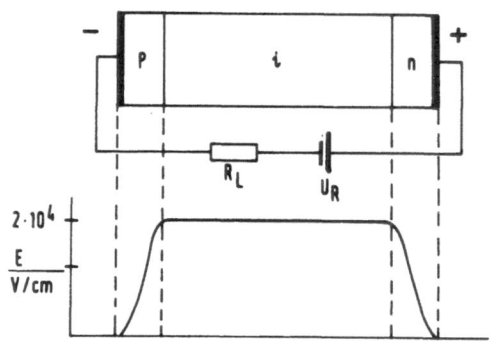

Fig. 39: Principle of a PIN-photodiode.

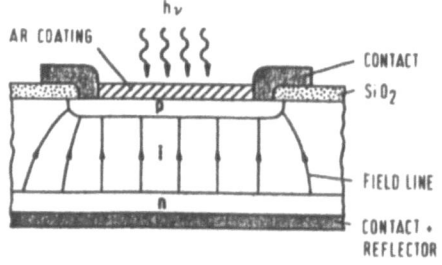

Fig. 40: Example of a PIN-photodiode structure.

Fig. 41 a) draws the simplest circuit for a PIN-photodiode with U_R = reverse bias voltage, U_d = voltage across the photodiode, i_d = photodiode current (U_d and i_d negative in the reversed direction) and R_L = load resistance. The analysis of this circuit can be easily performed with the diagram of Fig. 41 b) which shows the current-voltage characteristic along with the load line $U_R + U_d + i_d \cdot R_L = 0$ for the case of U_R = 20 V and R_L = 1 MΩ. This diagram allows to calculate the output voltage $U_o = -U_R - U_d$ as function of the optical input power P. For example P = 30 µW yields U_d = -5V and U_o = 15 V.

4.3   Avalanche Photodiode (APD)

For very small optical powers, i.e. levels in the nanowatt regime, the corresponding photocurrent in a PIN-photodiode may be less than the receiver noise current. In such cases application of an avalanche photodiode (APD) is adequate which provides multiplication of the primary electron-hole pairs created by photoabsorption.

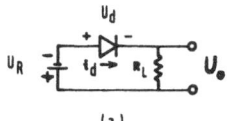

(a)

Fig. 41: (a) PIN-photodiode circuit and (b) diode charac-teristics with load line /23/.

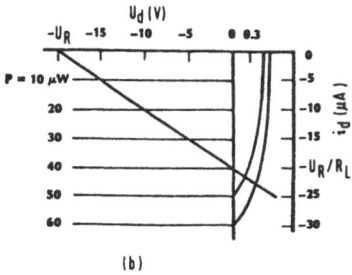

(b)

Fig. 42 shows the principle of an APD, in which photoelec-trons that are produced in the intrinsic layer, drift into the high electric field region of the p-n junction. There they are accelerated to energies that allow for secondary electron-hole production by collision. This can continue to yield an avalanche process with a current gain that may vary in Si from about 10 to over 100. The reverse bias voltage may be several hundred volts. Therefore APD structures are provided with a guard ring in order to prevent reverse breakdown. The following Table 1 summarizes the typical properties of semiconductor PIN and APD photodetectors /23/.

MATERIAL	STRUCTURE	τRISE ηs	λ μm	RESPONS. A/W	I_{DARK} nA	GAIN
Si	PIN	0.5	0.3-1.1	0.5	1	1
Ge	PIN	0.1	0.5-1.8	0.7	200	1
InGaAs	PIN	0.3	1.0-1.7	0.6	10	1
Si	APD	0.5	0.4-1.0	77	15	150
Ge	APD	1	1.0-1.6	30	700	50

Table 1: Comparison of PIN- and APD-pho-todetector characte-ristics.

The frequency response of a photodiode is the result of the transit times involved with the diffusion and drift of charge carriers as well as the diode capacitance C_d and the junction load resistance R_L (time constant $R_L \cdot C_d$). Fig. 43 shows a simplified equivalent photodiode circuit.

In most applications the PIN-photodiode is preferable since it is as fast as the APD, but works at much lower reverse bias voltages. The APD is needed in cases where low optical signal powers are to be detected.

288

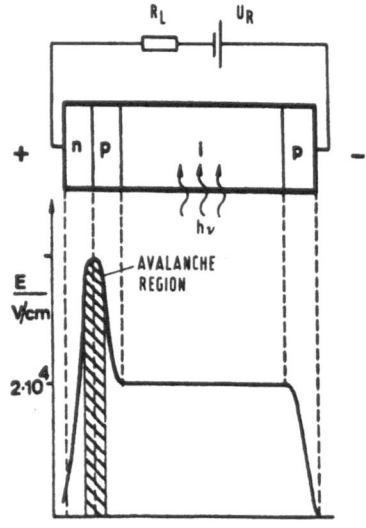

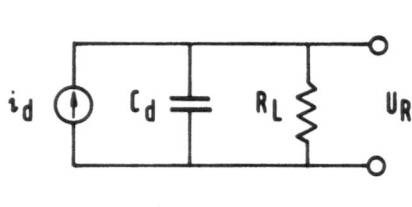

Fig. 42: Principle of a APD photodiode.

Fig. 43: Simplified equivalent photodiode circuit.

5   SOURCES AND DETECTORS FOR FIBER-OPTIC SENSORS

In the previous chapters we have reviewed the main properties of conventional light sources as well as semiconductor sources and detectors. The present chapter is devoted to the question of which source and detector aspects are to be considered for fiber-optic sensor technology. This will be done by looking at a set of representative fiber-optic sensor examples and discussing the suitable choice of sources and detectors in the light of the individual sensor requirements. These requirements follow from specific physical and technological sensor properties involving for example the kind of light wave modulation (intensity, phase, polarization, spectral distribution), the encoding concept (analog, digital, wavelength, frequency encoding) as well as the optical power budget (measuring range, resolution, sensor networking) and aspects of reliability and long term stability.

5.1   Intensity-modulated fiber-optic sensors

In case of intensity modulated sensors parasitic intensity changes may be caused by drift and aging of the source, losses on the fiber links due to environmental influences, bending and variations of the effective connector coupling.

Extended efforts have been and are being devoted to develop methods that make fiber-optic sensors independent on parasitic intensity variations. One basic approach relies on the measurement of the spectral radiance distribution of a black body radiator operating at a temperature T which is to be measured. By evaluating the wavelength of maximum radiance according to Planck's law the sensor's output signal is largely independent on intensity drifts and slow fluctuations along the fiber link and on variations in the detector characteristics. This concept has been successfully applied to the measurement of high temperatures /24/. Fig. 44 shows the principle of this sensor with the black body radiator realized as a passive element by sputtering an iridium film and a protective aluminum oxide film onto the tip of a high temperature saphire fiber.

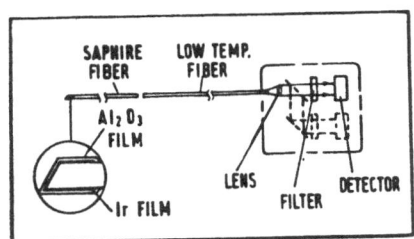

Fig. 44: High temperature fiber-optic sensor with black-body radiating sensor element /24/.

The well known LUXTRON ratioing fluoroptic temperature sensor uses a UV (200 - 400 nm) incandescent lamp to excite a phosphor at the probing fiber tip /25/. The intensity ratio of two separate lines of the returning fluorescent light which is a monotonous function of the temperature to be measured turns out to be largely independent on parasitic intensity variations. A somewhat similar concept is used in the ASEA temperature sensor in which the light of a LED emitting at around 750 nm excites a miniature crystal of GaAs/AsGaAs the photoluminescence of which is analyzed via two interference filters as shown in Fig. 45. The integral photoluminescence to which filter 1 is fully transparent acts as intensity reference whereas the optical signal transmitted via filter 2 depends on the spectral position of the luminescence spectrum which in turn is a unique function of temperature /26/. Both filters block the exciting LED light at 740 nm with a spectral bandwidth of about 50 nm. In the cases of LUXTRON and ASEA wavelength referencing relates to selected fluorescence lines and to the integral photolumines-cence spectrum (transmitted partially by filter 2), respectively.

There are also concepts for wavelength referencing related to two different source wavelengths. These might come from two separate LED's as in the case of a pressure sensor with schutter modulator that has been developed at UMIST/Manche-ster /27/. The shutter modulator consists of two holographic grids which are transparent to the reference wavelength λ_1 = 948 nm. The transmission for the measuring wavelength λ_2 = 820 nm is a periodic function of the relative grid position. This example demonstrates the advantage of the fact

290

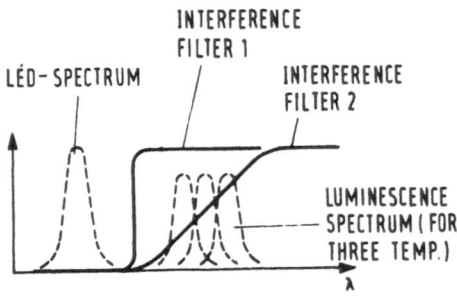

INTERFERENCE
FILTER 1

LÉD-SPECTRUM

INTERFERENCE
FILTER 2

LUMINESCENCE
SPECTRUM (FOR
THREE TEMP.)

λ

<u>Fig. 45:</u> Interference
filter spectral characte-
ristics for the ASEA tem-
perature sensor /26/.

that semiconductor sources can be fabricated via selection of
suitable compound ratios for various emission wavelengths.
Another advantage is the possibility to directly modulate a
LED via the injection current. In the case of the two
wavelength referencing pressure sensor LED 1 was modulated at
1,1 kHz, and LED 2 at 0.7 kHz. This allows to detect
selectively the corresponding AC components by Si-photodiodes
along with phase sensitive detection (PSD) and thus separate
the two intensities at λ_1 and λ_2.

Modulation of LED (and laserdiode) sources in fiber-optic
sensors is often used along with electronic band filters or
PSD to eliminate parasitic light components and drift
effects. Typical modulation frequencies are up to several kHz
which presents no problems for the use of LED's and
photodiodes (eventually PIN-diodes).

A further step in wavelength referencing is achieved by using
broad band sources, i.e. incandescent lamps and LED's, along
with birefringent Fabry-Perot comb filters /28/. Two comb
spectra, separated by their orthogonal states of polarization, are interleaved, and referencing is made of their
respective integral spectral intensities. This reduces the
effects of differential fiber losses that still may cause
errors in fiber-optic sensors with two (discrete) wavelength
referencing. Using for example a LED source at 845 nm and a
comb filter of finesse 14 each comb spectrum consists of
about 30 lines within the spectral half-width of the LED.
Another possibility of spectral encoding in order to become
independent on parasitic intensity changes is to use a
broadband source in coupled interferometers and spectrometers. As an example a coupled Michelson-Interferometer
as displacement sensor has been reported /29/ that uses a
tungsten lamp and a Si-PIN-diode receiver along with a
HeNe-laser interferometer as an absolute reference.

Broadband sources are also used in a rather large class of
wavelength-encoded fiber-optic sensors, in which the parameter to be measured causes a shift in the transmission (or
reflection) wavelength of a dispersive sensor element /30/.
Finally broadband sources play an important role in wave-

length-multiplexed fiber-optic sensor networks in which each
individual sensor operates and is identified in a small
wavelength interval that is filtered out of the broadband
source spectrum.

Concerning modulation requirements for LED's in fiber-optic
sensors higher modulation frequencies than usual may be
needed to resonantly excite micromechanical structures by
optical pulses that are absorbed by the surface material thus
causing periodic thermal strain. An LED (750 nm) has been
pulsed to excite resonantly oscillations of a 10 mm x
1,5 mm + 60 µm glass cantilever beam at around 590 Hz /31/.
Fig. 46 shows the schematics of the experiment. Only 100 µW
optical power are needed to excite the vibrational motion of
the cantilever which in turn is detected via reflection of
the light of a second LED. If much smaller structures
fabricated by anisotropic etching of silicon are used, the
resonant frequencies may be as high as several 100 kHz /32/.

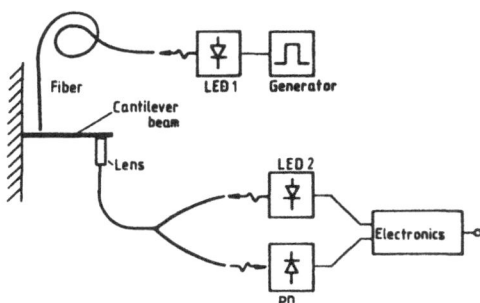

Fig. 46: Schematics of
the experiment for reso-
nant excitation of a
micromechanical struc-
ture /31/.

For this very interesting new class of fiber optic sensors
which provide a frequency as output signal, a more careful
selection of the LED as to its modulation bandwidth and the
use of PIN-diodes or even APD's is necessary.

5.2 Phase-modulated fiber-optic sensors

Additional source properties such as coherence, wavelength
stability and wavelength tunability are to be considered in
phase-modulated, i.e. interferometric fiber-optic sensors.
Fig. 47 shows three configurations of fiber-optic interferometers, the Michelson (a), the Fabry-Perot (b), and the
Sagnac (c) interferometer. Gas lasers and semiconductor
lasers are the sources appropriate for phase-modulated
fiber-optic sensors. The transmission type Mach-Zehnder is
topologically similar to the reflexion type Michelson
configuration. It is realized by replacing the two mirrors M
by a second coupler C in which the reference and measuring
arms of length L_R and L_M, respectively, are recombined for
interference. The phase $\emptyset = 2\pi n \cdot L/\lambda$ is changed by the parameter P to be measured which acts upon the optical length n·
L. P may be temperature, strain, pressure etc. The general
expression for the phase change is:

$$\Delta \Phi = \frac{2\pi}{\lambda} \Delta (n \cdot L) - \frac{2\pi n \cdot L}{\lambda^2} \Delta \lambda \quad .$$

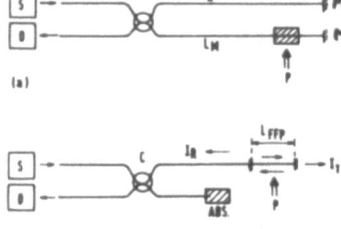

Fig. 47: Interferometric fiber
-optic sensor configurations,
Michelson (a),
Fabry-Perot (b), and
Sagnac (c).

In a Michelson/Mach-Zehnder interferometer the phase difference between the optical waves in the reference and measuring arms is detected:

$$\delta \Phi = \Delta \Phi_M - \Delta \Phi_R = \delta \Phi_{nL} - \delta \Phi_\lambda$$

$$= \frac{2\pi}{\lambda} [\Delta_M (n \cdot L_M) - \Delta_R (n \cdot L_R)] - \frac{2\pi \Delta \lambda n}{\lambda^2} (L_M - L_R) .$$

If there is an a priori length mismatch $\Delta L_0 = L_{M0} - L_{R0}$ of the interferometer arms then the optical path difference

$$\delta (nL) = \Delta_M (n L_M) - \Delta_R (n L_R)$$

includes a component $n \cdot \Delta L_0$. A first consequence of this mismatch is that the coherence length L_C of the source S used for the interferometric fiber-optic sensor must be larger than $2 \cdot n \cdot \Delta L_0$ in order to provide good fringe visibility. A second consequence is that any wavelength change $\Delta \lambda$ introduces a phase change $2\pi \Delta \lambda n \Delta L_0 / \lambda^2$ which will be erroneously interpreted as a parameter change by the sensor's signal evaluation system. A third consequence finally is a phase noise component that increases with the mismatch ΔL /1/ and adds to the detector and amplifier noise of the sensor.

The Fiber-Fabry-Perot(FFP-)-interferometer of Fig. 47b) needs a source with coherence length L_C of at least $2nL_{FFP}$, where L_{FFP} is the geometrical length of the FFP sensor element.

With decreasing L_c the effective finesse $F_{eff} = \Delta f / \delta f_{eff}$ (Δf = free spectral range, δf_{eff} = effective fringe width taken half way between maximum and minimum fringe intensity) and hence the fringe visibility of the FFP-resonator decreases. A FFP-sensor of high finesse (multiple-wave interference) needs a source with large L_c (typically 10 L_{FFP}) whereas FFP-sensors of low finesse (two-wave interference) require $L_c \gtrsim 2nL_{FFP}$. CSP-semiconductor lasers as described in section 3.3 meet these source requirements provided they are sufficiently temperature and injection current stabilized. Temperature stabilization to the mK and current stabilization to the µA provide a wavelength stabilization of about $5 \cdot 10^{-5}$ nm which is sufficient for most FFP-sensor applications.

The fiber-optic Michelson - as well as the FFP-interferometer require optical isolation of the source in order to prevent instable laser diode operation due to optical feedback. This isolation is provided by a polarizer plus a quarter-wave plate.

The wavelength tunability of laser diodes by tuning the injection current can be used to compensate the action of a measuring parameter by a suitable change of the laser wavelength. The compensation current is then calibrated against the measurand. This concept has been applied to a FFP-thermometer for biomedical applications /34/. Fig. 48 shows the functional scheme of this fiber-optic sensor that makes use of a CSP-laser diode operating at 780 nm.

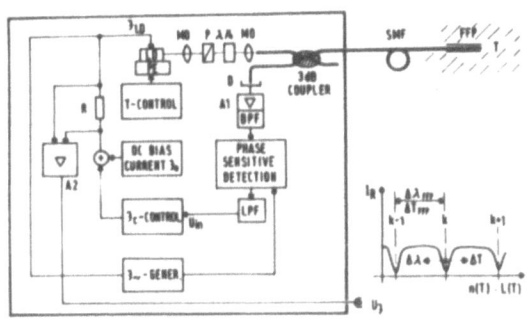

Fig. 48: FFP-thermometer for biomedical applications /34/.

The temperature and injection current prestabilized laser is tuned by current control such that a temperature change T at the FFP sensor element is compensated by a suitable wavelength change $\Delta\lambda$. This way the sensor remains locked to a fixed order of interference. A PIN-diode is used to detected the AC component of the reflected FFP-signal via phase sensitive detection. This AC component originates from a slight wavelength modulation of the laser diode optical output due to a superimposed injection current component oscillating at a few kHz. The signal is passed through a low pass filter (LPF) to provide an input voltage U_{in} for the control of the DC injection current. The laser diode can be tuned in wavelength over a range $\Delta\lambda = 0.1$ nm by varying the injection current over the range ΔI_{LD} = 15 mA without mode hop corresponding to a unique measuring range of 20 ^{o}C which can be covered with an accuracy of better than 0.1 ^{o}C.

In the fiber-optic gyroscope as depicted schematically by the Sagnac-interferometer in Fig. 47c) backscattered light originating at different positions within the fiber coil will interfere at the photodetector with fluctuating phase differences. The corresponding intensity fluctuations produce noise in the sensor output signal. Scatter events within an interval $[\frac{1}{2}(L-L_c), \frac{1}{2}(L+L_c)]$ in the fiber of length L with L_c being the coherence length produce phase noise which increases with increasing L_c. This perturbing effect can be eliminated by using a source with very small L_c, i.e. a superradiant diode (SRD) as described in section 3.4. The short SRD coherence length of the order of several μm is sufficient for this interferometer, since reference (clockwise light wave) and measuring (counterclockwise light wave) arm are accomodated in one and the same fiber and thus perfectly matched.

Recently fiber-optic sensors have been reported in which the optical source is RF-amplitude-modulated and the modulation phase difference in an interferometric configuration is measured /35,36/. LED as well as laser diodes may be used at modulation frequencies up to a few 100 MHz, which comes close to the modulation bandwidth requirements for optical communications sources.

Also serrodyning as a means to interrogate serial and parallel arrays of fiber-optic interferometric sensors /37/ requires high-frequency source modulation, in this case modulation of the wavelength or optical frequency. A laser source is needed in such cases that allows for wavelength ramping without procuding instable laser behaviour, in particular mode hops. The distributed feedback (DFB) laser will be particularly suited for this application, and successful efforts have been reported to develop an AlGaAs DFB laser at 800 nm in particular view of fiber-optic sensor applications /20/ as mentioned in section 3.3.

6   OUTLOOK: WHICH WAVELENGTH TO GO?

An important argument in favour of future fiber-optic sensor applications relates to the advantage of making use of components fabricated in large quantities for optical communications. This argument of "matched technology" promises low cost sensor developments by "riding on the back of communications technology".

Optical communications clearly go for the 1.3 μm and 1.55 μm low loss/low dispersion fiber windows. Is fiber-optic sensor technology likely to follow this development?

The outlook of using cheap mass produced components like LED's, laser diodes, couplers and connectors, optical isolators and compact source modules with built-in Peltier coolers, reference photodiodes and fiber pigtails is very appealing for following the long wavelength route. Also the large potential of integrated optics in the InGaAsP/InP material systems could be used in the future. In addition special integrated-optics active components such as phase modulators, switches, Bragg cells etc., fabricated in LiNbO$_3$ (titanium-indiffused waveguides) would provide advantages if operated at 1.3 or 1.55 µm including less stringent tolerances in photolithography and diffusion, negligeable photorefractive effect (optical damage) and easier fiber to integrated-optical chip coupling. Finally low loss optical isolators made in YIG materials by epitaxial techniques probably will be available in a forseeable future. At 800 nm these isolators have non-acceptable losses.

On the other hand, cheap and reliable sources are available now in GaAs/AlGaAs technology whereas InGaAsP-components are less available, less reliable and quite expensive. In addition at 800 nm the sensitivity of interferometric sensors is increased by a factor of about 1.6 as compared to operation at 1.3 µm. Short distance local area networks are expected to operate at 800 nm so that to a certain extent the argument of "matched technology" will be met in this wavelength window, too. Finally, DFB lasers for the 800 nm regime are being developed. They are, however, not yet available on the market and certainly will be rather expensive.

Thus within the years to come we most likely will see continuing work on fiber-optic sensors and sensor networks operating at 800 nm, but at the same time increasing sensor activities at 1.3 µm.

REFERENCES

1. Giallorenzi, Th.G. et al., "Optical Fiber Sensor Technology, IEEE J. Quantum Electronics, Vol. QE-18 (1982), 626 - 664.

2. Proceedings of the 1st Internat. Conf. on Optical Fiber Sensors, London (April 1983).

3. Proceedings of the 2nd Internat. Conf. on Optical Fiber Sensors, Stuttgart (Sept. 1984).

4. Proceedings of the 3rd Internat. Conf. on Optical Fiber Sensors, San Diego (Febr. 1985).

5. Medlock, R.S., "Review of modulating techniques for fibre optic sensors", Internat. J. of Optical Sensors, 1 (1986), 43 - 68.

6. Catalogue of the MICRO-GLÜHLAMPEN-GESELLSCHAFT, Hamburger Landstr. 1, D-2057 WENTORF/Germany.

7. Naese, C.J., "III-V-alloys for optoelectronic applications, J. Electron. Mat. 6 (1977), 253 - 293.

8. Hewlett-Packard Optoelectronics division, "Optoelectronics Applications Manual", McGraw Hill Book Company (1977).

9. Sharma, A.B., Halme, S.J., Butusov, M.M., "Optical fiber systems and their components", Springer-Verlag Berlin, Heidelberg, New York (1981).

10. Davis, Ch.M. et al., "Fiberoptic sensors technology handbook", Dynamic Systems, Inc. (1982).

11. Kersten, R.Th., "Einführung in die optische Nachrichtentechnik", Springer-Verlag Berlin, Heidelberg, New York (1983).

12. Siemens Datenbuch Opto-Halbleiter (1981/82).

13. Wagner, E., Gillessen, K., "Halbleiter-Emitter im sichtbaren und nahen infraroten Spektralbereich", GMR-Bericht Nr. 5, Meßverfahren mit optoelektronischen Halbleiter-Bauelementen, Lahnstein (Febr. 1985), 17 - 36.

14. Product information of LASER COMPONENTS (1986).

15. Marschall, P., Schlosser, E., Wolk, C., "A new type of diffused stripe geometry injection laser", Proc. 4th European Conf. Opt. Commun. (Sept. 1978), 94 - 97.

16. Chinone, N. et al., "Highly efficient (GaAl)As buried heterostructure lasers with buried optical guide", Appl. Phys. Lett. 35 (1979), 513 - 516.

17. Aiki, K. et al., "Transverse mode stabilized $Al_xGa_{1-x}As$ injection lasers with channeled-substrate-planar structure", IEEE J. Quant. Electr. Vol. QE-14 (1978), 89 - 94.

18. HITACHI Optoelectronic semiconductor products data book.

19. Itoh, M., Kimura, T., "Carrier density dependence of refractive index in AlGaAs semiconductor lasers", IEEE J. of Quantum Electron. QE-16 (1980), 910 - 911.

20. Kojima, K. et al., "Low threshold current AlGaAs/GaAs DFB laser grown by MBE", Proceedgs. IOOC-ECOC Venice (1985), 99 - 102.

21. Bludau, W., Rossberg, R.H., "Low-loss laser-to-fiber coupling with negligible optical feedback", J. Lightwave Technology, Vol. LT-3, (1985), 294 - 302.

22. Wang, C.S. et al., "High-power low-divergence superradiance diode", Appl. Phys. Lett. 41 (1982), 587 - 589.

23. Palais, J.C., "Fiber-optic communications", Prentice-Hall Inc. (1984).

24. "Award-Winning Fiber Temperature Sensor slated for commercial Production", Laser Focus/Electro-Optics (Dec. 1983), 66 - 67.

25. Wickersheim, K.A., Alves, R.B., "Recent advances in optical temperature measurement", Industrial Research/ Development (1979), 82 - 89.

26. Ovrén, C., Adolfsson, M., Hök, B., "Fiber-optic systems for Temperature and Vibration measurements in industrial applications", Optics and Lasers in Engineering 5 (1984), 155 - 172.

27. Jones, B.E., Spooncer, R.C., "Optical fiber pressure sensor using a shutter modulator and two-wavelength intensity referencing", 2nd Internat. Conf. on Opt. Fiber Sensors, Stuttgart (1984), 223 - 226.

28. Dabkiewicz, Ph., Ulrich, R., "Spectral encoding for fiber-optic industrial sensors", Proceedgs. EFOC/LAN 85, Montreux (June 1985), 212 - 217.

29. Bosselmann, Th., Ulrich, R., "High-accuracy position-sensing with fiber-coupled white-light interferometers", 2nd int. Conf. on Optical Fiber Sensors, OFS '84, Stuttgart (Sept. 1984), 361 - 364.

30. Hutley, M.C., "Wavelength encoded optical fiber sensors", 2nd Int. Conf. on Optical Fiber Sensors, OFS '84, Stuttgart (Sept. 1984), 111 - 116.

31. Hök, B., Gerritsen, S.J., "Multimode fibre-optic sensors with frequency output", Int. J. of Optical Sensors, Vol. 1 (1986), 89 - 93.

32. Venkatesh, S., Culshaw, B., "Optically activated vibrations in a micromachined silica structure, Electron. Lett. 21 (1985), 315 - 317.

33. Kist, R., Ramakrishnan, S., Wölfelschneider, H., "The Fiber-Fabry-Perot and its applications as a fiber-optic sensor element", 2nd Int. Techn. Sympos. on Optical and Electro-optical Applied Science and Engineering, Cannes (Nov. 1985).

34. Kist, R., Drope, S., Wölfelschneider, H., "Fiber-Fabry-Perot Thermometer for biomedical applications", 2nd Int. Conf. on Optical Fiber Sensors, Stuttgart (Sept. 1984), 165 - 170.

35. Wade, C.A., et al., "Optical fibre displacement sensor based on electrical subcarrier interferometry using a Mach-Zehnder configuration", 2nd Int. Techn. Sympos. on Optical and Electro-optical Applied Science and Engineering, Cannes (Nov. 1985).

36. Kotrotsios, G., Falco, L., Jeanneret, J.-P., Parriaux, O., "Radio frequency phase detection for intensity modulated fiber sensors", 2nd Int. Techn. Sympos. on Optical and Electro-optical Applied Science and Engineering, Cannes (Nov. 1985).

37. Giles, I.P., et al., "Coherent optical-fibre sensors with modulated laser sources", Electron. Lett., 19 (1983), 14 - 15.

ALL-FIBER GYROSCOPE: DESIGN AND PERFORMANCES

S. DONATI, V. ANNOVAZZI LODI, G. MARTINI
Università di Pavia, Pavia - Italy

SUMMARY

Principles and design philosophy of the fiber gyroscope are reviewed. The interest is focussed on the all-fiber configuration, of which performances and limitations are discussed.

1. INTRODUCTION

Fiber-gyro development has matured, since the first experimental demonstration by Vali and Shorthill ten years ago [1,2], to the point where the first commercially available units are now ready to appear for a number of applications in the industrial (e.g., robotics, oil well logging) and military (e.g., flight and guidance control, radar) areas.

An impressive progress in sensitivity and accuracy has been recorded in these years, as a result of a worldwide research effort devoted to overcome a series of relatively small but disturbing effects. Today that the fiber-gyro performance is close to the photon noise limit, the problems are with the engineering development of rugged, low cost units capable of competing with other technologies.

The first bulk-optics gyros had several hundred deg/h noise and even larger long-term drifts, but after exploiting the concept of reciprocity introduced by Ulrich [3,4] , gyro sensitivities of 1 deg/h and less were soon achieved. Further advance then came from the study of residual sources of errors, such as Rayleigh scattering and optical Kerr effect, with a corresponding improvement in the gyro performance, both in sensitivity and scale-factor accuracy.

Also, the progress in electro-optical technologies has made it feasible to develop the all-fiber gyro approach, where all the optical components (beamsplitters, polarizer, phase modulator, birefringent plates) are implemented by fibers [5,6] and, more recently, the integrated approach in which all the components but the fiber coil, the detector and the source, are fabricated on an electro-optical Li NbO_3 chip [7-9] , see fig. 1.

2. BASIC CONFIGURATION OF THE FIBER GYRO

As it is well known, the phase shift induced in the Sagnac ring interferometer between the two counterpropagating waves, due to an inertial angular velocity Ω perpendicular to the ring plane, is given by

$$\varphi_s = 8\pi \, NA\Omega/\lambda c \qquad (1)$$

where λ and c are the vacuum wavelength and speed of light, A is the coil area and N the number of turns. Using reasonable values for fiber length and

300

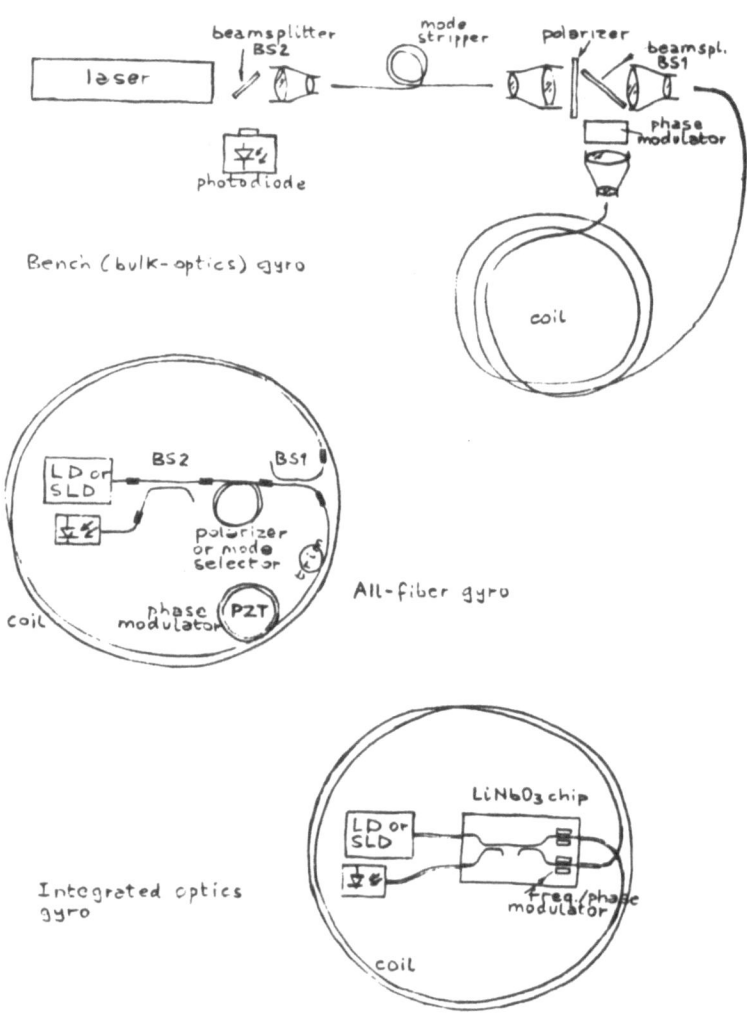

Fig. 1 - The fiber-gyro generations

coil size, e.g. L = 200 m and A = 100 cm^2, one has at λ = 0.85 μm a responsivity R,

$$R = \varphi_s/\Omega = 8\pi NA/\lambda c$$

of about 2.7 μr/(deg/h), calling for a high sensitivity measurement scheme of the phase φ_s, not only for the inertial-grade gyros (0.001 deg/h) but also for the less demanding applications to guidance and robotics (1-10 deg/h). Another key problem of the early scheme of Sagnac interferometer is that, after recombination of the propagated beams on the photodetector, the beating signal is of the form:

$$I = 2 I_0 + 2 \, \mathrm{Re} \, (\vec{E} \, \vec{E}^*) = I_o (1 - \cos \varphi_s) \tag{2}$$

i.e. the differential responsivity $dI/d\varphi_s$ is zero at rest, the sign of Ω is not detected, and the scale factor linearity is affected by a sine-function error. Further, it was soon recognized that, to achieve high sensitivity without disturbing effects from the environment, the optical pathlengths of the two waves should be accurately balanced and tracked a condition leading to the concept of reciprocal propagation [3,11].

Many different approaches for removing the $\cos\varphi_s$ dependence have been demonstrated, (see e.g. [10]) either by means of a two-frequency optical source or by adding a frequency or a phase modulator in the propagation loop. However, the most viable and generally accepted scheme for an all--fiber implementation is that reported in fig. 2, originally proposed by Ulrich [4] and subsequently analyzed and improved by several authors [5, 7,8].

In the scheme, the fiber of the main sensing coil shall have the property of allowing the propagation of a well-defined mode (or mode distribution) with a well-defined state of polarization, while keeping as low as possible the cross-coupling to other modes or polarization states. The selector acts as a mode and polarization filter matched to the fiber.

Usually, with high-birefringent fibers offering a good polarization decoupling, the selector is simply a fiber-polarizer and mode filtering is left to the coil itself; in this case it can be shown that the phase error φ_{sn} due to a finite extinction ratio ε (in field amplitude) of the polarizer is of the order of [12] :

$$\varphi_{sn} = 2\varepsilon \, t \, a \tag{3}$$

where t is the ratio of cross-to main-polarized amplitude transmission of the fiber and a is the ratio of cross-to main-polarized field amplitudes supplied by the source. Eq. (3) is valid down to about $\varphi_{sn} = 10^{-7}$ because the fundamental HE$_{11}$ mode is not strictly linearly-polarized [13] , and to obtain smaller values one should perform a true mode-filtering selection. Suitable reinterpretation of eq. (3) allows to explain the operation with unpolarized source [11,14] and to evaluate the performance limits for this particular state-variables selection.

Beam splitter reciprocity is ensured, in the scheme of fig. 2, by moving the detector to collect the beams going back to the source. This is

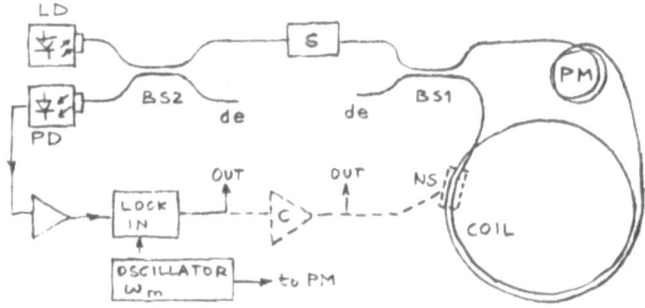

Fig. 2 - Basic configuration of reciprocal all-fiber gyro: BS1, BS2 = fiber
couplers; LD = laser diode or superluminescent diode; S = polarization and mode selector; PM = phase modulator; PD = photodiode
detector; de = dead ends. Dashed components are added for closed-
loop operation: C = feedback controller, NS = non reciprocal pha-
se-shifter.

accomplished with the fiber coupler BS2. The beams reaching the detector
thus pass the fiber coupler one time in straight-coupling and one time in
cross-coupling, only in interchanged order for each of the beams. This,the
configuration is reciprocal unless the phase shifts (associated with the
fiber coupler loss) suffered at a time t and t + L/c by the two beams in
straight and cross-coupling do not compensate exactly; in a good-quality
fiber coupler the error is negligible
The phase modulator, on the other hand, relies just on this time-dependent
source of nonreciprocity to allow the modulation of optical phase at a fre
quency ω_m. Usually implemented by several turns of fiber wound on a PZT
cylinder and located at an end side of the coil, it allows to impress a sub
stantial modulation phase amplitude ϕ_m even with modest voltage drives. The
counter-propagating waves then take a phase difference

$$\varphi = \varphi_s + \phi_m \left[\cos \omega_m (t + L/2c) - \cos \omega_m (t - L/2c) \right]$$

$$= \varphi_s + 2\phi_m \sin \omega_m L/2c \sin \omega_m t \qquad (4)$$

and the beating signal of eq. (2) now becomes:

$$I/I_o = \left[1 + J_o(\phi_m') \right] \cos \varphi_s - 2 J_1 (\phi_m') \sin \varphi_s \sin \omega_m t + \qquad (5)$$

$$+ 2 J_2 (\phi_m') \cos \varphi_s \sin 2 \omega_m t + \text{h.ord. harm.},$$

where $\phi' = 2\phi_m \sin \omega_m L/2c$ gives at 1.81 a maximum of the Bessel J_1 amplitude in eq. (5), at the fundamental frequency of modulation ω_m and with a
$\sin \varphi_s$ dependence of the Sagnac phase shift. Thus, by detection of the fundamental component 90° off the modulating signal through a lock-in amplifier, one has the sign detection of Ω and a fair linearity, if the sensitivity is high, since $I \propto \sin \varphi_s$.

Respect to this open-loop scheme, a further improvement of linearity
can be gained by nulling the gyro through a feedback loop (fig. 2).

This requires a non reciprocal effect in the propagation loop, by
which a phase shift equal and apposite to the Sagnac phase shift is added.
While in the integrated-optics gyro this is readily done with acusto-optic
modulation, leading also to a very convenient digital readout, in the all-
-fiber gyro it requires the combination of a Faraday (nonreciprocal) rotator, built in form of a solenoid wound around the sensing loop, and of a
pair of quarter-wave compensating fiber-loops (reciprocal) [15].

To summarize to above discussion on reciprocity in a more general framework, we may give the following picture. The optical signal propagating
through the fiber gyro can be regarded as a mixture of eigenstates of the
physical structure. Each eigenstate $\psi_k(x,y,z,t)$ can be expressed, e.g., in
term of a linear superposition of the orthonormal set of fiber modes,

$$\psi_k(x,y,z,t) = \Sigma\, a_{kjl}\, E_{jl}(x,y)\, \exp(i\beta_{jl}z + i\omega_{jl}t + i\varphi_{jl}) \qquad (6)$$

where the index l is for the mode order and the j index is for the polarization state. This implies a number of degrees of freedom of the optical
signal, connected to the modal composition, to the polarization state and
to the coherence length given by $\{\varphi_{jl}\}$ (see below). The eigenstates are found
by solving the characteristic equation of the propagation operator P,

$$P(\psi_k) = \lambda\, \psi_k \qquad (7)$$

where P is the cascade of the operator describing the optical elements in
the gyro setup, i.e. sensing loop, polarizers, beamsplitters, etc. The operator P and the eigenfunctions ψ can be broken down in a double set of counterpropagating waves,

$$P = \vec{P} + \overleftarrow{P}, \qquad \psi_i = \vec{\psi}_i + \overleftarrow{\psi}_i \qquad (8)$$

At the detector the photon-annihilation operator A gives the photodetector
current in form of:

$$I = \Sigma_{ij}\, \vec{a}_i \vec{a}_j\, \langle\vec{\psi}_i | A | \overleftarrow{\psi}_j\rangle \qquad (9)$$

and therefore, only if terms with $i \neq j$ are negligible, will the gyro exhibit a beating depending on the Sagnac phaseshift φ_s, in view of the dependence of the phase from ω_{il} and φ_{il} [eq. (6)]. As a consequence of eq. (9),
consider the case of limited coherence lenght of the source, expressed by
developing eq. (9) with the aid of eq. (6) in the form:

$$I = 2Re < \Sigma_{ij} \ \Sigma a_{ilp} \ E_{1p}(x,y) \ a^*_{jmq} \ E_{mq}(x,y) \cdot$$

$$\cdot \exp \left[i(\beta_{1p} - \beta_{mq})z + i(\omega_{1p} - \omega_{mq})t + i (\varphi_{1p} - \varphi_{mq}) \right] > \tag{10}$$

$$\cong 2 \ \Sigma_{ij} \ \Sigma a_{ilp} \ E_{1p}(x,y) \ a^*_{jmq} \ E_{mq}(x,y) \cdot$$

$$\cdot \cos \left[(\beta_{1p} - \beta_{mq})z + (\omega_{1p} - \omega_{mq})t \right] \quad <\cos (\varphi_{1p} - \varphi_{mq}) > \tag{11}$$

where the last factor $\mu = <\cos(\varphi_{1p} - \varphi_{mq})>$ represents the degree of coherence between eigenfunction 1p and mq. If a short coherence-length source is selected, $\mu = 0$ for $1,p \neq m,q$ and the error from different frequency modes is eliminated, as pointed out in Refs [6,11] . The coherence length can then be regarded as a further degree of freedom of the optical signal ψ, which allows an improvement of the discrimination selectivity. Lastly, going back to generalize eq. (3) we can state that, given a fundamental mode ψ_k selected by the gyro structure, the phase error due to residual imperfections can be expressed as:

$$\varphi_{sn} = 1 - \left[\vec{a}_k \vec{a}_k <\vec{\psi}_k \mid A \mid \vec{\psi}_k > \right] / \Sigma_{ij} \ \vec{a}_i \vec{a}_j <\vec{\psi}_i \mid A \mid \vec{\psi}_j > \tag{12}$$

3. FACTORS AFFECTING THE FIBER GYRO PERFOMANCE

Several effects limit the sensitivity performance of the fiber-optics gyro. Among these we shall consider:

- Photon noise. Assuming Poisson statistics for the number of photons detected in a given time interval [16-18] , the rms deviation of phase noise for an interferometric signal is, using eq. (5):

$$\sigma_\varphi = \frac{\sigma_I}{|dI/d\varphi|} = \frac{(2h\nu/\eta P\tau)^{\frac{1}{2}}}{2 \ J_1(\phi'_m)/ \left[1 + J_0(\phi'_m) \right]} \tag{13}$$

where ηP is the average detector power and η is quantum efficiency of the photodetector. Dividing σ_φ by the gyro responsivity R [eq. (1)] , we get the noise equivalent angular velocity NEΩ :

$$NE\Omega = \frac{\lambda c}{8\pi NA} \ \frac{(2h\nu/\eta P\tau)^{\frac{1}{2}}}{2J_1(\phi'_m)/ \left[1 + J_0(\phi'_m) \right]} \tag{14}$$

By inserting in eq. (14) typical values, i.e.:$R = 2.7 \ \mu r/(deg/h)$ and $\eta P = 200 \ \mu W$, we get as a result NE$\Omega = 0.01 \ (deg/h)s^{\frac{1}{2}}$, which indicates how the overall transmission efficiency of the fiber-gyro is important to achieve a reasonable photon noise limit at adequate integration times τ.
- Scattering noise. It originates in the sensing fiber under the form of

Rayleigh backscattering, and at the interfaces or splices between the various optical components. Light scattered in the forward or backward direction so as to superpose on the propagating beam is guided by the fiber and can beat with the useful signal, giving rise to a large source of error. A powerful cure is to use a small coherence-length source, such as the superluminescent diode, which reduces to a small length centered at the midpoint of the sensing coil the Rayleigh sensitive contribution,while backscattering from splices is virtually eliminated.

- Mode conversion noise. A large source of apparent nonreciprocal bias and noise is encountered when in the fiber sensing path, i.e. downward the counterpropagating wave splitting produced by BS1, one has splices between physically different fibers or relatively high loss splices. Then, because of mode conversion nonreciprocity, the attenuations $\vec{a}$ and $\overleftarrow{a}$ may be appreciably different, and similarly the associated phase shifts $\vec{\varphi}_a$ and $\overleftarrow{\varphi}_a$. The result is an unbalance of phase shifts of the two counterpropagating waves. Obviously, one should keep splice losses and any source of mode conversion to a minimum in the critical region of the sensing fiber, to avoid this additional contribution of noise.

- Faraday noise. A magnetic field intensity component, parallel to the sensing fiber axis, gives rise to a nonreciprocal phaseshift given by $\varphi = \oint V \, \overline{H} \cdot \overline{dl}$, where V is the Verdet constant of the fiber. While on a closed loop the closed line integral should be zero identically, the combined effect of reciprocal residual birifrangence and nonreciprocal Faraday birefringence [17] can give rise to a nonvanishing net effect, unless special care on the ratio of integer turns number versus the half-wave bending is provided in the loop design. Also, ac magnetic fields may be of concern, even though their effect can usually filtered out from the detected signal.

- Kerr effect bias. This effect alters the propagation constants $\vec{\beta}$ and $\overleftarrow{\beta}$ in dependance on the counterpropagating powers $\vec{P}$ and $\overleftarrow{P}$, as a result of third-order index of refraction nonlinearity [10]. In principle, the variation of propagation constants due to optical Kerr effect is given by:

$$\Delta\vec{\beta} = K(\vec{P} + 2 \overleftarrow{P}), \qquad \Delta\overleftarrow{\beta} = K(\overleftarrow{P} + 2 \vec{P})$$

where $K = 96\pi^3 \omega\chi_{(4)} r^2/n^2 c^2$ (where $\chi_{(4)}$ is the fourth order susceptibility and r is the core radius) is of the order of $4 \cdot 10^{-3}$ $W^{-1}cm^{-1}$ for a typical fiber [19]. While for a monochromatic polarized source the Kerr bias may amount to several tens of $\mu r/\mu W$ thus calling for a precise power balancing, either a 50 - percent duty cycle source modulation or the use of a broadband unpolarized source like the superluminescent diode tend to cancel the Kerr error efficiently for most practical applications [20].

4. DESIGN AND ENGINEERING

An all-fiber gyro, either in the open loop configuration for exploiting minimum part counts or in the closed loop configuration for maximum dynamic range, has the advantage of a relative easiness of engineering. However,

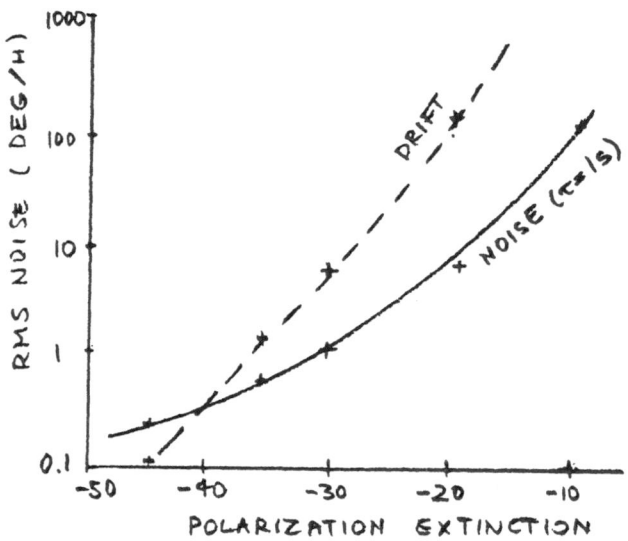

Fig. 3 - Baseline drift (1 - h period) and noise (1s integration time, 5 min sample time) as a function of the polarization extinction of the fiber-polarizer combination (fiber length is 200 m).

several issues shall be considered: (i) preserving the quality of mode/polarization discrimination of the components in the final assembly; (ii) ensuring a good ruggedness and immunity to vibrations and accelerations; (iii) hardening against electromagnetic interference; (iv) obtaining a good performance of zero drift and scale factor on a wide range of tempera ture. We briefly discuss the first two points, which impact more directly the fiber loop and optical components. Mechanical ruggednees calls for a rigid fixture of all components, with possible high stress. On the other hand, it is mandatory to avoid sharp bends and kinks in the winding of the fiber sensing loop, which can severely degrade the polarization mantaining properties. Indeed, in fig. 3 an experimental result is reported on the noise and drift performance versus the polarization extinction supplied by the fiber and polarizer combination. Also, the fusion splice of fiber coupler and main fiber loop has to be carefully performed, with proper alignment of the axes of the polarization mantaining fiber. Compensating loops could be used to trim small misalignments, but they are generally avoided in an engineered unit. The fusion splice technique, though well known, requires a substantial developmental work to allow low loss, reproducible splices at first trial, a constraint needed to have the correct fiber length, without any excess, to fit exactly in the gyro fixture. The phase modulator operation frequency shall be chosen carefully, evaluating the interferometric perfomance which reveals a number of minute risonances not found in the impedance vs frequency analysis. The relatively high temperature coefficient of the piezoelectric constant compels to use a compensation scheme if scale-factor accuracy better than 1% is required. Usually, a good compensation can be achieved by correcting the fundamental -frequency signal $2 I_o J_1(\phi'_m)$ [eq. (4)] by the aid of two auxiliary signals: the second-harmonic amplitude $2 I_o J_2(\phi'_m)$ and the intensity I_o, which can be taken by a photodiode placed at the dead-end of coupler BS2; a simple analog-ratio processing yields a correction also against the long term and warm-up drifts of the power launched by the source as well as of the optical attenuation of fusion splices and fiber components.

REFERENCES

1 V.Vali, R.W.Shorthill: Fiber Ring Interferometer, Appl. Optics, 15 (1976) pp. 1099-1100

2 V.Vali, R.W. Shorthill, M.F.Berg: Fresnel-Fizeau Effects in Optical Fiber Interferometers, Appl. Optics, 16 (1977) pp. 2605-2607

3 R. Ulrich, M Johnson: Fiber Ring Interferometer: Polarization Analysis, Opt. Lett., 4 (1979) pp. 152-154

4 R. Ulrich: Fiber-Optic Rotation Sensing with Low Drift, Opt. Lett. 5 (1980) pp. 173-175

5 R.A. Berg, H.C.Lefevre, H.J.Shaw: All-Single-Mode Fiber Gyroscope with Long-Term Stability. Opt. Lett. 6 (1981) pp. 502-504.

6 K.Bohm et al.: Low-Drift Fibre Gyro Using a Superluminescent Diode, Electr. Lett. 17 (1981) pp. 352-353

7 H.J.Arditty et al.: Integrated-Optics Fiber Gyroscope, Proc. Opt. Fiber Sensors Conf., Stuttegart 5-7 sept. 1984, pp. 321-325

8 R.A.Bergh, H.C. Lefevre, H.J.Shaw: An Overview of Fiber-Optic Gyroscope, J. Lightware Techn., LT-2 (1982) pp. 91-107

9 B.Y. Kim, H.J.Shaw: Fiber Optic Gyroscope, Spectrum IEEE, march 1986, pp. 54-60

10 S.Ezekiel, H.J.Arditty: Fiber-Optics Rotation Sensors, Springer-Verlag, Berlin 1982

11 R. Ulrich: Polarization and Depolarization in the Fiber-Optic Gyroscope, ibidem, pp. 52-77

12 E.C. Kintner : Polarization Control in Optical Fiber Gyroscope, Opt. Letters 6 (1981) pp. 154-156

13 M.P. Varnham, D.N.Payne, J.D.Love: Fundamental Limits to the Trasmission of Linearly Polarized Light by Birefringent Optical Fibres, Electr. Lett.

$\underline{20}$ (1984) pp. 55-56

14 G.A. Pavlath, H.J.Shaw: Multimode Fiber Gyroscope, in 10 , pp. 111-113

15 V. Annovazzi-Lodi, S.Donati: Combined Reciprocal and Nonreciprocal Birefringence in Optical Monomode Fibres; J. Opt. and Quant. El., $\underline{15}$ (1983) pp. 381-388

16 F.T.Arecchi, E.O. Schulz Du Bois: Laser Handbook, vol. 4, North Holland Press, Amsterdam 1985

17 V.Annovazzi-Lodi, S. Donati, G.Martini: Il Giroscopio Laser a Fibra Ottica, 84[a] Riunione Annuale AEI, Cagliari 1983, paper B 38 pp. 1-15

18 S.Donati, V.Annovazzi Lodi: Fiber Gyroscope with Dual Frequency Laser, Proc. LIA-ICALEO, $\underline{34}$ (1982) pp. 85-89

19 S. Ezekiel, J.L.Davis, R.W. Hellwerth: Observation of Intensity Induced Nonreciprocity in a Fiber Optic Gyroscope, Optics Lett. 1 (1982)pp.457-459

20 R.A.Berg, B.Culshaw, C.C.Cutler, H.C.Lefevre, H.J.Shaw: Source Statistics and the Kerr Effect in Fiber-Optic Gyroscopes, Optics Lett. $\underline{7}$ (1982) pp. 563-565

PHASE RECOVERY IN THE INTERFEROMETRIC FIBER-OPTIC SENSORS

M.Martinelli

CISE S.p.A., P.O.B.12081, 20134 Milano,Italy

1.INTRODUCTION

In the processing of the optical fiber interferometric signal a current technique refers to a Phase Tracking, PT, scheme[1], where a feedback loop permits to loock the interferometer in the point of maximum sensitivity, around $\pi/2$ of the phase shift. This approach, useful in several applications for its extreme high sensitivity, is inherently limited by the very small range of linearity offered. The extension of the range of linearity is one of the primary objective to reach in the development of industrial optical fiber sensors. In fact, considering for example, a gage length of 10 cm, typical engineering values of strain are in the order of 100-1000 $\mu\varepsilon$, which are very large if compared with the fraction of $\mu\varepsilon$ obtainable when working under fringe as imposed by the PT techniques. To overcome this limitation a Frequency Demodulation technique joints to the heterodyne operation can be adopted, as developed in several interferometric laser applications. The first part of the paper reviews the coherent detection process. Phase and Frequency modulations and homodyne and heterodyne schemes are then compared and a new method to directly extract the phase change values from a FM spectrum signal is pointed out.

2.COHERENT DETECTION

2.1.Interference

The coherent detection of a light wave consists in adding the signal wave to a reference phase correlated wave (local oscillator) and low-pass filtering the resulted signal[2,3]. The addition is often obtained by means of a semi-reflecting mirror that launches the signal and the reference waves parallel against a photodetector presenting a square-law response. The physical characteristics of the photon-current conversion of a typical solid-state photodetector make it possible to perform an average measurement of the optical signal only. As a matter of fact, this property operates a high-frequency cut-off of the detected signal.

Let $\theta(t)$ and $\Psi(t)$ denote the time-dependent phase of signal $E_s\cos\theta(t)$ and reference $E_r\cos\Psi(t)$ wave. It is

$$\theta(t) = \theta_o + \omega_s t \qquad (1)$$

$$\Psi(t) = \psi_o + \omega_r \tau \qquad (2)$$

where θ_o and ψ_o are the starting phases, and ω_s and ω_r the angular carrier frequencies of the two incident waves.

After the superimposition, the photodetector response r is

$$r = \beta \{ (E_s \cos \theta(t))^2 + (E_r \cos \Psi(t))^2 + E_r E_s$$

$$\cos[\theta(t) + \Psi(t)] + E_s E_r \cos[\theta(t) - \Psi(t)] \} \quad (3)$$

where β is a proportionality constant depending on the detector efficiency. Owing to the average property of the photodetector, the first two terms become 1/2, the third becomes null and (3) turns out:

$$r = \beta \{ \frac{E_s^2}{2} + \frac{E_r^2}{2} + E_s E_r \cos[\theta(t) - \Psi(t)] \} \quad (4)$$

which could be expressed as the sum of a d.c.term and an intermediate frequency, i.f., term containing the phase signal intelligence

$$r_{if} = \beta E_s E_r \cos[\theta_o + \omega_s t - \psi_o - \omega_r t] \quad (5)$$

The need for phase correlation between the two waves means that during the measurements the two signals are to be synchronized or, at least, show a definitive phase correlation. In other words, if the two signals are completely uncoherent also the last term of (3) becomes null. In an optical arrangement this condition is quite easily obtained when the local oscillator and the signal wave are both drawn out of the same light beam, as it occurs in a Michelson or Mach-Zehnder interferometer. Later on in the text it will be shown how the need for phase correlation practically determines the choice of the laser source. The r_{if} expression (5) develops into two principal branches according to the choice of the ω_s and ω_r values.
When $\omega_s = \omega_r$ a homodyne (here and after in the text used in the literally meaning: from the greek ομος = identical and δυναμις = power) detection is obtained and r_{if} becomes:

$$r_{if} = \beta E_s E_r \cos(\theta_o - \psi_o) \quad (6)$$

If a $\theta(t)$ phase change is added to the signal wave, putting $\theta_o - \psi_o = \theta'_o$, we obtain

$$r_{if} = \beta E_s E_r \cos[\theta'_o + \theta(t)] \quad (7)$$

which shows the modulation role played by $\theta(t)$ on the detected intensity. Consequently, the homodyne scheme seems appropriate for a coherent detection of a limited phase modulating signal, the limit being imposed by the polydrome character of the function cosine: $\theta(t)$ must be less than π radians.
Let's now assume that both reference and signal waves are drawn out of the same optical beam of carrier ω_o. If a frequency modulating signal acts on the signal beam causing a frequency

shift $\omega_m/2\pi$, we obtain : $\omega_s = \omega_o + \omega_m$ and $\omega_r = \omega_o$. In synchronized conditions and putting for simplicity $\theta_o' = 0$ expression (5) becomes:

$$r_{if} = E_s E_r \cos \omega_m t \qquad (8)$$

that is, the detected intensity is simply related to the frequency modulating signal. Because of the physical process that causes the modulation, the frequency changes introduced by the modulating signal usually scan from a plus or minus peak value, $+\Delta\omega_m/2\pi$. In these conditions, expression (8) would be ambiguous. To solve the ambiguity $|\omega_s - \omega_r|$ must be at least $>\Delta\omega_m$, and a frequency shift $\omega_m/2\pi$ must be introduced into the reference or into the signal beam. As a result the detection scheme operates in heterodyne (here and after in the text used in the literally meaning: from the greek ετεpoς = different and δυναμις = power) conditions: $\omega_r \neq \omega_s$.

To summarize, coherent detection is a general optical process permitting to detect the phase intelligence carried out by an optical signal. Although it is difficult to set a precise boundary, it is possible to split the coherent detection into two main branches, homodyne and heterodyne schemes, which seem appropriate to detect a phase modulating and a frequency modulating signal, respectively. Phase and frequency modulation are a particular type of angle modulation occurring in an optical wave, and the relation with the optical fiber sensor property will be investigated in the next chapter.

2.2. Phase and frequency modulation

To fix the idea without loss of generality let's refer to a Michelson interferometer scheme where the sensing part of the fiber is considered a device set in one arm of the interferometer causing a phase delay with integral peak value ϕ_o sinusoidally time-dependent at the frequency $\omega_m/2\pi$, $\phi(t) = \phi_o \sin \omega_m t$. For example this fiber-optic sensor could be a strain sensor like that presented in Ref.4, which detects a strain modulation.

"Phase modulation"[5] is the process occurring when the signal wave is modulated by a phase change linearly proportional to the modulating cause, $M(t)$. It is therefore possible to write (1) as

$$\theta(t) = \theta_o + \omega_s t + k_1 M(t) \qquad (9)$$

In our case $M(t)$ is directly the phase delay $\phi(t)$, and (9) becomes

$$\theta(t) = \theta_o + \omega_s t + 2\phi(t) \qquad (10)$$

where $k_1 = 2$ denotes the double passage of the light beam through the sensing fiber. A phase modulation can be detected by a homodyne scheme, and from expressions (7) and (10)

$$r_{if} = \beta E_s E_r \cos[\theta_o' + 2\phi_o \sin \omega_m t] \qquad (11)$$

312

The maximum of sensitivity of this detection is obtained around the value $\cos \theta = 0$, that is for $\theta' = \pi/2$. Moreover, in this condition, the sign ambiguity is also removed. When ϕ_o assumes large values, expression (11) becomes undetermined. In interferometric terms, a fringe shift is reached anytime ϕ_o equals $\pi/2$ radian and a method to make very large phase measurements could be the fringe counting. However, if ϕ_o suddenly changes, the fringe counting involves some detection problems and a powerful interpretative scheme must be introduced.

"Frequency modulation"[5] is the process where the radian frequancy ω_s of the signal wave is added by a time-dependent istantaneous radian frequency linearly proportional to the modulating cause $M(t)$, that is, in terms of phase change,

$$\frac{d\theta(t)}{dt} = \omega_s + k_2 M(t) \qquad (12)$$

If the modulating cause is the phase delay $\phi(t)$, (12) becomes

$$\frac{d\theta(t)}{dt} = \omega_s + k_2 \phi(t) \qquad (13)$$

where k_2 is to be computed.

To do this, let's consider the scheme in Fig.1 as a vibrometer where the "target" is a rest, but the "path" changes its velocity.

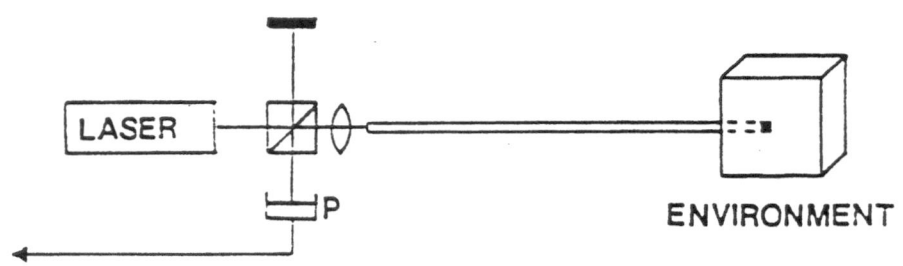

FIGURE 1. The Michelson scheme for the optical fiber interferometric sensor.

The Doppler-induced frequency swing will be

$$\Delta\omega_m = 2\pi \frac{2v}{\lambda} \qquad (14)$$

where v is the normal target velocity. This expression can be developed into a more general form as

$$\Delta\omega_m = 2kv = 2k\frac{dX}{dt} = 2\frac{dX}{dt} \qquad (15)$$

where k is the light avevector and χ a generic phase. To fix the idea, let's now assume that the optical fiber sensor is a strain sensor. Then, the indiced phase change $d\chi$ will be

$$d\chi = k \; dl - \frac{1}{2}lk_o n^3 d(\frac{1}{n^2})_z \qquad (16)$$

which, substituded into expression (15), gives

$$\Delta\omega_m = 2k\frac{dl}{dt} - lk_o n^3 \frac{d}{dt}(\frac{1}{n^2})_z \qquad (17)$$

The first term of this expression is the "classical" Doppler frequency shift due to the "distance" change dl. The second is an additional term due to an effective light path change introduced by the modulation of the refraction index. Then, considering $\phi(t)$ as phase change and taking into account its time-dependent behaviour,(15) becomes

$$\Delta\omega_m(t) = 2\omega_m\phi(t) \qquad (18)$$

and $\Delta\omega_m(t)$ can be considered the time-dependent frequency proportional to the modulating cause, $\phi(t)$. Consequently,(13) becomes

$$\frac{d\theta(t)}{dt} = \omega_s + \Delta\omega_m(t) = \omega_s + 2\,\omega_m\phi(t) \qquad (19)$$

where $2\omega_m$ is the wanted proportional term k_2.
A frequency modulation can be detected by a heterodyne scheme. If $\omega_s - \omega_r = \omega_c$, equal to the artificially introduced frequancy shift, the detected intensity becomes

$$r_{if} = \beta E_r E_s \cos\left[\int^t \frac{d\theta(t)}{dt}dt - \psi_o - \omega_r t\right] \qquad (20)$$

$$= \beta E_r E_s \cos\left[\omega_c t + 2\phi_o \cos \omega_m t\right] \qquad (21)$$

$$= \beta E_r E_s \cos\left[\omega_c t + \frac{\Delta\omega_m}{\omega_m} \cos \omega_m t\right] \qquad (22)$$

where the phase constant term has been disregarded. Term $\Delta\omega_m/\omega_m$ is called modulation index n of the frequency modulation process and results equal to $2\phi_o$.
Comparison between expression (21) and (11) enhances analogies and differences of the two modulating processes. In PM the r_{if} is related to the modulating signal, while in FM it is related to the integral of the modulating signal. Besides, the sign indetermination is removed in PM by working around the $\pi/2$ phase delay, and in FM by heterodyning the scheme. In both cases the integral value ϕ_o plays the important role of "modulation depth". In general, the PM approach seems to be more convenient for small fast time-dependent ϕ, while the

FM approach applies to larger ϕ.

2.3.Homodyne and heterodyne processes

The most used homodyne detection scheme is sketched in Fig.2 and termed Phase Tracking PT. An integrator operates a low-pass filtering of the signal coming from the photodiode ampli-fier and feeds an HV amplifier connected with an ADP crystal (or a generic phase modulator). If the working point of the interferometer moves from $\pi/2$, (because of slow phase drifts) a dc signal is generated by the integrator, which, amplified by the HV, generates a phase delay of opposite sign on the reference arm of the interferometer.

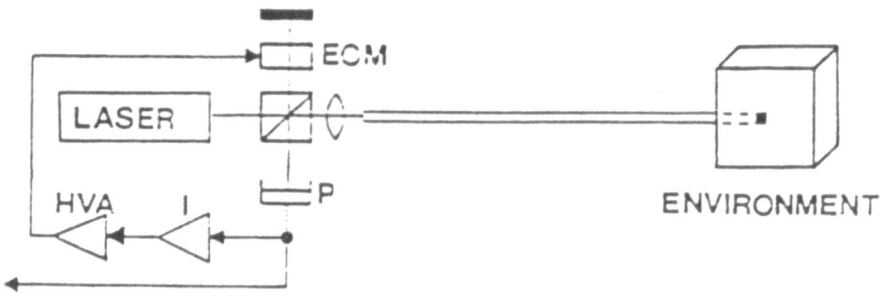

FIGURE 2.The Phase Tracking process.

Obviously, the constant time of the feedback loop is chosen in dependence on the frequency of the detected signal. The PT scheme is particularly fit for a vibration detection of very low amplitude. The $\pi/2$ phase delay coincides with the maximum of sensitivity of the system, defined as

$$ s = \frac{dr_{if}}{d\theta} = \beta E_s E_r \sin \theta \tag{23} $$

Besides, the function "sine " is quite linear around $\pi/2$, and this permits to obtain a simple linear expression linking the induced phase modulation ϕ to the signal

$$ r_{if} = \beta E_s E_r \cos \left[\pi/2 + 2\phi_0 \sin \omega_m t \right] = $$

$$ = \beta E_s E_r \sin 2\phi_0 \sin \omega_m t \tag{24} $$

$$r_{if} \simeq \beta E_s E_r \, 2\phi_o \sin \omega_m t, \qquad \text{for } \phi_o \text{ little} \qquad (25)$$

If amplitude or alignment fluctuations involve the detection scheme, the peak amplitude of the intermediate frequency term $(r_{ifo} = \beta E_s E_r)$ changes, thus causing a possible error in the phase measurement. Since r_{ifo} is a figure easily measurable, it can be taken as a reference value for the expression of ϕ_o and from (25) to obtain

$$\phi_c = \frac{1}{2} \frac{r_{if\ peak}}{r_{ifo}} \qquad (26)$$

If both r_{igo} and $r_{if\ peak}$ are expressed in the same unit (e.g. volts), the value of ϕ_o is read directly in radians.

A typical heterodyne arrangement is shown in Fig.3. The frequency shift is produced by means of a sawtooth generator feeding an electro-optical crystal.

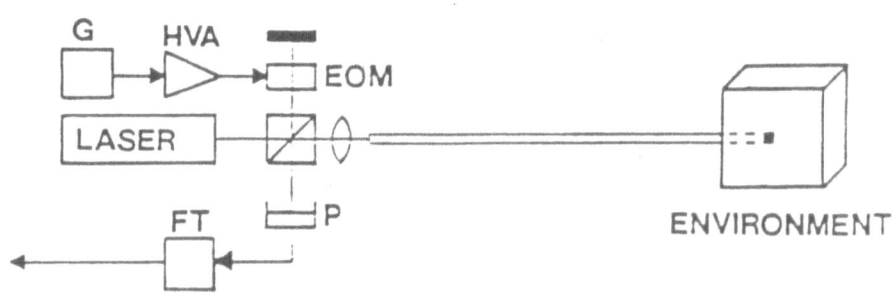

FIGURE 3.The Frequency Tracking process.

Many electronic schemes permit to operate the FM demodulation process[3,5,6]. One of the most advanced schemes refers to the phase-locked loop circuitry [7,11] and is known as Frequency Tracking, FT. In a FT, a feedback loop permits to restore an input FM signal by means of a comparison circuit based on a phase detector. In this process a signal proportional to the bandwidth B of the restored FM wave is available. As it will be shown, B is directly related to the modulation index m of the modulating wave. Consequently, the wanted value of ϕ_o can be found from expression (18).

The heart of the matter in a demodulation process is: how the bandwidth of the modulated signal is related to the modulating wave and, in particular, for a single sinusoidal modulating carrier, to the modulation index? Let's start considering that expression (22) writes as a function of modulation index m

316

$$r_{if} = 3E_r E_s \cos\left[\omega_c t + m \cos \omega_m t\right]$$

(27)

By means of the Fourier-series expansion expression (27) becomes

$$r_{if} = 3E_r E_s \sum_{n=-\infty}^{\infty} J_n(m) \cos(\omega_c + n\omega_m)t$$

(28)

where $J_n(m)$ is the Bessel function of the first kind of order n and argument m. Expression (28) defines a discrete frequency spectrum consisting in a central frequency of values equal to carrier frequency $\omega_c/2\pi$ (for n=0) surrounded by an infinite number of sidebands, spaced at frequencies $+\omega_m/2\pi$, $+2\omega_m/2\pi$, etc. Thus, even with a simple sinusoidal modulating signal, the FM modulated wave has an infinite bandwidth[3]. In other words, in order to completely restore the modulating signal a theoretically infinite wideband is needed. Practically, accepting a very small amount of non-linear distortion of the recorded signal, the required bandwidth depends on the modulation index m and is approximately given by the double of the frequency deviation $\Delta\omega_m$ [3]. To show this, let's consider the behaviour of the Bessel coefficients in expression (28), which give the amplitude of both carrier and side frequencies. For the upper sideband and for a modulation index m equal to 13, Fig.4,[11] shows how $J_n(12)$ rapidly approaches zero when n becomes larger than m: the maximum of the energy carried by the modulated wave is bound by a band just larger than the frequency corresponding to the last maximum.

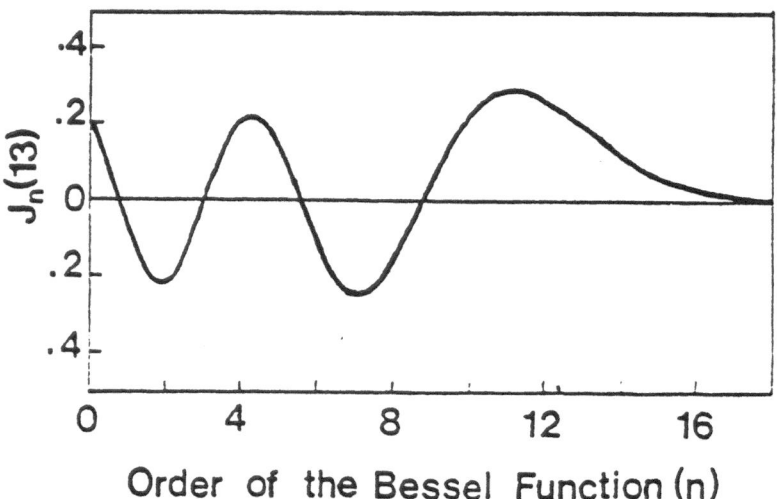

FIGURE 4.The Bessel function of argument 13 and variable order.

The larger is the modulation index, the righter is this approximation, and a bandwidth $\bar{B}$ equal to

$$B = 2m \; \omega_m = 2\Delta\omega_m \tag{29}$$

can be accepted.
A semiempirical rule-of-thumb is given to extend expression
(29) also to small m^5

$$B = 2 \; \omega_m \; (m + 1) \tag{30}$$

To measure the correctness of the last two expressions it
is necessary to define a minimum value accepted for J_n (m) or,
in other words, to define the last significant sideband. If
the last sideband has a magnitude of 10% at least of the unmodulated carrier magnitude, expression (30) agrees with the
effectiveness bandwidth very well for a large range of values
of m. Usually, a linear relationship[12] exists between the
analog output of the FT and the detected frequency f

$$V = T \; f \tag{31}$$

where T is an instrumental parameter. Using this expression it
is possible to find the wanted relationship between modulation
index and instrumental output. Obviously, the correctness of
this relationship depends on what the FT "choses" as the last
sideband and on the validity degree of (29). Hence, it is
possible to express the peak value of the FT output voltage as
a function of B

$$V_{peak} = \frac{1}{2} \; T \; \frac{B}{2\pi} \tag{32}$$

and to obtain the wanted value of ϕ_o from expression (18),(29),
(31)

$$\phi_o = \frac{1}{4} \; \frac{B}{\omega_m} = \frac{1}{2} \; \frac{V_{peak}}{T \; f_m} \tag{33}$$

A direct relationship between phase shift ϕ_o and modulated
signal is obtainable from a spectral display. From expression
(28) the spectrum of r_{if} is

$$r_{if}(\omega) = \beta E_s E_r \; J_{\omega-\omega_m} \; (m) \tag{34}$$

where $\omega = n\omega_m$. When ω_m is small compared to the full scale
frequency of the spectrum analyzer, r_{if} (ω) can be considered
a continuous function of ω. A typical power spectrum of r_{if}
is sketched in Fig.5 for a large value of the modulation
index. The power spectrum shows two well identifiable maxima,
the frequency position of which depends on the modulation index. The question is: at what frequency the maximum of function
(34) does it occur given a certain value of m? The answer
permits to define a relation between the frequency distances
betnween the two maxima and m, and consequently to obtain a
direct measurement of ϕ_o.

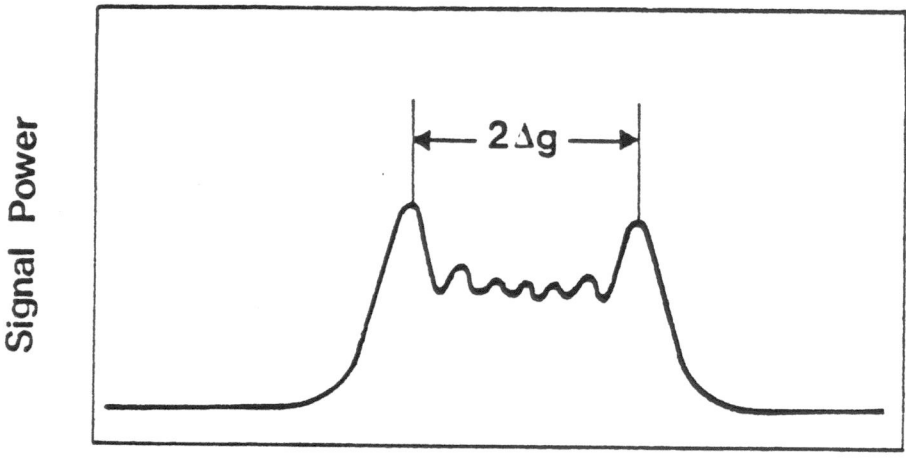

FIGURE 5. Power spectrum of a sinusoidal frequency modulated signal.

It is possible to demonstrate[14,15] that the absolute maximum of the spectrum function $r_{if}(\omega)$ is reached at the angular frequancy value ω_M

$$\omega_M = \omega_m \left[m - 0.80861651\ 7466\ \sqrt[3]{m} - \frac{0.0606\ 4998\ 7910}{\sqrt[3]{m}} - \frac{0.0316\ 7351\ 0263}{m} - \ldots \right]$$

(35)

and has the value

$$3E_s E_r\ J_{\omega_M - \omega_m}(m) = 3E_s E_r \left[\frac{0.6748\ 8509\ 6430}{\sqrt[3]{m}} - \frac{0.0727\ 6309\ 8182}{m} + \frac{0.0199\ 5975\ 0328}{\sqrt[3]{m^5}} + \ldots \right]$$

(36)

The interruption after the third term of the second member of expression (35) and (36), introduces a negligible error.

If ω_m and ω_M are known, we can solve equation (35) as a function of m. Thus, by means of a measurement of the frequency distance between the two maxima, $2\Delta g$, it is possible to obtain the value of ϕ_o directly. This measurement is affected by the reading error of $2\Delta g$, dependent on the resolution of the spectrum analyzer, and by the approximation of (35) and (36). Besides, it is applicable only when the spectrum pattern clearly shows the two peaks of the absolute maxima. Within these approximations, the method gives a direct evaluation of the modulation index m for any assumed value. It can be considered as an alternative method to measure m with respect to the "carrier null method" [16] that permits the measurement of m only for a discrete set of values.

REFERENCES

1. Giallorenzi TG et al.: Optical fiber sensor technology, IEEE J.of Quant.Electr.QE-18, 626, 1982.
2. Teich MC: Infrared heterodyne detection, Proc.IEEE 56, 37, 1968.
3. Schwartz M: Information transmission, modulation and noise, Chapter 4, Mac Graw-Hill, 19870.
4. Martinelli M: The dynamical behaviour of a single-mode optical fiber strain gauge, IEEE J.of Quant.Electr.QE-18, 666, 1982.
5. Black HS: Modulation Theory, D.Van Nostrand, 1953.
6. Terman FE: Electronic and Radio Engineering, Mc Graw-Hill, 1955.
7. Jamuar SS, Mullick SK: Noise-free analysis of a tracking filter FM demodulator, IEEE Trans. on Aer. and Electr. Syst.AES-15, 58,1979.
8. Ishigaki et al.: A phase-tracking loop detector for FM signals and its application to an FM receiver, IEEE Trans. on Cons.Electr. CE-24, 215, 1978.
9. Davies BR: Equivalent variable centre frequency amplifiers, Radio Electr.Eng.28, 381,1964.
10. Enloe LH: Decreasing the threshold in f.m.frequency feedback, Proc.I.R.E. 50,18,1962.
11. Ruthroff CL: FM demodulators with negative feedback, Bell Syst.Techn.J.40, 1149,1961.
12. Jahke-Emde: Tafeln Huherer Funktionen, B.G.Teubner Verlagsgesellschaft, 1952.
13. DISA 55N20 Doppler Frequency Tracker Instruction Manual, DISA Documentation Department,1982.
14. Meissel: Beitrag zur theorie der Bessel'schen functionen, Astronom.Nach.128, 435, 1981.
15. Carrington MS: Variation of bandwidth with modulation index in frequency modulation, Proc.IRE 35, 1013, 1947.
16. Oliver BM: Cage JM: Electronics measurements and instrumentation, Mc Graw-Hill, 1970.

OPTICAL FIBER SENSOR COATINGS

Joseph A. Bucaro

Naval Research Laboratory
Washington, DC 20375-5000

1. INTRODUCTION

Interferometric fiber optic sensors[1] are based upon the principle that environmental perturbations shift the phase of light, ϕ, propagating in a single mode fiber. This change of phase, $\Delta\phi$, compared to a phase reference (see Fig. 1) is ultimately converted by a demodulator to an electrical signal proportional to the environmental perturbation.

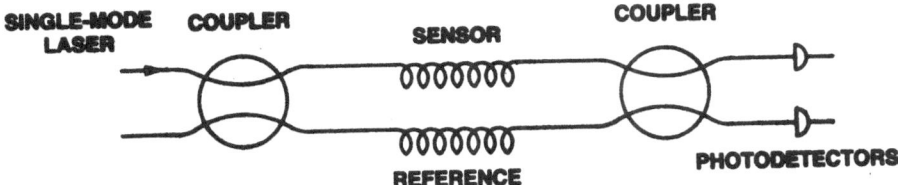

Figure 1. Mach-Zehnder Fiber Interferometer

An important attribute of interferometric fiber sensors is its generic applicability to sensing a broad range of environmental parameters. This derives from the fact that coating materials applied as jackets to the optical fibers usually play the major role in determining the fibers sensitivity and dynamic response to a particular field or parameter. This fact was first realized when experimental studies[2] on the acoustic response of fibers showed dramatic increases for fibers coated with plastic materials. Subsequent studies demonstrated that mechanical strains induced in the coating material communicate directly and in a very controllable manner to the slender glass fiber resulting in an optical phase shift which is subsequently used for detection.

In general, the normalized phase response of a fiber to some environmental perturbation, P, is given by:

$$\frac{\Delta\phi}{\phi} \simeq \epsilon_z(P) + \frac{\Delta n}{n}(P) \tag{1}$$

Here ϵ_z is the axial strain and $\Delta n/n$ is the refractive index change, both determined at the fiber core due to the perturbation, P. Thus, there are two major effects which contribute the fiber phase sensitivity. The first is a change in the physical length of a given section of fiber. The second is a change in the optical index of refraction of the fiberguide glass, part due to the strain optical effect, and (in the case of temperature) part due to temperature induced electronic polarizability changes.

322

The key to the widespread application of this sensor technology is the fact that this shift in optical phase caused by a particular environmental parameter - pressure, temperature, magnetic/electric field, etc. - can thus be effectively controlled and optimized by application of appropriately designed jacketing materials onto the optical fiber. (See Fig. 2)

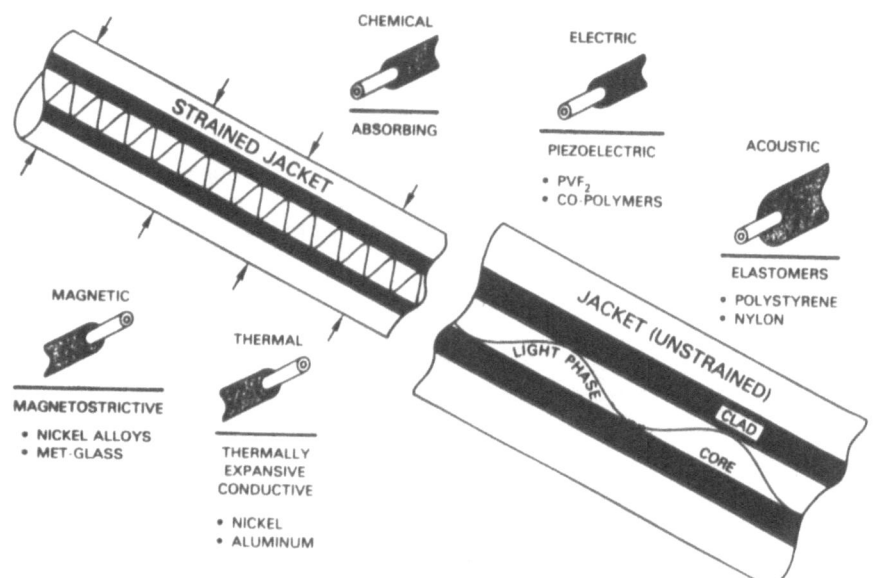

Figure 2. Special coatings for fiber optic sensors.

The application of these coatings to the fiber typically involves straight-forward procedures such as extrusion, dip-coating, and electroplating. More significantly, it does not alter the glass optical waveguide itself. This "dissociation" of the optical waveguide from the designer's attempt to control the fiber's sensitivity is perhaps the most important factor responsible for the rapid pace at which this generic sensor technology has been developed.[3]

2. ACOUSTIC COATINGS

A significant effort has been expended to develop optimized coatings for hydrophone applications.[4,5] Generally, the optical index contribution [see Eq. (1)] is opposite in sign to the length change effect, and each depends differently on fiber coating parameters such as thickness and elastic constants. Accordingly, a great deal of flexibility exists in designing fibers with well-defined acoustic responses.

Consider a glass fiber coated with several layers of elastic material. So long as the acoustic wavelength is large compared to the fiber diameter no shear strains are generated in the fiber. In this case Eq. (1) can be written as

$$\Delta\phi/\phi = \varepsilon_z - n^2/2 \left[(P_{11} + P_{12}) \varepsilon_R + P_{12} \varepsilon_z \right] \tag{2}$$

where ε_z and ε_r are the axial and radial strains at the fiber core and P_{11}, P_{12} are the glass Pockel's coefficients. The acoustic pressure in the fluid exerts stresses at the coating. The polar stresses σ_R, σ_ϕ, and σ_z in the various coated fiber layers are related to the strains ε_R, ε_ϕ, and ε_z as follows:

$$
\begin{bmatrix} \sigma_R{}^i \\ \sigma_\phi{}^i \\ \sigma_z{}^i \end{bmatrix} = \begin{bmatrix} (\lambda^i + 2\mu^i) & \lambda^i & \lambda^i \\ \lambda^i & (\lambda^i + 2\mu^i) & \lambda^i \\ \lambda^i & \lambda^i & (\lambda^i + 2\mu^i) \end{bmatrix} \begin{bmatrix} \varepsilon_R{}^i \\ \varepsilon_\phi{}^i \\ \varepsilon_z{}^i \end{bmatrix} \tag{3}
$$

where i is the layer index (0 for the fiber core, 1 for the fiber clad, etc.), and λ^i and μ^i are the Lame' parameters. These are related to the bulk and Young's modulus B and E by

$$
B^i = \frac{\lambda^i + 2\mu^i}{3} \tag{4}
$$

and

$$
E^i = \frac{3\lambda^i + 2\mu^i}{\mu^i + \lambda^i} \mu^i \tag{5}
$$

For this cylindrical geometry, the strains can be obtained from the Lame' solutions as calculated by Timoshenko and Goudier[4]:

$$
\begin{aligned}
\varepsilon_R{}^i &= U_0{}^i + (U_1{}^i/r^2) \\
\varepsilon_\theta{}^i &= U_0{}^i - (U_1{}^i/r^2) \\
\varepsilon_z{}^i &= W_0{}^i
\end{aligned} \tag{6}
$$

where $U_0{}^i$, $U_1{}^i$, and $W_0{}^i$ are constants to be determined. Since the strains must be finite at the center core, $U_1{}^0 = 0$. For a fiber with m layers the constants can be determined from the boundary conditions:

$$
\sigma_R{}^i|_{R=R_i} = \sigma_R{}^{i+1}|_{R=R_i} \quad (i = 0,1,\ldots m-1) \tag{7}
$$

$$
u_R{}^i|_{R=R_i} = u_R{}^{i+1}|_{R=R_i} \quad (i = 0,1,\ldots,m-1), \tag{8}
$$

$$
\sigma_R{}^m|_{R=R_m} = -P, \tag{9}
$$

$$
\sum_{i=0}^{m} \sigma_z{}^i A_i = -PA_m, \tag{10}
$$

$$
\varepsilon_z{}^0 = \varepsilon_z{}^1 = \ldots = \varepsilon_z{}^m, \tag{11}
$$

where $U_R{}^i$ ($= \int \varepsilon_R{}^i \, dR$) is the radial displacement in the ith layer, and R_i and A_i are the radius and cross-sectional area of the ith layer, respectively. Equations (7) and (8) describe the radial stress and displacement continuity condition across the boundary layers. Equations (9) and (10) assume the applied acoustic pressure is hydrostatic. Equation (11) is the plane strain approximation which ignores end effects. Using the above boundary conditions, the constants U_0^i, U_1^i, and W_0^i can be determined and ε_R^0, ε_θ^0, and ε_z^0 calculated. Equation (2) can then be used to give the acoustic sensitivity, $\Delta\phi/\phi\Delta P$, where ΔP is the acoustic pressure.

Figure 3 shows the calculated pressure sensitivity of a typical, commercially available single mode fiber as a function of the plastic coating thickness. As can be seen, the largest contribution to $\Delta\phi/\phi\Delta P$ is due to

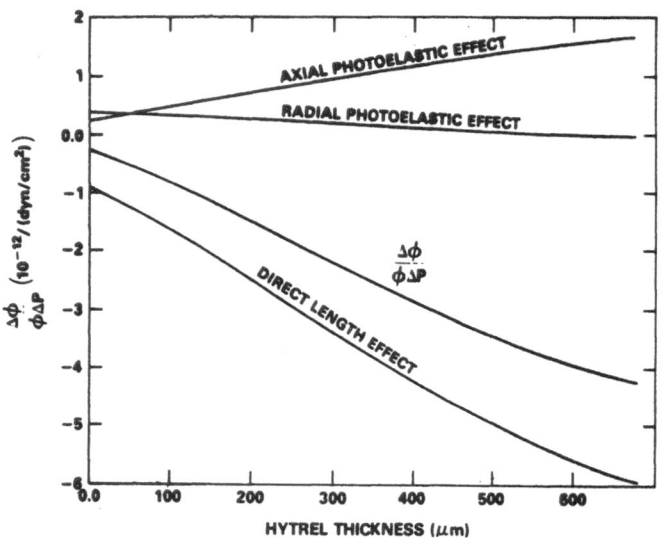

Figure 3. Calculated pressure sensitivity of a typical, commercially available single mode fiber versus plastic jacket thickness.

the pressure induced fiber length change [first term in Equation (2)]. The photoelastic terms [last two terms in Equation (2)] give smaller contributions of opposite polarity. As the plastic coating thickness increases, the magnitude of the pressure sensitivity increases rapidly due primarily to the length change.

Commonly used coating materials for optical fibers include rubbers, thermo-set plastics, and UV curable elastomers. The coating which is applied directly to the waveguide is typically a soft material such as rubber introduced for minimizing microbend optical loss. The outer coating is typically much harder. As is shown, the soft inner coating plays almost no role in determining the acoustic sensitivity. Accordingly, optimization of the fiber acoustic response involves selecting proper outer jacket materials.

Figure 4 shows the calculated acoustic response of a fiber of a typical fiber thickness (0.7 mm) coated with an elastic material whose bulk and Young's modulus are varied. As can be seen from this figure, for high Young's moduli, the fiber sensitivity is a strong function of the bulk modulus. This dependence becomes weaker as the Young's modulus decreases. This can be understood in the following way. For a composite fiber geometry, the axial stress carried by a particular layer is governed by the product of the cross-sectional area and the Young's modulus of that layer. Thus, for high Young's modulus materials, very little coating is required

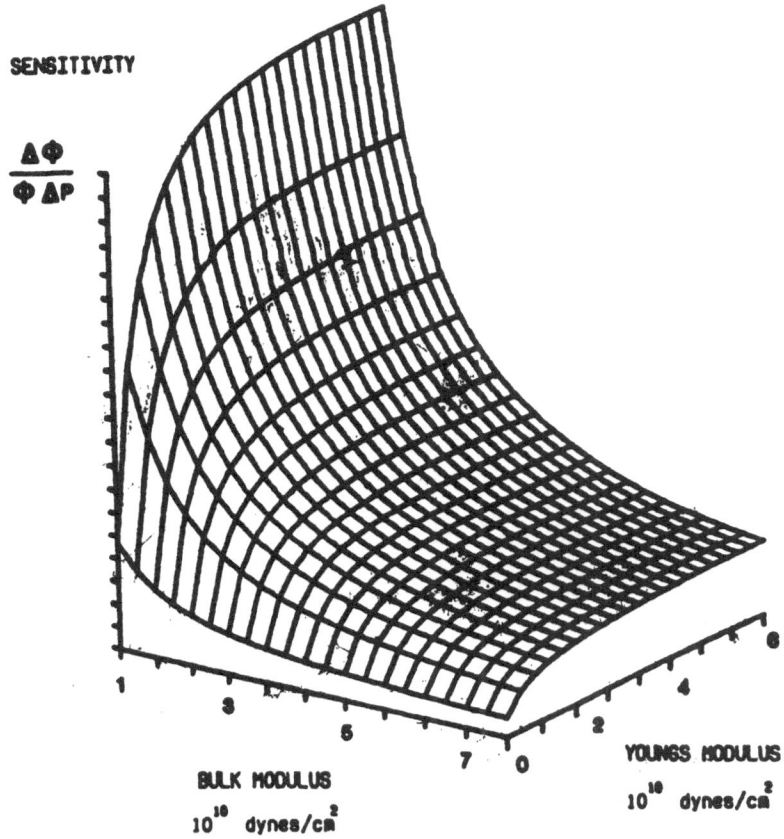

Figure 4. Acoustic sensitivity as a function of bulk and Young's moduli.

to reach the "thick coating limit" in which the sensitivity is governed
essentially by the compressibility of that layer. For low-Young's modulus
materials, however, the degree to which the coating contributes to the
axial strain is diminished and we begin moving toward the limit in which
the glass waveguide plays the major role in the sensitivity. Accordingly,
for typical fiber coating thickness, high acoustic sensitivity requires a
material with a high Young's modulus and low bulk modulus.

Often in designing an acoustic sensor, one wishes to maximize the
acoustic sensitivity for a particular packaging size. As can be seen
from Figure 3, for a typical plastic jacket, the sensitivity increases
monotonically with increasing coating thickness. However, if packaging
volume is a constraint, then an optimum thickness exists which results in
maximum sensitivity for that volume. This can be seen in Figure 5 where
the acoustic sensitivity normalized by the total fiber cross-sectional
area is plotted. As can be seen, this volume normalized sensitivity peaks
at moderately low coating thickness values on the order of 50 microns.

Another important consideration in choosing an optimum acoustic coating is the resulting variation in sensitivity as a function of frequency
and temperature. This variation is determined by the temperature and

SENSITIVITY

$$\frac{\Delta\Phi}{\Delta P\, D^2}$$

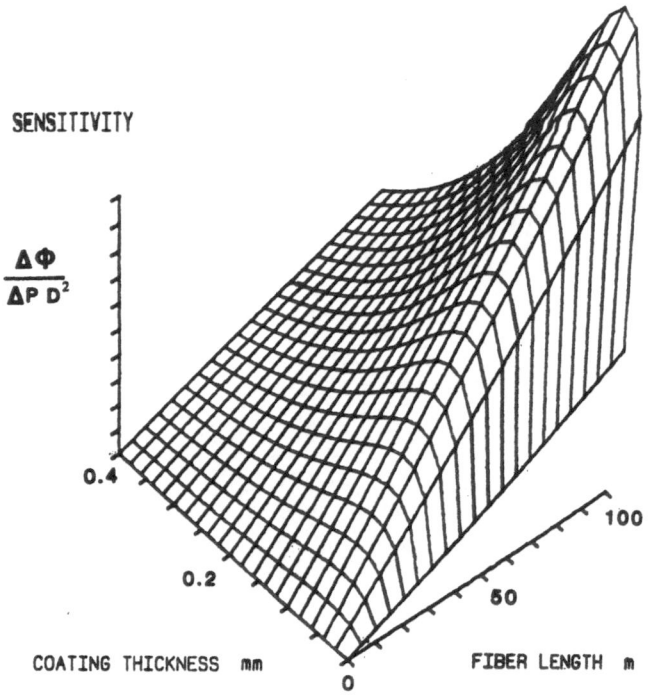

0.4

0.2

COATING THICKNESS mm

0

100

50

FIBER LENGTH m

Figure 5.   Volume normalized acoustic sensitivity versus
coating thickness and fiber length.

frequency dependence of E and B. Literally hundreds of plastics are
available with various behaviors which can be chosen to obtain specific
responses.

Minimizing the acoustic sensitivity of optical fibers is also important for acoustic fiber sensors. It is generally required that the fiber
acoustic sensor be localized in the sensing fiber and that the reference
and lead fibers be insensitive to acoustic signals.

The pressure sensitivity of an optical fiber, $\Delta\phi/\phi\Delta P$, is due to the
effect of the fiber length change and the effect of the refractive index
modulation, which effects are generally of opposite polarity. Thus, it is
possible to design coatings of the proper thickness in order to balance
these two effects and eliminate the acoustic sensitivity.

Figure 6 shows the calculated acoustic sensitivity of glass and metal
coated fibers as a function of coating thickness. As can be seen, for a
nickel jacket of about 13 microns, an aluminum jacket of about 95 microns,
or a calcium aluminate glass thickness of about 70 microns, the fiber has
zero acoustic sensitivity. For this case, nickel requires the smallest
jacket thickness. However, the sensitivity versus thickness curve is
rather steep, requiring critical control of thickness. The corresponding
slopes for aluminum, and in particular for the glass, are much less steep
making them more attractive from a dimensional tolerance point of view.

Figure 7 shows acoustic measurements which have been made on fibers
desensitized with aluminum and nickel jackets. The aluminum jacket thickness was 20 microns which is far from the exact thickness (95 microns)
required for zero sensitivity. The nickel jacket thickness (13 microns) is
within a few microns of the critical thickness.

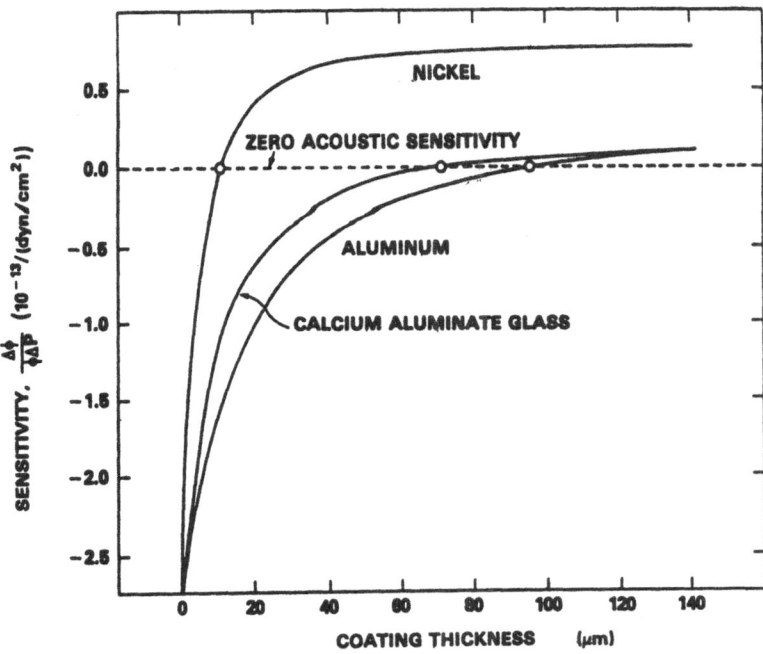

Figure 6. Desensitizing fibers with high Young's modulus coatings.

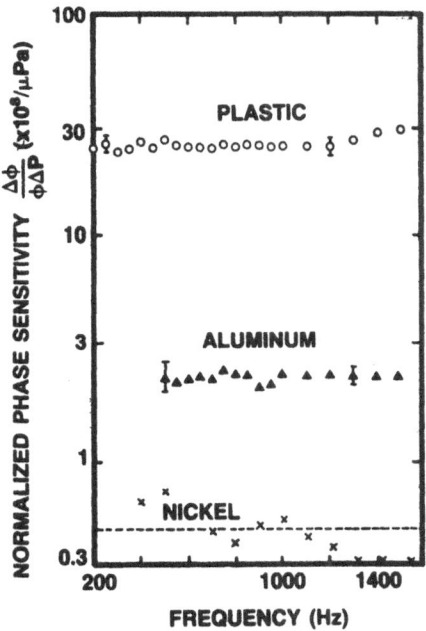

Figure 7. Acoustic measurements made on a fiber coated with a 20
micron aluminum jacket and one with a 13 micron nickel
jacket, compared to plastic jacketed fiber.

3. ULTRASONIC COATINGS

While coatings play the major role in fiber optic acoustic transduction, they find only limited application in higher frequency ultrasonic
sensors.[6] This can be understood by considering Figure 8 which shows the
broadband frequency response of a typical fiber immersed in a fluid and
subjected to an incident strain wave. As shown in the figure, the response

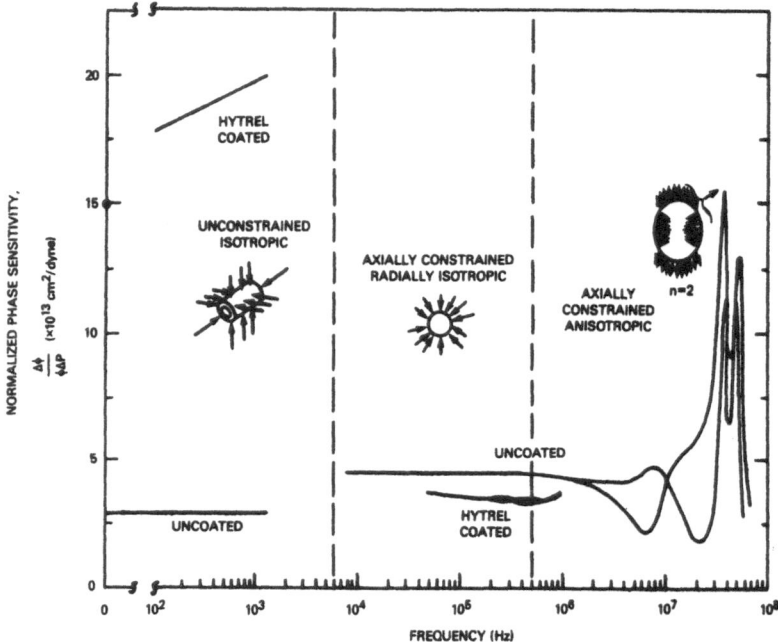

Figure 8. Sensitivity of a straight segment of fiber in a
fluid to an incident strain wave vs. frequency.

can be described in terms of three distinct frequency regimes. At the low
frequencies characteristic of acoustic fields, the fiber mechanical
response is "unconstrained" in the axial direction and isotropic in its
cross-sectional plane. As we have just discussed, in this regime coatings
can have a significant positive or negative effect on the axial strain and
can thus be used to control the acoustic sensitivity. At ultrasonic frequencies, however, the fiber becomes axially "constrained", i.e. due to
inertial effects no net length changes take place. Here transduction takes
place only through index of refraction changes caused by fiber radial
strain. In this case, Equation (2) becomes

$$\frac{\Delta\phi}{\phi} = \frac{-n^2}{2} (P_{11} + P_{12}) \, \epsilon_R \qquad (12)$$

which is valid as long as the ultrasonic wavelength is still large compared
to the fiber diameter. The radial strain, ϵ_R, can be found by again

utilizing Equations (3)-(11) with $\varepsilon_z{}^i = 0$. The result for a single coating layer is that

$$\frac{\Delta\phi}{\Delta\phi P} = \frac{n^2}{2}(P_{11} + P_{12})F_0 \frac{F_1 + (1/2\mu_1)}{F_0(1-r_{01}{}^2)+ F_1 r_{01}{}^2 + (1/2\mu_1)} \tag{13a}$$

where $F_0 = [2(B_0 + \mu_0/3]^{-1}$, $F_1 = [2(B_1 + \mu_1/3]^{-1}$, and $r_{01} = r_0/r_1$.

Here the zero subscript refers to the glass fiber and the one subscript to the coating.

For a fiber without a coating, F_1, μ_1, $\to 0$ and Equation (13a) becomes:

$$\frac{\Delta\phi}{\phi\Delta P} = \frac{n^2}{4}(P_{11} + P_{12}) \frac{1}{B + \mu/3} \tag{13b}$$

Analysis of Equation (13) will show that a coating softer than the glass ($F_1 > F_0$) will enhance the sensitivity of the bare fiber [Equation (13)]. However, this enhancement is limited owing to the fact that the bulk modulus (B) of coating materials is always comparable to, or higher than, their shear modulus. This can be seen from Figure 9 where the ultrasonic sensitivity is shown for various coating shear and bulk moduli for a typical silica fiber (84 µo.d.). The solid curves represent realizable materials where the bulk modulus is greater than the shear modulus and the dashed portions unrealizable materials for the opposite case. As is seen from this figure, it appears that sensitivity increases of only about a factor of two are possible over the uncoated fiber case (arrow). From this it can be concluded that substantial enhancement of ultrasonic sensitivity of conventional fibers is not likely to be achieved by coatings.

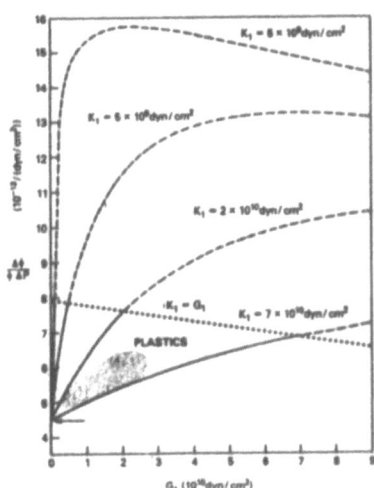

Figure 9. Calculated ultrasonic sensitivity vs. shear modulus (G_1) for various bulk moduli (K_1). Solid lines: $K_1 > G_1$; dotted lines: $K_1 < G_1$.

From Equation (13) it can be seen that coatings harder than glass ($F_1 < F_0$) can be used to reduce the fiber sensitivity when that is required. However, for moderate hard outer coating thicknesses (~ 200 microns), only moderate reductions can be achieved, i.e. factors of 2 or 3. However, increased ultrasonic desensitization of fibers can be obtained by employing more than one coating. In this case, the analytic expression for the sensitivity is too complicated to be given here. The analysis shows that fibers with a soft inner coating and a harder outer coating have less sensitivity than the bare fiber. Significant reduction of sensitivity is obtained when a fiber is first coated with a low bulk modulus material, such as silicone, and then surrounded by a high bulk modulus coating such a nickel. In this case, the radial pressure applied to the fiber will produce a relatively small radial strain in the outer coating due to its high bulk modulus. The small radial compression of the outer jacket results in a comparatively small strain in the soft inner coating. This is turn will communicate a very small pressure to the glass fiber due to the high compressibility of the soft coating. This can be seen in Figure 10 where the calculated sensitivity is shown as a function of the outer metal thickness for a silicone coated (100 μm o.d.) single mode fiber.

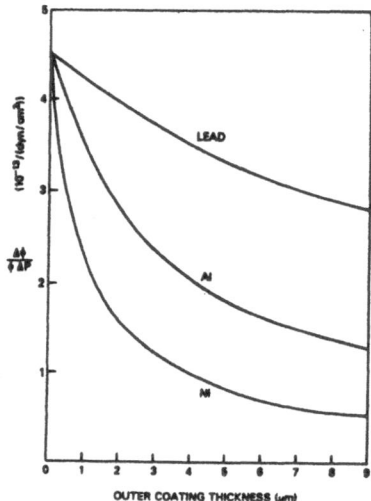

Figure 10. Calculated ultrasonic sensitivity versus thickness of outer coating for a silicone coated fiber.

Finally, as the frequency becomes high enough that the ultrasonic wavelength is comparable to the fiber diameter, we enter the third frequency regime (See Figure 8) in which the strains are not isotropic within the cross-sectional area of the fiber. This results in induced birefringence and the induced phase shifts now also depend upon the optical polarization state. In this regime, coatings can be used in three ways. The first is to weaken the induced birefringence. The second is to reduce the overall glass fiber strain through high frequency absorption effects. And the third is to enhance the mechanical response (ultrasonic sensitivity) through various resonances of the jacket structure.

4. ELECTRIC FIELD COATINGS.

Piezoelectric polymer materials have been successfully applied to optical fibers by first extruding the polymer from the melt onto the fiber and subsequently inducing strong piezoactivity by electric field poling under high field strengths.[7] To date, two poling states have been utilized as shown in Figure 11.

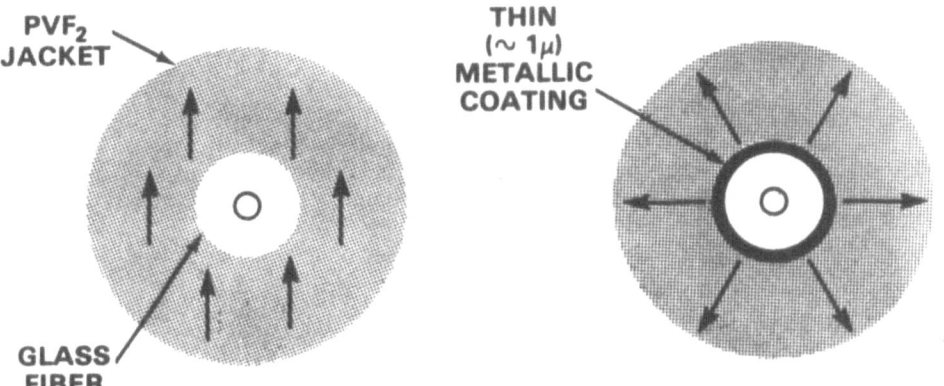

TRANSVERSLY POLED   **RADIALLY POLED**

$\uparrow$—DIRECTION OF POLARIZATION

Figure 11. Two configurations for fibers coated with piezo-electric material: Transversely poled for electric field sensing (left) and radially poled for characterization and modulator applications.

The transversely-poled coating on the left would yield a high sensitivity fiber for an electric field sensor. Along its length, the sensitivity of the fiber would, however, depend upon the relative orientation of the external signal field and polarization direction of the fiber. The radially poled configuration on the right would yield a much lower sensitivity fiber for electric field sensing. However, the sensitivity would be independent of the signal field, fiber orientation. In addition, the radially poled configuration is ideal for laboratory characterization studies and for electro-optic phase modulator applications. For the radially poled case, the stress-strain-electric displacement relations for piezoelectric coating are:

$$\sigma_R = C_{11}\,\epsilon_R + C_{12}\,\epsilon_\theta + C_{13}\,\epsilon_z - h_{11}D(r)$$

$$\sigma_\theta = C_{12}\,\epsilon_R + C_{11}\,\epsilon_\theta + C_{13}\,\epsilon_z - h_{12}D(r) \tag{14}$$

$$\sigma_z = C_{13}\,\epsilon_R + C_{13}\,\epsilon_\theta + C_{33}\,\epsilon_z - h_{13}D(r)$$

where the C are the appropriate elastic constants and h_{ij} is the piezoelectric constant. The stress-strain relation for the glass fiber is given by Equation (3) with $\sigma_\theta = 0$. The electric displacement in the jacket decreases with radial distance from the center:

$$D(r) = D_0/r \quad \text{and} \quad \Delta D = D(b)-D(a) \tag{15}$$

where a and b are the inner and outer jacket radii respectively. In the plane strain approximation, the axial displacement W is assumed to be equal in the fiber and jacket. The axial stress is different in these two regions, and the average axial stress is set to satisfy the axial boundary condition. In this approximation, the radial and axial displacements of the jacket material are

$$U^1 = U_0^1 r + U_1^1/r + U_2^1 \tag{16}$$

$$W = W_0 z$$

where again the superscript one refers to the jacket material and U^1, U_1^1, U_2^1 and W_0 are constants to determined. The corresponding strains in the jacket are

$$\epsilon_R^1 = \frac{\partial U^1}{\partial r}$$

$$\epsilon_\theta^1 = U^1/r \tag{17}$$

$$\epsilon_z^1 = \frac{\partial W}{\partial z} = W_0$$

and the stresses are obtained by substituting Equation (17) into Equation (13). If we confine ourselves to the two lower frequency regions in Figure 8, the stresses can be considered quasi-static and obey the conditions:

$$\frac{\partial \sigma_R}{\partial r} + \frac{(\sigma_R - \sigma_\theta)}{r} = 0$$

$$\frac{\partial \sigma_z}{\partial z} = 0 \quad . \tag{18}$$

This is satisfied if

$$U_2^1 = \frac{h_{12}}{C_{11}} D_0 \tag{19}$$

The displacements in the glass fiber are

$$U^0 = U_0^0 r \tag{20}$$

$$W = W_0 z \quad .$$

The boundary conditions are that $\sigma_R = 0$ at $r = b$, continuity of σ_R at $r = a$, continuity of U at $r = a$, and the average axial force = 0.

The boundary conditions lead to four simultaneous equations which can be solved for the displacement coefficients U_0^1, U_1^1, U_0^0, and W_0. From these and Equation (17), the strains in the jacket and fiber are determined. The electric field in the jacket is

$$E(r) = \frac{1}{K\epsilon_0} D(r) - h_{11}\sigma_R - h_{12}\sigma_\theta - h_{13}\sigma_z \qquad (21)$$

where $D(r)$ is given by Equation (15), ϵ_0 is the permitivity of free space, and K is the jacket dielectric constant. The voltage difference V between the outer and inner jacket electrodes is determined by

$$V = - \int_a^b \epsilon(r)dr \qquad (22)$$

Finally, the phase shift can be computed from Equation (2). In Figure 12 is shown the results of these calculations for a 120 micron jacket of PVF$_2$-TFE copolymer. The solid lines are calculated and the points experimental measurements. The calculated value shown at low frequencies

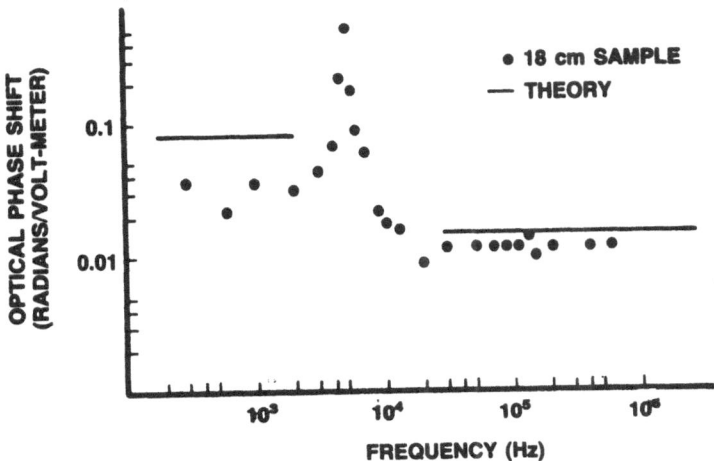

Figure 12. Measured and calculated optical phase shift vs frequency of applied voltage for fiber coated with piezopolymer jacket.

corresponds to the axially unconstrained region discussed in Figure 8. The calculated value at the higher frequencies corresponds to the axially constrained region in Figure 8 and was obtained by setting $\epsilon_z = 0$ in the above development. The peak located between these two regions is related to axial resonances of the finite length of the coated fiber.

Although we do not describe it here, it is possible to consider dynamic stresses and eliminate the constraint imposed by Equation (18). When this is done calculations can be carried out for much higher frequencies than those in Figure 12. Actual calculations of this type predict a number of large resonant peaks related to various coating and glass fiber resonances.

334

In Figure 13 is shown measurements made at NRL on a 20 cm length of fiber coated with a 40 micron piezoelectric polymer jacket. As can be seen, very high frequency performance can be achieved.

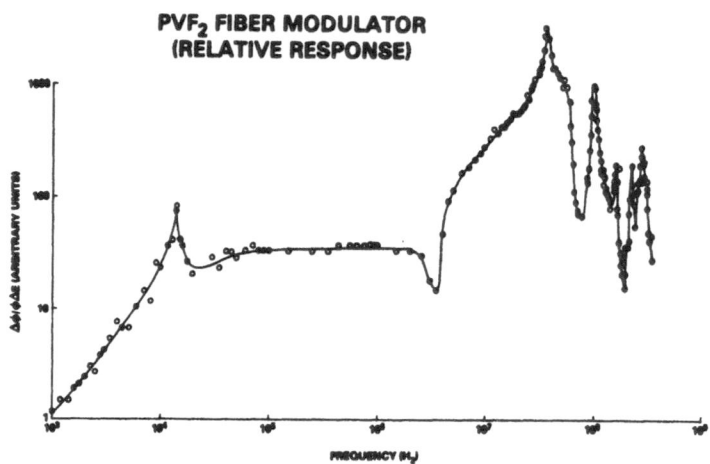

Figure 13. Experimentally measured phase shifts vs frequency for a piezopolymer coated fiber.

5. MAGNETIC FIELD COATINGS

The use of magnetostrictive coatings for magnetic field sensors is analogous to the use of piezoelectric coatings for electric field sensing. In most fiber optic magnetic sensor arrangements, a magnetic field in the direction of the fiber axis causes a strain in the magnetostrictive jacket which produces an optical phase shift in the fiber. In addition, most interest is related to the design of very high sensitivity devices. This latter requirement demands the use of long lengths of coated fiber.

The calculation[8] of optical phase shift for a magnetostrictively coated optical fiber can be made in the same manner as that for the electric field case with magnetostrictive co-efficients replacing the piezoelectric coefficients. However, the values of several of the elastic and magnetostrictive co-efficients appearing in such a formulation have not been measured for materials of interest here. Therefore, a somewhat different formulation is developed here which utilizes parameters already measured.

The stress strain relations are written in the following form:

$$\varepsilon_R = \frac{\sigma_R}{E} - \frac{\gamma \sigma_\theta}{E} - \frac{\gamma \sigma_z}{E}$$

$$\varepsilon_\theta = \frac{\gamma}{E} \sigma_R + \frac{\sigma_\theta}{E} - \frac{\gamma \sigma_z}{E} \tag{23}$$

$$\varepsilon_z = \frac{\gamma}{E} \sigma_R - \frac{\gamma \sigma_\theta}{E} + \frac{\sigma_z}{E} + d_{33} H_z$$

where ϵ and σ are the strain and stress respectively. H_z is a weak magnetic field superposed on a bias field (also along the fiber axis). In Equation (23) it is assumed that the magnetostrictive material is elastically isotropic and the magnetic field terms perpendicular to the axial direction are neglected. Equation (23) is an adequate approximation for calculations of the magnetically induced optical phase shift, which is due mainly to the physical elongation of the fiber driven by the $d_{33}H_z$ field term in the coating. The quantity γ is the Poisson ratio and $d_{33} = (\partial \epsilon_z)/(\partial H_z \sigma)$. Again assuming the plane strain condition, we have for the strains

$$\epsilon_R{}^i = U_0{}^i + U_R{}^i/r^2$$
$$\epsilon_\theta{}^i = U_0{}^i - U_R{}^i/r^2 \tag{24}$$
$$\epsilon_z{}^i = W_0$$

where the constants above are to be determined from the boundary conditions and again $U_1{}^0 = 0$. Solving Equations (23) and (24) for σR and σ_z in the coating material fields:

$$\sigma_R = \frac{E}{q} U_0{}^1 - \frac{E}{1+\gamma} \frac{U_1{}^1}{r^2} + \frac{\gamma E}{q} W_0 - \frac{\gamma E d_{33} H_z}{q} \tag{25}$$

$$\sigma_z = \frac{2\gamma E}{q} U_0{}^1 + \frac{(1-\gamma)E}{q} W_0 - \frac{(1-\gamma)E d_{33} H_z}{q}$$

where $q = (1-2\gamma)(1+\gamma)$ and γ is the Poisson ratio. The expressions for the strains and stresses in the glass fiber are similar except for the magnetic field terms. The boundary conditions are net axial force = 0, continuity of radial displacement, and continuity of radial stress.

The normalized phase shift calculated from this approach per length of coated fiber and per oersted of magnetic field is shown for several materials in Figure 14. In order to appreciate the levels involved it is useful to know that using one centimeter of coated fiber having a sensitivity of ten in these units when coupled to a state-of-the-art optical sensor demodulator would allow the detection of a 10^{-7} gauss magnetic field. State-of-the-art cryogenic superconducting magnetometers have a performance some four orders of magnitude better than this. Thus as can be seen from Figure 14, if nickel coated fiber were used with a thickness of 40 μ, 400 meters of fiber would be required for matching the superconducting magnetometer performance.

Thus for only magnetostrictive nickel has been deposited onto fibers and only in lengths of about one meter. Also, because of the relatively low magnetostrictive coefficient for nickel, relatively thick coatings are required which leads to long coating times. Figure 14 identifies more attractive magnetostrictive materials. In particular, with the metallic glass shown in Figure 13, the superconducting magnetometer performance could be reached with only about 100 meters of coated fiber with a three micron jacket thickness.

336

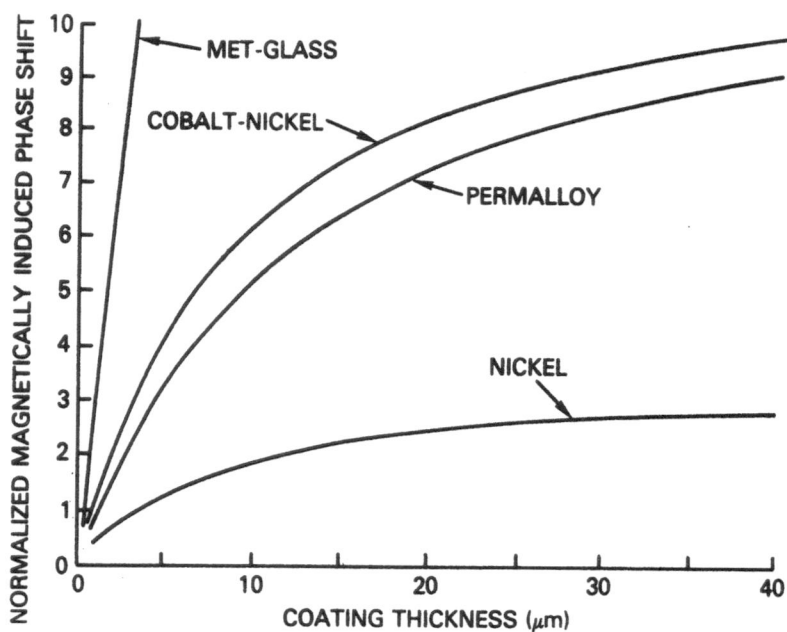

Figure 14. Normalized phase shift for a magnetostrictively coated
optical fiber per unit length per oersted.

6. THERMAL COATINGS
    Thus far, for all the cases considered the environmental field caused
phase shifts in the optical fiber via only two mechanisms: (1) a large
strain developed in the external coating or, (2) a strain developed in the
glass fiber directly. This is because although typical silica glass from
which fibers are made do have non-zero magneto-optic and electro-optic
coefficients, the index change caused by these effects are many orders of
magnitude lower than the direct strain effects. Temperature effects,
however, introduce the need to consider direct glass index of refraction
changes via the temperature effect on the electronic polarizability.
Thus, Equation (2) is no longer appropriate and the term

$$\frac{1}{n}\left(\frac{\partial n}{\partial T}\right)_{\varepsilon} \Delta T \text{ must be added to the left hand side.}$$

    For truly static temperature sensors, the addition of this term pre-
sents no particular analytic problem. Dynamically, however, heat transport
effects must be considered and this has been treated in detail by Shuetz
et. al..[9]
    The results of that work is presented in Figure 15 where the thermally
induced phase shift is shown versus frequency for an unjacketed fiber, a
plastic coated fiber, and a metal coated fiber.
    Consider first the uncoated fiber. Since the thermal expansion co-
efficient of silica glass is very low, the major effect seen here is in
fact due to the

$$\left(\frac{\partial n}{\partial T}\right)_\epsilon \Delta T$$

term. As the frequency of the temperature perturbation is increased, the
finite heat transport time into the glass fiber causes the effect to drop
off. In addition, the direct thermal expansion effect is also dropping off
owing to the fact that although at these frequencies the strain is communi-
cately instantaneously from the outer strained region to the glass core,
less and less of the outer region is being strained.

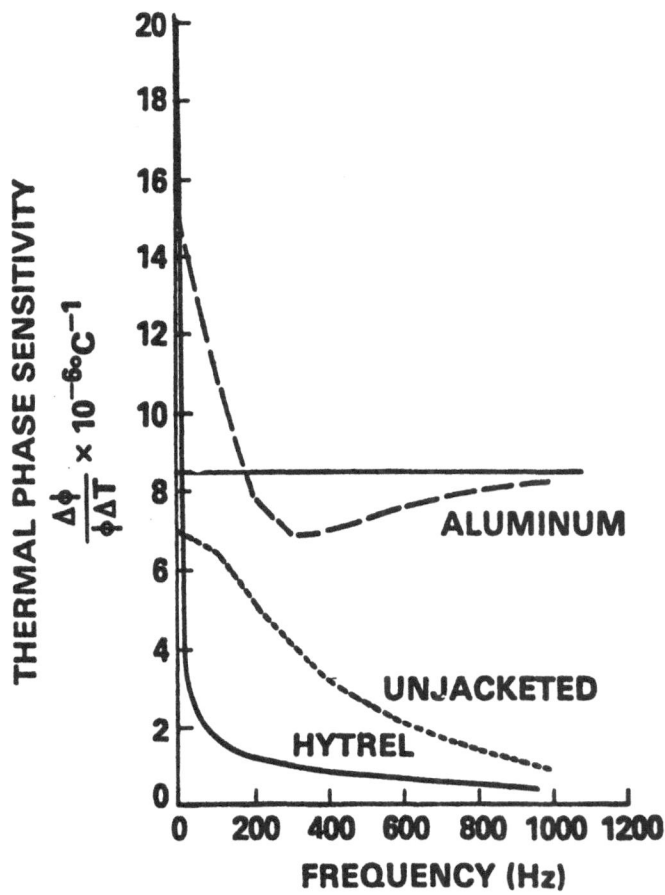

Figure 15. Thermal phase sensitivity vs frequency.

Consider next the plastic coated fiber. The plastic has a very large
expansion co-efficient and indeed one can see a very large response owing
to it at very low frequencies. However, the plastic's thermal conductivity
is relatively low so that this effect is lost almost immediately as the
thermal perturbation frequency is increased. Obviously, such a fiber would
make a very sensitive static thermometer.

Finally, consider the metal jacketed fiber response. The metal also has a relatively large thermal expansion co-efficient leading to a large "static" response. However, in addition the thermal conductivity is also large. Thus, a thermally induced strain generated in the metal coating (which accounts for more than half of the phase shift) can take place "instantaneously" even when considered at very high frequencies. Thus, as shown here, the response holds up beyond kilohertz frequencies (and in fact into the tens of kilohertz band). Obviously, such a fiber would be ideal for broadband applications. The initial fall off is of course related to the fall off in the $(\partial n/\partial T)_\varepsilon$ ΔT effect in the silica glass itself. Statically, this term is in phase with the coating expansion term. However, at a few hundred hertz, the transport times have also caused a phase lag in the $(\partial n/\partial T)_\varepsilon$ ΔT term and this leads to the minimum seen for aluminum at about 300 Hz.

7. REFERENCES

1. T. G. Giallorenzi, J. A. Bucaro, A. Dandridge, G. Sigel, J. H. Cole, S. C. Rashleigh, and R. G. Priest, "Optical Fiber Sensor Technology," IEEE J. Quantum Elect., 18, 626, 1982.

2. J. A. Bucaro and T. R. Hickman, "Measurement of Sensitivity of Optical Fibers for Acoustic Detection," Appl. Opt. 18, 938, 1979.

3. J. A. Bucaro, N. Lagakos, J. H. Cole, T. G. Giallorenzi, "Fiber Optic Acoustic Transduction," Physical Acoustics Vol XVI, edited by W. P Mason and R. N. Thurston, Academic Press, p. 385, NY, 1982.

4. N. Lagakos, E. V. Schnaus, J. H. Cole, J. Jarzynski, and J. A. Bucaro, "Optimizing Fiber Coatings for Interferometeric Acoustic Sensors," IEEE J. Quantum Electron 18, 683, 1982.

5. N. Lagakos and J. A. Bucaro, "Pressure Disensitization of Optical Fibers," Appl. Opt. 20, 2716, 1981.

6. N. Lagakos, J. H. Cole, and J. A. Bucaro, "Ultrasonic Sensitivity of Coated Fibers," J. Lightware Tech. LT1, 495, 1983.

7. J. Jarzynski, "Frequency Response of a Single Mode Optical Fiber Phase Modulator Utilizing a Piezoelectric Plastic Jacket," J. Appl. Phys. 55, 3243, 1984.

8. J. Jarzynski, J. H. Cole, J. A. Bucaro, and C. M. Davis, Jr., "Magnetic Field Sensitivity of an Optical Fiber with Magnetostrictive Jacket," Appl. Opt. 19, 3746

9. L. S. Schuetz, J. H. Cole, J. Jarzynski, N. Lagakos, and J. A. Bucaro, "Dynamic Thermal Response of Single-Mode Optical Fiber for Interfero-metric Sensors," Appl. Opt. 22, 478, 1983.

THERMODYNAMIC LIMITATIONS TO THE MEASUREMENT OF PHASE SHIFTS IN OPTICAL FIBERS

W. H. GLENN

United Technologies Research Center
Silver Lane
East Hartford, CT 06108
USA

1. INTRODUCTION

In discussions of the noise limitations of coherent optical sensors it is usually assumed that in absence of externally applied temperature or pressure variations, the optical path length of a single mode fiber has a perfectly well defined value. This is not strictly true. At any temperature above absolute zero, there will be statistical fluctuations of all the physical properties, including the optical path length, and this will set a lower limit to the measurable phase shift due to an external influence. These fluctuations are as unavoidable as Johnson Noise; they are significant for the fiber because of its small size. In this paper the magnitude and spectral distribution of these fluctuations are calculated. Some aspects of this problem have recently been discussed by Shelby, Levenson and Bayer (1). The present treatment approaches the problem in a slightly different way.

The optical path length of the fiber is:

$$L_o = \int_{-L}^{L} n(z,t)dz \qquad (1)$$

where z is the direction along the fiber axis and the fiber is of length 2L. We consider the index of refraction to be a function of any two thermodynamic variables. Pressure P, and entropy S will be chosen, for reasons that will become evident later. If the pressure and entropy undergo fluctuations, the index will fluctuate as

$$\Delta n(z,t) = \left(\frac{\partial n}{\partial P}\right)_S p + \left(\frac{\partial n}{\partial S}\right)_P s \qquad (2)$$

Here the capital letters P and S are the equilibrium values and the lower case letters p and s are the fluctuations. The temporal and spatial dependence of the fluctuations is contained in p and s. The coefficients are constants, characterizing the material. We may express the second term in a more convenient form by noting that

$$\left(\frac{\partial n}{\partial S}\right)_P = \left(\frac{\partial n}{\partial T}\right)_P \left(\frac{\partial T}{\partial S}\right)_P \tag{3}$$

Using the thermodynamic identity

$$\left(\frac{\partial T}{\partial S}\right)_P = \frac{T}{\rho c_p V} \tag{4}$$

where c_p is the specific heat (Joules/gram) of the material being considered. We obtain then

$$\Delta n = \left(\frac{\partial n}{\partial P}\right)_S p + \frac{T}{\rho c_p V} \left(\frac{\partial n}{\partial T}\right)_P s \tag{5}$$

$$\Delta n = \alpha p + \beta s$$

The coefficients of the fluctuating quantities are now expressed in terms of readily measurable properties of the material

$$\left(\frac{\partial n}{\partial P}\right)_S = \frac{n^3 K}{2} \frac{1}{\rho v^2} \tag{6}$$

where K is the photoelastic constant, ρ the density and v the acoustic velocity. ($\rho v^2 = Y = $ Young's modulus).

The mean square values of the pressure and entropy fluctuation, averaged over a small volume, may be calculated from statistical thermodynamics (see, for example, Ref. (2), Section 111). They are

$$\overline{s^2} = k\rho V c_p$$

$$\overline{p^2} = -kT \left(\frac{\partial P}{\partial V}\right)_S$$

$$= kT \frac{\rho v^2}{V} \tag{7}$$

The average values are by definition zero. It should be noted that in the cited reference, the quantity C_p, the heat capacity of the whole body is used (Joules/degree). Here we will use the specific heat c_p (Joules/gm). The relation is $C_p = \rho V c_p$. In the second result, use has been made of the relation

$$v^2 = \left(\frac{\partial P}{\partial \rho}\right)_S = -\frac{V}{\rho} \left(\frac{\partial P}{\partial V}\right)_S \tag{8}$$

It is also shown in the same reference that s and p are statistically independent, i.e., sp = 0. This was the reason why they were chosen. The fluctuations in n are thus seen to consist of two independent processes. In the adiabatic, "p" processes all quantaties., density, pressure, temperature, etc. fluctuate but the entropy is constant. In the isobaric "s" processes, all quantities except the pressure fluctuate. The use of the variables p and s allows the separation of these two types of fluctuations. We would now like to calculate the frequency distribution of the fluctuations. This cannot be done from thermodynamics, the detailed dynamics of the medium (fiber) must be considered. The calculation can proceed in a way similar to that used in the Debye theory of specific heat. The treatment is similar to that discussed by Fabelinskii(3).

2. PRESSURE FLUCTUATIONS

We consider first a pressure fluctuation. The pressure, to a first approximation, satisfies the wave equation.

$$\nabla^2 p = \frac{1}{v^2} \frac{\partial^2 p}{\partial t^2} \tag{9}$$

where v is the velocity of sound. For pedagogical purposes and to minimize unnecessary mathematical complexity we will consider a square fiber of side 2a, a square core of side 2b and length 2L (see Fig. 1). Extension of the approach to a round fiber is straightforward. The core-cladding boundary presents a negligible acoustic discontinuity. The solution of the wave equation is readily found to be

$$p = p_0 \begin{Bmatrix} \sin \beta_x x \\ \cos \beta_x x \end{Bmatrix} \begin{Bmatrix} \sin \beta_y y \\ \cos \beta_y y \end{Bmatrix} \begin{Bmatrix} \sin \beta_z z \\ \cos \beta_z z \end{Bmatrix} e^{i\omega t} \tag{10}$$

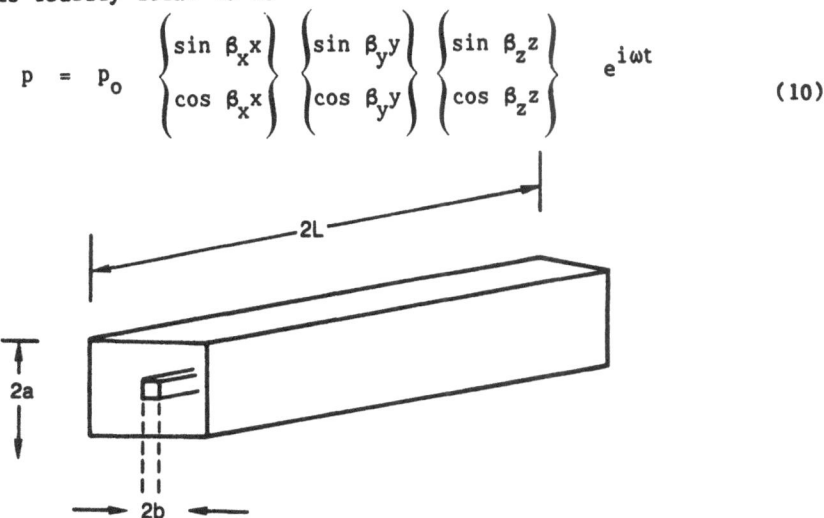

NOMINAL DIMENSIONS

2a = 100 microns

2b = 5 microns

2L = 1 meter

FIGURE 1. Simplified fiber geometry.

and the dispersion relation is

$$\left[\beta x^2 + \beta y^2 + \beta z^2\right]^{\frac{1}{2}} v = \omega \qquad (11)$$

We require that the pressure fluctuation be zero at the fiber boundaries. (This is a simplification; some modifications will be mentioned later). This requires that

$$\beta_x = \frac{m\pi}{a} \quad \text{for the sine modes}$$

$$(12)$$

$$\beta_x = \frac{(2m+1)\pi}{2a} \quad \text{for the cosine modes}$$

Here m is an integer running from 0 to ∞. A similar expression holds for βy and βz, (with a replaced by L). We now calculate the spatial average of p over the fiber. This average is not zero, it is the time average that makes p = 0. In the spatial average the antisymmetric (sine) modes do not contribute so

$$\langle P \rangle_{s_p} = \frac{1}{V} P_0 \int \cos \beta_x x \cos \beta_y y \cos \beta_z z \ dV$$

$$= P_0 \left(\frac{2}{\pi}\right)^3 \frac{1}{(2m+1)} \frac{1}{(2q+1)} \frac{1}{(2r+1)} e^{i\omega t} \qquad (13)$$

Here q and r are the wave vector indices in the y and z directions.

So far, the fluctuation has been treated as a coherent excitation while it is actually a stochastic process and can be described only by its Power Spectral Density (PSD). The PSD of Eq. (13) is

$$|\langle p(\omega) \rangle_{sp}|^2 = po^2 \left(\frac{2}{\pi}\right)^6 \frac{1}{(2m+1)^2 (2q+1)^2 (2p+1)^2} \delta(\omega-\omega_{mqr}) \qquad (14)$$

where ω_{mqr} is given by the dispersion relation Eq. (11) and Eq. (12). The mean square value $\overline{p2}$ is obtained by integrating over all frequencies

$$\overline{p^2} = po^2 \left(\frac{2}{\pi}\right)^6 \sum \frac{1}{(2m+1)^2 (2q+1)^2 (2r+1)^2} \qquad (15)$$

$$= \frac{1}{8} po^2$$

This may now be equated to the value of $\overline{p2}$ calculated earlier from thermodynamics

$$\frac{1}{8} \, po^2 \;=\; kT \, \frac{\rho v^2}{V} \tag{16}$$

We note that the energy in a standing acoustic wave is

$$E \;=\; \frac{1}{8} \, \frac{po^2}{\rho v^2} \, V \tag{17}$$

so that each mode has an energy of kT.

3.0 ISOBARIC FLUCTUATIONS

The isobaric fluctuations satisfy a diffusion equation

$$\theta \nabla^2 s \;=\; \frac{\delta s}{\delta t} \tag{18}$$

$$\theta \;=\; \text{thermal diffusivity} \;=\; \text{thermal conductivity} \,/ \rho c_p$$

The normal modes may be found for this equation. They are identical to those of Eq. (10) except that the complex $\exp(i\omega t)$ is replaced by the real $\exp{-\delta t}$ where

$$\delta \;=\; \theta \left[\beta_x^2 \;+\; \beta_y^2 \;+\; \beta_z^2 \right] \tag{19}$$

The components of the wave vector satisfy the same conditions as in Eq. (12). The β_z^2 term may be neglected in this expression since it is of the form $[(2r + 1) \, /2L]^2$ while the others are of the form and $[(2m + 1) \, /2a]^2$ and $L^2 >> a^2$. We take the spatial average as before to obtain

$$\langle s \rangle_{sp} \;=\; s_o \left(\frac{2}{\pi} \right)^3 \frac{1}{(2m+1)} \frac{1}{(2q+1)} \frac{1}{(2r+1)} \, e^{-\delta t} \tag{20}$$

This expression decays exponentially in time and represents the decay of a single "isolated" fluctuation. In reality, we have a continuous random process consisting of an incoherent superposition of many such elementary events. In this case we can only speak of the power spectral density. If we have a signal, as a voltage $A(t)$, that is a random superposition of such decays

$$A(t) \;=\; \sum_k a(t-t_k)$$

$$a(t) \;=\; a_o e^{-\delta t} \quad t > 0 \tag{21}$$

$$0 \qquad t < 0$$

Then its power spectral density is

$$|A(\omega)|^2 = A_o^2 \frac{\delta}{\omega^2 + \delta^2} \tag{22}$$

where A_o^2 has the dimensions of $(Volts)^2$ and the mean square value

$$\overline{A(t)^2} = \int A_o^2 \frac{\delta}{\omega^2 + \delta} d\omega = A_o^2 \tag{23}$$

Similarily, the power spectral density for s is

$$|s(\omega)|^2 = \rho_o^2 \left(\frac{2}{\pi}\right)^6 \frac{1}{[(2m+1)(2q+1)(2r+1)]^2} \frac{\delta}{\omega^2 + \delta^2} \tag{24}$$

for each fluctuation mode and the mean square value is the integral over all ω,

$$|S|^2 = \rho_o^2 \left(\frac{2}{\pi}\right)^6 \frac{1}{[(2m+1)(2q+1)(2r+1)]^2} \tag{25}$$

The total mean square value is the sum over all modes and is identical to the sum in Eq. (15).

$$\overline{s^2} = \frac{1}{8} \rho_o^2 = k\rho Vc_p$$

$$\overline{s_o^2} = 8k\rho Vc_p \tag{26}$$

Having calculated the magnitude and spectum of the fluctuations, we may now investigate the effect on the optical signal.

4.0 OPTICAL SIGNAL MODULATION BY FLUCTUATIONS

In a rigorous treatment a coupled wave approach to evaluating the interaction of an optical wave with the fluctuations could be taken. Here it is sufficient to consider the optical fiber as a lumped phase modulator where optical path length varies as

$$\Delta L_o = \int_{-L}^{L} \Delta n(z,t)dz \tag{27}$$

This will be evaluated along the center line of the fiber since the core is assumed very small with respect to the entire fiber. The x and y variations can be replaced by their values of x = y > 0.

For the p fluctuations, the change in phase is

$$\phi_p = \frac{2\pi}{\lambda} \alpha \int_{-L}^{L} pdz = \left(\frac{2\pi}{\lambda}\right)\left(\frac{2}{\pi}\right) \alpha \frac{2Lp}{(2r+1)} e^{i\omega t} \qquad (28)$$

Here α is the constant from Eq. (5) relating the index change to the pressure. By the same reasoning leading to Eq. (14), the PSD is

$$|\Phi_p(\omega)|^2 = \left(\frac{8L}{\lambda}\right)^2 \alpha^2 p_o^2 \sum \frac{\delta\,(\omega-\omega_{mqr})}{(2r+1)^2} \qquad (29)$$

This is a line spectrum and will be dicussed below.

For the isobaric fluctuations

$$\phi_s = \frac{2\pi}{\lambda} \beta \int_{o}^{L} sdz = \left(\frac{2\pi}{\lambda}\right)\left(\frac{2}{\pi}\right) \beta \frac{2L}{(2r+1)} s_o e^{-\delta t} \qquad (30)$$

And by same the reasoning leading to Eq. (14), the PSD is

$$|\Phi_s(\omega)|^2 = \left(\frac{8L}{\lambda}\right)^2 \left(\frac{2}{\pi}\right) S_o^2 \beta^2 \sum \frac{1}{(2r+1)^2} \frac{\delta}{\omega^2+\delta^2} \qquad (31)$$

The sum over all modes in this case is complicated by the fact that δ depends on the mode indices

$$\delta = \theta \left[\frac{(2m+1)^2}{(2a)^2} \pi^2 + \frac{(2q+1)^2}{2a^2} \pi^2 \right]$$

$$\sum \frac{1}{(2r+1)^2} \frac{\delta}{\omega^2+q^2} = \sum \frac{1}{(2r+1)^2} \frac{(2a)^2}{\theta\pi^2} \frac{\left[(2m+1)^2 + (2q+1)^2\right]}{(u')^2 + \left[(2m+1)^2 + (2q+1)^2\right]^2} \qquad (32)$$

$$= \left(\frac{2}{\pi}\right) \frac{(2a)^2}{\theta\pi^2} F(u)$$

Here u is a normalized frequency

$$u = \frac{(2a)^2\,\omega}{\theta\pi^2} = \frac{\omega}{\omega_o} \qquad \omega_o = \frac{\theta\pi^2}{(2a)^2} \qquad (33)$$

346

and F(u) is the function plotted in Fig. 2. Finally we obtain

$$|\Phi_g(\omega')|^2 = \left(\frac{8L}{\lambda}\right)^2 s_0^2 \beta^2 \left(\frac{2}{\pi}\right) \frac{1}{\omega_0} F(u) \tag{34}$$

The expression for the mean square phase fluctuations constitutes the frequency dependent Optical Nyquist Theorem. It is somewhat more complex than the electrical version for the voltage across a resistor

$$\overline{V^2} = 4kTB \tag{35}$$

since the latter has a PSD that is independent of frequency, at least as long as h /kT << 1. We may obtain an integrated version for the total mean square phase fluctuation by integrating over all frequencies.

$$\overline{\phi_t^2} = \overline{\phi_p^2} + \overline{\phi_s^2}$$

$$= \frac{8L}{\lambda}^2 \frac{2}{\pi} \left(\alpha^2 \overline{p_0^2} + \beta^2 s_0^2\right) \tag{36}$$

If an optical signal $\exp(i\omega_0 t)$ is injected into the fiber the output will be

$$A e^{i(\omega_0 t + \phi(t))} = A e^{i\omega_0 t} \left[1 + i \phi(t)\right] \tag{37}$$

with a corresponding PSD

$$A^2 \left(\delta(\omega-\omega_0) + \Phi^2(\omega-\omega_0) \right) \tag{38}$$

So that the optical PSD has symmetrical sidebands about the center frequency.

The spectrum due to the "s" processes is continuous and proportional to the function F(u). The spectrum of the "p" processes is a line spectrum with lines at frequencies given by the dispersion relation Eq. (11). The lowest of them is for m = q = r = 0.

$$\omega_0 \simeq \left[2\left(\frac{\pi}{2a}\right)^2\right]^{\frac{1}{2}} v \tag{39}$$

For typical fiber dimensions, these frequencies start in the tens of MHz and extend upward. Around each of the frequencies there is a much more closely spaced line spectrum due to the various axial (r index) modes. The entire spectrum is shown schematically in Fig. 3. (For the nominal dimensions of the fiber shown in Fig. 1, the spectrum starts at a somewhat higher frequency, ~ 42 MHz). We must now consider a refinement of the boundary conditions given in Eq. 12. A more rigorous treatment of the vibrations of a cylinder shows that there is a set of quasi-longitudinal modes whose frequencies are much lower and are given by

$$\omega^2 = \left(\frac{(2r+1)\pi}{2L}\right)^2 \tag{40}$$

These occur at much lower frequencies. They are longitudinal compressional waves. These should have the same mean square phase fluctuations and are also shown in Fig. 3.

The quantities α and β are defined in Eq. 5. From this we may calculate

$$\alpha^2 \overline{P_0{}^2} = \frac{8kT}{v}\left(\frac{n^2k}{2}\right)\frac{1}{\rho v^2} \approx 4 \times 10^{-24} \text{ for silica}$$

$$\beta^2 \overline{S_0{}^2} = \frac{8kT^2}{\rho C_p V}\left(\frac{2n}{2T}\right)_p^2 \approx 2 \times 10^{-24} \text{ for silica}$$

$$\omega_0 \approx 715 \text{ rad/sec}; \quad f_0 \approx 113 \text{ Hz} \tag{41}$$

Finally we obtain, for $\lambda = 600$nm, $L = 1$ meter

$$\left(\frac{8L}{\lambda}\right)^2 \overline{P_0}{}^2 \alpha^2 = .72 \times 10^{-9} \text{ rad}^2/\text{Hz} \quad \sim L, T, (1/a)^2$$

$$\text{rms} = 2.8 \times 10^{-5} \text{ rad}/\sqrt{\text{Hz}} \quad \sim L^{\frac{1}{2}}, T^{\frac{1}{2}}, (1/a)$$

and

$$\left(\frac{8L}{\lambda}\right)^2 \frac{\overline{S_0{}^2}\,\beta^2}{\omega_0} = 5.2 \times 10^{-13} \text{ rad}^2/\text{Hz} \quad \sim L, T^2, (1/a)^2$$

$$\text{rms} = 7.2 \times 10^{-7} \text{ rad}/\sqrt{\text{Hz}} \quad \sim L^{\frac{1}{2}}, T, (1/a)$$

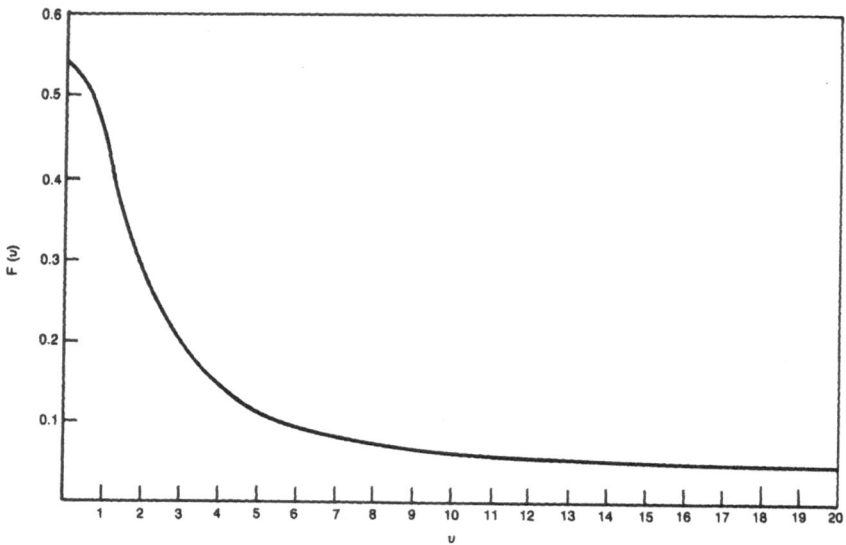

FIGURE 2. Spectrum of isobaric fluctuations.

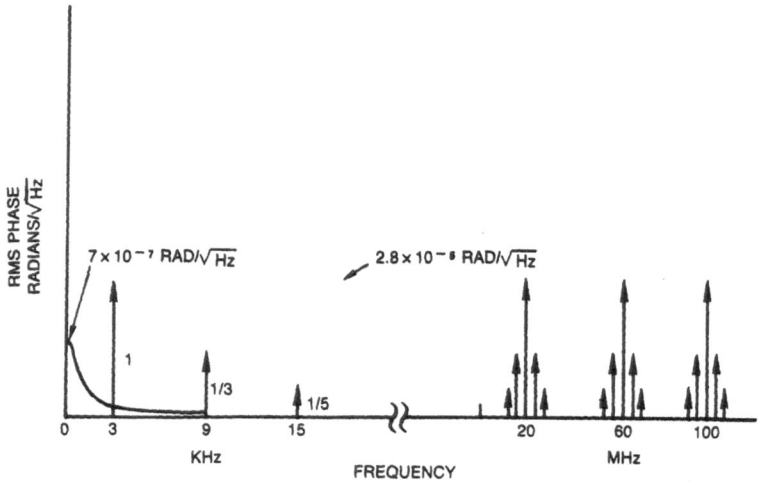

FIGURE 3. Total optical spectrum due to fluctuations

5.0 EXPERIMENTAL RESULTS AND DISCUSSION

Experimental observatins of the "p" process have recently been reported by Shelby et al. (1). Their experimental arrangement is illustrated in Fig. 4 and consists of a Mach Zehnder interferometer with a 1 meter length of fiber. The output was detected and analyzed in a spectrum analyzer. Signal levels were adjusted so that the dominant noise level was the shot noise level, as shown in Fig. 5. Multiple peaks, extending from about 20 MHz to beyond 400 MHz, the bandwidth of the electronics. These authors carried out a much more rigorous analysis of the vibrational modes, including torsional modes, and found excellent agreement between the calculated and observed values of the peaks.

One factor that has not been considered in the effect of acoustic damping. This could be included by appropriate modification of Eq. (9) describing the wave propagation. This damping will have the effect of lowering and broadening the peaks although the total area under the peaks will remain the same. In the work cited (1) it was found necessary to remove the buffer coating to reduce the acoustic damping.

In conclusion, this paper has presented a simplified treatment of statistical fluctuations in an optical fiber and their effect on the phase of an optical signal. An optical version of the Nyquist Thereom has been presented. The effect of these fluctuations can be larger than the shot noist limit and they should be considered in systems based on the measurement of extremely small phase shifts.

REFERENCES

1. R. M. Shelby, M. O. Levenson and P. W. Bayer; Guided Acoustic-Wave Brillouin Scattering, Phys. Rev. B 31 pp. 5244 - 5252, 15 April 1985.

2. L. D. Landau and E.M. Lifshitz; Statitical Physics, Pergamon Press LTD, London - Paris, 1958, distributed by Addison - Wesley Publishing Company, Inc., Reading, Massachusetts, U.S.A.

3. I. L. Fabelinskii, Molecular Scattering of Light, Plenum Press, New York, 1968.

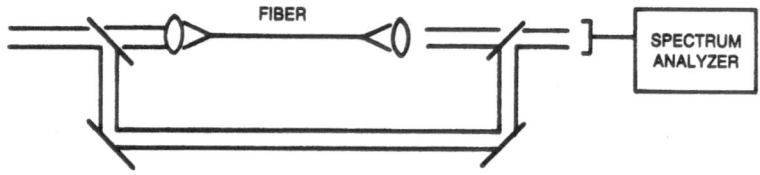

FIGURE 4. Experimental arrangement to observe fluctuations.

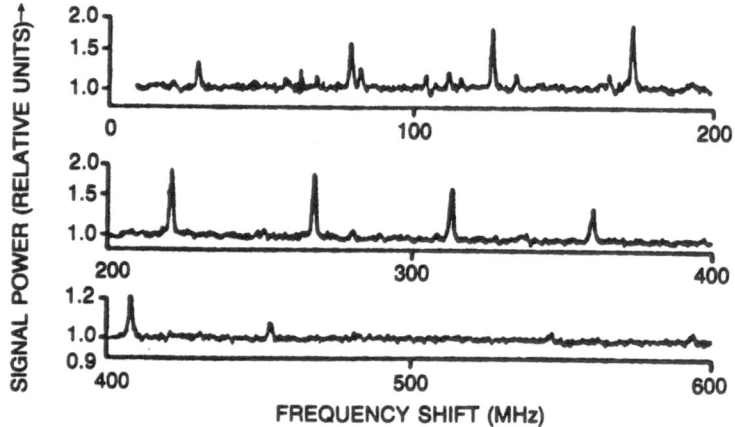

FIGURE 5. Measured spectrum due to pressure fluctuations
(From Shelby, Levenson and Bayer (1.)).

POLARIMETRIC OPTICAL FIBER PRESSURE SENSOR WITH LOW TEMPERATURE EFFECTS

S.J. HUARD

INSTITUT D'OPTIQUE THEORIQUE ET APPLIQUEE - Centre Universitaire d'Orsay - Bât. 503 - B.P. 43 - 91406 ORSAY Cédex FRANCE

1. INTRODUCTION

In order to measure with high resolution a parameter such as the temperature or the pressure, the interferometric optical fibers sensors (O.F.S.) have been proposed during the last years {1}. While the resolution is quite good, they present the disadvantage of a reference arm in which perturbations can occur. Such situations appear in the Mach-Zehnder or in the Michelson optical fiber configuration but not in the Fabry-Pérot configuration {2} in which the two arms are in the same fiber. Other configurations using single-mode fiber in which the external parameter modifies the evolution of the state of polarization (S.O.P.) have been developed in connection with interferometric O.F.S. {3}.

A perfect isotropic single mode fiber supports two degenerated orthogonal modes and it is impossible to use such a configuration as a polarimetric sensor. Two approaches can be done to elaborate an OFS using polarization properties. In the first one, high birefringent single-mode fibers (see the paper of D.N. PAYNE) can be used to left the mode degeneracy and so to work on the state of polarization. In the second one, very low birefringence single-mode fibers are wrapped around cylinder {4}; because of the photoelastic effect a linear birefringence appears. In the two cases the sensing signal is the SOP at the output of the fiber.

That second way will be followed in this paper. Moreover in single mode fiber with well-polished end faces, interferences can occur between the reflected light. This phenomena has been used recently to define a high resolution and high dynamic temperature sensor {5}.

2. THEORITICAL BACKGROUND

Because of two different phenomena can be involved in this kind of sensors a preliminary theoritical approach must be carried out. If a single-mode optical is wrapped around a cylinder in order to produce a static birefringence, the possible interference pattern induced by reflection onto the fiber-ends must not be forgotten. The total round phase trip ϕ and the differential phase ψ between the two eigenstates have been evaluated. Their variations under some parameters such as temperature or pressure can be derived in order to define a single mode optical fiber sensor {5}. Nevertheless, for the transmitted light and using a relatively low coherence-length sources, interferences phenomena can be neglected for the considered application. If must be quoted that polarization and interference phenomenas can be turned into account in order to conceive new sensors.

2.1. Birefringence induced in a tension coiled fiber

An isotropic weakly guiding monomode fiber with length L is considered. Its outer radius is ρ and its refractive index is n. It is stretched in order to induce a small axial strain ε. When N turns of the fiber are wrapped,

without any twist, around a cylinder with a radius b, it can be shown {6},
in the LP-modes approximation {7}, that a linear birefringence with two local
principal axis is induced. The fast axis (x-axis) and the slow axis (y-axis)
are, respectively, normal to the cylinder and parallel to its axis (Fig. 1).

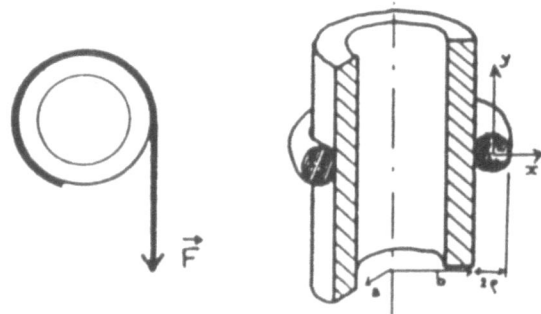

Fig. 1 : Local axis of the birefringent coil

In this case the principal indices become $n_x = n + \delta n_x$ and $n_y = n + \delta n_y$.
The differential phase shift Ψ between the two eigenstates has been evaluated by Rasleigh {6} :

(1)
$$\psi = C_s \left[\frac{1}{2} \left(\frac{\rho}{b} \right)^2 + A \frac{\rho}{b} \varepsilon \right] . L$$

where

$$C_s = k_o \frac{n^3}{2} (p_{11} - p_{12}) (1 + \nu_o)$$

and :

$$A = \frac{2 - 3 \nu_o}{1 - \nu_o}$$

In those expressions $k_o = 2\pi / \lambda_o$ (λ_o is the vacuum wavelength) and ν_o is
the fiber Poisson's ratio. p_{11} and p_{12} are the well known photoelastic constants {8}. It must be quoted that no torsion of the fiber (inducing optical
rotary power) is taken into account and that strains are supposed uniform
inside the core cross section in order to derive (1). In the bracket the first
term depends only on a pure bending which the second term is related to the
initial axial strain ε. When an external parameter such as temperature (or
pressure) is exerted onto the system (cylinder + fiber) the external radius
of the cylinder varies. In order to evaluate the variation $\Delta \psi$ of the initial

differential phase shift Ψ, the action of the parameter on the fiber but also on the cylinder must be considered. Such an approach is now carried out with temperature and pressure as external parameters.

2.2. Differential phase shifts
2.2.1. Temperature
If the temperature is increased, because of the thermal expansion of the cylinder, b is modified. Moreover the thermooptic effect induces a change of the index of refraction. Assuming no radial expansion of the fiber and a small variation ΔT of the temperature, the variation $\Delta\Psi_T$ is :

$$(2) \qquad \Delta\Psi_T = 2\pi \, N \, C_s \frac{db}{dT} \frac{\rho}{b} - \left[\frac{1}{2}(\frac{\rho}{b})(1 - \frac{3}{n}\frac{dn}{dT}) + A(1 + \frac{3}{n}\frac{dn}{dT})\varepsilon\right] . \Delta T$$

Considering a silica fiber $\left[\frac{dn}{dn} \approx 10^{-5} \text{°} . C^{-1}\right]$, a small enough initial tension $[\varepsilon \approx 0.01]$ and the smallness of $\frac{\rho}{b}$ (typically 0.01), the expression (2) can be reduced to :

$$(3) \qquad \Delta\Psi_T = 2\pi \, N \, C_s . A . \frac{\rho}{b} \frac{db}{dT} . \Delta T.$$

It must be emphased that $\Delta\Psi_T$ does not depend on the initial strain ε. However, if only pure bending is considered, the variation of the differential phase shift is given by :

$$(4) \qquad \Delta\Psi_T^b = \Delta\Psi_T \frac{\rho}{2b} \frac{1}{A} \, ,$$

which is hundred time weaker than $\Delta\Psi$. The only factor to be determined is now $\frac{1}{b}\frac{db}{dT}\Delta T$. According the well-know linear thermoelasticity theory {9}, this factor reduces to $\Delta b/b$. We assume a long enough hollow cylinder with an inner and outer radii a and b. If the inner and outer surfaces are respectively submitted to temperature variations ΔT_i and ΔT_o the result is {10} :

$$(5) \qquad (\frac{\Delta b}{b})_T = \alpha(1 + \nu)\left[\frac{\Delta T_o - r^2 \Delta T_i}{1 - r^2} + \frac{\Delta T_o - \Delta T_i}{2 \, Log \, r}\right] ,$$

where α is the thermal expansion coefficient, ν the cylinder Poisson's ratio and $r = a/b$. When the thickness of the cylinder is small enough and if the thermal conductivity of the material is high, $\Delta T_i = \Delta T_o = \Delta T$ and $(\Delta b/b)_T$ becomes :

$$(6) \qquad (\frac{\Delta b}{b})_T = \alpha(1 + \nu) \, \Delta T$$

The most important result derived from (6) is that the thickness of the cylinder does not affect $(\Delta b/b)_T$. This particularity will be used widely in the next sections in order to compensate temperature disturbances in a polarimetric pressure sensor.

A more accurate elasticity analysis, including the clamping effect of the fiber on the cylinder has been carried out {10}. Its leads to a relative decreasing of $(\Delta b/b)_T$ by 5 %.

2.2.2. Pressure

In that case an over-pressure ΔP is applied inside the cylinder. Assuming a long enough cylinder in order to use only plane strains {10}, the relative variation $(\Delta b/b)_P$ of the outer radius has been evaluated.

$$(7) \qquad (\frac{\Delta b}{b})_P = \frac{2(1 - \nu^2)}{E} \ \frac{r^2}{1 - r^2} \ \Delta P,$$

where E is the Young's modulus of the former cylinder. That variation leads to a variation of the differential phase shift $\Delta \Psi_P$ such as :

$$(8) \qquad \Delta \Psi_P = 4\pi \ N \ C_s . \ A. \ \rho . \ \frac{1 - \nu^2}{E} \ \frac{r^2}{1 - r^2} \ \Delta P$$

It must be quoted that the sensitivity increases when the thickness of the cylinder decreases and when the outer radius decreases. Practically the thickness is down-limited by the mechanical properties of the material. Moreover the curvature 1/b is also down-limited by curvature losses of the coiled single-mode fiber. A small Poisson's ratio ν is also suitable.

2.2.3. Temperature and pressure

When the temperature and pressure variations are very slow (static case) the two effects are not coupled according the thermoelasticity theory {9}. So the total variation $\Delta \Psi$ of the differential phase shift is simply :

$$(9) \qquad \Delta \Psi = \Delta \Psi_T + \Delta \Psi_P$$

In order to compare the two contributions on $\Delta \Psi$ an expandable copper pipe with an inner radius of 7 mm and an outer radius of 8 mm on which is coiled a low birefringence 105 µm diameter single-mode fiber is considered. The same phase variation is produced by either 1°C or 6 Atm. These values show clearly that the temperature effect is much more important that the temperature effect is much more important than the pressure effect. The reduction of thermal influences can be obtained by reducing the expansion coefficient α (low expansible material such as silica or INVAR) or more efficiently by using the porperty of $\Delta \Psi_T$. This variation is independant of the thickness of the tube according (6). So if two identical coils are wrapped around the same outer radius cylinder on two sections with a different inner radius the two signals will be same for a temperature variation but not for pressure variation (Fig. 2) {11}.

If, using a suitable devide optical device, the phase difference $\Delta \Psi = \Delta \Psi_1 - \Delta \Psi_2$ can be obtained the sensor is, in principle, temperature independant. While the final pressure sensitivity is reduced it is possible to conceive a differential pressure sensor with an overpressure ΔP_1 and ΔP_2 in the two separated sections of the pipe.

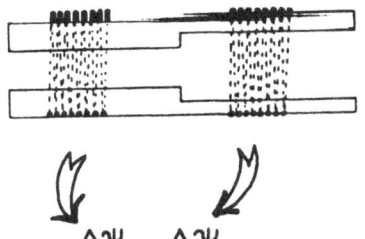

Fig. 2 : Two coils on two
    section of a pipe.

$$\Delta \Psi_1 - \Delta \Psi_2$$

2.2.4. Principles of a temperature compensated polarimetric optical fiber sensor

It can be shown that in order to perform the differential phase shift difference $\Delta \Psi$ between two linear retarder plates characterized by their respective differential phase shift $\Delta \Psi_1$ and $\Delta \Psi_2$, and half-wave device correctly adjusted is needed. Consider two coils with so-called principal axis lying at 45° with respect Ox and Oy. An immediate Jones calculus {12} leads to an equivalent matrix M for the entire device (Fig. 3).

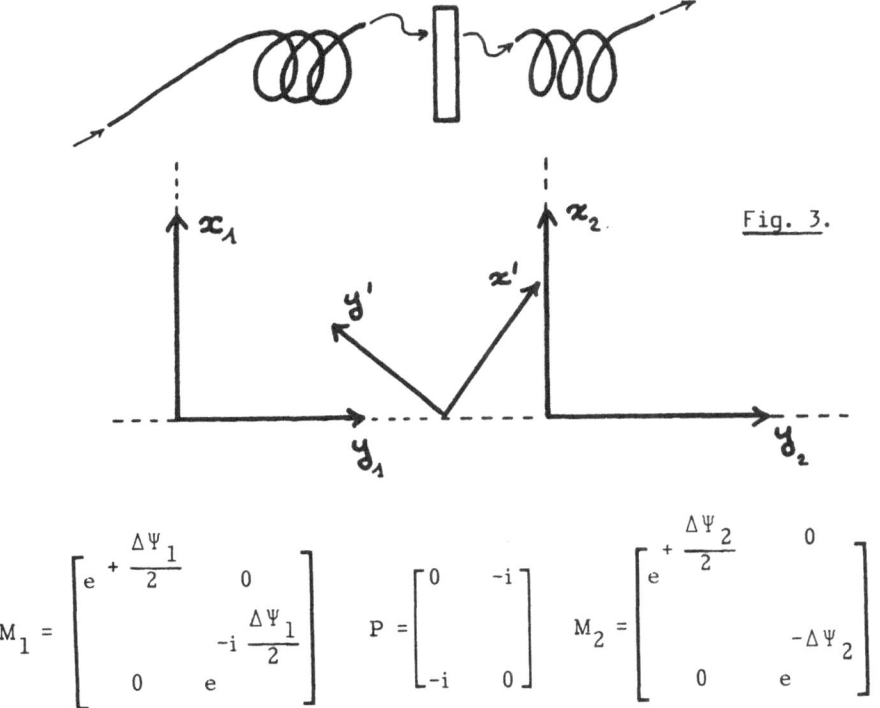

Fig. 3.

$$M_1 = \begin{bmatrix} e^{+\frac{\Delta \Psi_1}{2}} & 0 \\ 0 & e^{-i\frac{\Delta \Psi_1}{2}} \end{bmatrix} \quad P = \begin{bmatrix} 0 & -i \\ -i & 0 \end{bmatrix} \quad M_2 = \begin{bmatrix} e^{+\frac{\Delta \Psi_2}{2}} & 0 \\ 0 & e^{-\Delta \Psi_2} \end{bmatrix}$$

M_1, M_2 and P are respectively the Jones matrix of the two coils and the half-wave device.

$$(10) \qquad M = M_2 \, P \, M_1 = -i \begin{bmatrix} 0 & e^{i(\Delta\Psi_2 - \Delta\Psi_1)/2} \\ e^{-i(\Delta\Psi_2 - \Delta\Psi_1)/2} & 0 \end{bmatrix}$$

M is clearly the matrix of an equivalent linear retarder with a differential phase shift $(\Delta\Psi_2 - \Delta\Psi_1)$ followed by an halfwave plate adjusted to 45° with respect to the retarder axis. This concept is applied to our system : two identical coild wrapped around two sections of the same tube with differents inner radii the total differential shift is then :

$$(11) \qquad \Delta\Psi = (\Delta\Psi_1)_P - (\Delta\Psi_2)_P$$

provided the temperature variations are the same on the two sections of the tube. This device is described on the experimental point of are in the next section.

3. EXPERIMENTAL VERIFICATIONS

3.1. Halfwave device
Before the description of the sensor it is necessary to recall how to make an half wave device without any cutting and/or splicing between the two coils fibers. If a free loop of single mode with a radius R is consider it is easy to see, according to (1) that :

$$(12) \qquad \Delta\Psi_L = \frac{2\pi}{\lambda_o} \frac{n^3}{4} (P_{11} - P_{12}) (1 + \nu_o) \frac{P^2}{R^2} \times 2\pi R$$

So for a known single-mode fiber and a well defined wavelength it is possible that $\Delta\Psi_L = \pi + 2m\pi$. That device is equivalent to an half wave plate with its local principal axis rectively along the loop radius and perpendicular to the plane of the loop. Lefevre {13} have been the first to propose these kind of device in order to control the polarisation state along a single mode fiber system.

If the plane of the loop is adjusted at 45°C with respect the common coils axis, the final result is the interchange of the slow and fast axis of the two coils. It must be quoted that Dakin (see his paper) have also proposed an analog device using two high birefringent fibers spliced with their principal axis crossed.

3.2. Experimental set-up

The typical experimental scheme is described in this section Fig. 4.

A linearly polarized He-Ne laser beam is launched into a monomode optical fiber coiled on a copper hollow cylinder with a = 7 mm and b = 8 mm. Its polarization is adjusted at 45° with respect to the principal axis of the coil. The S.O.P. of the transmitted light is analyzed by a quaterwave plate Q and an analysor A. In order to control the temperature, two electric heating

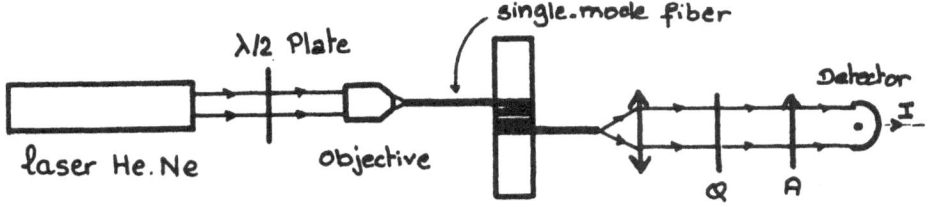

Fig. 4 : Experiment set-up

wire coils are also wrapped on the pipe. A thermocouple give the variations of the surface temperature.

For a given bias point (P_O and/or T_O) Q and A are adjusted in order to cancel the transmitted light. So if any variations $\Delta\Psi$ on the differential phase Ψ occur, the output intensity is given by {14}.

$$(13) \qquad I = I_o \sin^2 \frac{\Delta\Psi}{2}$$

Because of $\Delta\Psi$ is either proportionnal to ΔP or to ΔT the two experiments have been carried out.

3.3. Temperature and pressure sensitivity
With the same pipe it is possible to increase the inner pressure or to increase the temperature. Because of high thermal distrubances, for pressure measurement sensitivity the pipe is thermo-regulated. Results are given on the following alrawings (Fig. 5) :

The measured sensitivities are $2.8 \ 10^{-7}$ rd/Pa/m for pressure and 0.2 rad/°C/m for temperature. Those values are about seven time weaker than the theoritical predictions. This difference is attributed to the soft jacket of the fiber which does not transmit all the strains. Nevertheless those results show clearly the great influence of temperature on future pressure sensor.

3.4. Temperature independant pressure sensor
In this case the cylinder outer diameter is 15 mm and the two sections have an inner diameter of 10 and14 mm respectively. The two coils device separated by an halfwave optical fiber system appear on Figure 6.

It must be quoted that the $\lambda/2$ device is temperature insensitive because of no initial strain before curvature. A 100°C temperature increasing leads to 0.8° on the 180° static differential phase shift of the loop device. In order to verify the influence of the halfwave device, preliminary experiments have been carried out by heating the tube with the loop adjusted to 45° versus the cylinder axis and cooling when the loop in out of order.

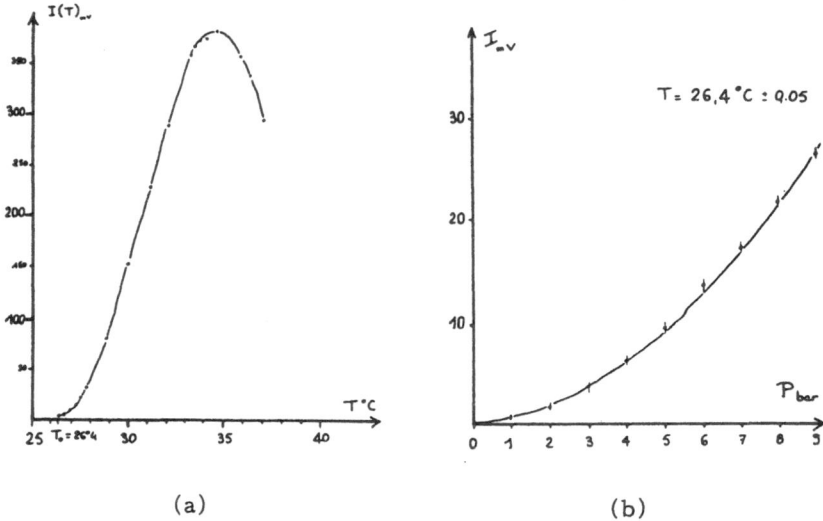

(a)   (b)

Fig. 5 : (a) Temperature response ; (b) Pressure response.

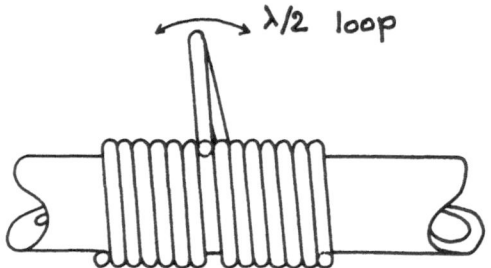

Fig. 6 : Two coils device on
the pipe.

   The experimental results (Fig. 7) clearly show that temperature effects
are reduced by a factor higher than 50 {15}. A slight difference in the two
coils or in a small imperfection on the half wave loop device can explain the
residual influence of the temperature. Nevertheless such a configuration
leads to a good improvement but pressure sensitivity is reduced as it can be
see on Fig. 8. This last one $\Delta\Psi/P$ is $3.1 \ 10^{-2}$ rad/bar while the sensitivity
with n_0 device compensation but with a thermo-regulated temperature is
$6.8 \ 10^{-2}$ rad/bar. The loss in sensitivity is higher than the predictions
issued from (11). This fact can be explain by the high jump of the inner
radius between the two coils.

4. CONCLUSION
   While temperature disturbances on all the optical fiber sensors are very
high, it has been proved that by means of a suitable device they can be
highly reduced. A single mode polarimetric optical fiber pressure sensor has
been developped for medium and high pressure into pipes. Using two fiber

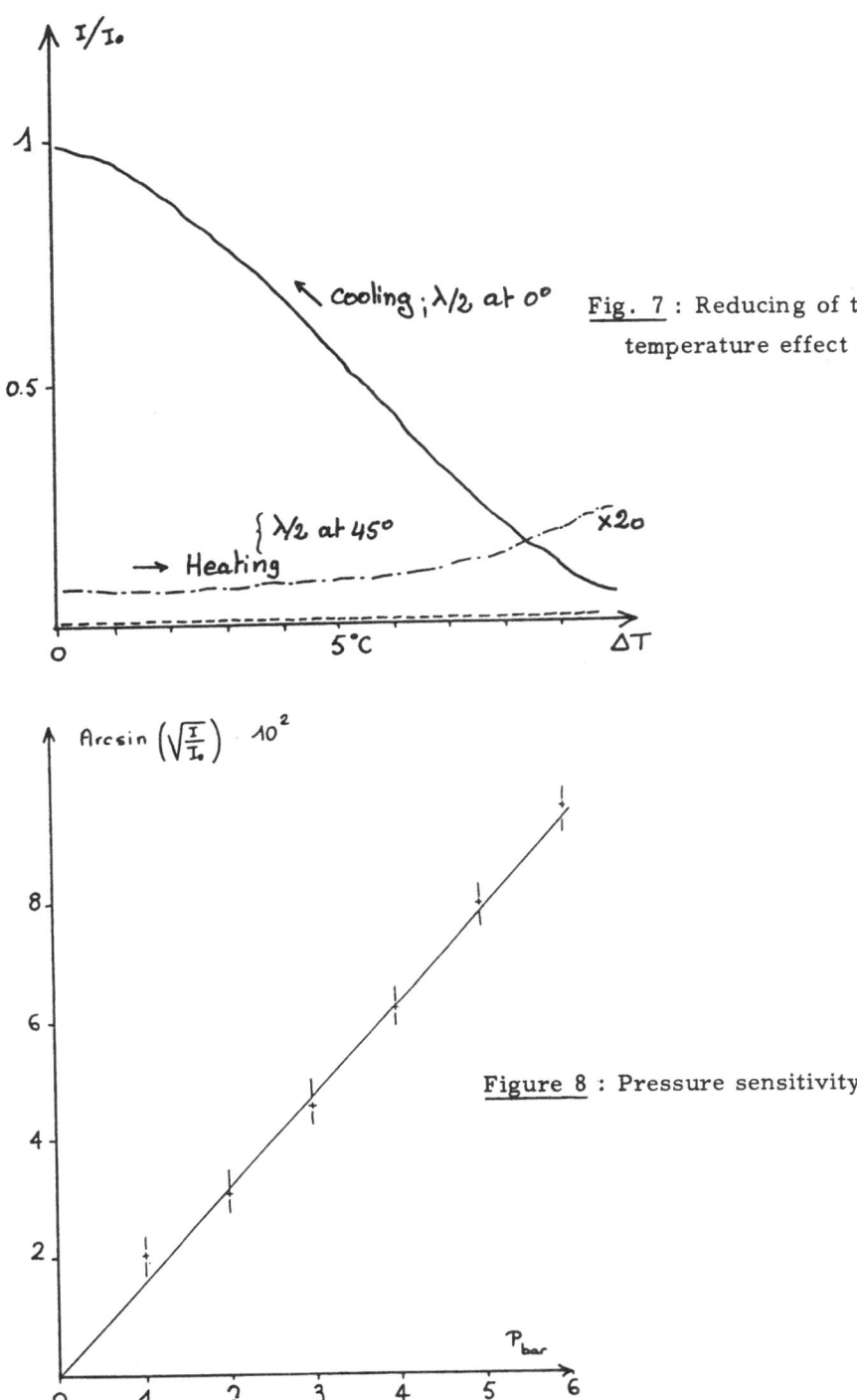

Fig. 7 : Reducing of the temperature effect .

Figure 8 : Pressure sensitivity

coils separated by an in-line half wave loop device it is possible to desensitized the sensor for temperature variation. Moreover if the thicknesses of the pipe under the two coils are different a small reduction in pressure sensitivity is obtained, and a differential pressure manometer is now under development. But this first approach can be developed in order to add interference phenomena and so to have a second information on the transducer. While in transmission such interference are not well adapted because of the poor contrast, in the reflection beam the contrast is high enough to leads to a fringes counting method.

REFERENCES

1. J.A. BUCARO, H.D. HARDY and E.F. CAROME, J. Acoust. Soc. Am. 62, 1977, 1302.
    G.B. HOCKER, Opt. Lett. 4, (1979), 320.

2. T. YOSHINO, K. KUROSAWA, K. ITOH and T. OSE, IEEE QE-18, (1982) 1624.

3. M. CORKE, J.D.C. JONES, A.D. KERSEY and D.A. JACKSON, Electron. Lett. 21, 148

4. S.C. RASHLEIGH and R. ULRICH, Opt. Lett. 5, (1980), 87.

5. D. CHARDON and S.J. HUARD, J. Light. Techn. (to be published).

6. S.C. RASHLEIGH, Opt. Lett. 5, (1980), 392.

7. A.W. SNYDER and J.D. LOVE, "Optical Waveguide Theory". Chapman and Hall London, New-York (1983).

8. J.F. NYE, "Physical properties of Crystals" Clarendon Press, Oxford (1955).

9. W. NOWACKI, "Thermoelasticity" Pergamon Press Oxford, (1962).

10. D. CHARDON and S.J. HUARD, Capteurs 84, Paris, June 1984.

11. D. CHARDON and G.R. ROGER, EFOC-LAN 85 Montreux, (1985), I.G.I., BOSTON.

12. H.G. JERRARD, Optics and Lasers Technology (1982), 304.

13. H.C. LEFEVRE, Electron. Lett. 16, (1980), 778

14. J.M. BORN and E. WOLF. "Principles of Optics" Pergamon Press, New-York, (1980).

15. D. CHARDON and S.J. HUARD, Horizons 85, Besançon (1985).

FIBEROPTIC TEMPERATURE PROBE UTILIZING A SEMICONDUCTOR SENSOR

D.A. CHRISTENSEN AND J.T. IVES

DEPARTMENT OF ELECTRICAL ENGINEERING, DEPARTMENT OF BIOENGINEERING
UNIVERSITY OF UTAH
SALT LAKE CITY, UTAH, 84112

1. INTRODUCTION

There are several applications for fiberoptic temperature sensors in biological and clinical studies.   For example, in thermodilution measurement of cardiac output, their use instead of thermistors in heart catheters precludes the possibility of patient electrical shock.   They are also finding application in multipurpose fiberoptic catheters for the monitoring of blood PO_2, PCO_2, pH, and pressure in critical care patients.

The fiberoptic temperature probe described in this paper was developed for use during hyperthermia therapy for cancer, an experimental treatment in which the region containing the cancer is heated from normal body temperatures (37°C) to approximately 43-44° C for about 30 minutes, followed by radiation therapy or chemotherapy. This procedure leads to increased regression rates in some types of tumors [1].   Careful in vivo temperature monitoring of both healthy and cancerous tissue during patient treatment is essential for safe and effective application of heat [2].

Accurate temperature measurement is especially difficult when electromagnetic means are used for hyperthermia production, since conventional thermometers such as thermistors and thermocouples may produce large measurement errors due to interactions from scattering, internal ohmic heating, and noise pickup between the electromagnetic fields and the device's associated conductive wires.   This has led to the need for a temperature monitoring probe with the following characteristics:

1. It should be nonperturbing in the presence of electromagnetic fields strong enough to heat large tissue volumes, and should eliminate shock hazard.
2. It should be small in diameter to minimize tissue trauma and temperature errors.
3. The system should be relatively accurate (±0.2°C), possess good resolution and stability, and be easy to calibrate.

To date, two basic approaches have been followed to reduce the conductivity of the probe's leads and minimize interference: high-resistivity lead wires (typically carbon-impregnated plastic) attached to a small thermistor bead as sensor [3,4] and optical fibers attached to one of several possible optical temperature sensors.   Reported so far as optical sensing schemes have been liquid crystals [5,6], polarization rotation by a birefringent crystal [7], band-edge absorption in a semiconductor crystal [8], fluorescence [9,10], and thermally-sensitive cladding material on a section of the fiber's core [11,12,13].   In addition, a viscometric method based upon temperature-related changes in fluid viscosity has been suggested [14].   Each of the above techniques has interesting features and potential advantages.

We have developed in our laboratory a complete temperature measuring system which utilizes the semiconductor band-edge absorption type of sensor and optical fibers.   The system measures and displays four probe tem-

peratures simultaneously, employs a microprocessor for linearization of
the readings and convenient calibration, and has interfacing capabilities
to other systems. The small size of the sensor is a major feature of this
design. A description of the instrument and its components follows.

2. SENSOR AND OPTICS MODULE DESIGN

2.1 Sensor material

The sensors are fabricated from bulk single-crystal gallium arsenide,
obtained in intrinsic purity as a lapped and polished wafer of 350 μm
thickness in 1-0-0 orientation intended for integrated circuits manufacturing. A simplified drawing of our sensor configuration and dimensions
is shown in Fig. 1a; many such sensors can be obtained at one time in a
batch fabrication process. No special reflection coatings are needed on
the rear faces of the sensor since total internal reflection provides
reflection of the radiation incident on the GaAs/air interfaces due to the
high index of refraction of GaAs -- 3.34. The simplicity of sensor design
results in very small possible sensor sizes, approximately 125 x 125 x
250 μm.

GaAs possesses favorable optical and physical characteristics for use
as a sensor [15,16]. The most important property for this application is
the nominal energy gap spacing (1.43 eV, corresponding to a near infrared
wavelength of about 905 nm) and the temperature coefficient of this energy
gap (-6.4×10^{-4} eV/°C).

2.2. Temperature measurements

The measurement of temperature changes in the GaAs is made possible
by utilizing the shift in wavelength of the optical band absorption edge,
which is the wavelength of radiation for which the incident photons have
sufficient energy to excite valence band electrons across the energy gap
into the conduction band and thereby become absorbed. This process is
shown in the energy level diagram of Fig. 1b. As the temperature of the
semiconductor increases, the magnitude of the energy gap decreases, thus
increasing the wavelength at the position of the optical absorption edge.

Figure 2 shows the experimentally-derived absorption edge for a
250 μm thick sample of GaAs at two different temperatures. As can be
seen, higher temperatures shift the edge toward longer wavelengths. If a

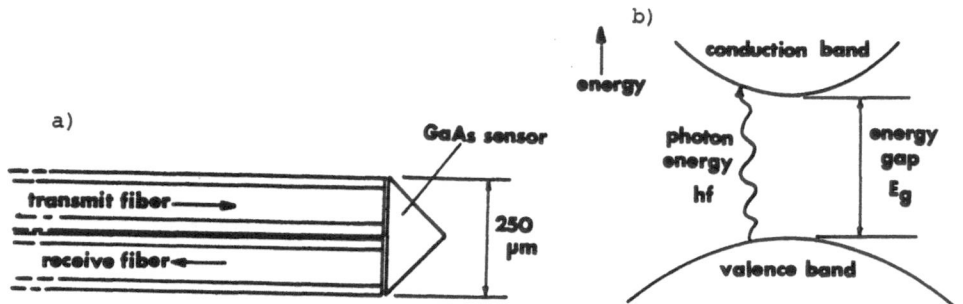

Figure 1. a) A drawing of the semiconductor sensor and its attachment to
the probe fibers. b) A schematic diagram of the energy bands in a direct-
gap semiconductor.

suitably monochromatic source is positioned within the steep portion of the absorption edge, a change of temperature will result in a variation of the optical power absorbed (and, therefore, of the amount transmitted) by the sensor. For the source wavelength shown in Fig. 2, a change of sensor temperature from 25°C to 40°C results in a 33 percent reduction in transmitted power, yielding an average sensitivity on the order of -2%/°C.

There is also a very slight pressure effect on the energy gap of GaAs (-9×10^{-10} eVm^2/kg), but for the pressure changes expected to be encountered in the environment where the sensor will be used, this effect is entirely negligible. Other than the influence of temperature and pressure, the energy gap of a semiconductor is stable.

2.3. Optical source

As the source of optical power to the sensors, we have chosen to use light emitting diodes (LEDs). These are GaAs LEDs with a peak emission wavelength of about 905 nm matching the absorption edge of the sensors and a half-power bandwidth of approximately 35 nm. Although this does not represent a perfectly monochromatic source, the concepts shown in Fig. 2 still apply, and temperature sensitivity with an LED is only slightly lower than with a theoretical monochromatic source.

The wavelength and amplitude stability of the optical source is of prime importance in the practical implementation of this idea since a drift in either parameter will be indistinguishable from a change in

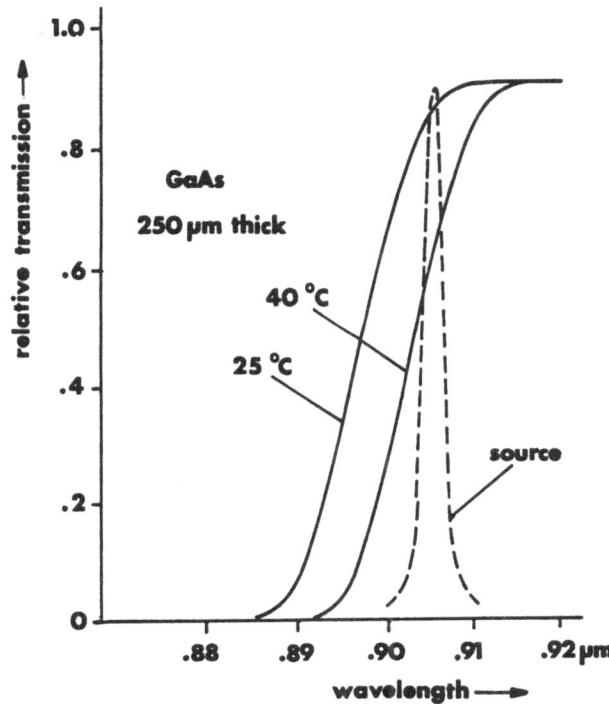

Figure 2. The optical absorption edge of GaAs at two temperatures, 25°C and 40°C. The dotted line shows the spectrum of a narrowband source.

sensor temperature (unless a separate channel is added to independently monitor LED power and wavelength). Therefore, the LEDs are placed in a temperature-controlled and thermally-insulated optics module to minimize temperature effects on output wavelength and power. A Peltier thermoelectric cell holds the module temperature to approximately 20 ±.1°C. The silicon photodiode employed to detect returned power from each probe is also housed in this module to reduce temperature drifts in its sensitivity, although this is not as critical as with the LEDs. The first stage of the electronic amplification (an op amp in a transconductance configuration) is located in the module in close proximity to the detector, thereby reducing noise pickup at the high gains employed.

The source LEDs and photodetector are optically connected to the individual probes via short plastic fiber pigtails of 400 μm clad diameter terminating in Amphenol Series 915 optical connectors (two per probe) placed in feedthroughs on the front panel of the instrument. In the case of the LED, the pigtail is epoxied directly in contact with the emitting surface after removal of the protective can. The pigtail to the photodetector is epoxied on the outer window of the detector's can; the detector area is sufficiently large -- approximately 1 x 1 cm -- to intercept all light exiting from the fiber's end face.

3. PROBE CONFIGURATION AND FIBERS
3.1 Fiber choice

The selection of the fibers used in the probes was guided by many considerations and trade-offs, including the following:

1. It is very advantageous to use two side-by-side fibers per probe -- one for carrying the incident light from LED to sensor, and one for transmitting the light from sensor to detector. This effectively isolates the received power to only radiation which has passed through the sensor, and avoids alignment difficulties and stray scattering associated with beam splitters. The minimum diameter of the fiber portion of the probe is then determined by twice the clad (outside) diameter of the individual fibers.

2. The larger the numerical aperture (NA) of the fibers, related to the overall angle of rays accepted by the fibers, the higher the power which will be coupled from a broad-angled radiation source such as an LED [17]. In general, all-plastic fibers possess higher NAs than either glass core/glass clad fibers, or glass core/plastic clad fibers.

3. The attenuation characteristic of the fibers, which is the sum of the optical absorption and scattering losses, is not significant for glass fibers but may limit the possible length of the probe if plastic fibers are chosen since most plastic fibers suffer from strong absorption near 905 nm from an OH^- resonance.

4. Bending losses in the fibers must be minimized; otherwise tight bending of the probe will decrease returned power and will be interpreted as a change in sensor temperature. Gloge [18] has shown that, for a given bend radius, bending losses are lower in fibers with large NA, small core diameter, and with a step rather than a graded index profile.

5. Finally, the fibers must be rugged enough to withstand the repeated flexing which the probe is likely to encounter in actual use. They should also be easy to handle, cut, and polish.

There is no single fiber type which is optimum in every regard. Glass fibers have very low absorption loss and are readily available in the small diameters desired, but they generally suffer from a fragility unacceptable for the intended use of the probes, and their low NA results in relatively high bending losses. Therefore, plastic fibers (polystyrene core, PMMA clad) were chosen for this application. These step-index fibers have a clad diameter of 125 µm, and an NA = .58. Unfortunately, their high absorption at 905 nm limits the length of present probes to about 1.5 meters.

3.2. Sheathing

When unsheathed, the sensor exhibits a very fast thermal time constant due to its small size and the low thermal conductivity of the optical fibers. When exposed to calm room air, the display of a bare sensor's temperature will appear erratic as it tracks the fluctuations in temperature at that point in space. Such a rapid time response is not needed for hyperthermia monitoring, so in order to mechanically protect the sensor and fibers and to restrict water-layer formulation around the sensor, we have sheathed the probe in thin-walled 0.61 mm OD Teflon tubing sealed at the distal tip. The sheathing increases the probe's thermal time constant to several hundred milliseconds, but this is still fast enough for hyperthermia applications; in fact, to improve the signal-to-noise ratio of the received optical pulses, low-pass filtering is purposely added in the electronic sections which further increases the response time constant to approximately 2 seconds.

Figure 3 is a photograph of a typical probe, showing the small sheathing at the distal end and optical connections at the proximal end.

4. SYSTEM DESCRIPTION AND PERFORMANCE
4.1. Analog and digital electronics

The analog portion of the prototype system has the task of supplying pulses to the LEDs, receiving and amplifying the returned signal pulses from the probes, rectifying these pulses by synchronous detection to reduce interference by noise which is nonsynchronous with the source, and low-pass filtering the rectified voltages prior to analog-to-digital conversion. Each probe channel has a dedicated source LED, but all channels share a common photodiode detector; thus the probe signals are time-multiplexed at the front-end receiver and are later separated in the syn-

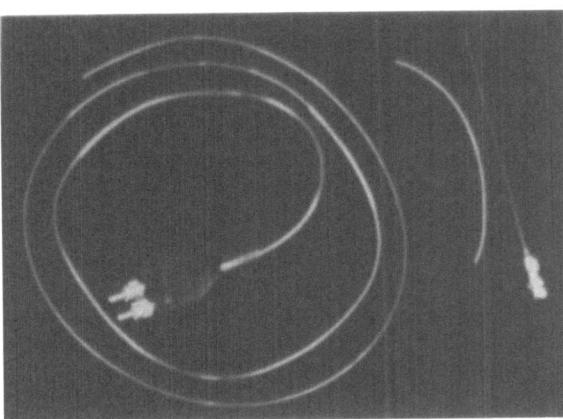

Figure 3. A photograph of a fiberoptic temperature probe, showing sheathing tubing and optical connectors.

chronous detection stage. An auto-zero stage cancels the effects of dc drift in the receiver amplifiers.

The digital portion is controlled by an Intel 8085A microprocessor. This section includes the analog-to-digital converter, a 12-bit successive approximation device chosen with the speed necessary to convert each 80 μs-wide LED pulse and with the resolution necessary for 0.01 to 0.02°C temperature resolution. The microprocessor, its software stored in a 2K-byte read-only memory, controls the A/D converter, linearizes the four temperature readings according to calibration curves stored in random access memory during a previous calibration procedure, and transmits the temperature readings to the liquid crystal displays (LCDs).

4.2. Calibration

Another major task for the microprocessor is automatic calibration of the probes. Full-scale internal calibration is accomplished by simultaneously placing all probes in the built-in calibration well, composed of a 1" x 1" x 2" copper block for heat uniformity, a Peltier module to provide cooling and heating upon microprocessor command, an accurate electronic temperature sensor for reference, and a small diameter copper tube for insertion of the probes. During calibration, the temperature of the well is first cooled to the lowest temperature of the chosen range, followed by slower heating to the upper temperature of the range. The entire procedure, which lasts about ten minutes, is automatic and convenient enough to be employed before every experiment.

Immediately after full-scale calibration, the accuracy of the system is approximately ±0.2°C. Due to small drifts of the optical and electronic components, the accuracy six hours after calibration is normally ±0.4°C. Resolution on the displays is either 0.01-0.02°C or 0.1°C, depending upon the position of an option switch inside the instrument. Two ranges are selectable by another switch; the normal range is from 21.72°C to 49.88°C, and an extended range spans from 16.60°C to 55.00°C.

4.3. Interface

We have provided several modes of possible interfacing between the system and other data recording and display units. Analong ouput voltages corresponding linearly to the temperature readings of the four probes are available at rear-panel BNC connectors for analog stripchart recording. Also, parallel digital outputs are presented in a "talker-only" IEEE-488 format on a 25-pin rear-panel connector. The interface that has proven to be most useful, however, is a serial RS-232 digital output utilizing an optical transmitter inside the unit, a fiberoptic cable link, and an optical receiver with TTL-level output. This fiberoptic link avoids any possibility of the pickup of stray electromagnetic radiation by wires when used in an environment of large leakage fields and has successfully interfaced the system with an Apple computer and also a small thermal printer for permanent data logging.

ACKNOWLEDGMENTS
    This work was supported by a grant from the National Cancer Institute, Public Health Service. The contributions of Richard Volz, Pi Chang Ko, Chris Burkes, Tom East, and Steve Rockhold are deeply appreciated.

REFERENCES

1. Rossi-Fanelli, A., Cavaliere, R., Mondovi, B., and Moricca, G., eds: Selective Heat Sensitivity of Cancer Cells. Berlin: Springer-Verlag, 1977.
2. Cheung, A. Y. and Samaras, G. M., eds. Special edition on hyperthermia. J Microwave Power, 16:85-233, 1981.
3. Bowman, R. R. A probe for Measuring Temperature in Radio-Frequency-Heated Material. IEEE Trans Microwave Theory Tech, MTT-24:43-45, 1976.
4. Larsen, L. E., Moore, R. A., and Acevedo, J. A Microwave Decoupled Brain Temperature Transducer. IEEE Trans Microwave Theory Tech, MTT-22:438-444, 1974.
5. Johnson, C. C., Gandhi, O. P., and Rozzell, T. C. A Prototype Liquid Crystal Fiberoptic Probe for Temperature and Power Measurements in RF Fields. Microwave J, 18:55-59, 1975.
6. Rozzell, T. C., Johnson, C. C., Durney, C. H., Lords, J. L., and Olsen, R. G. A Nonperturbing Temperature Sensor for Measurements in Electromagnetics Fields. J Microwave Power, 9:241-249, 1974.
7. Cetas, T. C. A Birefringent Crystal Optical Thermometer for Measurements in Electromagnetically Induced Heating. In Proc 1975 USNC/URSI Symposium (Bureau of Radiological Health),eds. C. C. Johnson and J. L. Shore. Rockville, Maryland:1976.
8. Christensen, D. A. A New Nonperturbing Temperature Probe Using Semiconductor Band Edge Shift. J Bioeng, 1:541-545, 1977.
9. Wickersheim, K. A. and Alves, R. V. Recent Advances in Optical Temperature Measurement. Indust Res and Dev, 21:82, 1979.
10. Samulski, T. and Shrivastava, P. N. Photoluminescent Thermometer Probes: Temperature Measurements in Microwave Fields. Science 208:193-194, 1980.
11. Scheggi, A. M. Brenci, M., Conforti, G., and Falciai, R. Optical-Fibre Thermometer for Medical Use. IEE Proc, 131:370-372, 1984.
12. Gottlieb, M. and Prandt, G. B. Temperature Sensing in Optical Fibers Using Cladding and Jacket Loss Effects. App Optics, 20:3867-3873, 1981.
13. Kopera, P. M., Melinger, J., and Tekippe, V. J. Modified Cladding Wavelength Dependent Fiber Optic Temperature Sensor. Proceedings of SPIE, Fiberoptic and Laser Sensors, 412:82-89, 1983.
14. Chen, M. M., Cain, C. A., Laun, K. L., and Mullin, J. The Viscometric Thermometer: A Nonperturbing Instrument for Measuring Temperature in Tissues under Electromagnetic Radiation. J Bioeng, 1:547-554, 1977.
15. Sturge, M. D. Optical Absorption of Gallium Arsenide Between .6 and 2.75 eV. Phys Rev, 127:768-777, 1963.
16. Pankove, J. I. Optical Processes in Semiconductors. New York: Dover, 1975.
17. Yang, K. H. and Kingsley, J. D. Calculation of Coupling Losses between Light Emitting Diodes and Low-Loss Optical Fibers. App Optics, 14:288-293, 1975.
18. Gloge, D. Bending Loss in Multimode Fibers with Graded and Ungraded Core Index. App Optics, 11:2506-2513, 1972.

LASER INJECTION MODULATION SENSORS

S. DONATI, T. TAMBOSSO
Dipartimento di Elettronica
Università di Pavia, Pavia, Italy

SUMMARY

Using optical feedback effect, a fiber interferometric sensor can be implemented. We describe the feedback into the laser as a injection modula tion, both in amplitude and frequency, of the cavity field, and outline the development of this class of fiber sensors.

1. INTRODUCTION

Among the various fiber-optic sensors, those based on the interfero-metric readout are particularly attractive because of their very high sensitivity. Commonly, fiberoptic interferometric sensors employ an optical configuration in which the fiber and the interferometer are physically external to the laser source, and the detected signal has the well-known expression:

$$I = I_o \left[1 + \cos (\Delta\omega t + \varphi) \right] \tag{1}$$

where $\Delta\omega$ is a possible frequency difference between the recombined beams, and $\varphi = \varphi(A)$ is the optical pathlength depending on the physical quantity A to be measured.

A second class of interferometers, referred to as the internal one, is that of the configurations where the laser resonator itself is the opti cal interferometer (fig. 1): a well known example is the laser ring gyro, while e.g. the fiber-optic gyro belongs to the class of external interfero-meters. The counterpart of eq. (1) for internal interferometers is:

$$I = I_o \left[1 + \cos (c/2L)\varphi t \right] \tag{2}$$

where $c/2L$ is the longitudinal mode spacing of the cavity.

A third class of interferometers, intermediate between internal and external ones, is the so-called induced-modulation configuration which is encountered whenever a mirror external to the laser cavity re-injects even a minute fraction of the propagating wave back into the cavity (fig. 1). The effect has been first analyzed by Spencer and Lamb [1] in the frame-work of a three-mirror laser structure, and successively regarded as a self-injection phenomena [2], useful for performing an interferometric measure ment [2,3] since the perturbed cavity field exhibits amplitude and frequen cy modulations whose indexes are proportional to the $\cos \varphi$ and $\sin \varphi$ interferometric terms:

$$E = E_0(1 + m \cos\varphi) \exp i \left[(\omega_0 + \Delta\omega \sin\varphi)t + \psi \right] \qquad (3)$$

being

$$m = (c/2L)\alpha T^2/2(g_0 - \Gamma), \qquad \Delta\omega = (c/2L)\,\alpha T^2 \qquad (4)$$

where αT^2 is the relative amplitude of re-injected field and $g_0 - \Gamma$ is an effective gain parameter [1].

An induced-modulation interferometer was demonstrated in [2], using a HeNe Zeeman Laser supplying two orthogonally polarized modes with a slight frequency difference. One mode was allowed to propagate, the other mode was blocked by a polarizer and served as the local oscillator for heterodyne detection of AM and FM modulations carried by the propagated mode. Used as a vibration or displacement sensor, i.e. with

$$\varphi = 2k\ \Delta s$$

the interferometer had a noise-equivalent displacement s_n of about 5pm/ /$(Hz)^{\frac{1}{2}}$ and was able to operate also on diffusing surfaces without additio nal optics, in view of the inherent matching of the speckle projected back into the cavity to the cavity mode, with the obvious limitations due to the speckle pattern statistics [4].

When an optical fiber is used in connection with the interferometer, the phase φ becomes a function of an intrinsic or extrinsic physical quan- tity - e.g. vibration, acoustic field, strain, etc.; thus a very simple and compact interferometric fiber sensor can be implemented.

The amplitude modulation due to injection effects has been observed in virtually any laser, from CO_2 and IR - HeNe to dye lasers [5].

In laser diode, it has been early recognized both as a useful effect for detecting diffusivity variations [6], such as those of the optical disk, and as a disturbing effect of the emitted intensity and noise [7], e.g. in optical communications. An additional feature of the induced modulation in laser diode is that even the drive current exhibits the effect, i.e. the laser functions simultaneously as the light source and detector of its own radiation [6].

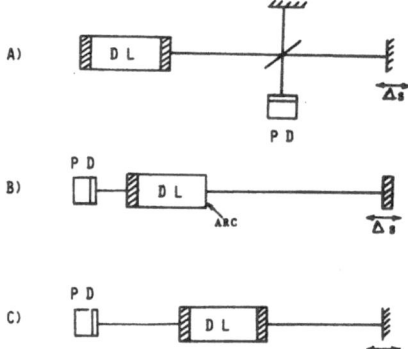

Fig. 1 - External (A), internal (B), and induced-modulation (C), basic interferometers.

2. ANALYSIS OF THE INJECTION MODULATION

The injection modulation can be analyzed with the aid of the semiclas sical theory of the laser [8,1]. For the cavity-field amplitude E and phase φ, in the approximation of slowly varying quantities, we have:

$$\dot{E} - \left[g_o(1 - \beta E^2) - \Gamma\right]E = (c/2L)\alpha T^2 E \cos(\varphi + \psi) \tag{5}$$

$$\dot{\varphi} - \zeta g_o(1 - \beta E^2) + \Delta = (c/2L)\alpha T^2 \sin(\varphi + \psi) \tag{6}$$

where $T \exp i \psi$ is the complex transmittance (field) of the output mirror, $\alpha \exp i\varphi$ is the roundtrip propagation attenuation and phase, and the other terms have the usual meaning [8]. By solving eqs. (5) and (6) one readily obtains [2] the AM and FM expression given by eqs.(3) and (4).

For the laser diode, it is $1/\beta = I/I_{th}|E_{sat}|^2$, where I and I_{th} are the diode current and its threshould value, and $|E_{sat}|$ is the field satu- ration amplitude; then eqs. (3) and (4) can also be written in the form [7]:

$$I_{th} = I_{th(o)} (1 - \alpha T^2 \cos\varphi) \tag{7}$$

where $I_{th}(o)$ represents the unperturbed threshold current. Eq. (7) shows how the laser diode current variations carry the interferometric term $\cos\varphi$ (fig 2). The other interferometric term $\sin\varphi$ is impressed as FM of the optical frequency and is much more difficult to recover, in a laser diode, because a two-frequency counter-part of the gas lasers is not easily im- plemented. An attractive alternative to a heterodyne detection is that proposed in [9], of an optical FM to AM conversion performed with the aid of the dispersion of a long fiber.

In the above discussion, full coherence of injections phenomena is tacitly implied. Interesting locking phenomena occur when the coherence length L_c is shorter than the optical pathlength, $L_c < s$; this case has been treated in detail in Ref. [7], showing self-synochromization and frequency-attraction [10] effects.

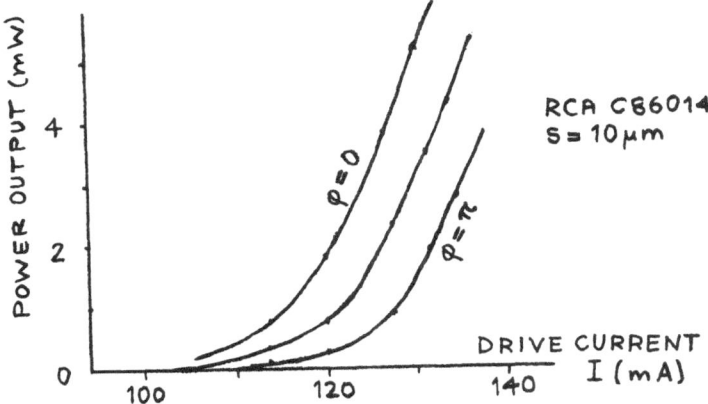

Fig. 2 - Power-current characteristic of unperturbed diode (center) and of diode with induced modulation ($\varphi = 0, \pi$).

3. SENSOR CONFIGURATIONS

Laser diode sensors can be implemented by means of two main configurations. When a reflective membrane sensitive to the quantity to be measured (e.g. acoustic pressure, piezomagnetic strain, etc.) is placed close to the laser output mirror, a direct induced modulation is obtained, with minimum disturbances from extraneous sources. This kind of sensor has been demonstrated in Ref.[11] to achieve interferometric-grade sensitivity using the AM signal.

However, several applications demand the use of an optical fiber, inserted between the laser diode (and the associated electronic circuits) and the sensing region, for the well-known advantages of this arrangement. The experimental setup utilized for this kind of optical fiber laser sensor based on the induced modulation is that shown schematically in fig. 3.

Some problems arise from the use of the fiber. Firstly, the efficiency of laser to fiber and fiber to laser coupling is relatively low, resulting in a reduction of the modulation indexes [eq.(4)] by a factor of the order of 0.05. Secondly, all retroreflection contributions such as those from scattering and joints, add to the useful signal and therefore shall be kept as small as possible. Actually, fixed added contributions are insignificant respect to AM and FM signals because they correspond to a bias, while their fluctuations are sources of error. Reflections from optics (e.g., a launch objective) would give rise to an intolerable level of disturbance in the induced modulation signal, also because of the ψ-dependence expressed by eqs. (5) and (6). An index-matched butt-joint on the laser was found to be an adequate solution. Lastly, since the optical pathlength of the fiber adds to the phaseshift φ under measurement, it must be controlled by means of a feedback so as to stabilize, e.g., the AM signal $\cos\varphi$ at the optimum bias $\varphi = \pi/2$.

To implement the stabilization, we employ (fig. 3) a phase modulator driven by the low-frequency component of the interferometric signal, which is picked either by the laser drive circuit or by a separate photodiode. A few turns coiled on a 1" PZT piezoceramic, driven at $\pm$ 200V, suffice for compensation over several tens of wavelengths, with high ($\sim$ 50 dB) dc loop gain. Since the stabilization loop introduces a low-frequency cutoff, the sensor is only suited for ac signals, as in the applications reported in fig. 3, i.e. vibrometers, acoustical hydrophones and ac magnetic field sensors. Present measured sensitivity of the interferometer section is 50 pm/$\sqrt{\text{Hz}}$ at 1 KHz which, though adequate for most applications, appears to be limited by the fiber. However, the laser injection modulation sensor is attractive because of its very simple setup, suitable for a compact layout, and the inherent self-alignment with minimum optical and mechanical adjustment operations. A last, distinctive feature of the injection modulation sensor is that the AM/FM interferometric signal is carried by the beam, and is therefore available for detection at any point, including e.g. the far end of the fiber.

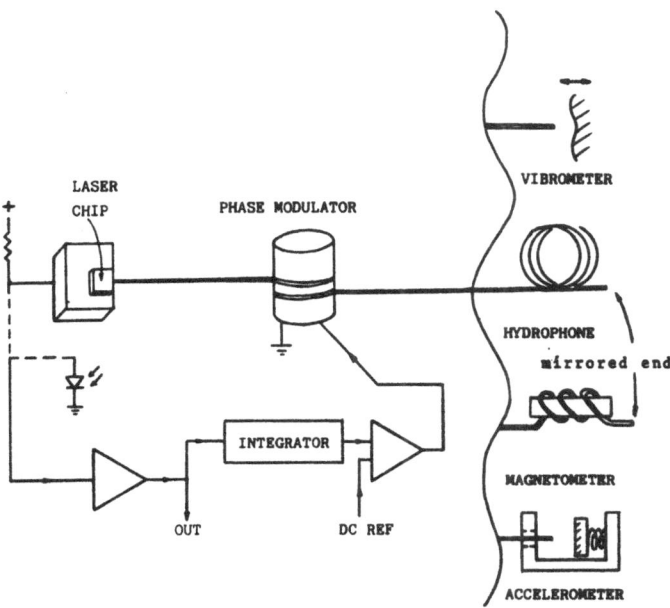

Fig. 3 - Schematic of induced modulation laser-diode sensors.

REFERENCES

1 M.B.Spencer, W.E. Lamb jr.: Laser with a Transmitting Window, Phys. Rev.
   A 5 (1972) pp. 884-892

2 S. Donati: Laser Interferometry by Induced Modulation of Cavity Field,
   J.Appl. Phys. 49 (2) (1978) pp. 495-497

3 Donati: A Novel Laser Interferometer for Distance Measurements, Proc. Conf.
   Prec. Electromagn. Measur., Ottawa June 1978, pp. 75-77

4 S. Donati, G.Martini: Speckle Pattern Intensity and Phase: Second Order
   Statistics, J. Opt. Soc. of Am. 69 (1979) pp. 1690-1694

5 see e.g.: Honeycutt and Otto, J.Quant. El. QE-8 (1972), 91; Doyle and
   Gerber, QE-3 (1967), 479; Rudd, J.Sci. Instr. 1 (1968), 723; Matsumoto,
   Appl. Opt. 19 (1980), 1; Schroter and Kuhlke , J. Opt. Quant. El. 13
   (1981), 251

6 W.J. Burke, M.Ettenberg and H.Kressel: Optical Feedback Effects in CW
   Injection Lasers, App. Opt. 17 (14) 1978 pp. 2233-2238

7 O. Hirota and Y. Suematsu: Noise Properties, of Injection Lasers Due
   to Reflected Waves, IEEE J. Quant. El. QE-15 (1979) pp. 142-149

8 W.E.Lamb jr., Theory of an Optical maser, "Phys. Rev. 134 (1964) pp.
   1429-1450

9 P.Brosson, C.A.Riberio and J.E. Ripper: Frequency Demodulation of a
   Semiconductor Laser by FM/AM Conversion in an Optical Fiber, IEEE J.
   Quant. El., QE-17 (1981) pp. 689-693

10 V.Annovazzi Lodi, S.Donati: Injection Modulation in Coupled Laser
   Oscillators, IEEE J. Quant. El. QE-16 (1980) pp. 859-864

11 R.O.Miles, A. Dandridge, A.B.Tveten, T.G.Giallorenzi: An External Cavity
   Diode Laser Sensor, J.Lightw. Tec. LT-1 (1983) pp. 81-93

CHEMICAL SENDUCERS

A.D'AMICO
Università degli Studi di L'Aquila
Dipartimento di Ingegneria Elettrica, 67100 L'Aquila - Italy
G.PETROCCO
IESS-CNR, Via Cineto Romano 42, 00156 Roma - Italy

1.INTRODUCTION

A senducer is a device which senses and transduces a physical or chemical parameter into an electrical or optical signal suitable for electrical or optical processing. In practice the physical parameters usually monitored are the following: temperature, pressure,electric or magnetic field, strain stress etc., while the chemical parameters most widely considered are: chemical activity, chemical potential, ion or gas concentration, etc.

The measurement of the above mentioned parameters has proved extremely important in the framework of many industrial and scientific problems at the instrumentation and control levels. Merely to give an idea of the impor tance of the development of chemical senducers let us list a few areas which deal with the detection of chemical species and the measurement of their concentration.

1.1 Biomedics: a) Cardiopulmonary bypass(N^+, Na^+, Ca^{++},P_H, P_{O2}, P_{CO2},CN^-, Cl^-, F),b) blood analysis(P_{O2},P_{CO2}, N^+,K^+, Ca^{++}, Cl^-, H^+), 3) renal dialysis and renal hepatic disfunctions(uric acid, creatinine, phosphates,P_H, acetates, K^+, bicarbonates, bilirubin, etc.

1.2. Microbiology: enzymes, markers, P_H, etc.

1.3.Neuropharmacology: P_{O2}, K^+, etc.

1.4 Industrial control: a) Combustion(CO, CO_2, SO_2, H_2, etc. b) central biogas sysstems(CH_4,CO_2, H_2, etc., c) wastes(CO, CO_2,O_2, NH, etc.,c)pollution in operating theatre(CO_2, N_2O, etc.

1.5. Atmospheric control: a) Primary atmospheric pollution sources(H_g,Pb, Cr, Zn, SO_2, NO, CO, etc.,b)secondary atmospheric pollution sources(NO_2, CO_2, O_3), aldehydes, H_2SO_4, etc.

1.6. Security systems: high concentration of H_2 in energy storage.

An ideal chemical senducer must be:a) higly sensitive,b)reversible,c) chemically selective,d)reliable, e)small-sized,f)low in noise in the low frequency region, g)noncontaminating, h)nonpoisonous, i)nonsensitive to radiations,l)nonsensitive to temperature variations, m)simple to prepare,n)ro-

bust and solid in construction, o)simple to calibrate, p)compatible with lo
cal preprocessing, q)quik responding, r)low in cost.

Since it is obviously impossible to satisfy all these constraints for
every senducer imaginable, a compromise must be reached at the design level: the particular needs, the characteristics of the measurement ambience,
and the level of fabrication technology available will together dictate the
most suitable sensducer characteristics to be chosen.

One of the most difficult tasks is that of developing gas senducers with
sufficiently high sensitivity, selectivity and reversibility to measure low
concentrations at very low exposure limits with a minimum power dissipation.

Despite all these problems involved, the current state of the art is one
of considerable vitality not only in Universities, private and public research institutes but also, and more importantly, in those industries which
see in chemical senducers the promise of new and more lucrative markets.

This paper reviews some of the recent developments which should have
long term consequences on the study of chemically sensitive sensucers.

The particular topics to be presented and discussed are as follows:
a) Senducers based on metal-oxide-semiconductor capacitors (MOS) and
   metal-oxide-semiconductor transistors(MOSFET),
b) Senducers based on surface acoustic wave (SAW) device,
c) Senducers based on pyroelectric devices.

In order to give an idea of the three different detection procedure relative to the same gas,i.e. hydrogen, the catalytic metal considered is in
all cases palladium. The causes of the H_2 sensitivity in the first, second,
and third cases respectively are: the charge variation developed at the
Pd-SiO$_2$ interface, a change in both the surface and bulk density of the Pd
due to H_2 adsorption and desorption and the heat of reaction developed between H_2 and Pd.

2. THE MOS AND MOSFET STRUCTURES: MOS AND MOSFET HYDROGEN SENDUCERS

Metal-oxide-semiconductors(MOS) and metal-oxide-semiconductors field
effect transistors(MOSFET) with palladium as a metal have both been extensively investigated in research into chemical(hydrogen) senducers (1,2) .

The Pd-MOS capacitor is schematically shown in fig.1

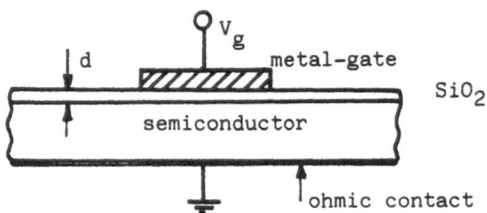

Fig.1 Schematic of the Pd-MOS capacitor.

Its working principle is as follows:

$$N(H_2) \longrightarrow H+H \longrightarrow \Delta Q \longrightarrow \Delta C$$
$$\text{(on Pd)} \qquad \text{(diffusion) dipole}$$

A change in the H_2 concentration in the ambient, after the catalytic reaction on the Pd surface, produces at the Pd-SiO$_2$ interface a dipole layer and
or a change in the surface state density which causes an easily detectable
capacitance variation ΔC. The most straithforward method for investigating
adsorption and desorption processes in this particular case is represented
by C/V measurements. Fig. 2 shows a typical result obtainable by C/V measurements performed at 1MHz.

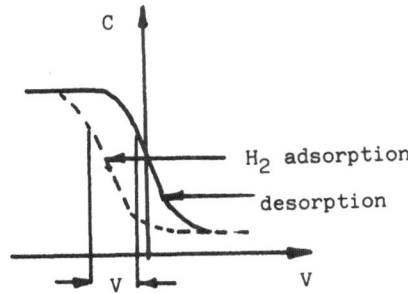

Fig. 2 Capacitance versus voltage behaviour for a MOS capacitor.

Of particular interest is the flat band voltage capacitance related to
the flat band voltage given by the following relationship:

$$V_{FB} = \phi_m - \chi_s - (E_C - E_F) - Q_s \frac{d}{\epsilon_i}$$

where ϕ_m is the Pd work function, χ_s is the electronic affinity of the semiconductor, $E_C - E_F$ is the energy interval between the conduction band and
the Fermi level, Q_s is the charge at the oxide interface per unity area, d
is the oxide thickness and ϵ_i is the dielectric constant of the oxide.

In practical experiment the shift of the C/V curve is measured as ΔV_{FB}
and is then correlated to the hydrogen concentration.

The simplified structure of a MOSFET is shown in fig.3

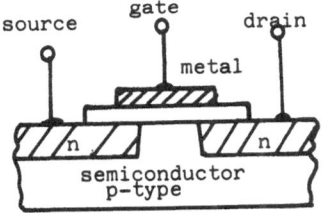

Fig.3 Schematic of a Pd-MOSFET.

When the gate is represented by a Pd metal film a hydrogen MOSFET senducer is obtained. In this case the flow diagram of the basic operating
principle is as follows:

$$N(H_2) \longrightarrow H+H \longrightarrow \overset{\text{dipole}}{\Delta Q\ (V_g)} \longrightarrow \Delta I \longrightarrow \Delta V$$

(on the Pd)   (diffusion to SiO$_2$)   output

A variation in the H_2 concentration causes a dipole layer(or a change in the surface state density N_s) at the Pd–SiO$_2$ interface which is due to the catalytic behaviour of the Pd film deposited on the SiO$_2$.

The dipole or/and ΔN_s causes a variation ΔI_d in the drain current which produces, by a load resistance, an output voltage variation related to the hydrogen concentration. The drain current, in the nonsaturated region of the MOSFET is given by:

$$I_d = \mu \left(\frac{w}{l}\right) C_{ox} \left(V_d(V_g - V_t) - \frac{1}{2} V_d^2\right)$$

where μ is the mobility of the carriers in the FET channel, w and l are the gate dimensions, C_{ox} is the oxide capacity per unity area, V_d and V_g are the DC drain to source and gate to source voltage respectively and V_t is the threshold or turn on voltage. 3

A variation in the drain current is caused by a threshold voltage change which is dependent on the electrical state at the metal–SiO$_2$ interface; and this state is strongly related to the adsorption and/or desorption processes of chemical species from the ambient. By applying a specific membrane onto the gate a MOSFET senducer can became CHEMFET senducer.

3. THE SURFACE ACOUSTIC WAVE SYSTEM: THE HYDROGEN SAW SENDUCER

This senducer is based on the velocity changes in surface acoustic waves (SAW) which propagate on a piezoelectric substrate(LiTaO$_3$,quartz,ZnO,etc.) coated with an absorbing film under gas exposure.

In the case of a hydrogen senducer, palladium can be used as the adsorbing film.

The basic structure of the SAW senducer consists of a SAW delay-line whose propagating path is coated by the catalytic film. Adsorption and desorption of hydrogen producechanges in density and elastic properties of the film which produces variations in the surface wave velocity. These can be detected in different ways according to the type of detecting circuit used.(4)

The working principle of the SAW senducer is shown in the following flow diagram:

$$\Delta N(H_2) \longrightarrow \text{surface} \longrightarrow \Delta v \longrightarrow \Delta \tau$$
$$\text{bulk}$$

$$\Delta N(H_2) \longrightarrow \text{surface} \longrightarrow \Delta v \longrightarrow \Delta \tau \longrightarrow \Delta f$$
$$\text{bulk}$$

A variation in the number of the impinging molecules produces a change in density which gives rise to a SAW velocity variation. This variation induces a phase shift of the wave which can be detected by the circuit represented in fig. 4. According to this figure a radio frequency is injected into the interdigital transducer T_I which generates SAW's along two opposite directions. Along one of the two propagation path there is a thin film of Pd.(5). The propagating SAW's are collected by T_2 and T_3 which generate two voltage signals which are then mixed and filtered to give the output voltage. For a given velocity variation Δv in the SAW velocity v, the output voltage variation V_o is given by:

$$V_o = 2V_M FL \frac{\Delta v}{v^2}$$

where V_M is the maximum value of the input signal, f is the frequency, L is the length of the film.

This configuration, which is also called a differential configuration, is characterized by a specific common mode rejection versus temperature and pressure variations due to variable ambient conditions.

In another configuration (fig.5) the SAW delay-line is connected with an amplifier to form an oscillator whose frequency f is compared with a fixed frequency f_o.

A frequency to voltage converter translates $f-f_o$ into the useful analog signal.

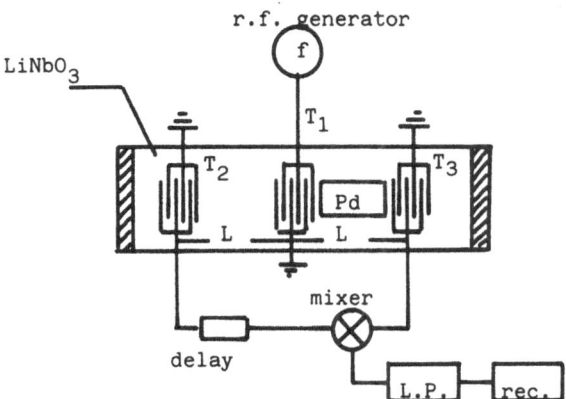

Fig.4 A differential SAW senducer

Assuming that, due to the adsorbtion desorbtion processes, the variation of the relative velocity $\Delta v/v$ is small, the relative variation of the frequency is given by:

$$\frac{\Delta f}{f} \simeq K \frac{\Delta v}{v} \qquad \text{where } K = \frac{L}{L_p}$$

Fig. 6 shows a typical response of the hydrogen SAW senducer in the case of the oscillator configuration.

The use of suitable catalytic films enables this kind of structure to be also used to construct senducers for gases of other kinds.

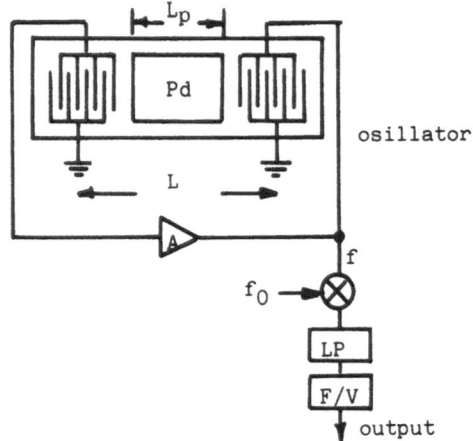

Fig.5 An osillator SAW senducer for H_2

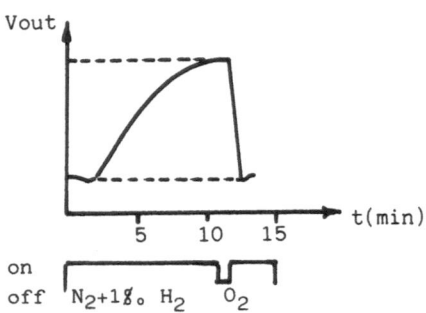

Fig. 6 Typical response to H_2 of a SAW senducer

4. THE PYROELECTRIC SYSTEM: THE HYDROGEN PYROELECTRIC SENDUCER

Pyroelectricity is the thermal equivalent of piezoelectricity.

In a given pezoelectric crystal a net strain produces a net polarization along suitable crystallographic directions. In a pyroelectric crystal a change in thermal energy causes a change in temperature producing a net polarization(again along particular crystallographic directions) which is easy to measure by suitable electronic circuit. (6)

Pyroelectric materials belong to the class of ferroelectrics and are cha racterized by having a Curie temperature T_c above which the pyroelectric properties disappear. Below the ferroelectric Curie temperature, ferroelec tric crystals posses domains of a spontaneous but randomly oriented electri cal polarization which can be aligned by applying a suitable electric field during a slow-varying cooling procedure(-2 °C/min) from T_c plus 50 °C down to room temperature(pooling method). It is essential that the resulting net polarization is preserved for all temperatures below T_c.

The working principle of the pyroelectric senducer can be observed in the following flow-diagram:

$$\Delta Q \longrightarrow \Delta T \longrightarrow \Delta p \longrightarrow \Delta Q_{el} \underset{C}{\overset{}{\longrightarrow}} \begin{array}{l} \Delta V \\ \Delta I \end{array}$$

This means that a heat variation ΔQ, causes a temperature variation ΔT which causes an internal polarization change Δp. Δp will produce an electrical charge variation ΔQ_{el} which can develop a current ΔI or a voltage ΔV on an appropriate external circuit.

The sensitivity value of a pyroelectric senducer can be deduced from the fact that a typical noise equivalent power for a pyroelectric material such as $LiTaO_3$ is 10^{-9} Watts/cm^2, which corresponds to an equivalent thermal input of 0.239×10^{-12} Kcal/cm^{2}sec. or about 6×10^{10} eV/cm^{2}sec.

In fact it has been shown , from experiments, that under appropriate conditions, extremely small temperature changes down to less than 10^{-6} °C can be detected and measured.

A practical pyroelectric senducer which can sense hydrogen is schematically shown in fig. 7

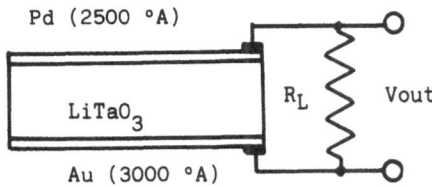

Pd (2500 °A)

LiTaO$_3$

R_L Vout

Au (3000 °A)

Fig.7 Schematic of a pyroelectric senducer.

We point out that the above considerations on the sensitivity of this kind of senducer suggest the theoretical possibility that nanocalorimetry can be studied.

The net charge arising from a polarization depending on temperature variation can be written as:

$$Q = CV - p(T)A (\langle T \rangle - T_0)$$

where $C = \epsilon \epsilon_0 \dfrac{A}{d}$ is the capacitance of the structure, ϵ is the dielectric constant of the pyroelectric material, ϵ_0 is the free space dielectric permittivity, d is the thickness of the pyroelectric material, V is the voltage across the capacitance and p(T) is the pyroelectric parameter which is temperature independent till 200 °C (for $LiTaO_3$) and temperature dependent from 200 °C up to T_c.

The current induced in an external load R_1 when the average temperature of the material is changing, is given by:

$$i(t) = -C \frac{dV}{dt} + \frac{d}{dt}(p(T) A \langle T \rangle)$$

The above relationship is the basic equation describing the pyroelectric effect.

In practice, when low level temperature variations are involved, it is important to consider the noise behaviour in order to define the accuracy of the measurement.

A possible equivalent sennducer circuit where an amplifier is used to detect the voltage signal, is represented in fig. 8

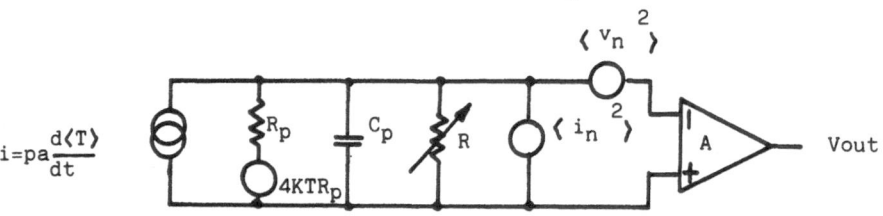

$$i=pa\frac{d\langle T\rangle}{dt}$$

Fig. 8: R is the matching resistor between the senducer and the amplifier.

The noise factor F is expressed as:

$$F=\frac{\langle V^2{}_{tot}\rangle}{4KTR_{eq}} = 1 + \frac{\langle V^2\rangle}{4KTR_{eq}} + \frac{\langle i^2\rangle R_{eq}}{4KT}$$

which has a minimum when:

$$R_{eq} = R_p//R = (\langle V_n^2\rangle / \langle i^2\rangle)^{1/2}$$

The optimum resistance is obtained when:

$$R_{opt} = \frac{(\langle V^2\rangle /\langle i^2\rangle)^{1/2}}{1-(\langle V^2\rangle/\langle i^2\rangle)/\ R_p}$$

Particular attention must be paied to the problem of selscting the most appropriate R value. A high R value produces both a high output signal level and a high electric time constant. At the design level a compromise between the two conditions must always be reached.

Fig.9 shows the response of a pyroelectric senducer during adsorbtion and desorbtion of hydrogen.(7)   The dotted lines represent variations in temperature.

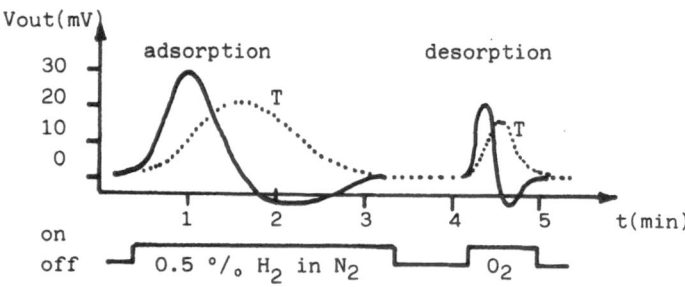

Fig.9

It must be observed that the pyroelectric senducer can either detect or measure the concentration of other gases, when an appropriate catalytic material is used.

CONCLUSIONS

Three kinds of senducers have been introduced and discussed. The MOS and MOSFET structures are sensitive to charge variations at the $Pd-SiO_2$ interface. The SAW senducer responds to variations of both the density and elastic properties of a film deposited on the acoustic path of the SAW.

The pyroelectric senducer is sensitive to small temperature variations due to adsorption and/or desorption processes.

Much research work must still be done to improve these different structures from the point of view of both the technological compatibility with adsorbing membranes and the cost/performance ratio.

The first kind of structure is the most advanced due to the fact that it benefits from the rapid development of microelectronics.

REFERENCES

1) I.Lundstrom, Hydrogen Sensitive MOS Structure, Part 1, Sensors and Actuators, 1, 403, 1981.

2) I.Lundstrom and D.Soderberg, Hydrogen Sensitive MOS Structure, Part 2, I05, 1981.

3) S.M.Sze, Physics of Semiconductor Devices, Wiley Interscience, 1969.

4) A.D'Amico, A.Palma, E.Verona, Palladium Surface Acoustic Wave Interaction for Hydrogen Detection, Appl. Phys. Lett. vol. 41, N°3 1982.

5) A.D'Amico, A.Palma, E.Verona, Surface Acoustic Wave Hydrogen Sensor, Sensors and Actuators, 3 1983.

6) A.D'Amico, J.N.Zemel, Pyroelectric enthalpimetric Detection, J.Appl. Phys. 57(7) 1985.

7) A.D'Amico, G.Fortunato, Sensori Chimici, Fisica e Tecnologia, vol.4 N°2,1983.

A VERY SMALL VOLUME UV ABSORBANCE DETECTOR FOR CAPILLARY SEPARATION SYSTEMS

K. Ogan[a], F.M. Everaerts[b], Th.P.E.M. Verheggen[b]

[a] Applied Research Department, Perkin-Elmer Corporation, Norwalk, CT 06859 , U.S.A.
[b] Dept. of Instrumental Analysis, Eindhoven University of Technology, Eindhoven, Netherlands

INTRODUCTION

The introduction of the capillary column has certainly revolutionized the practice of gas chromatography, particularly when high resolving power is required. Attempts are being made to achieve similar benefits in liquid chromatography and in other separation techniques. These efforts have been less than totally successful, in part due to the dimensional limits imposed by the magnitude of solute diffusivity in liquids relative to gases, and in part due to the lack of detectors having the sensitivity and small volume required for use in such systems. Approximate values of the capillary diameters required for the different separation systems are given in Table I. One interesting approach to the detector problem for spectroscopic detectors has been demonstrated

TABLE I
Capillary Dimensions for Separation Systems

Separation System	Capillary Diameter
Gas chromatography	75 - 250 µm
Liquid chromatography	3 - 20 µm (theor.)
	10 - 50 µm (exp'tl.)
Super-critical fluid chromatography	20 - 50 µm (theor.)
	75 - 200 µm (exp'tl.)
Capillary zone electrophoresis	50 - 200 µm (exp'tl.)
Isotachophoresis	200-500 µm (exp'tl.)

by Yang , et al.[1], for capillary liquid chromatography, by Jorgenson, et al.[2], for capillary zone electrophoresis, and by Everaerts, et al.[3] for capillary isotachophoresis. These authors have used the separation capillary directly as the flow cell for optical detection, thereby avoiding problems of unwanted extra peak dispersion which arise in coupling a detector system to the separation system. This approach usually requires conventional lens arrangements in order to focus the incident light down to the small scale (75 to 250 µm diameter capillaries). Naturally, care must be taken to ensure that all of the incident light passes through the capillary, rather than around it, in order to achieve full sensitivity and linearity. One of the few disadvantages to this approach is the requirement that the material used for the separation capillary be optically transparent. This is a very severe restriction in the case of UV spectroscopic detection, in which case the capillary must be fused-silica. Different types of stationary phases are required for the different separation techniques, and the application of these stationary phases to the surface of fused-silica is often not as easily achieved as it could be to with other surfaces.

This preliminary report describes a novel approach to such detectors, one which makes use of optical fibers for the introduction and collection of the light. The optical fibers are cast in a thermoset resin (typically, an epoxy resin) which serves several purposes. The polymer casting (1)provides a means of directly forming the separation capillary (or that section of the capillary involved in the detector), (2)enables the ready selection of the chemical and physical characteristics required for the particular separation technique being used, (3)permanently positions the fibers in direct contact with the contents of the capillary, and (4) forms a slit which limits the incident light to just the central portion of the capillary.

EXPERIMENTAL

The fused-silica optical fibers were supplied by Fiberguide Industries, (Stirling, NJ). These fibers were 200 μm core diameter, with a 20 μm thick cladding, and a vinyl jacket (ca. 370 μm final diameter). The ends of approximately 10 cm lengths of the fiber were polished manually, using a series of alumnium oxide paper with grits spanning the range of 40 to 0.3 μm . An optical fiber was positioned on either side of a 200 μm diameter stainless steel wire, as shown in Figure 1. The wire had been previously treated with a silicone-based release agent (MR-5002, Conap Inc., Olean, NY).

Once the optical fibers were tightly positioned perpendicular to the stainless steel wire, an epoxy resin (RP- 4032-A, Ren Plastics, Inc., Lansing, MI) was then applied to the junction region and allowed to cure. This sub-assembly was then carefully placed in a casting mold which incorporated suitable coupling fixtures to allow the detector to be connected to the rest of the capillary system. A different epoxy (Insulcast 510, Pergamon-Salmon, Ltd., Plainview, NY) was used to pot this assembly. After this epoxy had cured, the stainless steel wire was carefully pulled out of the casting. The resulting channel was rinsed with acetone, then toluene, then acetone, and finally water.

The fiber detector cell was tested in the simple source/cell/detector configuration shown in Figure 2. A low pressure mercury arc lamp (Hamamatsu, 81-1025-01 "pen lamp"), driven by a modular high voltage source (SHV-10, MIL Electronics, Inc., Dracut, MA) served as the source. The pen lamp was mounted in a plastic (Delrin) holder which was fitted with a plastic pressure nut such as used in LC instruments. This LC coupling nut served to hold the end of the fused-silica optical fiber directly against the lamp. (The optical fiber was also sheathed in a light-tight jacket by means of opaque heat shrink tubing.) A solid-state photodiode (Hamamatsu S1226-5BQ) was used as the detector, and was monitored with a standard current-to-voltage converter utilizing a high-impedance FET amplifier (AD41411) and a 100K feedback resistor. This pre-amplifier was packaged in a metal box with a small hole through which the photodiode protruded. A Delrin holder over the photodiode was fitted with a plastic LC coupling nut such that the fiber (again jacketed with heat-shrink tubing) was directly impingent on the face of the photodiode. The pre-amp output was fed to a second-order Butterworth filter (12 Hz cut-off frequency), then to a second gain stage (x 40), and finally to a voltage offset stage.

Sodium iodide was used for the test solute in the initial tests of the detector. [This solute is recommended for use in evaluating the stray light characterisitics of spectrophotometers (ASTM Method E 387-72) since it has a single strong absorbance band between 200 and 280 nm.] Solutions of NaI were prepared by dilutions of a saturated solution, and spanned the range of 24 μM to 1230 μM. These solutions were prepared in a sodium phosphate buffer (0.1 M, pH 7).

RESULTS AND DISCUSSION

The general concept of the optical fiber detector is shown in **Figure 1**. Two optical fibers with polished ends are positioned axially opposite each other, with a "template" wire fixed perpendicularly in between them. An opaque thermoset resin is next cast around this system. Forcing the optical fibers tightly against the template wire will prevent the polymer from filling the contact region between the optical fiber and the template wire. (This also requires that the

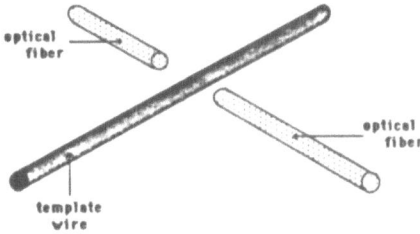

Figure 1. Configuring of the optical fibers and the template wire for the detector.

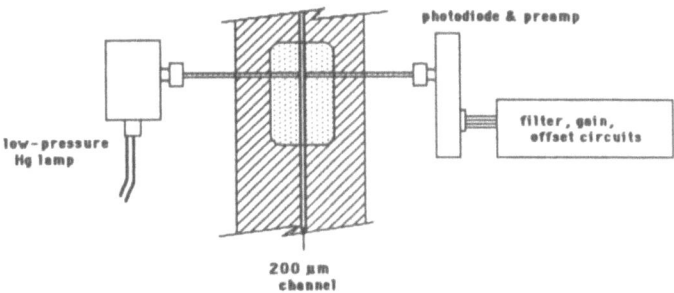

Figure 2. Experimental system for evaluation of the optical fiber detector.

polymeric material be chosen to have appropriate values of viscosity and surface tension.) This restricted space is illustrated in **Figure 3**, viewed along the optical fiber axis, and iin **Figure 4**, viewed along the template wire axis. The polymer is cured, and then the template wire is removed, thereby creating a capillary channel with the optical fibers rigidly positioned in direct contact with the channel, as shown in **Figure 4**B. The non-occluded space directly against the fiber face in **Figure 4**B now serves as an optical slit for each of the optical fibers, such that only that light passing directly across the center of the capillary channel is monitored.

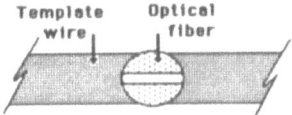

Figure 3. Detector configuration before casting, as viewed along the axis of the optical fiber. The clear area represents the region of contact between the fiber face and the template wire.

388

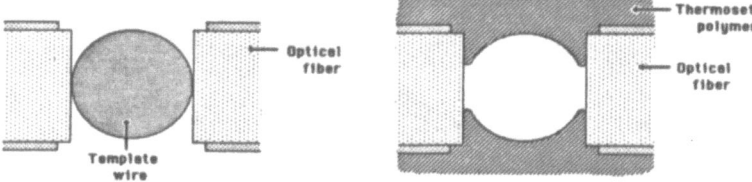

Figure 4. Detector configuration as viewed along the axis of the template wire. A. Before casting. B. After casting and removal of the template wire. The empty clear area is now the separation capillary.

This approach to the construction of the UV detector offers several advantages. First, the fiber directly forms the window into the separation channel; consequently, the separation channel can now be constructed from materials other than fused-silica. This makes possible the selection of materials which would be more suitable for the particular separation technique being used. Second, the partial filling of the polymer around the end of the fiber, as shown in **Figure 4B**, directly forms an optical slit that restricts the transmitted light to just that which transverses the center of the separation channel. In principle, this should improve the linearity relative to a system in which the light can be refracted around the edge of the capillary. Also, light can only pass through the capillary in this system, and not around the outside of the capillary, as is possible in many other conventional designs. Thus, an expanded linear range is possible. Finally, the optical elements are permanently fixed relative to the capillary in this approach, thereby providing a rugged and reliable detector module.

The need for UV detection requires that the optical fibers be fused-silica, with high transmission levels in the UV wavelength range. (The use of fused-silica fibers in and of itself does not guarantee that they will have high transmission in the ultraviolet range. It was found that fused-silica optical fibers from several different manufacturers absorbed strongly at 254 nm.) The optical fibers supplied by Fiberguide Industries were found to have the good transmission at the 254 nm line of the mercury arc lamp.

The first test systems were constructed using 200 μm diameter stainless steel wire as the template for the separation channel, together with the 200 μm core diameter optical fibers. These dimensions yield an approximate detector volume of 6 nL (6×10^{-3} μl). The initial evaluation of this detector design was directed toward the development of an inexpensive UV detector, and drew upon elements and components from other successful systems. The source was an inexpensive low pressure mercury arc, and the detector was a photodiode as described in the Experimental section. The NaI test solutions were introduced directly into the detector module (no separation system), and the change in light in light intensity recorded on a strip-chart recorder.

The change in light intensity, ΔI, is proportional to the exponential of solute concentation,

$$\Delta I = I_0 - I_t = I_0(1 - 10^{-\beta c})$$

where I_0 and I_t are the incident and transmitted light intensities respectively, c is the solute concentration, and β is the product of the molar absorbtivity and pathlength as usual. The ΔI values for the lowest three NaI concentrations were used to determine β and I_0, from which the absorbance values for all the ΔI measurements were then calculated. These results are plotted in **Figure 5** against NaI concentration. Good linearity is observed for the lower absorbances, with a usable linear range up to approximately 1.0 A. Non-linearity does begin to appear beyond this

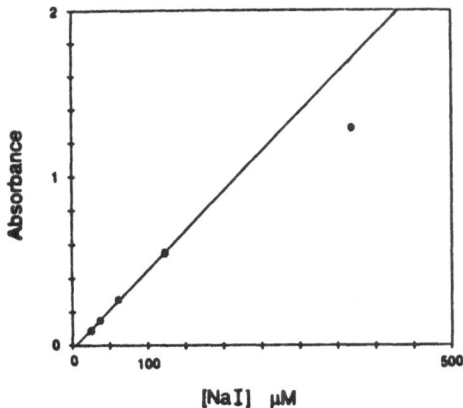

Figure 5. Absorbance vs. concentration for NaI in 0.1 M NaH_2PO_4 (pH 7.0). The absorbance was monitored at 254 nm using the configuration in Figure 2. The solid line is the least square fit to the lower four data points.

point, but 1.0 A is typically the upper limit found in commercial LC detectors having cell volumes orders of magnitude larger. The detection limit with this prototype system is approximately 2 μM NaI, corresponding to 0.017 A. The electronic circuits used in these experiments were simple breadboards, and it is expected that the system noise can be reduced at least another order of magnitude, thereby increasing the dynamic linear range.

The results of this initial evaluation of the optical fiber based UV detector are promising; however, the system is not without its problems. First, the apparent absorbance of a highly concentrated NaI solution (1230 μM) is smaller than that of the 370 μM solution (which is the highest point in Figure 5). Refractive index changes (which are a recurrent problem for standard LC detectors) offer one likely explanation for this effect. Light from the incident fiber that is ordinarily too divergent to be collected by the output fiber might be refracted into the collection fiber by the increased refractive index of the highly concentrated NaI solution. Small changes in baseline in response to changes in the concentration of the non-absorbing buffer solution suggest that this is a good possibility. Another troublesome point is that the apparent absorbances of the NaI solutions are about five times too small in comparison to absorbances measured in a conventional UV spectrophotometer. Further investigations are underway.

SUMMARY
Modern separation systems require very small volume detectors, and optical fibers offer an excellent means of introducing and collecting light on the such small scales for optical detectors. Casting of the fibers in a polymeric matrix offers several benefits, among these being;
1) The faces of the optical fibers serve as "windows" through separation capillaries made of material compatible with the particular technique being used.
2) A natural optical aperture or slit is formed which helps restrict the light monitored to that which traverses the central part of the capillary.

3) The optical fibers are permanently fixed in place, such that the detector sub-assembly becomes a compact and rugged unit.

This approach also appears to offer the benefits of further reduction in scale, by simply reducing the size of the template wire (and perhaps the optical fibers). Sub-nanoliter detector volumes thus appear feasible. It also appears that other configurations of the fibers could be constructed for other types of optical detectors.

---

[1] F.J. Yang, J. High. Resol. Chromatogr. 4 (1981) 83.

[2] J.W. Jorgenson and K.D. Lukacs, Anal. Chem. 53 (1981) 1298.

[3] F.M. Everaerts, J.L. Beckers, and Th.P.E.M. Verheggen, Isotachophoresis, (Elsevier, Amsterdam, 1976), p. 164.

Total Internal Reflection Fluorescence Surface Sensors.

J.T. Ives, W.M. Reichert, J.N. Lin, V. Hlady[1], D. Reinecke, P.A. Suci, R.A. VanWagenen, K. Newby, J. Herron, P. Dryden and J.D. Andrade

University of Utah Department of Bioengineering, Salt Lake City, Utah 84112 USA
[1]Institute "Ruder Boskovic", Zagreb, Yugoslavia

1. Introduction.

Total internal reflection fluorescence (TIRF) technique shows potential as the basis of a remote fluoroimmunoassay design (1,2). Evanescent excitation of fluorescently labelled antigens (Ag) complexed with surface immobilized antibodies (Ab), or vice versa, significantly simplifies the rinsing required in standard immunoassay techniques, and allows smaller sample volumes to be measured. The development of integrated waveguide optics as an evanescent spectroscopic technique has opened the possibility of optically detecting interfacially bound biological molecules in remote environments. This paper will review the research at the University of Utah on a remote fiber optic immunosensor and a polymer thin film evanescent sensor. We will also discuss the basic TIRF system as applied to the study of proteins at interfaces because it represents evanescently excited spectroscopy in its simplest form.

2. Fixed Angle TIRF.

Several groups have used total internal reflection in the fluorescence mode to study protein adsorption-desorption reactions (3). The total internal reflection fluorescence of interfacially bound proteins may be monitored intrinsically in the ultraviolet by exciting the tryptophan moieties or extrinsically in the visible using fluorescent labeling techniques. In either case, the fluorescence of the adsorbed protein is excited by an exponentially decaying optical field created immediately adjacent to a solid-liquid interface oriented at such an angle to totally reflect an incident light beam. This exponentially decaying field, commonly called the "evanescent wave", produces a fluorescence excitation volume extending from the

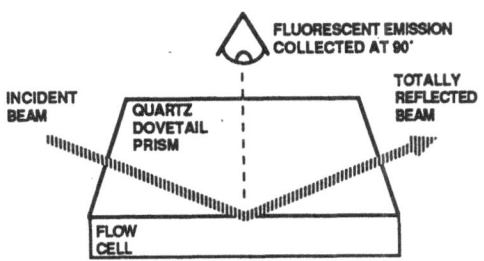

Fig. 1. SCHEMATIC OF FIXED ANGLE TIRF SYSTEM USED TO STUDY PROTIENS ADSORBED TO THE SOLID-LIQUID INTERFACE.

surface of the total reflection element (typically a 70 degree quartz dovetail prism) into the adjacent liquid medium to an effective depth of 1/3 of the exciting wavelength. The liquid phase is generally contained within a flow cell for the controlled introduction of biological molecules to the totally reflecting interface. The basic TIRF design is displayed in Figure 1 and described in detail in (3,4).

TIRF has been used extensively in our laboratory for several years to study the fluorescence of interfacially bound proteins(3,4). Recently, we have used TIRF as an immunosensor to follow interfacial antigen (Ag)-antibody (Ab) reactions (5).

3. Variable Angle TIRF

Variable angle TIRF (VA-TIRF) is a method to obtain the concentration and thickness of an adsorbed fluorescent layer. The advantage of VA-TIRF over fixed angle TIRF is that one is not required to assume a thickness or refractive index of the adsorbed film when attempting to obtain quantitative data. The concentration-distance profile of fluors at the interface results directly from the numerical inverse Laplace transform of the fluorescence angular spectrum representation where one of the two angles (observation or incident) is varied while the other angle is held fixed (6). In our experimental design a fluorescence curve is obtained by holding the angle of incidence constant and collecting the angular distribution of the interfacial fluorescence by varying the angle of observation (Fig. 2). The VA-TIRF technique in the variable observation angle mode was demonstrated by the collection and numerical analysis of data from an immunoglobulin (IgG) protein film adsorbed to a quartz hemi-cylinder (7). Currently we are testing the accuracy and reliability of the optics and inversion software by collecting spectra of dye impregnated Langmuir-Blodgett monolayers of known thickness and dye concentration (8). In addition to determining the concentration-distance profiles of adsorbed protein layers we are also planning to use VA-TIRF to detect the step increase in thickness when an Ag from solution is complexed with a preadsorbed Ab monolayer producing a bilayer at the prism surface.

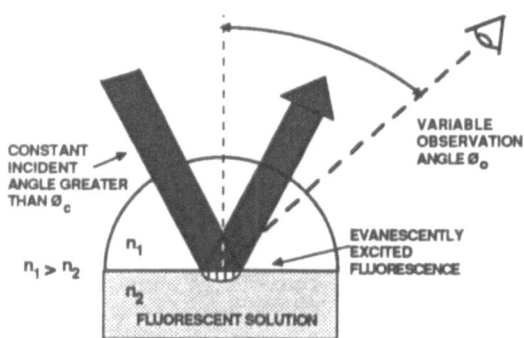

FIG. 2. VARIABLE OBSERVATION ANGLE TIRF (VA-TIRF) SYSTEM FOR DETERMINING THE CONCENTRATION-DISTANCE PROFILES OF ADSORBED PROTIEN MOLECULES ADSORBED TO THE SOLID-LIQUID INTERFACE.

4. Waveguide TIRF.

Waveguide TIRF shows promise as a remote surface sensor to monitor protein adsorption-desorption reactions and Ag-Ab complex formation. The waveguides used in our laboratories to date have been cylindrical glass optical fibers and 1-3 um transparent polymer films spun-cast onto pyrex or glass substrates. Both of these systems utilize an integrated optics modification of the evanescent excitation principle of the conventional total internal reflection geometry.

The evanescent field at the surface of approximately 2 um thick spun-cast poly(styrene) films have been used by our group to excite the fluorescence of Langmuir-Blodgett deposited cyanine dye-fatty acid derivative monolayers (9) and surface adsorbed films of dye labeled IgG (10). Figure 3 is an illustration of the prism technique used to couple guided waves of light into thin polymer film. The fluorescence from the surface deposited fluorescent monolayers is collected at 90° to the polymer film surface.

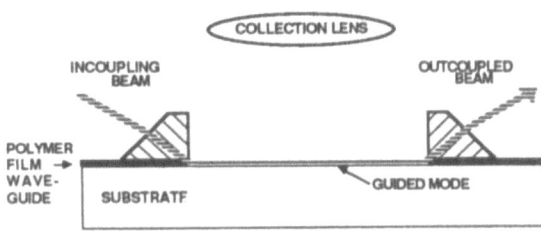

FIG. 3. PRISM TECHNIQUE FOR COUPLING GUIDED MODES OF LIGHT INTO SPUN-CAST POLYMER THIN FILM WAVEGUIDES. THE EVANESCENT FIELD AT THE POLYMER SURFACE IS USED TO EXCITE THE FLUORESCENCE OF SURFACE DEPOSITED LAYERS.

Cylindrical glass waveguides have also been used in our labs as an evanescent surface sensor. A sensor tip is formed by stripping the cladding from one end of a multi mode fiber (600 um) and capping the terminal end with an opaque epoxy to prevent light leakage out of the fiber tip into the sample solution (Fig. 4). In this design only the evanescent field of the stripped fiber tip is exposed to the surrounding liquid sample solution. The remote operation of the fiber tip sensor has been demonstrated with fluorescent dye solutions (11). The application of this system to the remote detection of protein adsorption was demonstrated with dye labeled IgG (12).

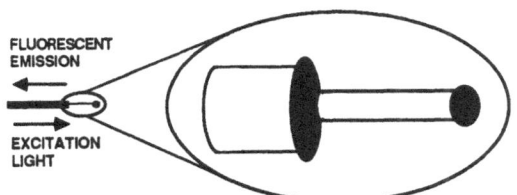

FIG. 4. FIBER SENSOR TIP PRODUCED BY STRIPPING THE CLADING FROM THE DISTAL END OF AN OPTICAL FIBER. THE EXPOSED FIELD AT THE STRIPPED FIBER CORE SURFACE IS USED TO EVANESCENTLY EXCITE FLUORESCENCE OF MOLECULES ADSORBED TO THE SENSOR TIP.

5. Integrated Optics Based Immunoassay.

Immunoassay systems are widely employed due to their high sensitivity and, with monoclonal antibody techniques, high selectivity (13). In brief, the immunoassay concept is based on the pre-immobilization of either Ab or Ag to a totally reflecting surface which is exposed to a solution containing the complementary immunospecific species. The observed fluorescence is proportional to the surface concentration of the formed Ag-Ab complexes which is in turn proportional to the concentration of the unbound species in solution. Figure 5. shows a typical fluorescence signal response to pre-immobilized Ab binding with free Ag. When the immobilized Ag is in turn exposed to free Ab the fluorescence signal intensity increases further.

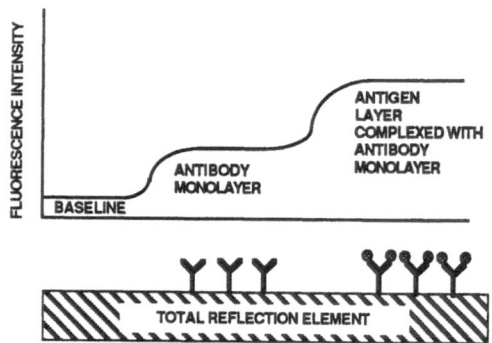

FIG. 5. TIRF IMMUNOASSAY CONCEPT FOR FLUORESCENT ANTIBODY AND ANTIGEN MOLECULES. THE OBSERVED FLUORESCENCE WILL INCREASE ONE STEP WITH THE ADSORPTION OF THE ANTIBODY MONOLAYER FOLLOWED BY A SECOND STEP INCREASE AFTER THE COMPLEXATION OF THE ANTIGEN FROM SOLUTION WITH THE PRE-IMMOBILIZED ANTIBODY LAYER.

Fiber Sensors.

The attractiveness of fiber optic sensors are their small size and ability to deliver and collect light to and from remote locations. The concept of a semi-continuous fiber optic immunosensor utilizing a competitive assay scheme is presented in more detail elsewhere (2). Advancing the fiber optic sensor to an immunosensor requires the stable pre-immobilization of the Ab to the surface of the quartz fiber and a mechanism for regenerating the sensing surface in situ.

Surface immobilized Ab molecules which are stable, active, and selective have been obtained using two different techniques. A technique which uses a 3-aminopropyl-triethoxysilane prepared surface has been described by Weetall, et al. (13), and covalent binding on a dimethyldichlorosilane prepared surface has been developed by Lin, et al (14). Work is being performed on these pre-immobilized surfaces in the TIRF mode to test their potential as affinity surfaces suitable for fluoro-immunoassay.

The current immunoassay techniques are essentially single use techniques. A remote fiber optic immunosensor should include a mechansim for semicontinuous monitoring. We are investigating two methods for long term use based on remote surface regeneration. One method is a photoinduced conformational change in the Ab which causes the Ag to be released. Light at a wavelength other than the fluorescence excitation wavelength is propagated down the optical fiber to induce a change in a photosensitive segment of the Ab. Preliminary work with azobenzene rings on poly-L-glutamic acid has shown a photoinduced conformational change. The second method involves a permanent modification of the Ab binding site to reduce the binding constant. A reduced binding constant would yield a faster response time of the sensor to fluctuations in the solution Ag concentration.

A number of optical improvements also need to be considered before a suitable immunosensor is developed. Currently, the system is standardized with fluorescent dyes in the test chamber, but future referencing should probably utilize two wavelength or broad band comparisons. The coupling and detection systems are large and expensive (lasers, microscopes, spectrometers, photon counting), and much simpler components (diode lasers operating in the visible wavelengths, fiber optic spectrometers) are needed. Fluorescence is currently used because of its large emission signal offsets (Stokes shift) and quantum efficiency. Improved sensitivity and a tunable source/detector would allow chromophore absorption to be studied with perhaps wider applications than fluorescence. Another important parameter in fluorescence spectroscopy is polarization anisotropy which compares light absorption and emission in two orthogonal polarizations. Only with polarization maintaining fibers can such studies even be considered, but the probability of success is low due the small amount of bound energy in these single mode fibers.

Polymer Thin Film Waveguides.

An alternative to fiber optics are slab waveguides made from spun-cast polymer films. Polymer waveguides potentially have several advantages over optical fibers: 1) the large surface area of polymer waveguides is easily adapted to different surface chemistries (i.e., polar vs. nonpolar, hydrophilic vs. hydrophobic, mobile vs. rigid) through the use of different polymer films or surface derivatizaition, 2) the mode and polarization can be selected for different effects at the polymer film surface, and 3) the linear shape of the waveguide streak is efficiently imaged onto the spectrometer slit (3). The disadvantages of a thin film sensor are the scatter loss along the polymer film and the bulky prism system used to couple light into the waveguide.

Fluorescence excitation caused by the scattered field represents a large noise factor in the detected signal. Waveguide features of evanescent surface selectivity and mode selection of different excitation volumes become difficult to distinguish above the background scatter. This problem is being approached with calculated energy profiles of four layer waveguide systems (substrate-waveguide-film-superstrate), theoretical absorption and scatter losses and experimental work measuring the decay of light along the waveguide. Scatter caused by the Langmuir-Blodgett surface films and the ability to vary their absorption will be an important part of this study.

The first generation polymer film sensor involves a flow cell designed for the waveguide surface. This design allows protein adsorption and desorption and Ag-Ab reactions to be studied at the polymer surface. The second generation sensor will incorporate the remote sensing advantages of the fiber optics while retaining the surface chemistry and optics advantages of polymer waveguides to produce the optimum immersible sensor. In this design, fiber to waveguide coupling (15) will be used to create an evanescent streak along the polymer waveguide sensing surface (Fig. 6) .

6. Summary.

Evanescent wave spectroscopy in the TIRF mode has several advantages for biomedical surface studies (1-3). When coupled with optical fibers and polymer waveguides, remote surface sensitive sensors can be developed for immunoassay measurements and protein adsorption studies. With the techniques reviewed in this paper, information has been obtained about protein adsorption and desorption kinetics, conformational changes of adsorbed proteins, surface immobilized fluorescent immunoassays, and surface monolayer fluorescence.

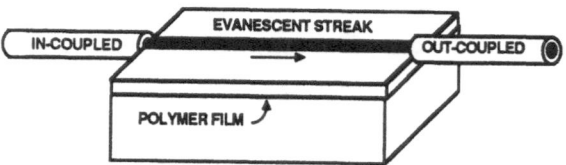

FIG. 6. SCHEMATIC OF FIBER TO POLYMER FILM WAVEGUIDE COUPLING WHICH WOULD INCREASE THE SUITABILITY OF THIN FILM WAVEGUIDES FOR REMOTE SENSOR APPLICATIONS.

7. References.

1. Place, J.F., R.M. Sutherland and C. Dahne, "Opto-Electronic Immunosensors: A Review of Optical Immunoassay at Continuous Surfaces", *Biosensors*, 1, 321-353 (1985).

2. Andrade, J.D., R.A. Van Wagenen, D.E. Gregonis, K. Newby and J-N Lin, "Remote Fiber Optic Biosensors Based on Evanescent Excited Fluoro-immunoassay: Concept and Progress," IEEE Trans Electr Dev, ED-32, 1175-1179 (1985).

3. J.D. Andrade, ed., <u>Surface and Interfacial Aspects of Biomedical Polymers;</u> Vol 1: <u>Surface Chemisrty and Physics</u>, Vol. 2: <u>Protein Adsorption.</u> Plenum Press, New York (1985).

4. Hlady, V., Reinecke, D.R., and J.D. Andrade, "Fluorescence of Adsorbed Protein Layers. 1. Quantitation of Total Internal Reflection Fluorescence," *J Coll Interface Sci*, 111, 555-569 (1986).

5. Lin, J-N., V. Hlady, W.M. Reichert and J.D. Andrade, "Immunosensors Based on Evanescent-Excited Fluorescence", *Annual Meeting of the Electrochemical Society*, Las Vegas, Oct.,1985.

6. Sansone,M., F. Rondelz, D.G. Peiffer, P. Pinus, M.W. Kim and P.M. Eisenberger, "Concentration Profile of a Dissolved Polymer near the Air-Liquid Interface: X-Ray Fluorescence Study", Phys Rev Lett, 54, 1039-1042 (1985).

7. Suci, P.A. "Variable Angle Total Internal Reflection Fluorescent Spectroscopy", Masters Thesis University of Utah 1984.

8. Suci. P.A., W.M. Reichert, J.T. Ives and J.D. Andrade, "Variable Angle Total Internal Reflection Fluorescence", *Annual Meeting of the Society for Biomaterials*, Mpls, May, 1986.

9. Ives, J.T. W.M. Reichert, P.A.Suci and J.D. Andrade," Waveguide Evanescent Streak Excitation of Surface Deposited Dye Monolayer Fluorscence," *J. Opt. Soc. Am.*, A 2(13) (1985),p.53

10. Reichert, W.M., K. Newby and J.D. Andrade, "Waveguide Evanescent Streak Excitation of Adsorbed Protein Fluorescence," *Annual Meeting of the Society for Biomaterials*, April 1985.

11. Newby, K., W.M. Reichert, J.D. Andrade and R.E. Benner, "Remote Spectroscopic Sensing of Chemical Adsorption Using a Single Multimode Optical Fiber," *Appl Optics*, 23 1812 (1984).

12. Newby, K., J.D. Andrade, R.E. Benner and W.M. Reichert, "Remote Sensing of Protein Adsorption Using a Single Mode Optical Fiber," *J. Colloid Interface Sci.*, 111 (1986),pp 280-283.

13. H.H. Weetall, ed., Immobilized Enzymes, Antigens, Antibodies and Peptides, , Plenum Press (1975).

14. Lin, J-N., J.N. Herron and J.D. Andrade, in preparation.

15. Bear, P.D. "Microlenses for Coupling Single-Mode Fibers to Thin Film Waveguides", *Appl. Optics*, 19, 2906-2909 (1980).

IMMOBILIZED ANTIBODIES - FIBER OPTIC SENSORS FOR BIOCHEMICAL MEASUREMENTS

D. DE ROSSI, A. NANNINI, M. MONICI

CENTRO "E.PIAGGIO", FACOLTA' DI INGEGNERIA, UNIVERSITA' DI PISA
ISTITUTO DI FISIOLOGIA CLINICA DEL C.N.R., PISA, ITALY

SUMMARY

The possibility to implement fluoro-immuno assay and competitive binding methods on fiber optic devices to remotely measure or monitor concentration of complex organic molecules is receiving considerable attention in the scientific community.

In this paper, biochemical sensors proposed in the literature based on Near Total Internal Reflection (NTIR), Total Internal Reflection Spectroscopy (TIRS) and Total Internal Reflection Fluorescence (TIRF) are described in some detail.

Problems related to the effects of antigen-antibody affinity on the expected sensitivity of the sensor and its eventual reversibility are examined.

In addition, details are given on a new TIR-based configuration which may potentially improve existing configurations.

Finally, the combined use of monoclonal antibodies and photoactive compounds is proposed as a possible mean to perform continuous monitoring of antigen concentration.

1. INTRODUCTION: MEASURE OF BIOCHEMICAL SPECIES INTO MULTI-COMPONENT MEDIA.

The growing interest in measuring concentration of molecular species into multicomponent media is common in several fields such as basic science, medicine, environment protection and industrial biotechnology.

The investigation of thermodynamics and kinetics of interfacial reactions involving biological macromolecules, monitoring of concentration of harmful compounds (pesticides, pollutants) or viruses, controlling biological parameters connected to operations in bioreactors, performing clinical diagnosis in human liquids, have lead to the development of sensitive and selective analytical essays based on biochemical, spettroscopical, immunological and nuclear techniques (1).

More recently, considerable research efforts have been devoted to the development of methods and devices combining essays techniques with suitable sensor elements to provide useful measuring tools for automating clinical and biochemical analytical procedures (2).

The main goal of this activity consists in the development of a sensor capable to on-line monitor the concentration of complex biochemical species in presence of interfering compounds; however, since it may be a matter of great difficulty to obtain full reversibility, especially using immuno-assay, an intermediate objective would consist in developing an integral detector (dosimeter type), possibly reusable after some sort of reconditioning procedure.

Related to the high complexity of the biochemical target to be monitored and the requested high specificity of the sensor, is the need for a very selective, sensitive transducing element.

Chemical sensing in living systems is performed by very specific sensor elements: antibodies and membrane receptors. Recent progress in immunology and biotechnology have made available a large wealth of basic understanding and practical technique related to the use of immobilized antibodies (3).

Basically, antibodies are proteins that selectively bind (in vivo and in vitro) to a specific substance (antigen) to produce an antigen-antibody complex (4).

Antibodies are naturally produced by living organisms under the stimulus of the antigens. An antigen is in general a "not-too-small molecule" but in principle it is possible to induce the production of a specific antibody for a small molecule (hapten) by conjugating it to a suitable carrier protein. At present biotechnology can provide a large spectrum of antibodies that are capable to selectively bind to their natural or "artificial" antigen producing the antigen-antibody complex.

Biologically relevant antigens are typically complex molecules, characterized by several binding sites and animal-produced correspondent antibodies result in an etherogeneous population of proteins reacting with the different sites of the antigen. These broad spectrum antibodies are called polyclonal. It is possible to obtain monoclonal antibodies that are an omogeneous population of antibodies reacting with the antigen by selectively binding to a particular site. Monoclonal antibodies are characterized by the same affinity for a particular active site of the antigen while polyclonal ones present a spread spectrum of affinity constants (5).

So, monoclonal antibodies may represent the solution to the sensitivity and selectivity problems if methods and techniques are conceived to detect the complex formation and to transfer this information to the elaboration stage.

2. READING IMMUNOCOMPLEX FORMATION: FIBER OPTIC SENSORS.

Among the various techniques which have been recently proposed for chemical sensing (6, 7), optical sensing techniques (8) appear particularly suited to monitor antigen concentration, through induced variations of refractive index, light scattering, absorption, intrinsic or extrinsic fluorescence and spectroscopy.

Optical techniques are by far the most powerful tool to investigate structure and functions of macromolecules and fiber optics can provide some additional advantages such as immunity from electro-magnetic interference and miniature dimensions; moreover their functionality is not impared when they operate in aqueous media. Using fiber optics as "reading" tools, methods need to be developed to immobilize antibodies onto optical grade surfaces. In addition, provisions should be adopted to define a measuring volume closed to the optical probe to confine the chemical species under measure and to avoid interference from the bulk solution.

Two possible approaches are reported in literature related to this last aspect; the first one (9) uses membranes to confine immobilized antibodies and optically labelled antigens and to separate them from the solution containing the antigens to monitor.

The external antigen which diffuses into the optically probed volume through the membrane, enters in competition with the confined labelled antigen for the binding sites of the immobilized antibody inducing a variation of the optical signal due to the optically active label. This technique is based on the well known principle of competitive binding; the main limitation of this approach may be related to the effective possibility of certain antigens, relatively large molecules, to cross the membrane. A different approach exists to create a sampling volume probed by evanescent waves (10). At the interface between two media light is completely reflected if the angle of the incident light ray is larger than the critical angle $\vartheta = \arcsin (n_2/n_1)$, n_1 and n_2 being the refractive index of the media. In the rarer medium evanescent wave only propagates exponentiality damped and interesting only a very thin volume, whose thickness is of the order of the wavelength of the impinging light, close to the interface.

Evanescent wave behaves as an ordinary light ray and it can be used to excite fluorescent substances, to make absorption spectroscopy or to study light scattering in a well defined and confined sampling volume, avoiding interferences from the bulk (11).

Several configurations operating on the basis of evanescent wave principles have been proposed and some of them are reported in figure 1.

Figure 1 shows the configuration used at the University of Utah (12) based on an exposed core fiber optic; on the exposed core a layer of antibodies is absorbed and the optical probe so constructed can excite and detect intrinsic fluorescence and Raman spectra of antigen-antibody complex.

The device proposed by Parriaux et al. (13) is reported in figure 1b. A tapered fiber can be coated with antibodies in the region in which the core is partially exposed. This configuration, although, intrinsically more robust than the former one, is still very fragile and has a very reduced active area (wich is obviously strictly related to the transducer sensitivity and dynamic range).

Extensive studies on evanescent wave based configurations are performed by Batelle in Geneva (14); their device is capable to detect signals due to antigen-induced increase of light scattering in the evanescent volume.

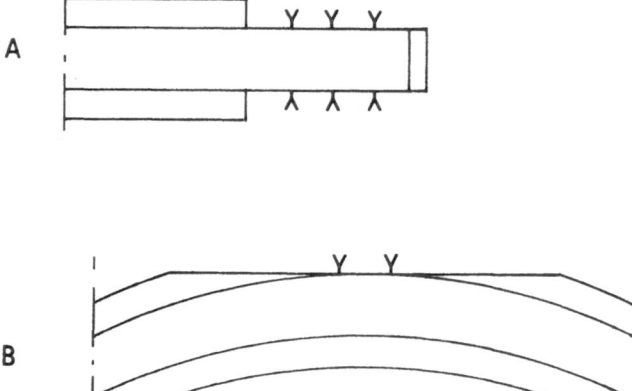

FIGURE 1. A) Exposed core fiber optic immuno-sensor probe (Andrade et al. 1985). B) Tapered fiber configuration (Parriaux et al. 1983).

In figure 2 the "near total internal reflection (NTIR)" configuration originally proposed by Phillips (15) for the realization of hydrophones is depicted. The tip of an optical fiber is polished at an angle very close to the critical angle and pressure induced variations of the refractive index of the rarer medium (water) modulate the intensity of the optical reflected signal.

It is possible, in principle, to modify the Phillip's configuration by coating with a layer of antibodies polished fiber tip surface to detect the intensity modulated signal produced by the index variations due to the antigen-antibody complex formation.

There are two basic problems connected with this idea:

i) the presence in the front of the fiber of a stratified medium;

ii) the reduced area of the layer that it is possible to coat on the fiber because of the very little dimensions of the optical interface.

A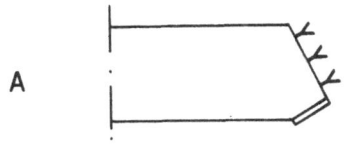

B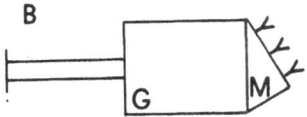

FIGURE 2. A) Modified Phillips configuration. B) Configuration with GRIN lens (G) and microprism (M).

In figure 2b a configuration, employing a GRIN lens as beam expander and collimator and a microprism cut near to the critical angle is presented, capable in principle to overcome the difficulty related to the ii) point. For what the i) problem is concerned, theoretical means are available (16) to evaluate the effective sensitivity of the device.

3. UNANSWERED QUESTIONS (17).

Several questions need to find an answer before fiber optic biosensors can eventually prove to offer a viable solution to measurement problems.

A technological problem concerns the coupling of the sensing molecules to the optical substrate. Two possibilities exist:

i) to physically absorb antibodies on the optical surface; this procedure is quite simple to perform but not fully satisfactory from the point of view of stability and uniformity of the absorbed layer;

ii) to covalently bind antibodies to the previously treated optical surface.

Solution ii) is better than i) for what stability is concerned, but an additional problem arise related with possible changes of structural and functional properties of the sensing molecules due to the covalent immobilization. It is easy to conclude that stronger the binding, the more probable will be the antibody denaturation. On the other hand absorption, being an almost umpredictable process in the definition of the binding site and attachment procedure, can lead to a loss of activity of the absorbed layer if the active site of the antibody is somehow sterically masked because of the binding modality.

Using covalent binding this problem can be avoided, at least in principle, if functional groups grafted onto the optical surface induce a sort of steric order in the successive antigen binding.

The second problem is to define the optimum optical geometry for best sensitivity and it is strongly connected with the need for theoretical modeling of electromagnetic propagation in thin macromolecular layers (18).

This study can provide an answer to questions as: is better a fluorescence based technique or light scattering or an intensity modulated devices? Or: is the sensitivity of the NTIR based device as good for a biosensor as it seem to be for an hydrophone or the reduced thickness of the antigen layer is not able to modulate the intensity of internally reflected ray?

However, the most crucial issue in the design of biosensors for continuous monitoring is related to the reversibility of the sensor, or, in other terms, the value of the binding time constant.

Difficulties to induce release of antigen by the monoclonal

antibody are intrinsically connected to the high affinity constant between the two molecules. It is possible, in general, to recondition the antibody layer by changing chemico-physical parameters as pH or temperature, but these are solutions suitable only with integral sensors.

A much more ambitious goal would consist in the design of an analogic sensor able to follow in time variations in concentration of the target molecule.

Photocontrol of macromolecule conformations is an area of active research (19) and it may eventually provide a possibility to confer reversibility and reuse to immunosensors.

Many photosensitive molecules are known (20) to be transformed under photoirradiation into different isomers, which can return to the initial state after thermal or photochemical excitation.

Control of peptide chains conformation by photoisomerising chromophores has been achieved and extensively used to modulate enzymes catalytic activity (21).

It appears conceivable to label or contour antibodies immobilized on the optical substrate with suitable photoactive molecules (chromophores) which under specific light irradiation may induce transient structural rearrangements in the antibody-antigen complex leading to the release of the antigen and reactivation of the antibody binding capabilities.

Hence, at least in principle, an immunosensor with limited saturation capacity and integral functioning mode could be reconditioned without removing it from the measuring site by using the fiber optic either as a "reading" light and "reconditioning" light carrier.

REFERENCES

1. Shuurman et al.: Physical models of radioimmuno assay, Annal Chem. 51 (1979) 2.
2. Lowe CR: An introduction to the concept and technology of biosensors, Biosensors 1, 1985, 3.
3. Andrade JD: Principles of protein adsorption in "Surface and Interfacial Aspect of Biomedical Polymers, New York, Plenum, 1985.
4. Pressman D, Grossberg AL: The structural bases of antibody specificity, New York, W.A. Benjamin, 1968.
5. Goding JW: Monoclonal antibodies, New York Academic, 1983.
6. Bergveld P, De Rooij NF: The history of chemically sensitive semiconductor devices, Sensors and Actuators 1, 1981, 5.
7. Zemel NJ et al.: Non-fet chemical sensors, Sensors and Actuators, 1, 1981, 427.

8. Place JF et al.: Opto-Electronic Immunosensors at continuous surfaces, Biosensors 1, 1985, 321.
9. Liu BL, Schultz JS: Equilibrium binding in Immunosensors, IEEE BME 33, 1986, 133.
10. Born M: Principles of Optics 6th Ed., New York Pergamon, 1980.
11. Harrik NJ: Internal reflection spectroscopy, New York, Wiley, 1968.
12. Newby K et al.: Remote spectroscopic sensing of chemical adsorption using a single multimode optical fiber, Applied optics 23, 1984, 1812.
13. Lew A, Depeursinge C, Cochet F, Berthou H, Parriaux O: Single made fiber evanescent wave spectroscopy, Proc. 2nd Int. Conference on Optical fibre sensors, Stuttgard, 1984.
14. Dähne C, Sutherland RM, Place JF: Detection of Antibody-antigen reactions at glass-liquid interface: A novel fiber optic sensor concept, Proc. 2nd Int. Conference on optical fibre sensors, Stuttgard, 1984.
15. Phillips RL: Proposed fiber-optic acoustical probe, Optics Letters 5, 1980, 318.
16. Marcuse D: Theory of dielectric optical waveguides, New York, Academic Press, 1974.
17. Andrade JD, VanWagenen RA, Gregoris DE, Newby K, Lin JN: Remote fiber optic biosensors based on evanescent-excited fluoro immunoassay: concept and progress, IEEE ED. 32, 1985, 1175.
18. Berreman DW: Optics in stratified and anisotropic media: 4x4 matrix formulation, SOSA 62, 1972, 502.
19. Montagnoli G, Erlanger BF Ed.: Molecular Models of Photoresposiveness, New York, Plenum, 1983.
20. Irie M: Photoresponsive Synthetic Polymers, Ibid, 291.
21. Montagnoli G, Pieroni O, Suzuki S: Control of peptide chain conformation by photoisomerizing chromophores: Enzymes and model Compounds, Polymer Photochemistry 3, 1983, 279.

OPTICAL FIBER SENSORS IN MEDICINE

A.M.SCHEGGI

Istituto di Ricerca sulle Onde Elettromagnetiche of C.N.R.
Florence, Italy

1. INTRODUCTION

Sensors of physiological parameters represent an important aspect of modem biomedical instrumentation and their use may result in a better diagnosis and better quality of health care.

Medical diagnosis is mainly based on chemical analysis of a variety of biological parameters but physical parameters such as flow, pressure and temperature are also important in assessing a patient's condition and in guiding therapy. An incredibly relevant number of tests are performed in clinical chemistry laboratories and consistent blood samples are often drawn from patients and particularly frequently from patients in critical care ward. The costs for this type of analysis is relevant while delays and potential errors may occur; miniaturized semiportable versions of the common analytical instruments which could allow "to bring the laboratory to the patient" would enable more rapid and better quality analysis with consequent more prompt therapeutic decisions and at the same time reduced costs.

The use of optical fiber sensors could in principle solve several medical problems owing to the peculiar advantages of optical fibers: small size, flexibility and immunity to e.m. interference. For instance the small size of the fiber could lead to miniaturized sensors for ex vivo tests which would need smaller amounts of sample material. A more difficult task is related to the construction of invasive sensors to be inserted in ancillary channels in endoscopes or in hypodermic needles to perform measurements in the tissue, in organs or within vessels. However the mechanical flexibility and again the small size make optical fiber sensors fulfilling all these requirements, while e.m. immunity is a precious quality whenever the therapy requires electronic apparatus (e.g. high voltage defribillation, or microwave hyperthermia).

Biomedical optical fiber sensors are now investigated and developed in many industry and academic laboratories while several review papers on the subject have appeared in the literature (1-5). In the present paper the attention will be focused to wards two types of sensors: temperature and pH sensors.

Precise spatial temperature measurements are required to con
trol heating of biological tissues in microwave or R.F. hyperthermia for treatment of some cancers, where the use of conventional thermocouples or thermistors can perturb the incident e.
m. field and may also cause localized heating spots. There are
also particular fields of application of optical fiber thermome
ters, for instance to determine thermal distribution during pho
toradiation therapy of malignant tumors or for blood flow measu
rements. In fact, apart from flow measurements obtained through
laser Doppler velocity technique, a more commonly used method
is based on thermodilution, where a known change in the heat con
tent of the blood is induced at one point of the circulation by
injection of a cold saline and the resultant change in temperature is detected at one point downstream.

Blood gas analysis include PO_2, PCO_2, as well as pH measurements which are important for metabolic and respiratory problems, in particular for continuous monotoring during intensive
care. Optical fiber sensors of such chemical parameters are a-
mong the most recent ones to appear: their basic design consists of a reversible indicator system fixed   into an appropriate permeable container at the fiber end, which changes its optical properties in response to changes in the measurand material.

2. TEMPERATURE SENSORS

Optical fiber temperature sensors developed for medical applications in general involve a temperature sensitive optical
material constituting the transducer joint to a bundle or to a
single optical fiber. Resolution of $0.1^{\circ}C$ over a range of a few
degrees ($35 \div 50^{\circ}C$) are the main measurement requirements. Several approaches have been suggested, however this paper will be
limited to a few significant systems which are now well establi
shed or are in the research stage. Two thermometers have been
developed by Luxtron Co. (USA) and by ASEA (Sweden): the first
one based on the work of Wikersheim and Alves (6,7) makes use of
the fluorescence quenching with rising temperature of a rare e-
arth phosphors mixture bonded at the end of an optical fiber
(400 μm core). The sharp emission spectra of the phosphors exci
ted with u.v. radiation can be easily insulated by narrow-band
interferencial filters: the ratio of the two emission lines is
utilized for temperature measurements which result independent
of short term problems of source fluctuations and long term pro
blems of degradation of optical contacts (Fig.1). Luxtron thermometer has a very wide working range (from -50 to above $200^{\circ}C$):
a specific medical version is now on the market, working over a
temperature range $0 \div 80^{\circ}C$ with resolution of $0.1^{\circ}C$ and best accuracy of $\pm 0.2^{\circ}C$.

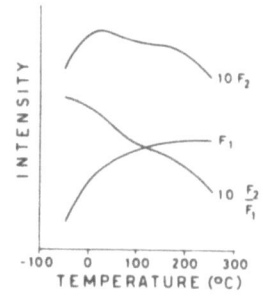

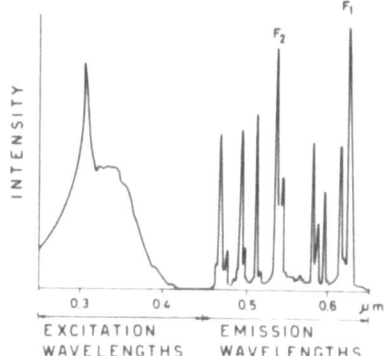

FIGURE 1. Upper: Temperature dependence of two lines and their ration of the phosphor compound.
Lower: Excitation and emission spectra.

A technique based upon light absorption and emission in semiconductors which are characterized by a forbidden gap between the valence and the conduction bands is utilized in the second sensor. Christensen has pioneered in this type of device utilizing the absorption of a small GaAs prism used as sensing element (8). The sensor developed at ASEA utilizes the photoluminescence of a GaAs crystal sandwiched between two confinement AlGaAs layers (9,10). Light from a LED is transmitted through the optical fiber (100 μm core) to the sensor: here the light is converted and returned at different wavelengths depending on temperature. The intensity of the returned light is detected by two photodiodes and ratioed to give a temperature measurement (Fig.2). The operative range is $0 \div 200°C$ with sensitivity of $0.1°C$ and accuracy of $\pm 0.1°C$ in the physiological range.

Recently a new temperature sensor has been proposed and tested in its laboratory version (11). The sensor consists of a FFP (with finesse = 10) made of a short piece of single mode

410

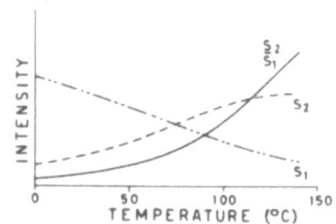

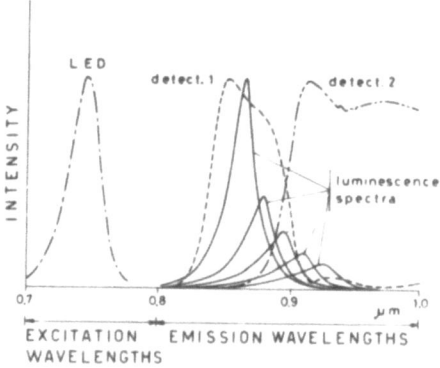

FIGURE 2. Upper: Output singlas and their ratio versus tempera
ture.

Lower: Photoluminescence spectra emitted from the sen
sor and spectral curves of the source and photodetectors.

fiber (26 mm long) with the terminal end faces polished and die
lectrically coated; the light from a laser diode is transported
along a monomode fiber to the FFP and back reflected along the
same fiber through a directional coupler to the detecting system. The working principle is based on the compensation of the
temperature change ΔT which would shift the FFP output to the
neighbouring order ($K \to K+1$) by a suitable wavelength change
$\Delta\lambda$ so that the reflected signal is continuously kept at the
reflection minimum of order K. The sensor which has been envisa
ged for continuous monitoring in hyperthermia systems covers
the temperature range 25 ÷ 45°C with a resolution better than
0.1°C.

Two optical fiber thermometers have been proposed and tested
at IROE-CNR, Florence. The first one (12,14,15) is based on the
light intensity modulation induced by a thermosensitive clad-

ding applied on the distal end of a PCS fiber. Such a clad is
an oil exhibiting a temperature dependent refractive index so
that the temperature modulated NA of the liquid clad fiber por
tion gives rise to a temperature modulated backscattered signal
which is ratioed with a reference signal thus eliminating sour
ce fluctuations.

The probe experimental configuration is shown in Fig.3.

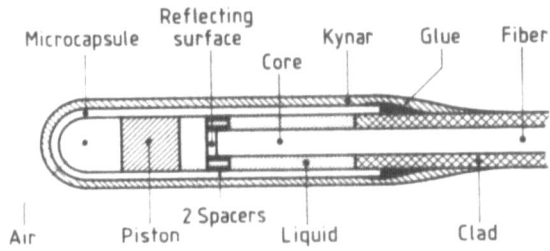

FIGURE 3.   Probe configuration

The end face is mirrored and the liquid is contained in a glass
microcapsule cemented to the fiber. In order to avoid deteriora
tion of the cement, due to pressure by the liquid clad expansion with increasing temperature, the bottom of the capsule is
left empty. A piston separating the empty from the liquid filled region allows the liquid expansion towards the empty region
and an open ring plastic spacer provides fiber centering. The
capsule and a portion of the fiber are coated with a plastic
layer for protection. By utilizing a 200 μm core diameter fiber
it was possible to construct miniaturized probes (1 cm long,and
with overall external diameter less than 1.5 mm) suitable for
medical use. The use of different liquids (mineral oils) with
different refraction index at room temperature would allow to
make the device sensitive over different temperature ranges.
However this sensor has been mainly finalized to work in the
physiological interval where it exhibits a sensitivity of $0.1^{\circ}C$
with accuracy of ± 0.3 $^{\circ}C$ and response time of ∿ 1 sec. Fig. 4
shows the package of the optoelectronic system connected to the
sensor and to the processing and display units.

The second thermometer is based on the thermochromic properties of a Cobalt salt solution in iso-propyl alcohol used as a
temperature transducer (15,16,17). The optical spectrum of such

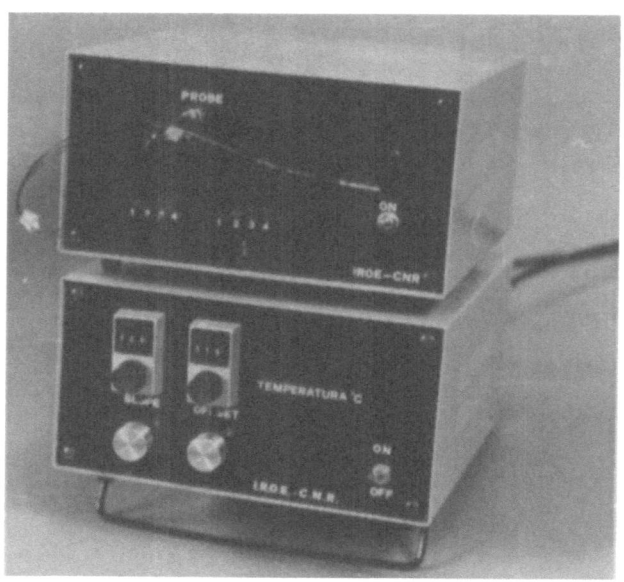

FIGURE 4. Thermosensitive-clad optical fiber temperature sensor
prototype.

solution results strongly modulated by temperature variations
at certain wavelengths and temperature independent at other wavelengths (Fig.5). By choosing two suitable wavelengths, one obtains a sensing and a reference signal: their intensity ratio
is a temperature function practically insensitive to fluctuations and transmission losses not strictly related to temperature
variations. The device has been set up and tested in the labora
tory version: the source is constituted by two LEDs at $\lambda = 660$
nm (sensing signal) and $\lambda = 840$ nm (reference signal) respectively, which are used in time multiplexing; the probe is constituted by a thin glass capillary with mirrored bottom filled
with the thermochromic solution and containing two optical fiber terminal portions: one for conducting the light to the tran
sducer and the other to collect the backscattered light. Its
overall dimensions are 1 cm length, 1.5 mm external diameter.
The resolution is of $0.1^{\circ}C$ over the temperature range 2.5 - 50
$^{\circ}C$ with a response time of ~ 2 sec.

3. pH SENSORS
    Activity in optical biochemical sensors is mainly in the re-
search stage: the number of parameters in the human body (proteins, pH, pO_2, pCO_2 etc.) are so high to require a number of

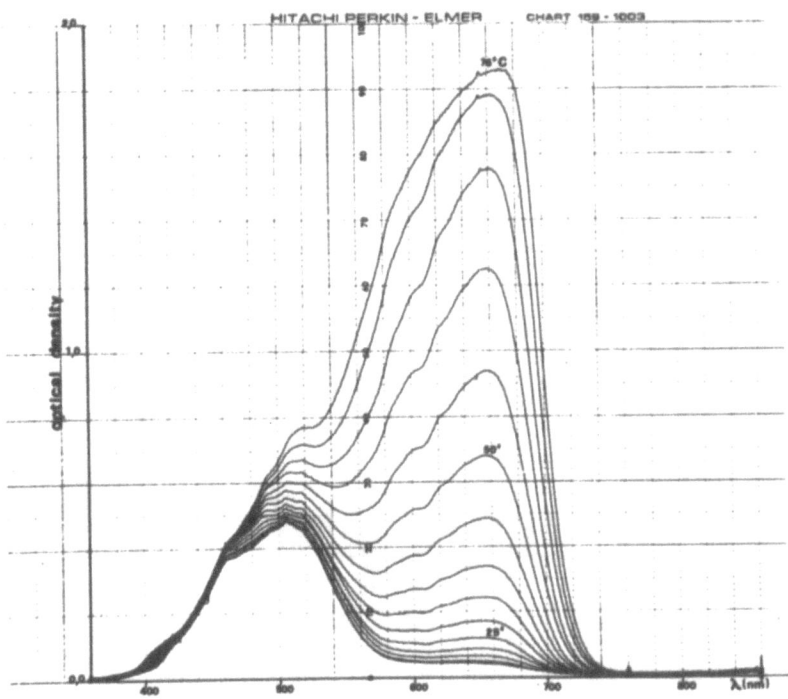

FIGURE 5. Optical absorption spectra at different temperatures
of the thermochromic transducer.

specific and selective sensors. However such sensors are substantially based on the extension of classical spectrophetome-
tric methods. While sensors of physical parameters (temperature, pressure, etc.) make in general use of a transducer which
can be hermetically encapsulated, in the chemical sensor, the
transducer must in general interfere with the measurand and hen
ce its realization results more complicate. In order to allow
interaction of the light with surrounding medium, the fiber is
terminated by an "optrode" (combination of optical and electrode in analogy with chemical electrode sensors), in general consisting of a reagent (in solid or liquid form) in a suitable
membrane enclosure, through which it is exposed to the chemical
being analyzed, by measuring changes in reflactance, absorbance
or luminescence.

This chapter is dedicated to pH optical fiber sensor: pO_2,
pCO_2 sensors are equally important for methabolic and respirato
ry problems, however they are very   similar in construction
to pH sensors and based on analogous principles.

The knowledge of pH in blood and tissues is desirable in a wide variety of chemical and biological studies such as respiration studies including blood and tissue oxygen content and oxygenéhemoglobin dissociation curve.
The pH of a solution is generally defined as (18)

$$pH = - \log H^+ \tag{1}$$

However a more exact definition must take into account the ionic strength of the solution. Accordingly pH must be defined as

$$pH = - \log a_{H+} \tag{2}$$

a_{H+} denotes the activity of H^+ which is related to H^+ by the expression

$$a_{H+} = f_{H+} H^+ \tag{3}$$

where f_{H+} (activity coefficient) varies with the ionic strength. For very dilute solutions f_{H+} may be assumed equal to unity so that expression (2) reduces to expression (1). Otherwise corrections due to ionic strength are of the order of a few percents.
Such considerations are important when absolute measurements are made: however the sensors here considered perform relative measurements by suitable calibration. Different calibrations must be done for different measurand substances.

Traditional techniques for pH measuring are based on electro metric or on colorimetric methods. The electrometric method is the most traditional and makes use of glass electrodes. The colorimetric method is based on absorption or on fluorescence of a suitable indicator. In the first case a dye (organic compound) is used which reveals the acid or basic character of the solution by a change of color with a consequent light absorption variation ; in the second case fluorescence variations of particular substances are observed when the acidity of the solution varies.

Peterson et alii (19) designed and developed a fiber optic pH sensor based on dye chemistry. The used dye indicator (phenol red) is weakly acid and has two tautomeric forms, each one having a different light absorption spectrum. As the pH of the solution varies, the relative size of each form, optical absorption peak varies in proportion to the changing relative concentrations of the acid and base form of the dye. A hydrophilic gel structure of polyacrylamide microspheres covalently bound to the dye for providing a fixed concentration, containing also smaller microspheres for light scattering, is packed in an envelope of cellulose hydrogen ion permeable dyalisis tubing at the end of a pair of large NA plastic fibers (150 μm core). The light from a high intensity tungsten lamp, injected into one fiber is sent to the sensor package backreflected and scattered into the second fiber and then selected into two wavelengths by filters. Green light (λ = 560 nm) is absorbed by the base form of the dye as a function of the pH, while the red light is not absorbed thus giving a reference signal. The ratio R between green and red light is measured by the photo detection and signal processing system connected to the fibers. It is related to pH by the expression

$$R = k \ 10^{(-C/ \ 1+10^{-\Delta})} \tag{4}$$

where k and C are constants of the probe and Δ = pH - pK (pK = -log K dissociation constant of the dye).

This results in an S curve with an appropriate linear region near pH = pK. (Fig.6)

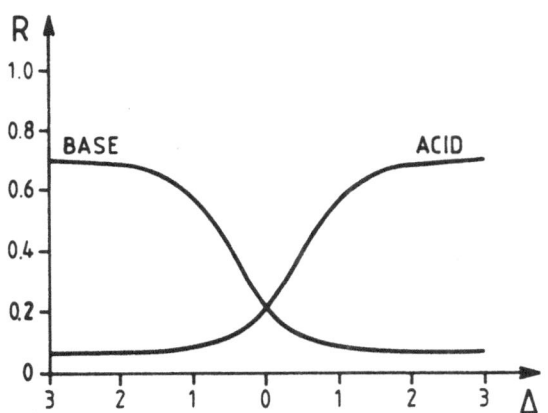

FIGURE 6. Reflectance curves of the dye indicator versus pH-pK

This sensor is capable of measuring pH over the physiological range 7÷7.4 with an accuracy of ±0.01 pH units. In vivo tests on animals have been successfully performed with this device.

Another pH sensor more recently proposed and tested is based on two wavelengths fluorescence (20). It makes use of a trisodium salt (HOPSA) immobilized on an anion exchange membrane. Fig.7 shows the absorption spectra for acid (HOPSA) and base-forms (OPSA^{-*}) as well as the fluorescence emission of the base form (OPSA^{-*}). Excitation wavelengths are 470 nm and 405 nm for base and acid forms respectively; as the excited state of the acid form (HOPSA*) dissociates more quickly than it returns to the ground state the observed fluorescence at λ = 510 nm is

416

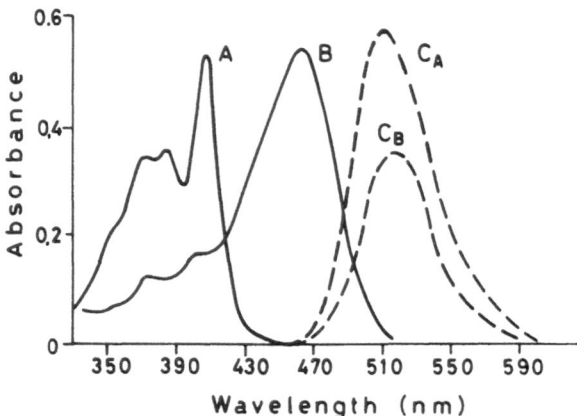

FIGURE 7. Absorption and emission spectra of acid and base form of HOPSA.

proper of the excited state of the base form (OPSA^{-*}). The ratio of the fluorescence intensities excited at λ = 470 and 405 nm respectively gives a pH   measurement   which results independent of several factors such as source fluctuations, temperature, quenching by oxygen and in particular ionic strength. The working range and best accuracy reached with a set of probe samples are 6 ÷ 8.5 and 0.02 pH units respectively.

At IROE-CNR, Florence an investigation has been tackled to set up a model of pH sensors for physiological use based on the absorption method and using a liquid optrode coupled with the fiber. The liquid optrode, while not requiring the immobilization of the dye on a polymeric matrix, needs a suitable membrane selective to H$^+$ and able to prevent dye exit. Accordingly attention was dedicated to the choice of the most suitable dye and membrane. The work is in progress with particular attention to the problem of miniaturizing the probe and shortening   the response time of the sensor.

It is to be observed that a problem concerning in general all these miniaturized probes for medical sensors is that their construction is done by hand and microfabrication technology for these devices has to be developed. Much work remains to be done in the field of biological sensors because the parameters to be measured in the human body require a large variety of sensors and also because of the need of technology implemen tation so to achieve miniaturized, reliable and low cost expendable devices.

REFERENCES

1. Peterson JI, Vurek GG: Fiber Optic Sensors for Biomedical Ap plications, Science, vol. 224, pp. 123-127, 1984
2. Seitz WR: Chemical Sensors Based on Fiber Optics, Anal.Chem. vol. 56, pp. 16A-28A, 1984
3. Vurek GG: In Vivo Optical Chemical Sensors, Proc. of SPIE, vol. 494, pp.2-8, 1984
4. Scheggi AM: Optical Fiber Sensors in Medicine, 2nd Int. Conf. on Optical Fiber Sensors (Stuttgart, September 1984), Conf. Proc. pp.93-106,VDE-VERLAG GmbH (Berlin), 1984
5. Scheggi AM: Optical Fiber Sensing in Medicine, Journal of Optical Sensors, vol. 1, pp.69-77, 1986
6. Wickersheim KA, Alves RV: Fluoroptic Thermometry: A New RF-Immune Technology, in Biomedical Thermology, Alan R.L.Liss, Inc. N.Y., pp. 547-554, 1982
7. Alves RV, Wickersheim KA: Fluoroptic Thermometry: Temperature Sensing Using Optical Fibers, Proc. SPIE, vol. 403, pp. 146-150, 1983
8. Christensen DA: A New Non Perturbing Temperature Probe Using Semiconductor Band Edge Shift, J. Bioeng. vol.1, pp.541-545 1977
9. Ovrén C , Broyàrdh T , Hidman T , Adolfsson M : A System for Temperature Measurements Using Fiber Optics, Proc. IOOC'83, Japan 1983
10.Ovrén C , Adolfsson M , Hök B : Fiber Optic Systems for Temperature and Vibration Measurements in Industrial Applications, Intern. Conf. on Optical Techniques in Process Control, The Hague, June 1983, paper B2, pp.67-81
11.Kist R , Drope G , Wölfelscheneider H : Fiber Fabry Perot Thermometer for Medical Applications, 2nd Int. Conf. on Optical Fiber Sensors, Conf. Proc. pp.165-170,VDE-VERLAG GmbH (Berlin)
12.Scheggi AM, Brenci M , Conforti G , Falciai R , Preti GP: 1st Int. Conf. on Optical Fiber Sensors (London, April 1983) IEE Conf. Publication n. 221, pp.13-16, 1983
13.Scheggi AM, Brenci M , Conforti G , Falciai R : Optical Fiber Thermometer for Medical Use, IEE Proc. vol. 13, pp.270-272, 1984
14.Brenci M , Conforti G , Falciai R , Scheggi AM: Optical Fiber Thermometer for Medical Use, Proc. SPIE, vol. 494, pp. 13-17, 1984
15.Brenci M , Conforti G , Falciai R , Mignani AG, Scheggi AM: Thermochromic Transducer Optical Fiber Temperature Sensor, 2nd Int. Conf. on Optical Fiber Sensors (Stuttgart, Sept. 1984), Conf. Proc. pp.155-160, VDE-VERLAG GmbH (Berlin),1984
16.Conforti G , Bacci M , Brenci M , Falciai R , Mignani AG,

418

Scheggi AM: Fiber Optic Thermometer for Biomedical Applications, Proc. SPIE, vol. 576, 1985, in press

17. Bacci M , Brenci M , Conforti G , Falciai R , Mignani AG, Scheggi AM: Thermochromic Transducer Optical Fiber Thermometer, in press in Applied Optics

18. Covington AK, Bates RG, Durst RA: Definition of pH Scales Standard Reference Values, Measurement of pH and Related Terminology, Pure and Applied Chem., vol. 55, pp. 1467-1476 1983

19. Peterson JI, Goldstein SR, Fitzgerald RV, Buckold DK: Fiber Optic pH Probe for Physiological Use, Anal. Chem. vol. 52, pp. 864-869, 1980

20. Markle DR, McGuire DA, Goldstein GR, Patterson RE, Watson RM A pH Measurement System for Use in Tissue and Blood Employing Miniature Fiber Optic Probes, Advances in Bioeng. D.C. Viaro Ed. (Am.Soc.Mech.Eng., New York, 1981)

21. Zhujun Z , Seitz WR: A Fluorescence Sensor for Quantifying pH in the Range from 6.5 to 8.5, Anal. Chem. Acta, vol.160, pp.47-55, 1984.

THE PRESENT AND FUTURE STATUS OF FIBRE OPTIC SENSORS IN INDUSTRY

R.S. MEDLOCK

BROWN BOVERI KENT PLC

1.DEFINITIONS

The term "sensor", and "transducer" often lack clear definition.
Sometimes the terms are regarded as being synonymous whilst at other
times they are considered as referring to different functions. The
author prefers to regard a sensor as a device which undergoes a
reversible physical or chemical change in response to a stimulus but
does not necessarily carry out an energy conversion. For example a
thermistor senses temperature in terms of a resistance change, but needs
a bridge circuit and electrical supply to provide the energy required for
the output signal. The more general case is a device which performs a
direct energy conversion to provide a measurable signal, (eg. a
thermocouple) in which case the term "transducer" seems to describe the
device more accurately. Many measuring devices need to combine the
functions of sensing, transducing and signal processing before they can
be usefully applied. In this paper the term "sensor" will be used
exclusively in order to be consistent with the title of the Course. It
can be regarded as synonymous with the term "transducer".

2.INTRODUCTION

Sensors for measurement are used in every facet of science and industry
as well as in the home. Their applicational requirements, their design
features, their specifications and their cost are highly dependent on the
market sector they serve. At least 12 sectors can be identified
(i)   Process (on-line measurement)
(ii)   Utilities (power, water, sewage)
(iii)   Food
(iv)   General industrial (inspection, test, metrology, robotics)
(v)   Environment (pollution, safety)
(vi)   Agriculture
(vii)   Laboratory and scientific
(viii) Health care, medical and biological
(ix)   Aeronautics and aerospace
(x)   Military
(xi)   Automobile
(xii)   Domestic
This classificiation is arbitrary as the sectors cannot be precisely
defined and because some sensors can serve more than one sector. There
can be common threads of technology in the operating principles of fibre
optic sensors for different sectors but the commercial products demand
individual designs and specifications in order to meet operating and cost
requirements. It is quite possible that some new sensors will make their
first appearance in laboratory or scientific instruments which later on

could undergo design modification and adaption to suit other markets.
This could apply particularly in the analytical field if fibre optic
sensors follow the same development path as their electronic
counterparts.

In this paper the emphasis will be on sensors for the industrial and
process control sectors which currently enjoy a market of US$ 2.8 billion
(excluding China and the Soviet bloc).

Process on-line measurement carries some severe problems:

. Measurement has to be continuous - often 24 hours each day without
  attention other than for routine maintenance
. If a sensor fails or measures inaccurately enormous plant product
  losses may be incurred, or worse, personnel and plant safety may be
  endangered
. Sensors suffer exposure to severe environmental conditions $-20^{o}C$ to
  $+80^{o}C$, dirt, corrosive influences, vibration, mechanical shock, high
  humidity
. Long transmission distances (up to 1km) may be necessary
. Multiplexing of sensor signals is frequently required necessitating
  compatibility of output signals from all sensors installed on a
  process plant.

The majority of process variables which require measurement can be
classified into five groups:

. Flow
. Pressure
. Temperature
. Level
. Analytical

This list can look disarmingly simple but the applicational variations
and the customer preferences can only be met by multitudinous design
variations. For example there are about sixty different operating
priciples available to meet all the applicational requirements of flow
measurement and each principle needs to be incorporated into a range of
sizes and ratings including pressure, temperature and chemical
compatibility to name just a few. As an inspired "guestimate" there are
about 2000 basic types of sensors and about 60,000 commercially available
products, mainly with electrical output signals. The implication for
fibre optic sensor manufacturers wishing to promote competitive products
is obvious - how can such a powerful, acceptable and established range of
electronic sensors be displaced by fibre optic equivalents without a huge
capital investment in development, production and marketing? History
shows that although the task is difficult it is not fundamentally
impossible. In 1958 a transition from pneumatic to electronic sensors
commenced and has proceded slowly but surely for the past 30 years and is
now about 80% complete.

When considering possible future trends in sensor technology in the
process industries, it has to be remembered that market conservatism is a
dominant feature - rightly so because no-one responsible for plant output
and safety can afford to take risks by rapid adoption of new sensors
without a thorough proving procedure. Also the industry has a huge
investment in diagnostic and maintenance procedures together with large
stocks of spare parts for its existing range of sensors, and this can put
a brake on the adoption of new developments until crystal clear economic
and reliability factors are tested and proved. Unlike the consumer
market the large users are not influenced by novelty or new technology
unless it can be economically justified.

This introduction is intended to provide the scenario for a more detailed discussion of the present and future status of fibre optic sensors in industry.

3. THE PRESENT SITUATION

3.1. Research and Development

The biggest spenders on fibre optic sensor R & D are undoubtedly in the military field, especially those involved in work on hydrophones and gyroscopes.

For a few years this market sector is likely to lead others in production output.

A considerable amount of R & D activity is proceeding in academic institutions, in medical research and in industry involving all five modulation techniques, i.e. intensity, wavelength, phase, polarisation and time.

In the industrial sphere, intensity modulation appears to have a promising future for alarms, switches and two state sensing such as level. This form of modulation is the subject of numerous patent specifications but for the reasons given in an earlier paper, the simple analogue forms of modulation without referencing seem incapable of meeting the industrial standards of accuracy. Nevertheless there are several existing process instruments which can use intensity modulated sensors in a switching or digital mode without any form of referencing. Examples include vortex and turbine flowmeters, tachometers, fibre optic microswitches, thermostats and proximity switches. There is a huge build up of patents and inventions on intensity sensing which apart from the examples given above are waiting the development of a cheap, accurate flexible and reliable referencing system. One of the contenders for the prize is the optical wheatstone bridge described in an earlier paper. Research work is continuing on the alternative types of modulation but these either involve costly optical components or involve components such as gas lasers which are not appropriate for the process control conditions referred to earlier or do not provide long-term calibration stability. However it would show lack of foresight to assume that these disadvantages will prevail for ever.

Because the applicational requirements of the laboratory and scientific sector are less severe than those of the process industries the first successful production of sensors using sophisticated modulating techniques is likely to be aimed at the former sector rather than the latter.

3.2 Sensors for the Process Industries - Advantages

The euphoria which developed when optical fibre sensing became viable a decade ago overlooked some of the practical and special problems associated with the requirements of the process industries referred to in the introduction. Enthusiasm tended to accentuate the advantages but remained silent on the disadvantages. The frequently quoted advantages include the following:

    3.2.1 Some sensors can be potentially cheap, simple and reliable
Examples: miscroswitches, intensity modulated reflective sensors, liquid level switches and microbend force or displacement sensors.

    3.2.2 Some sensors can be used in applications which are difficult, expensive or impossible with electronic techniques.
Examples: a) Temperature measurement in high voltage electrical equipment such as alternators and transformers

     b) Current and voltage measurement in EHT power
       distribution
     c) Rapid temperature measurement of small surfaces
       having low thermal conduction and variable emissivity
     d) Temperature measurement at positions having low
       accessibility such as transformer winding interspaces
       or in medical and biological applications

3.2.3 Some sensors can outperform electronic types.
In the short space of time since development began on fibre optic
sensors, some good examples have emerged which show a marked superiority
in performance over their contemporaries.

Examples: a) Hydrophones with improved sensitivity
      b) Fibre optic gyros
      c) The Accufibre temperature sensor which has a performance
        exceeding that of the best temperature standard as the
        following specification will show:
        Range: $500^{\circ}C-2000^{\circ}C$
        Resolution: $7.5 \times 10^{-6\circ}C$
        Accuracy: 0.0025%
        Bandwidth: 50,000Hz
        By way of comparison the Accufibre temperature sensor has
        2,000 times the resolution of a thermocouple, 100 times the
        accuracy and 5,000 times the bandwidth.

3.2.4 Safety. The absence of electrical power in the sensor and the
substitution of optical energy at levels not exceeding 1mW is almost
certain to ensure safety of sensors operating in the most hazardous
environments although this view still needs official approval. The most
dangerous environment could be one in which small dust particles exist in
the 1-100 micron mean diameter size range, which are either easily
combustible in air or are mixed with highly combustible vapours. The
actual safe limit of optical energy under the worst conditions has still
to be established but clearly high power gas lasers are not desirable for
operation with sensors in hazardous locations. It has been shown
theoretically that an optical power of 1mW absorbed by a single dust
particle of 10 micron diameter could raise its temperature by $332^{\circ}C$

3.2.5 Fibre optic sensors normally are free from interference arising
from high frequency electromagnetic and EMP fields and from earth
currents. This virtue can be exploited in many applications of
measurement, particularly temperature, when such forms of interference
exist e.g. in medical diathermy and in ducts carrying high power cables.
Very large savings are possible in the process control field by running
both power and measurement cables in a single duct. On high voltage
electrical grid systems this freedom from interference enables
communications and signalling to be undertaken by fibres running parallel
with the high voltage conductors.

3.2.6 Other advantages
. The application of fibre optic sensors removes the dangers of
  electrical shock. For this reason they are preferable to electrical
  sensors for medical examinations
. High data transmission rates are possible
. A unique advantage of fibre optic intrinsic sensors over the electronic
  types is their ability to provide distributed sensing over distances of
  the order of a kilometer using a single fibre.
Examples: a) Cryogenic leak sensing has been applied to tanks for

storing liquids at very low temperatures. The technique
is to employ a fibre with a cladding whose refractive
index exceeds that of the core when cooled to the
temperature of the cryogenic liquid. The fibre is laid
in a position which ensures that it will come in contact
with any spillage. When this occurs the light in the core
transfers to the cladding and becomes attenuated and this
can be detected in the usual way

b) Temperature sensing can be made in large tanks and
reactors to establish temperature profiles and peak
values. This type of distributed sensing requires an
intrinsic fibre optic temperature sensor used in
conjunction with OTDR techniques (Optical Time Duration
Reflectometry).

3.3 Sensors for the Process Industries - Disadvantages

In spite of the R & D investment in fibre optic sensors and the stir of
interest created in the worldwide sensor community, there are very few
commercially available sensors suitable for the process industries or
for that matter any other market sector except perhaps that of defence.
This section of the paper analyzes the underlying causes of this
situation.

3.3.1 Industrial conservatism and resistance to change

As mentioned in the introduction this conservatism of instrument
purchasers in the process industry is justified provided it is based on
economic considerations. At the present time there are very few
industrial fibre optic sensors available to purchase and not a lot is
known about their cost of ownership, functional performance, reliability,
installation requirements or safety certification.

3.3.2 Lack of compatible actuators

The logic of installing fibre obtic sensors is weakened if electrical
signals are still required to operate control elements like E to P
convertors and pneumatically actuated valves. The problem of controlling
actuators by light is a difficult one. At least one system has been
demonstrated but its cost and complexity is unlikely to make it
competitive with existing systems.

3.3.3. Multiplexing

Many systems have been described in the literature but in the opinion of
the author these are unlikely to be competitive with existing electrical
systems in terms of cost, reliability, flexibility and compatibility.
One of the problems encountered with optical multiplexing is the lack of
optical energy available for modulation if it has to be channelled into
more than about ten passive sensors connected to a single fibre highway.
The problem is immediately eased if each sensor is provided with its own
separate energy source but this would immediately undermine the cost
advantages of multiplexing. The problem is considered later on in
connection with the possibility of using battery operated sensors.

3.3.4 Field Connections

In spite of the ingenious equipment available for making excellent fibre
splices in the field, the cost of the equipment and the complexity of the
process compares unfavourably with the simplicity of making electrical
connections. The problem of making T branches is even worse. A new and
ingenious system of making branches with polymer fibres has been
developed but cannot be applied to glass or silica fibres.
Unfortunately the attenuation of polymeric materials is too high for the
distances involved in the process industries.

3.3.5 Incomplete sensor range

One of the most severe problems facing manufacturers of sensors for the process industries is the huge investment needed for the development, the engineering, the production and the marketing of an the enormous range of sensors and their variants which are required by the industry. This problem was highlighted in the Introduction.

The manufacturers will have already amassed a huge investment in electronic sensors through development, designing, tooling, stocks, publicity, training etc. so that only the harsh discipline of competition will force them to write off the investment and start a new product family. The users will almost certainly be unwilling to install new systems with a hybrid composition, i.e. a mixture of electrical and fibre optic sensors with little compatibility between them. This means that a time extended introduction of the complete new range is probably not acceptable.

3.3.6 Lack of standardization

A fair degree of standardization exists in respect of pneumatic and electrical sensors, particularly with regard to their output signal and this makes it possible for users to shop around to select individual sensors from several manufacturers knowing that regardless of source of manufacture they will all meet the system requirements. At the present time standardization of optical output signals and system connections is a future hope rather than a present reality and this will be an unfavourable aspect for potential purchasers.

3.3.7 Passive fibre optic sensors are not "smart"

In the last few years the process industries have become interested in the so called "smart" sensor particularly for differential pressure measurement but there is every reason to expect the technology to extend to other electronic sensors. This "smartness" or "intelligence" as some would call it, has been made possible by the advent of digital techniques, microprocessors and memories. A typical differential pressure sensor can boast of the following achievements:

. It can identify itself remotely by transmitting its code and can then follow up with its range, output signal and damping factor
. A remote hand held device can change any of the parameters
. Diagnostic facilities can provide trouble shooting assistance and the calibration can be checked
. All data can be displayed on the remote hand held device
. The calibration can be corrected for variable pressure and termperature conditions
. The output signal can be modified to suit the requirements of the display equipment

Unfortunately all these features are unavailable with passive fibre optic sensors. If equivalent optical 'chips' were available at a low price and could operate on a few microwatts of power the situation could be transformed but this development is much to far away for anybody to make a forecast.

It is therefore difficult to see optical sensors completely replacing electronic sensors where smartness is a key requirement.

3.3.8 Research and Development expenditure on "optics" is less than that on "electronics"

The disparity in expenditure is so great that it is difficult to accept that optical sensor technology can keep pace with electronic competition except in certain areas previously mentioned.

3.3.9 <u>Some important</u> optical components such as semiconductor lasers, couplers, monochromators, Bragg cells, diffraction gratings, polarization maintaining fibres etc. are currently expensive and impose a severe cost penalty. Whilst it is reasonable to accept price levels falling in the future with increased demand, it is unlikely that the fall will be as dramatic as the fall in prices of electronic components over recent years.

4. THE FUTURE

The future of fibre optic sensors can be read from this analysis of their advantages and disadvantages from which the following conclusions can be drawn:

4.1 <u>Discrete instruments</u>

For the reasons given in Section 3.3; complete integrated systems of fibre optic measurement and control of the type required by large process plants in the oil, chemical and petrochemical industries, are unlikely to be installed for many years other than for experimental or pilot plant purposes. The way ahead for manufacturers therefore is to exploit the quite large potential industrial markets for discrete sensors in which large integrated systems of sensors are not required and the problems of 'multiplexing, system compatibility and hybrid assortment, are therefore not important factors. Discrete sensors are also in demand in other market sectors listed in the Introduction and these hopefully could be acceptable or adaptable for the industrial sector.

The sales potential for discrete sensors could be enhanced by concentrating development on the very simple intensity modulated switching sensors for level, flow, pressure and temperature as well as on simple referencing techniques. These could have application in several industrial sectors including the utilities,(especially gas supply), food, the environment and the smaller industries. Great stress must be given to the importance of developing low cost intensity modulated sensors with a simple but adequate referencing system.

4.2 <u>Sensors with superior performance</u>

Some examples were given in Section 3.2.3 to 3.2.6. inclusive, of fibre optic sensors having special advantages in performance and application. This list was by no means exhaustive and there is undoubtedly a great opportunity for finding other applications requiring a better performance and applicability than that currently available from electronic sensors. As opposed to the merits of the low cost simple intensity sensors mentioned in the previous section, the attraction to the manufacturer of these superior sensors will be the opening up of new markets free from 'electronic' competition.

In order to reach this goal, development trends will need to be directed to the more sophisticated technology involving modulation by phase (interferometry), polarisation and time (frequency). This development will also call for resources to be applied to optical and electro-optical components to make them simpler and cheaper to produce and it will also be important to develop special fibres, including coated fibres, which have intrinsic sensor properties. The special capability of fibres to provide distributed sensing is a powerful weapon in the battle for supremacy over electronic sensors.

4.3 <u>An interim solution for the process industries</u>

As mentioned previously there is an enormous investment in electronic sensor designs and systems for the process industries and there seems to be no way of invading this market to any extent even in the next ten years with a complete range of passive fibre optic sensors. What is needed is an interim introduction which might be possible if the users were prepared to accept sensors containing long life batteries. Lithium batteries have now been available for some years for powering microcomputer systems in the event of power failure and they have been demonstrated to have a long shelf life and a high capacity/size ratio. It is conceivable that many existing sensors can be modified using a battery to transmit weak light pulses through a fibre to the receiving equipment either as a frequency or serial digital signal. The techniques for generating these signals by adaptation from existing sensors should be available on a relatively short time scale. The presence of the battery should also enable the sensor to be endowed with a modicum of intelligence or "smartness". One or two commercial systems have already made a modest entry into the process control market but it will need more manufacturers and more development and design for production to convince the large users of the practicability and advantage of this approach.

4.4 <u>Integrated optics</u>

Ten years ago there was some highly optimistic claims that optical "chips" would create a technical revolution even greater than that of the electronic microchips. At the present time there is very little sign of this photonic revolution coming to pass despite the fact that from time to time interesting announcements on the development of optical computers and memories are made by the press. However, optical chips have been demonstrated which are able to perform many of the functions required by sensing techniques e.g.

      Modulation
      Switching and multiplexing
      Light guiding
      Polarisation
      Frequency changing
      Diffraction
      Spectral Analysis
      Photoelectric conversion
      Reflection
      Memory
      Lasing
      Light amplification

It is certain that integrated optics will play an important role in the long term future but will not make a significant contribution within the time scale relevant to this paper

5. CONCLUSION

The problem facing the sensor manufacturing industry is caused not only by the variety of measurements required by industrial users but by the bewildering range of design variations in terms of operating spans, pressure ratings, materials of construction, fixings and special applicational requirements. Acquiring this variety may not involve much development work but it would impose a staggering burden of design and production engineering expense if an attempt was made to provide a complete competitive alternative to the existing electronic range in a short space of time.

The alternative short term strategy could be one of developing fibre optic sensors:
. as discrete devices where the application involves measurement rather than control
. for applications where electronic sensors are unavailable or unsuitable
. for applications where they out-perform electronic sensors in price, performance reliability, ease of maintenance, sensitivity, specificity etc.
. where standardization is not crucial
. where large systems of measurement and control are not involved
. where battery assisted sensors would be acceptable.

"MULTIMODE-FIBER COUPLED WHITE-LIGHT INTERFEROMETRIC POSITION SENSOR"

T. Bosselmann

Technische Universität Hamburg-Harburg, Harburger Schloßstr. 20, D-2100 Hamburg 90, FRG, present adress, Siemens AG, ZFE TPH 42, D-8520 Erlangen, FRG

Summary Multimode fibers interconnect two Michelson interferometers, white light source and detector. The mirror position of the transmitting interferometer is reconstructed and measured in the receiving interferometer by periodically scanning through three "white-light" fringes (laser reference, 22mm measuring range, 0.00001mm resolution).

Introduction One of the most accurate optical measuring instruments determining precisely position and distances is the interferometer. Conventionally operated with a monochromatic coherent source, its output signal has a positional ambiguity of $x=m\lambda/2$ (λ= light wavelength, m= 0,1,2,....) and lacks of a fixed internal zero point. The use of a broad-band fairly incoherent source and a second interferometer avoids both problems in the "white-light interferometry" /1-4/.

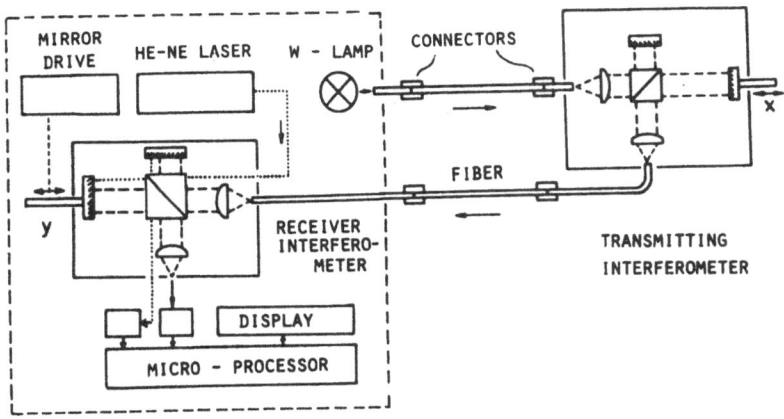

Fig. 1: Position sensor system, based on white-light interferometry.

Fig. 1 shows the fiber optics system which we have constructed for high accuracy position sensing. White light from a W-lamp is fed via multimode fibers consecutively through two Michelson interferometer acting as transmitter and receiver to a detector. The element whose position x is to be sensed displaces one mirror of the transmitter. In the remote controlling receiver interferometer one mirror is scanned periodically. During one receiver scan two times white light fringes do appear at the output of the receiver when the optical path differences OPD of receiver and transmitter coincide and one time with higher intensity at zero OPD of the receiver, Fig. 2. Recording a complete interferogram during one scan, the system provides a precise reproduction or transmission of positions. For measuring the position as the distance between two white light fringes (signatures) light of a He-Ne laser is fed additionally through the receiving interferometer. The laser fringes are detected by a separate detector and counted by a transient recorder. By this combination of white light and monochromatic interferometry, the system provides the usual high accuracy of laser interferometers, yet it does not forget its zero point when interrupted.

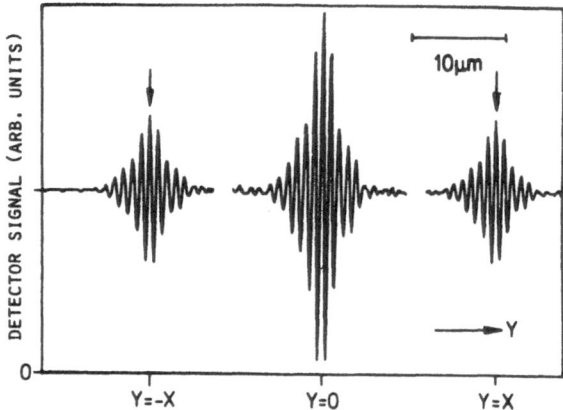

Fig. 2: Detector signal during receiver scan (interferogram). The arrows mark the central fringes of the side signatures.

The use of multimode fibers yields a sufficiently large optical power level to permit large transmission distances between transmitter and receiver of up to several kilometers and simple fiber connections of the system components.

Principle of operation Transmitter and receiver consist mainly of two beam Michelson interferometers, constructed using conventional optical components, Fig. 3. The OPD of the transmitter is denoted by 2x, representing twice the displacement of the moving mirror from its zero OPD position. The operation of the transmitter can be understood as a spectral filter acting on the incoming light passing a finely channeled comblike output power spectrum to the receiver. The periodicity of the output spectrum $\Delta \nu_T = 1/2|x|$, (optical frequency ν measured in wavenumbers) is inversely proportional to the measurand x. This output spectrum is filtered a second time in the receiver, where 2y denotes the OPD and determines the spectral periodicity $\Delta \nu_R = 1/2|y|$ of the interferometric filter. In general $|x| \neq |y|$ each interferometer has a mean power transmissivity of .5, so a small optical power I reaches the detector. During one scan the situation $|y| = |x|$ occurs twice. The filters become matched, that means their transmission spectra are identical and the detector signal reaches a maximum (side signature, fig. 2). These two side signatures are located symmetrically at $y = \pm x$ in the interferogram. Additionally a higher maximum (main signature, fig. 2) exists at y=0, where the receiver filter is fully transparent. By analyzing the interferogram (detector signal) at the receiver location it is possible to recover the OPD 2 |x| of the transmitter, i.e. the position x of the sensing mirror. A more detailed analysis shows, that the form of each of the signatures represents the Fouriertransform of the power spectrum of the light source, as seen by the detector /5/. The signatures do have a central fringe at $y = \pm x$ and y=0 respectively, which can not be exceeded by any other fringe of the same signature, fig. 2. According to Fourier theory the central fringe of a signature becomes strongly pronounced with increasing optical bandwidth of the light source. So a W-lamp and an optically broad band Si-detector are used here. By determining the central fringes of the three signatures and counting the number of laser fringes between two of these fringes with increased resolution by electronic subdivision the transmitted position x is measured.

The information of position x is encoded into the periodicity Δy_T of the optical power spectrum transmitted and decoded by a matched receiver tuned to the same filter periodicity Δy_R as the transmitter. Compared to other encoding methods employed in fiber optic sensors (e.g. intensity), the spectral encoding is advantageous, because it is highly immune to almost all conceivable pertubations and variations of the fiber optic transmission path. As the principle of spectral encoding can be explained fully by considering only the transmitted optical power spectrum, it is recognized that the transmission of information is independent of the modal structure of the fiber employed. Regardless of dispersion and polarisation properties, inexpensive multimode fibers can be used offering large optical throughput and easy handling. The spectral transmission loss of the employed fibers may influence the form of a signature, but do not effect the positions $y=\pm x$, $y=0$ of the central fringes in the interferogram. Consequently, the accuracy of the transmission does not depend on kind and length of fiber used, nor on connector losses, as long as the detector signal is large enough to evaluate from the interferogram the positions of its fringes. The position x is measured two times as the separation of the central signature ($y=0$) and either side signature ($y=\pm x$). Hence it is possible by comparing those two values for x to check for consistency of the measurement.

    <u>Realisation</u> Two types of compact transmitter interferometers for different tasks have been constructed, fig. 3.

<u>Fig. 3:</u> Photos of two types of transmitter interferometers (left) and receiver base plate (right).

Fig. 4 shows one transmitter interferometer with single mechanical input assembled in a stainless steel tube of 30mm diameter and 130mm length. Achromatic lenses, a light deflecting prisma and a beamsplitting cube are used. A precision linear ball bearing holding the movable mirror provides 25mm travel path.

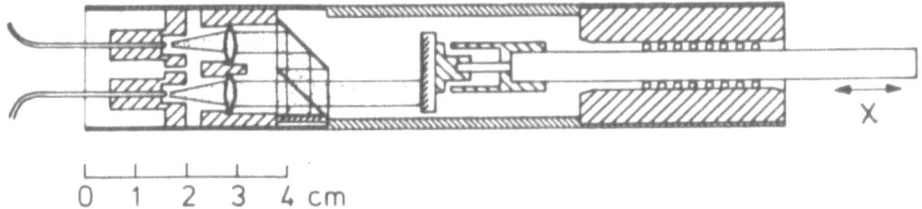

<u>Fig. 4:</u> Construction of one transmitter interferometer with single mechanical input.

The other transmitter interferometer in fig. 3 with a differential mechanical input is constructed in a similar way. The receiver consists of a base plate, fig. 3, and a special developed
transient recorder. On the base plate are mounted the receiver interferometer, a HeNe
laser producing calibrational reference fringes, a halogen white-light source, a photodetector for the white-light fringes and a second photodetector combined with a frequency
multiplier for generating laser reference pulses. One pulse corresponds to a shift in OPD
of 10nm. The receiver interferometer is dynamically balanced to reduce sensitivity against
mechanical shocks of the base plate. It uses two movable mirrors moving at any moment
in opposite directions with equal velocities driven by a crank drive. The transient recorder
determines the relative positions of a number of the more pronounced fringes in each of
the three signatures by means of the laser reference. It identifies the three central fringes,
evaluates their separations, checks for consistency and displays the transmitted position.
From the W-lamp (effective bandwidth as seen by the Si-detector 0.7-1μm) an optical
power of 100μW is coupled into a 80μm core fiber. From the noise equivalent power of
the detector (300pW/300kHz), the insertion losses of the interferometer (7dB each) and
a 13dB signal/noise ratio required for unambiguous fringe detection, a total power margin
of 28dB is available for fiber transmission losses. This should be enough for link lengths
up to several kilometers. Save operation has been demonstrated with 800m of fiber. The
achieved resolution is 10nm over a measuring range of 22mm at a scan repetition rate of
0.3Hz giving a total resolution of 10^6 resolved spots.

Conclusion A system has been presented for sensing and transmitting positions. It has
the attributes of interferometric accuracy, spectral encoding and absolute transmission.
It is independent of the transmission characteristics of the employed fiber. It requires a
sophisticated hardware but gives high performance of 10^6 resolved spots.

The author would like to thank R. Ulrich for inspiring this work and many helpful discussions and the Deutsche Forschungsgemeinschat for financial support.

References
1. Patten RA: Michelson Interferometer as a Remote Gauge. Appl. Opt. 10,
   p. 2217 1971
2. Delisle C, Cielo P: Application de la modulation spectrale à la transmission de
   l'information. Can. J. Phys. 53, p. 1047 1975
3. Al-Chalabi SA, Culshaw B, Davies DEN: Partially coherent sources in
   interferometer sensors. Proceed. 1st Conf. OFS, London, p. 132 1983
4. Bosselmann T, Ulrich R: High-accuracy position-sensing with fiber-coupled
   white-light interferometers., Proceed. 2nd Conf. OFS, Stuttgart, p. 361 1984
5. Bosselmann T: Spektral-kodierte Positionsübertragung mittels fasergekoppelter
   Weißlichtinterferometer. Thesis, Techn. Univ. Hamburg-Harburg, Hamburg 1985

MACH-ZEHNDER SYSTEMS FOR HETERODYNE FIBRE POLARIMETRY IN   DIFFERENT   COHERENCE
CONDITIONS

R. CALVANI, R. CAPONI, F. CISTERNINO

CSELT - VIA G. REISS ROMOLI 274 - TORINO - ITALY

1.INTRODUCTION
In this contribution two distinct types of heterodyne systems are presented
for the fast detection of polarization states at the output of a single -mode
fibre . Based on the same measurement principle, the two versions   differ   as
to source coherence properties, in order to cover a wider range of application
conditions. Though devoted to fibre characterization purposes in the area of
optical communications, the present work can be of considerable interest also
for the development of new polarimetric fibre optic sensors.

2.MEASUREMENT PRINCIPLE
In polarimetric interferometry for sensor applications, the   state   of
polarization of a light wave (S.O.P.) is currently specified by the electric
field vector (Jones vector), since the unpolarized radiation is negligible.
In this case the S.O.P. could be recognized directly through a detection of
the orthogonal field components E_x , E_y , with their ratio R and relative phase
ϕ . Such a measurement cannot be performed, of course, at the optical
frequency ω , i.e. in real-time. However an heterodyne mixing process of E_x
(E_y) with a reference field E_{Rx} (E_{Ry}) at a slightly different frequency
$\omega+\Omega$, conveys all the relevant information on the intermediate R.F. Ω .   The
two waves after superposition are analyzed in polarization, and the two
resulting beating signals separately detected , to get the R and ϕ
polarization   parameters (Fig.1). So, through a heterodyne frequency
rescaling, the electric field rotation is detected and can be directly watched
with an oscilloscope operated in x-y mode.
In formulas, the field to be measured can be represented as follows ·

$$E_{Mx} = E_{0x} \cdot \exp (i\omega t)$$
(1)
$$E_{My} = E_{0y} \cdot \exp [i(\omega t + \phi)]$$

while the reference field must be linearly polarized at 45 degrees for a
correct weighting of the two amplitudes with no extra phase added:

(2) $$E_{Rx} = (E_0/\sqrt{2}) \cdot \exp i[(\omega+\Omega) t - \phi_R] = E_{Ry}$$

After superposition and polarization analysis along x,y axes, the optical
intensities I_x , I_y separately detected are

$$I_x = I_{0x} + (E_0 E_{0x}/\sqrt{2}) \cdot \cos (\Omega t - \phi_R)$$
(3)
$$I_y = I_{0y} + (E_0 E_{0y}/\sqrt{2}) \cdot \cos (\Omega t - \phi_R - \phi)$$

where the two beating signals, filtered from d.c., carry the correct amplitude
ratio $R = E_{0x}/E_{0y}$ and phase difference ϕ of the S.O.P. to be evaluated. So,
beside the oscilloscope observation, precise measurements of the S.O.P.
parameters can be obtained with a vector voltmeter which assumes either of the
two x,y oscillating voltages as reference signal.

3.FIBRE POLARIMETRY WITH NARROW LINEWIDTH SOURCES
     The simplest way of generating a reference field is to take it from the
same optical source used for fibre excitation. Then, to achieve

beam-splitting, frequency-shifting, beam recombining and polarization analyzing, as sketched in Fig.1, the straightforward solution is a Mach-Zehnder interferometric configuration, illustrated in Fig.2. The source is a single-frequency laser, followed by an acousto-optic device (A.O.D.), which provides both the launching and the reference beam, being operated at a reduced Bragg efficiency with a relative frequency-shift $\Omega / 2\pi = 40$ MHz. The fibre to be measured is inserted in either of the two paths, while the reference arm includes an optical compensation device to obtain the required 45 degrees linear S.O.P. A beam-splitter and a Glan-Taylor prism are used for recombination and polarization analysis respectively. A separate detection of the two linearly polarized outputs provides the R.F. input signals for the oscilloscope and/or vector voltmeter.

The precision achievable in relative phase ϕ and amplitude ratio R has been checked using a quarter wave-plate in place of the fibre, and comparing the output data at different plate azimuths with computed results. The agreement was within .05 dB in R and .1 in ϕ [1].

A key feature of the system is that polarization measurements are feasible also on long fibre spans, even exceeding the source coherence length, provided that the laser linewidth ($\delta\nu$) is much less than $\Omega / 2\pi$. This incoherent operating mode is precisely the same as in heterodyne transmission systems. In both cases the phase noise bandwidth is much narrower than the intermediate frequency, so that the beating signal can be recognized in spite of the E_M, E_R mutual incoherence. To fulfil this particular requirement we used a stabilized single-mode He-Ne laser ($\lambda = 632.8$ nm) with $\delta\nu = 50$ kHz. The tested fibre was of single-mode type, 3.3 km long, cut-off wavelength $\lambda_C = 850$ nm.

The system performances have been tested through two sets of quantitative observations, in time and frequency domain. Fig.4a shows a typical x,y display of the polarization ellipse, representing the first Lissajous picture of the two R.F. signals. In Fig.5 three different spectrum analyzer measurements are presented. The first two traces a,b refer to x,y detected signals mixed with the external R.F. supplying the A.O.D. The presence of phase jitter in both channels is clearly displayed near the d.c. component, while the same beating signals mixed together exhibit an output current spectrum cleaned from phase noise (Fig.5c). Such cancellation experimentally demonstrates the subtractive character of relative phase measurements in our interferometric apparatus. This effect is an important feature which improves measurement precision and stability.

4. FIBRE POLARIMETRY WITH BROAD LINEWIDTH LASERS

The previous measurement scheme can be applied also to broad linewidth sources using a different configuration with balanced optical paths in which the fibre is placed outside the interferometer. In this case both arms contain the same S.O.P. as the fibre output. So the reference beam needs an S.O.P. reset, achieved with a polarizer at 45 degrees inserted in place of the compensator.

From a theoretical viewpoint, the present configuration differs from the previous one only with respect to the amplitude E_0 and phase ϕ_R of the reference field. These parameters are now dependent on the S.O.P. to be measured in the form

$$(4) \qquad E_0 = \sqrt{E^2_{0x} + E^2_{0y} + 2\, E_{0x}\, E_{0y} \cos \phi}$$

$$(5) \qquad \phi_R = \text{arctg}\, [E_{0y} \sin \phi/(E_{0x} + E_{0y} \cos \phi)]$$

Anyway, the amplitude ratio R and relative phase ϕ in intensities (3) remain unchanged. An experimental implementation of this scheme is illustrated in Fig.3. The launch section, the fibre under measurement and the output collimator are assembled in a standard set-up outside the interferometer. The output beam is sent to an A.O.D., which operates the same way as the previous system. The Mach-Zehnder arrangement includes in the reference arm a corner cube delay line, for path equalization, followed by a reset polarizer. Recombination and polarization analysis are unchanged.

To demonstrate the excellent quality of the fibre S.O.P. observation, an x-y polarization display is reported in Fig.4b. Moreover, a series of dynamical tests have been carried out using a standard (low coherence) semiconductor laser at λ =850 nm and a single-mode fibre with λ_C =800 nm, 750 m long, attenuation 2.95 dB/km at λ =850 nm. The fibre behavior in a real environment has been simulated with a 60 Hz shaker on which 20 turns of fibre were placed. The measurement consists of a spectral analysis of the output phase given by the vector voltmeter. Fig.6 shows: a,b noise spectra of x,y channels separately , assuming the R.F. of the A.O.D. driver as reference, and c, the relative phase ϕ spectrum, using either of the two channels as reference. The traces in Fig.6a,b exhibit many harmonics of the shaking frequency. These additional lines disappear in Fig.6c, where only the polarization phase noise is present. Such cleaning is another experimental proof, in a lower frequency range, of the same subtractive effect observable in Fig.5.

5.ADVANTAGES AND APPLICATIONS

The advantages of heterodyne polarimetry can be summarized in three different features:

1) as in any interferometric polarimeter, the detection of field amplitudes instead of optical intensities involves a scale halving in logarithmic units, so that the explorable S.O.P. range for a given instrumental dynamics is doubled.

2) The polarization phase measurement is immune from environmental noise, regardless of the interferometric stability, due to the subtractive character of the system.

3) The high frequency shift produced by the Bragg cell offers a fast S.O.P. recognition time (equal, in principle, to 1 R.F. period).

These three characteristics are interesting for sensor applications. In particular, the first item can improve the explorable S.O.P. range of a Faraday rotation sensor [2], while feature 3 would allow the detection of fast current transients. The Mach-Zehnder configuration can be also applied in phase polarimetric sensors, in which the measurands are strain, vibrations or temperature [3,4], owing to the ϕ stability recalled in item 2. The broad linewidth version is especially suited for such applications, being free from the usual constraints on the source coherence length in comparison with fibre length.

6.CONCLUSIONS

The feasibility of a Mach-Zehnder heterodyne polarimeter both with narrow- and broad- linewidth sources, has been extensively demonstrated through various kinds of experimental tests. The inherent advantages of the two system versions, and related applications to fibre optic sensors have been discussed.

REFERENCES
1.Calvani R.,Caponi R.,Cisternino F., Opt.Comm., 54, No. 2, 1985, pp.63-67
2.Culshaw B.: Optical fibre sensing . P. Peregrinus Ltd., 1984.
3.Akhavan Leilabady P. et al., J. Phys. E, 19, 1986, pp. 143-146.
4.Akhavan Leilabady P. et al., Opt. Lett., 10, No. 11, 1985, pp. 576-578.

436

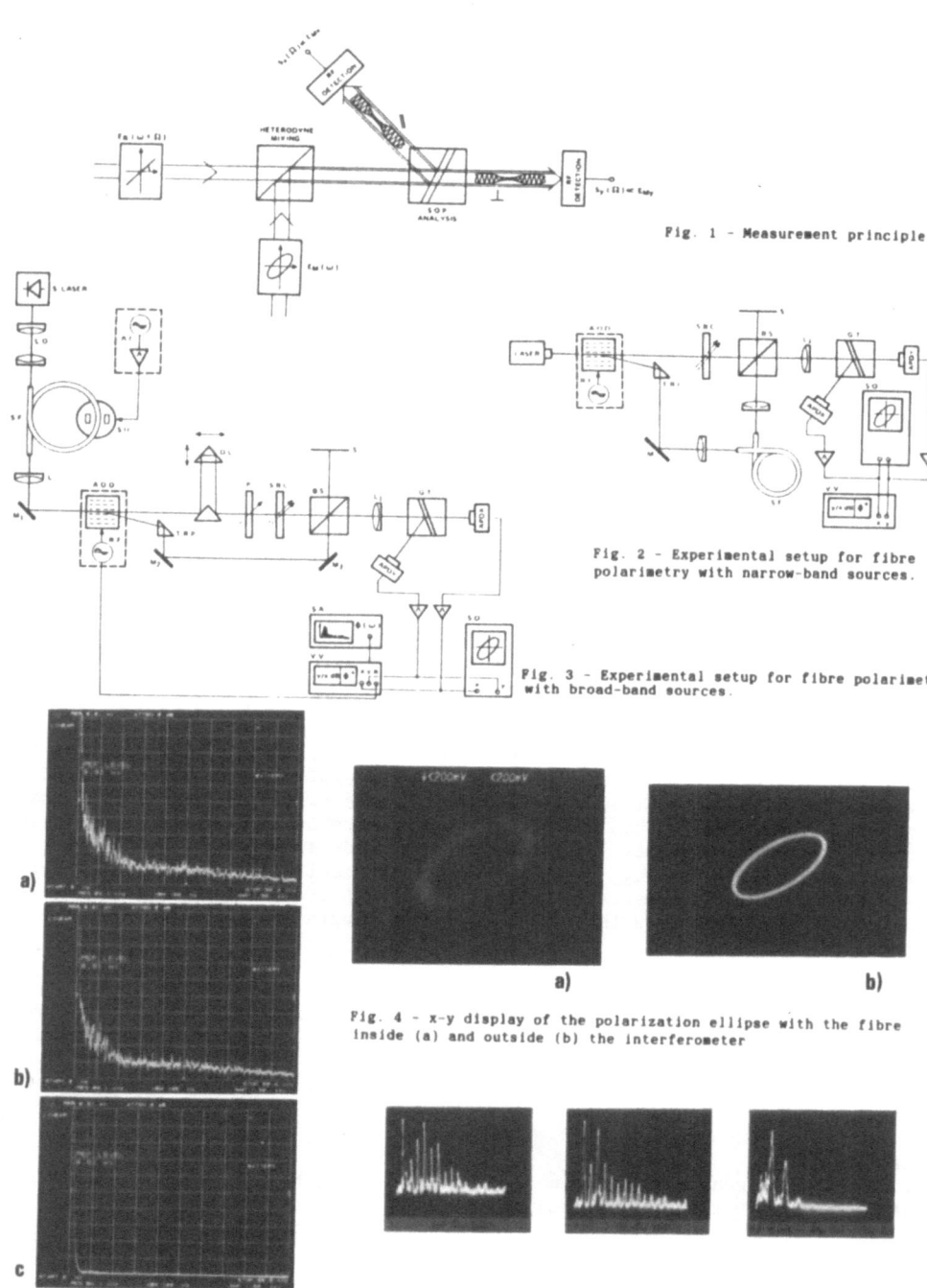

Fig. 1 - Measurement principle.

Fig. 2 - Experimental setup for fibre polarimetry with narrow-band sources.

Fig. 3 - Experimental setup for fibre polarimetry with broad-band sources.

Fig. 4 - x-y display of the polarization ellipse with the fibre inside (a) and outside (b) the interferometer

a)

b)

c

Fig. 5 - Phase noise spectra or output beating signals. (a) x-channel mixed with RF external source; (b) y-channel mixed with RF external source; (c) (x,y)-channel mixing.

Fig. 6 - Low-frequency phase noise spectra of output beating signals.

(a) $\phi_x(\omega)$, span 1 kHz; (b) $\phi_y(\omega)$, span 1 kHz;
(c) $\phi(\omega) = \phi_x(\omega) - \phi_y(\omega)$, span 0.5 kHz.

MODEL FOR AN OPTICAL FIBER pH SENSOR

F.BALDINI, M.BRENCI, G.CONFORTI, R.FALCIAI, A.G.MIGNANI

Istituto di Ricerca sulle Onde Elettromagnetiche del C.N.R.
Via Panciatichi, 64 - 50127 Firenze, Italy
phone +39-55-4378512

The traditional methods for the pH determination are electrometric and colorimetric methods. In the first case, pH determination by glass electrodes, is based on a measurement of the electromotive force of a cell which has a reversible electrode, whose electrochemical potential depends on activity of hydrogen ions and, hence, on pH. In the second case indicators are used, that reveal the acid or base character of a solution through a change of colour, or fluorescent substances whose fluorescence varies with the acidity of the solution.

As for the absorption [1,2], if we consider an indicator in solution, its equilibrium reaction may be written as:

$$HIn \leftrightarrows H^+ + In^-$$
(1)

where HIn denotes the indissociated form (or acid form) of the indicator, In^- the dissociated form (or base form) and H^+ the hydrogen ion.

By assuming a very dilute solution, it is possible to relate the pH of the solution directly to the concentration of hydrogen ions:

$$pH = - Log \left[H^+\right]$$
(2)

By using the equilibrium equation of the reaction (1), it turns out:

$$pH = pK - Log \left\{ \frac{[T]}{[In^-]} - 1 \right\}$$
(3)

where pK is related to the dissociation constant of reaction (1) and $[T]$ and $[In^-]$ denote the total concentration of the dye, which is constant, and the concentration of basic form, respectively.

The Lambert-Beer law relates the light absorption with the

concentration of the absorbent substances:

$$\text{Log} \frac{I_o}{I} = \left[In^-\right] L\varepsilon \tag{4}$$

Here I_o and I denote the incident and transmitted light intensities at the absorption wavelength of the base form of the dye, L is the optical path and ε the absorption coefficient. From (3) and (4) one obtains:

$$pH = pK - \log \left[\frac{C}{\text{Log } I_o/I} - 1 \right] \tag{5}$$

where C is equal to the product $L\varepsilon\left[T\right]$, which results constant for a fixed optical path.

Relation (5) shows how the pH of a solution is univocally related to the intensity I transmitted by the dye.

A few laboratory prototypes of optical fiber pH sensors, which make use of different colorimetric techniques [2-9], have been developed up to now.

The present paper describes the studies carried out to realize an optical fiber pH sensor model, working in the physiological range (pH = 7.0 ÷ 7.5), based on the absorption method which makes use of a liquid optrode joint to the fiber. The liquid optrode avoids the problem of fixing the dye on a polymeric support, which generally leads to a shift of the measurement dynamic range of the indicator and to a long response time. However a suitable membrane, which must be highly selective to hydrogen ions and able to prevent the dye exit, must be used. Accordingly two lines have been followed: choice of the most proper dye and choice of a membrane with the required properties.

Spectrophotometric measurements have been made on indicators whose response is in the physiological range: phenol red, bromothymol blue, rosolic acid and neutral red. Bromothymol blue is preferably chosen among the other indicators mainly for two reasons: very high sensitivity to pH changes and very high molecular weight which is important in order to fix a solution dye at the end of the fiber with a membrane. Absorption curves obtained at the spectrophometer with an aqueous solution of bromothymol blue for different values of pH are shown in Fig. 1. The concentration of the dye is 1.49 x 10^{-5} M and the absorption wavelength for base form is 616 nm while the absorption wavelenght of acid form is not constant but shows a shift towards longer wavelengths with the increase of pH.

As for the membrane, first we performed several tests on different kinds of filters (with porosity smaller than 1μ), which

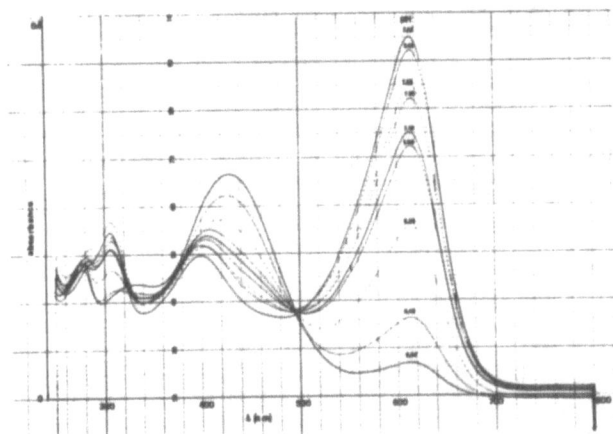

FIGURE 1. Absorption spectra of bromothymol blue for different pH values (dye concentration = 1.49×10^{-5} M).

resulted permeable to hydrogen ions, but allowing also the exit of dye molecules. The attention was focalized onto semipermeable membranes; two membranes based on polymers of piperazine and one made of cellulose acetate were tested; the best permeability to hydrogen ions and the capacity in retaining the dye after several hours were exhibited by the cellulose acetate membrane. It has an asymmetrical structure made by a very thin layer on a porous base, with progressive porosity, and has a thickness of 70 μ.

Systematic tests have been carried out at the spectrophotometer using two cuvettes, arranged as in Fig.2, one containing

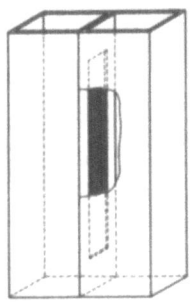

FIGURE 2. Cuvettes with intermediate membrane, used to test the permeability of the membrane.

the dye solution and the other the buffer while openings in the
lateral surfaces of the cells covered with membrane allow contact between the two solutions. Sodium chloride, which is a neu
tral salt with respect to the concentration of hydrogen ions,is
added in the dye solution in order to estabilish an equilibrium
for the osmotic pressure, which is quite different for the dye
solution and for the buffer (about three orders of magnitude),
thus avoiding membrane deformation due to pressure difference.

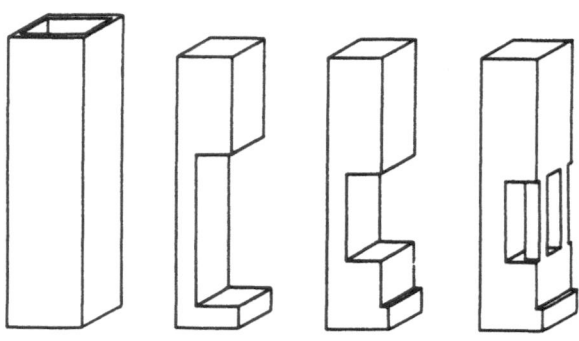

FIGURE 3. Cuvette and volume reducers.

Afterwards volume reducers (Fig.3), placed into the cuvette with
dye solution, allow to reduce the ratio (dye solution volume)/
(membrane surface) and consequently the response time.
Fig . 4a, b show the response time curves, for different volume reducers, when the dye solution is placed in contact,

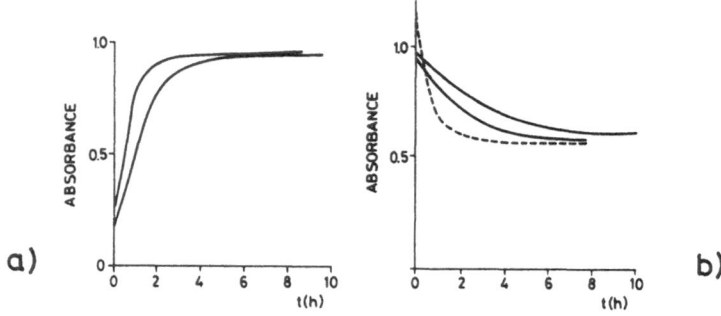

FIGURES 4a and 4b. Response curves corresponding to two different volume reducers, placed in the dye cuvette, with the dye
solution in contact with pH 7.5 and 6.75 respectively.

through the membrane, with two buffered solutions at pH 7.50 and pH 6.75 respectively. The longer response time of the second buffer (Fig. 4b) is probably due to an exceedingly protracted use of the membrane. This is confirmed by the dashed curve in Fig. 4b, corresponding to the response obtained by starting with a dye solution at pH 7.40 in contact with a buffered solution at pH 6.75 through a regenerated membrane;the response is quicker and comparable with those shown in Fig. 4a.

Fig. 5 shows the calibration curve that relates the absor-

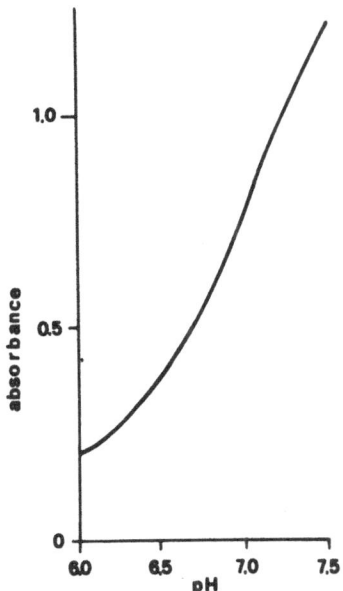

FIGURE 5. Calibration curve.

bance of the dye at 615 nm with its pH value. For base values of the buffer, the dye solution does not reach the same value of the buffered solution with which it is in contact. This can be seen in Fig. 4a, where the absorbance value at the steady state corresponds to a pH value 7.2 according to the calibration curve and not to the pH value of the buffer solution; this is probably due to the carbon dioxide in the air which attacks no buffered solutions whose pH is near neutrality.

By assuming that the diffusion velocity in liquid is larger than that through the membrane, one may expect a quasi-linear relation between the (dye solution volume)/(membrane surface) ratio and the response time. Fig. 6 shows the diagram of the response time as a function of such ratio;   the reciprocal of

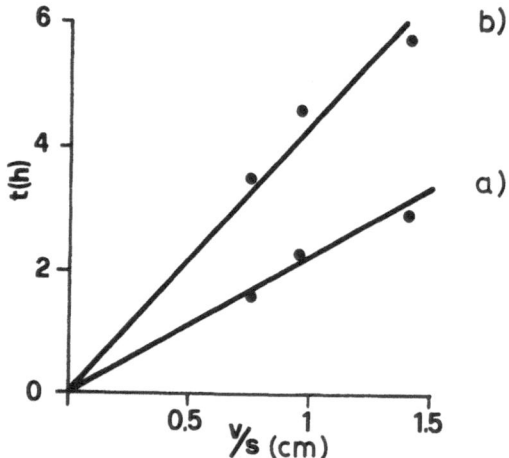

FIGURE 6. Time response versus volume/surface ratio.

the slope of the curves may be interpreted as membrane permeabi
lity to the hydrogen ion.

The results obtained by spectrophotometric measurements show
that a further reduction of the volume/surface ratio lead to a
sensible reduction of response time, eliminating also the problem of membrane deterioration for protracted use. This is possible in a miniaturized case with optical fibers, where the pro
be will be a membrane cylinder with specular bottom filled with
the dye solution and coupled with optical fibers.

REFERENCES

1. Bates RG: Determination of pH. Theory and Practice. John Wiley & Sons. New York, Cap. IV, 135, 1973.
2. Peterson JI, Goldstein SR, Fitzgerald RV, Buckhold DK: Fiber
   Optic pH Probe for Physiological Use, Anal. Chem. 52 (6),864
   1980
3. Goldstein SR, Peterson JI, Fitzgerald RV: A Miniature Fiber
   Optic pH Probe for Physiological Use, J.Biom.Eng. 102 (2),
   141, 1980
4. Markle DR, McGuire DA, Goldstein SR, Patterson RE, Watson RM:
   A pH Measurement System for Use in Tissue and Blood, Employing Miniature Fiber Optic Probes, Adv. in Bioeng., Amer.Soc.
   Mechan.Engin., New York, 1981
5. Kirkbright GF, Narayanaswamy R , Welti NA: Fiber Optic pH
   Probe Based on the Use of an Immobilized Colorimetric Indicator, Analyst 109, 1025, 1984

6. Ruzicka J , Hansen EH: Optosensing at Active Surfaces - a New Detection Principle in Flow Injection Analysis, Anal. Chim.Acta <u>173</u>, 3, 1985
7. Saari LA, Seitz WR: pH Sensor Based on Immobilized Fluoresceinamine, Anal.Chem. <u>54</u>, (4), 821, 1982
8. Zhujun Z, Seitz WR: A Fluorescence Sensor for Quantifying pH in the Range from 6.5 to 8.5, Anal.Chim.Acta <u>160</u>, 47, 1984
9. Gehrich JL, Lubbers DW, Opitz N, Hansmann DR, Miller WW, Tusa JK, Yafuso M: Optical Fluorescence and its Application to an Intravascular Blood Gas Monitoring System, IEEE Trans. Biom. Eng. <u>BME-33</u> (2) 117, 1986.

INTEGRATED OPTICS FOR SENSORS: A REVIEW OF THE ACTIVITY IN
ITALY

GIANCARLO C.RIGHINI

Gruppo Nazionale di Elettronica Quantistica e Plasmi e
Istituto di Ricerca sulle Onde Elettromagnetiche of C.N.R.
50127 Firenze, Italy

1. INTRODUCTION

Integrated optics has grown continuously since the very beginning of this field at the end of Sixties. However, it always follows with a certain delay the development of the companion field of fiber optics.

Thus, while optical fiber sensors are already commercially
available, and new or improved versions are under test in the
research laboratories, integrated optical sensors are still in
a research stage (see the paper by R.Th.Kersten in this issue).

Italian activity in integrated optics reflects such a general trend: in addition, due to the fact that the main interest
has been devoted so far to possible applications of planar optical circuits in analog signal processing, the area of integrated optic sensors has been scarcely considered. However,the
progress made in the design and fabrication of basic components
(lasers, modulators, couplers, lenses, etc.) should allow one
to easily build up a planar sensing device when required.

A summary of companies and institutes working in the field
of integrated optics is given in the following, together with
a short presentation of some special efforts.

2. INTEGRATED OPTICS ACTIVITIES

First investigations on optical guided-wave components began
at the Istituto di Ricerca sulle Onde Elettromagnetiche (IROE),
an institute of the National Research Council of Italy (CNR),in
1971/72 [1], and one of the first Italian patents in this field
was filed in 1977 [2]. Other groups had also started working on
related topics, and a first attempt to coordinate the national
activity in integrated optics, with more ambitious aims, was done just in 1977 [3]. Unfortunately, despite some support by companies like Telettra and Selenia, the project failed and the
researchers in this area had to wait till the beginning of the
current year (1986) to see the start of a 5-year Finalized Project of CNR,namely "Materials and devices for solid-state electronics", which, by including integrated optical devices among

its goals, has recognized the importance of this area and has provided adequate funding.

At present, some ten groups are actually involved in integrated optics, with possible occasional contributions from other groups. Their activity in the last years is briefly reviewed in the following, in a geographical order, from north to south of Italy (Fig. 1). The reader interested in details about previous works is referred to a national review published in 1979 [4].

FIGURE 1. Sketch of the geographical distribution of institutes and companies concerned with integrated optics research in Italy

TORINO

Politecnico di Torino and CESPA-CNR: Mainly work on various aspects of the propagation in planar optical waveguides had been done [5-8]. Recent investigations have concerned the design of specific devices, such as optical couplers [6] and optical parametric oscillators [7]. A project on design and optimization of laser diode structures has been started in cooperation with CSELT [8].

Centro Studi e Laboratori Telecomunicazioni (CSELT): In this institution, which is the main research center for telecommunications of
the Government-owned STET, most of the activity in the optical
field is devoted to fiber optic communication systems, but studies and developments of integrated optical devices have always
been considered with interest [8-13]. It is worth to underline
that CSELT is at the moment the only research laboratory in Italy using molecular beam epitaxy of III-V compound materials to
build sources and detectors for optical communications [12,13].
These components obviously can be employed in integrated optic
sensing circuits as well: Figures 2 and 3 show the cross section of a shallow mesa oxide insulated laser device and of a
detector integrated with a guiding layer respectively [13].Both

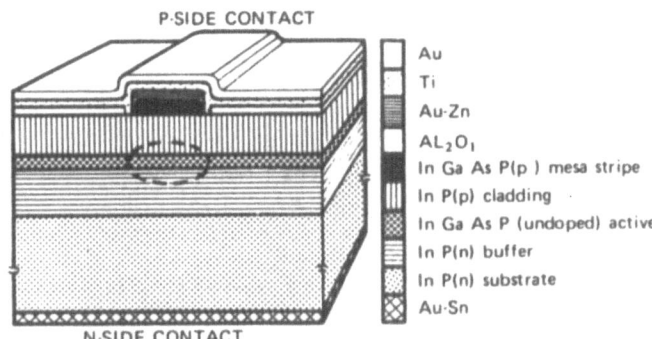

FIGURE 2. Cross section of a laser diode for 1.3 or 1.55 micron
operation (from Ref. 13).

these devices are suitable for multimode operation, either at
1.3 or at 1.55 micrometer wavelength, but work is in progress
to obtain single-mode index-guide devices.

PADOVA
Dipartimento di Fisica and Istituto di Elettrotecnica e di Elettronica:
Particular attention has been paid to the investigation of materials and material processing, in close cooperation with other
research groups [11,14-16]. Their work with the group at Bari
University has brought to a better understanding of the tita-
nium-diffusion process in lithium niobate, [17] which is an important step in the fabrication of optical waveguides.

BOLOGNA
Dipartimento di Elettronica : This research group has been concerned

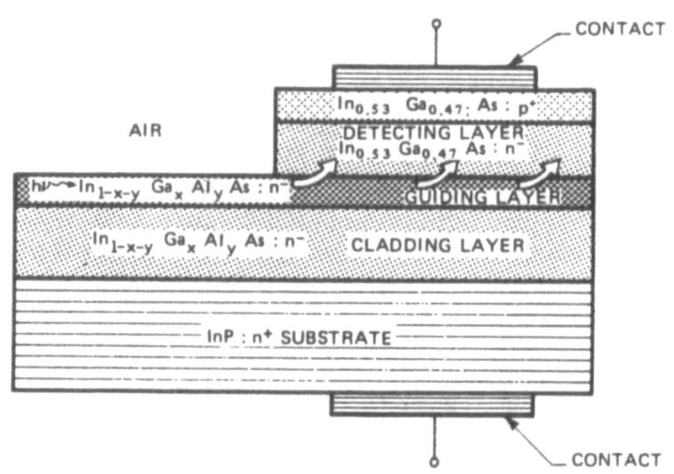

FIGURE 3. Cross section of a proposed PIN detector integrated with the optical waveguide (from Ref. 13).

mainly with theoretical activity; in particular, studies and characterization of tapered waveguides have been performed [18,19].

FIRENZE

I.R.O.E. - C.N.R.: A major goal of the research programs in integrated optics at IROE has been the design and development of thin-film geodesic components [20-23], since their first introduction in guided optics in 1972 [1]. Particular attention has been paid to signal-processing applications [24,25]. A recent improvement is represented by a new design of the lens profile that should reduce radiation losses and make easier the fabrication [26]: typical profile of a perfectly focusing lens with focal length of 25 mm, and corresponding radii of curvature are depicted in Fig. 4. The use of geodesic components in multiplexing devices has been also investigated [27-29]. At the same time, work on fabrication and characterization of glass waveguides has been carried out [30,31]. Recent activity, in cooperation with a research group at *Istituto Nazionale di Ottica*, has also concerned the investigation of grating [32] and refractive [33] integrated optical components. Figure 5 shows the ray-tracing trough a novel refractive thin-film lens [34], having aperture about f/3.

ANCONA

Dipartimento di Elettronica e Automatica: The activity so far has been mainly concerned with fiber optics [35], but there is some

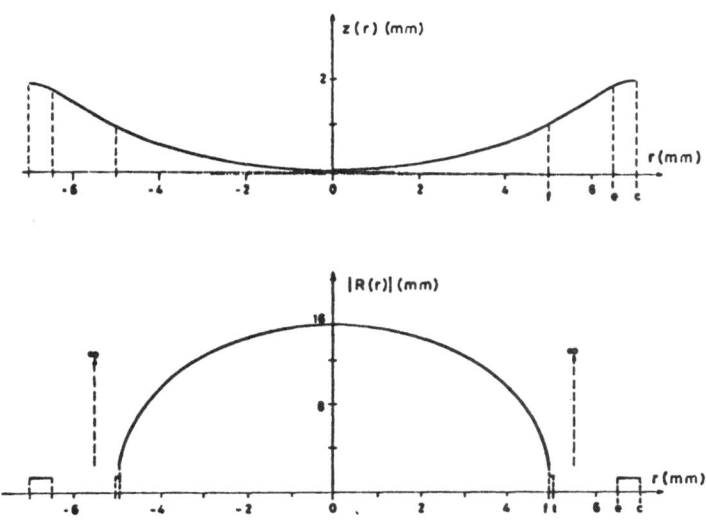

FIGURE 4. Typical profile (top) and radius of curvature (bottom) of a focusing geodesic lens, adopting a new connecting region to the plane. (See Ref. 26).

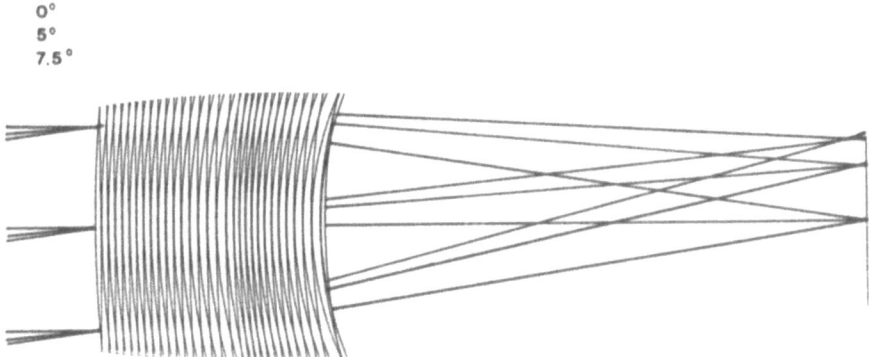

FIGURE 5. Ray-tracing for different input beam angles in a new lens, constituted by multiple refractive elements with circular boundaries.(See Ref. 33).

interest in starting a cooperative program on integrated optical devices.

ROMA

Fondazione Ugo Bordoni: This is a non-profit private institution working in close cooperation with the Italian Post Office (Poste e Telecomunicazioni). Due to the interest in optical communication systems, most of the work has been concerned with fiber optics, and with the investigation of laser diodes and detectors performance ([36-38]).

Dipartimento di Energetica, Università di Roma "La Sapienza": The main topics dealing with integrated optics have been laser annealing of thin films ([39]), in order to reduce scattering and absorption losses, and investigation of nonlinear phenomena in optical waveguides, with special concern about optical bistability ([40-42]).

Istituto di Acustica "Corbino" - C.N.R.: Activity is devoted mainly to the design and development of surface-acoustic-wave and acousto-optics devices ([43]); as far as the integrated optics is concerned, particular attention has been devoted to the use of ZnO films as piezoelectric material to be deposited on glass or silicon ([39,44]).

Selenia S.p.A.: This company, besides the general interest on integrated optical devices for signal processing applications, has a research group which is making use of its own expertise in optical thin-films for facing the problems related to the fabrication of hybrid optical circuits and in particular of Luneburg lenses on high-index materials ([45]).

BARI

Dipartimento di Elettrotecnica ed Elettronica: First investigations had concerned mainly with propagation in anisotropic waveguides for the design of integrated optical modulators ([46-48]). More recent work has dealt with characterization of lithium niobate ([11,14,17,49,50]) and with the design of specific devices in it ([51]). As an example of the experimental measurements carried out on diffused and proton-exchanged lithium niobate waveguides, Fig.6 shows the scattering intensity in plane for proton-exchanged z-cut waveguides as a function of the exchange time.

PALERMO

Dipartimento di Ingegneria Elettrica: One of the pioneering groups in integrated optics in Italy ([3,52]), together with IROE-CNR, CESPA-CNR and University of Bari. Among the research topics which have been investigated, thin film waveguide technology has attracted particular attention ([53]). It is worth to outline the

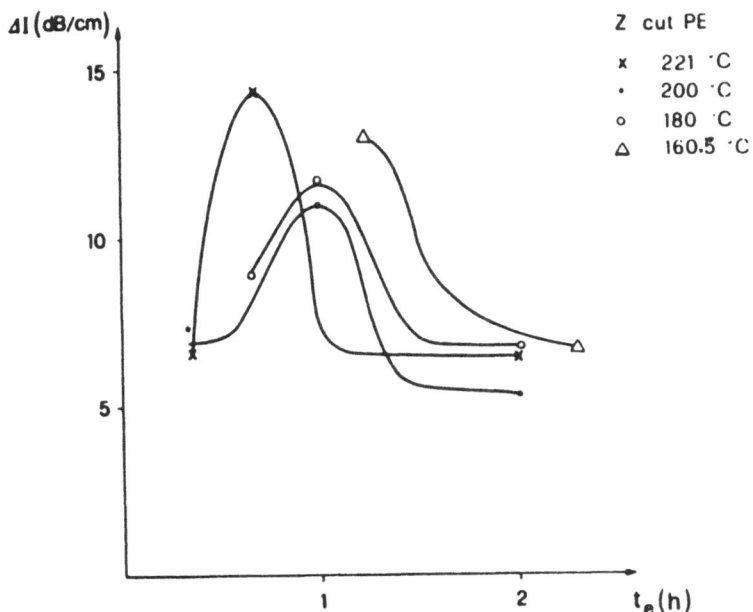

FIGURE 6. In-plane scattering for PE z-cut waveguides vs. exchange time. (See Ref. 51).

very recent works on selective photodeposition and laser patterning of different materials [54-58]. Fig.7 refers to the experimental arrangement used for testing light-induced localized growth of ZnS thin film on a previously deposited film of CdS [55].

INDUSTRIAL COMPANIES: As mentioned before, Telettra S.p.A. in Milano and Selenia S.p.A. in Roma have already been involved, even if in a different degree, in integrated optics activities. Other companies which are interested, but - at least to my knowledge- are not directly operating in this area are Pirelli and Italtel in Milano, and Face Standard in Pomezia (Roma). Some of them are participating in EEC projects, such as BRITE and ESPRIT, and some greater involvement in integrated optics activities can be expected in a relatively short time.

3. SUMMARY

Integrated optics work is mainly done at public research institutes and universities, but industrial attention to this area is rapidly growing. A short description of main research

452

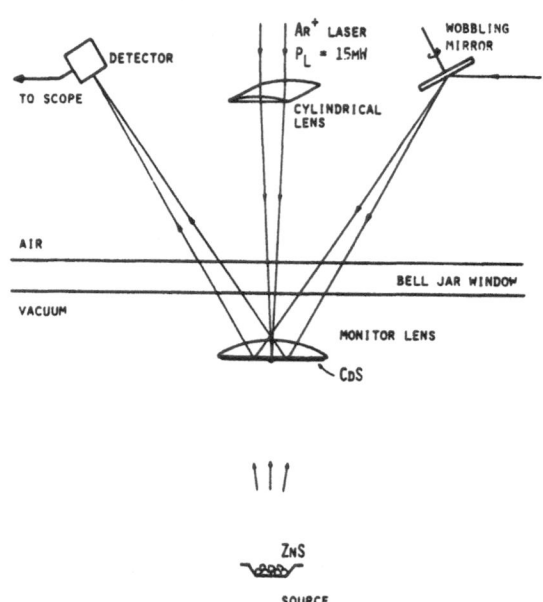

FIGURE 7. Experimental setup to observe photo-induced thin-film growth (from Ref. 55).

interests has been given, and a selected (certainly not exhaustive) list of references done. Integrated optical sensors are not included at present among the devices which are under investigation and development, but the advances in waveguide technologies and in components design which have been obtained recently can make relatively easy the approach to these devices as well.

REFERENCES

1. Righini GC, Russo V, Sottini S, Toraldo di Francia G: Thin Film Geodesic Lens; Appl. Opt., vol.11, 1442-1443, 1972
2. Righini GC, Russo V, Sottini S: Dispositivo a strato sottile per l'elaborazione di segnali unidimensionali, Italian Patent Appl. N.9534 (1/8/1977) - U.S.Patent 4,222,628 (Optical Thin Film Processor for Unidimensional Signals)
3. Righini GC, Riva Sanseverino S, Tomassini M: Possibilità di sviluppo di una attività coordinata di ottica integrata in Italia, IROE Report CMR.194-8.12, 1977
4. Scheggi AM: A National Review on Fiber and Integrated Optics Activity in

Italy, in "Fiber and Integrated Optics", D.B.Ostrowsky Ed. (Plenum Press, N.York, 1979), 389-405

5. Daniele V, Montrosset I, Zich R: Boundary Formulation of Propagation Problems in Guiding Structures for Integrated Optics, 1980 Intern. Symp. on Antennas and Propagation, 2-6 June, Québec, vol. I, 240-243

6. Bava GP, Ghione G: Exact Synthesis of Optical Couplers, Tech. Digest WIORT'84 (IEEE Catalogue n. CH2077-6), 145-148, 1984

7. Bava GP, Montrosset I, Sohler W, Suche H: Optimized Structure of $Ti:LiNbO_3$ Channel Waveguides for Optical Parametric Oscillators, in "Integrated Optics", H.P.Nolting and R.Ulrich Eds., (Springer, Berlin, 1985),196-201

8. Bava GP, Montrosset I, Potenza M: Threshold Current Optimization of Ridge-Waveguide Lasers, CSELT Tech. Reports vol. XIV, 127-131, 1986

9. De Bernardi C, Loffredo A, Morasca S: Optimization of Waveguide Attenuation Measurements by Scattered Radiation Detection, Proc. IWOCS, (Rome, 1984)

10. Coppa G, Di Vita P, Potenza M, Rossi U: Stratification Methods in Waveguiding Structures, in "Hybrid Formulation of Wave Propagation and Scattering", L.B.Felsen Ed. (Martinus Nijhoff, Dordrecht, 1984), 409-417

11. Canali C, Mazzi G, De Bernardi C, Loffredo A, Morasca S, De Sario M: Effects of Water Vapour on Refractive Index Profiles in $Ti:L_iNbO_3$ Planar Waveguides, Proc. IOOC-ECOC'85 (Venezia, 1985), 47-50

12. Genova F, Rigo C, Stano A: Molecular Beam Epitaxy Growth of InGaAs and InAlAs Layers for Photodiodes, CSELT Tech. Reports Vol. XIII, N.2, 127-131, 1985

13. Destefanis G: Lasers and Detector for 1.3 and 1.55 μm Optical Fibre Communication, CSELT Tecnical Reports Vol. XIII, N.6, 391-396, 1985

14. Armenise MN, Canali C, De Sario M, Carnera A, Mazzoldi P, Celotti G: Evaluation of the Ti Diffusion Process During Fabrication of $Ti:LiNbO_3$ Optical Waveguides, Journal of Non Crystalline Solids Vol. 47,2, 255-258, 1982

15. Canali C, Carnera A, Mazzi G, Mazzoldi P, De La Rue RM: Fabrication Process, Performances and Stability of Ti-Indiffused and Proton Exchanged $LiNbO_3$ Optical Waveguides, Tech. Digest WIORT'84 (IEEE Catalogue n.CH2077-6) 17-20, 1984

16. Canali C, Carnera A, Mazzi G, Mazzoldi P, Arnold G: Li and H Concentration Profiles in $LiNbO_3$ Waveguides Measured by Nuclear Techniques, Proc. IOOC-ECOC'85 (Venezia, 1985), 43-46

17. Armenise MN, Canali C, De Sario M, Carnera A, Mazzoldi P, Celotti G: Ti Diffusion Process in $LiNbO_3$, in "New Directions in Guided Wave and Coherent Optics", D.B.Ostrowsky and E.Spitz Eds., (Martinus Nijhoff, The Hague, 1984), 623-637

18. Bassi P, Zang DY, Ostrowsky DB: Near Field Radiation Pattern of Tapered Monomode Optical Waveguides, Optics Commun. vol. 41, 95-98, 1982

19. Bassi P: Tapered Monomode Optical Waveguides, in "New Directions in Guided Wave and Coherent Optics", D.B.Ostrowsky and E.Spitz Eds., (Martinus Nijhoff, The Hague, 1984), 577-587

20. Righini GC, Russo V, Sottini S: A Family of Perfect Aspherical Geodesic

Lenses for Integrated Optical Circuits, IEEE J.Quantum Electron. vol. QE-15, 1-4, 1979

21.Sottini S, Russo V, Righini GC: General Solution of the Problem of Perfect Geodesic Lenses for Integrated Optics, J.Opt.Soc.Am. vol.69, 1248-1254, 1979

22.Sottini S, Russo V, Righini GC: Fabrication Tolerances in Geodesic Lenses: a Rule of Thumb, IEEE Trans. Circuits & Systems vol. CAS-26, 1036-1040, 1979

23.Sottini S, Russo V, Righini GC: Geodesic Optics: New Components, J.Opt. Soc.Am., vol.70, 1230-1234, 1980

24.Sottini S, Russo V, Righini GC: FT Geodesic System for High-Resolution Spectrum Analyzer, IEE Conf. Publication N. 201, 95-98, 1981

25.Righini GC: Integrated Optical Signal Processors: Design and Implementation Criteria, in "Integrated Optics: Physics and Applications", S.Martellucci et al. Eds., Plenum Press (London, 1983), 247-265

26.Sottini S, Russo V, Giorgetti E: A New Geodesic Lens for Integrated Optics Processors, Proc. IOOC-ECOC'85 (Venezia, 1985), 183-186

27.Russo V, Sottini S, Righini GC, Trigari S: Spherical Waveguide Multiplexing Device, in "Integrated Optics", H.P.Nolting and R.Ulrich Eds. Springer Verlag (Berlin, 1985), 16-20

28.Russo V, Sottini S, Righini GC, Trigari S: Demultiplexing and Tapping Device Using a Spherical Geodesic Lens, Opt. Communications, vol.54, 87-90, 1985

29.Righini GC, Sottini S, Russo V: Aberration Analysis in a Waveguide Hemispheric Geodesic Lens, Optica Acta, vol.33, 771-780, 1986

30.Righini GC, Tecniche di fabbricazione di guide e componenti per ottica integrata in substrati di vetro, Vuoto, vol. XV, 183-189, 1985

31.Righini GC: GRIN Structures in Integrated Optics: Materials and Fabrication Techniques, in "Digest of GRIN'85", AEI (Milano, 1985), 53-60

32.Righini GC, Molesini G: Grating Structures in Integrated Optics, Proc. SPIE, vol.473, 21-30, 1984

33.Laybourn PJR, Righini GC: A New Design of Thin Film Lens, Electronics Letters, vol.22, 343-345, 1986

34.Laybourn PJR, Righini GC: Multielement Homogeneous Thin-Film Lens Design, to appear in Proc. SPIE, vol. 700, 1986

35.Cancellieri G, Orfei A: Effective Cut-off Condition in Single-Mode Optical Fibres, Proc. IOOC-ECOC'85 (Venezia, 1985), 659-662

36.Daino B: Integrated Optical Receivers for Telecommunications: a Critical Discussion, in "Integrated Optics, Physics and Applications", S.Martellucci and A.N.Chester Eds, (Plenum Press, N.York, 1983) 111-121

37.Tamburrini M, Spano P, Piazzolla S: Frequency Noise in Injection-Locked Single-Mode Semiconductor Lasers, Proc. IOOC-ECOC'85 (Venezia, 1985), 713-716

38. Spano P, Piazzolla S, Tamburrini M: IEEE J.Quantum Electron., QE-22, 427-435, 1986

39.Bertolotti M, Ferrari A, Jaskow A, Palma A, Verona E: Laser Annealing of ZnO Thin Films, J.Appl.Phys., Vol.56, 2943, 1984

40.Bertolotti M, Sibilia C, Sette D: Bistable Behaviour of Light Waves in a Graded-Index Planar Waveguide with Nonlinear Substrate, Phil.Trans. R. Soc.London, A313, 361, 1984

41.Sibilia C, Bertolotti M: Guided Waves in a Nonlinear Slab, Tech. Digest WIORT'84 (IEEE Catalogue n. CH2077-6), 55-59, 1984

42.Bertolotti M, Sibilia C, Anselmi I: Bistability in Nonlinear Waveguides, AGARD Conf. Proc. N.383, 17, 1985

43.Alippi A, Palma A, Palmieri L, Socino G, Verona E, Acousto-optics with Surface Acoustic Waves. Devices and Applications, Optica Acta vol.27, 1061-1076, 1980

44.Palma A, Verona E, Lagomarsino S: Zinc-Oxide Piezoelectric Films for Integrated Acoustooptics, Alta Frequenza vol. LII, 209 1983

45.Varasi M, Lenti per ottica integrata realizzate con tecnologie di film sottile, Vuoto vol. XV, 19-29, 1985

46.Armenise MN, De Sario M, Modal Characteristics of Diffused Rectangular Anisotropic Waveguides to Design Integrated Optics Devices, Proc. 8th European Microwave Conf., 808, 1978

47.Armenise MN, De Sario M: Investigation of the Guided Modes in Anisotropic Diffused Slab Waveguide with Ambedded Metal Layer, Fiber and Integrated Optics, vol.3, n.2, 197-219, 1980

48.De Sario M, Design of Microstrip Integrated Optical and Microwave Devices, Alta Frequenza, vol. XLIX, 298-303 N.4, 1980

49.Canali C, Armenise MN, Carnera A, De Sario M, Mazzoldi P, Celotti G: Effects of Water Vapour on TiO2,Linb3O8 and (TixNb1-x)O2 Compound Kinetics during Ti:LiNbO3 Waveguide Fabrication, Proc. SPIE, vol.460, 34-42, 1984

50.Armenise MN: Guide d'onda ottiche in LiNbO3: preparazione e caratterizzazione, Vuoto vol. XV, 5-10, 1985

51.De Sario M, D'Orazio A: Realistic Design of Ti-Diffused LiNbO3 Directional Couplers, Alta Frequenza vol.53, 221-30, 1984

52.Tien PK, Riva Sanseverino S, Martin RJ, et al: Two-Layered Construction of Integrated Optical Circuits and Formation of Thin-Film Prisms, Lenses, and Reflectors, Appl. Phys.Lett. vol.24, 547, 1974

53.Calì C, Daneu V, Riva Sanseverino S: Use of a Scanning Laser Beam for Thin Film Control and Characterization, Optica Acta vol.27, 1267-1274, 1980

54.Arnone C, Assanto G, Calì C, Riva Sanseverino S: Applicazioni di fotochimica laser per la realizzazione di microstrutture, Vuoto Vol. XV, 11-17, 1985

55.Arnone C, Calì C, Riva Sanseverino S: Study of Photo-Induced Thin Film Growth on CdS Substrates, Mat. Res. Soc. Symp. Proc. vol.29, 275-281 1984

56.Arnone C, Calì C, Riva Sanseverino S: Laser Evaporation Technique for CdS Thin Film Deposition, Tech. Digest WIORT'84 (IEEE Catalogue n.CH2077-6), 33-36, 1984

57.Arnone C, Rothschild M, Ehrlich DJ, Laser Etching of 0.4 μm Structures in CdTe by Dynamic Light Guiding, Appl. Phys. Lett. vol.48, 736-738,1986

58. Arnone C, Rothschild M, Black JG, Ehrlich DJ: Visible-laser Photodeposition of Chromium Oxide Films and Single Crystals, Appl. Phys. Lett. vol. 48, 1018 - 1020, 1986.

LIST OF PARTICIPANTS

F. BALDINI
C/O I.R.O.E. - C.N.R.
VIA PANCIATICHI, 64
50127 FIRENZE - ITALY

J. BANG
ELAB SINTEF GROUP
ELEKTRONIKKLABORATORIET VED NTH
N-7034 TRONDHEIM-NTH, NORWAY

G. BARILE
C/O I.R.O.E. - C.N.R.
VIA PANCIATICHI, 64
50127 FIRENZE - ITALY

A. BERTHOLDS
INSTITUTE DE MICROTECHNIQUE DE
L'UNIVERSITE' DE NEUCHATEL
RUE DE LA MALADIERE 71
CH-2000 NEUCHATEL 7, SWITZERLAND

A. BINI
C/O I.R.O.E. - C.N.R.
VIA PANCIATICHI, 64
50127 FIRENZE - ITALY

F. BLOISI
ISTITUTO DI FISICA
FACOLTA' DI INGEGNERIA
P.LE V. TECCHIO, 80
80125 NAPOLI - ITALY

P. BOFFI
CENTRO LASER
SOCIETA' CONSORTILE S.R.L.
VIA F. BLASIO, 1 (ZONA INDUSTR.)
70123 BARI   ITALY

T. BOSSELMANN
SIEMENS AG   ZFE THP 42
POSTFACH 32 40
D-8520 ERLANGER   F.R.G.

A.J.A. BRUINSMA
INST. OF APPLIED PHYSICS TNO-TH
P.O. BOX 155
2600 AD DELFT - THE NETHERLANDS

R. CAPONI
CSELT
VIA G. REISS ROMOLI, 274
10148 TORINO   ITALY

G. CONFORTI
I.R.O.E. - C.N.R.
VIA PANCIATICHI, 64
50127 FIRENZE   ITALY

B. CROSIGNANI
DIPARTIMENTO DI FISICA
UNIVERSITA' DI ROMA
P.LE ALDO MORO, 5
00185 ROMA - ITALY

B. CULSHAW
DEPT.ELECTRONIC & ELECTRICAL ENG
UNIVERSITY OF STRATHCLYDE
ROYAL COLLEGE BUILDING
204 GEORGE STREET
GLASGOW G1 1XW SCOTLAND - U.K.

J.P. DAKIN
PLESSEY E.S.R. LTD
ROKE MANOR
ROMSEY, HANTS - U.K.

A. D'AMICO
IESS - CNR
VIA CINETO ROMANO, 42
00156 ROMA - ITALY

H. DANIGEL
CIBA-GEIGY AG, R-1062-107
CH-4002 BASEL   SWITZERLAND

D. DE ROSSI
CENTRO "E. PIAGGIO"
UNIVERSITA' DI PISA
FACOLTA' DI INGEGNERIA
VIA DIOTISALVI, 2
56100 PISA - ITALY

HADRAWNIK DETLEV
HEWLETT-PACKARD
ANALITICAL DIVISION
WALDBRONN - F.R.G.

S. DONATI
UNIVERSITA' DI PAVIA
DIPARTIMENTO DI ELETTRONICA
VIA ABBIATEGRASSO, 209
27100 PAVIA - ITALY

C. EDWARDS
PHYSICS DEPT.
UNIVERSITY OF WESTERN AUSTRALIA
NETLANDS   WESTERN AUSTRALIA

R. FALCIAI
I.R.O.E. - C.N.R.
VIA PANCIATICHI, 64
50127 FIRENZE - ITALY

G.W. FEHRENBACH
DEGUSSA WOLFGANG, ABT. FM-P
POSTFACH 13 45
D-6450 HANAU - F.R.G.

M. FENN
POLYSENS S.P.A.
UFFICIO RICERCHE E SVILUPPO
VIA B. CELLINI, 32
50020 SAMBUCA VAL DI PESA (FI)
ITALY

K. FESLER
STANFORD UNIVERSITY
GINSTON LAB.
STANFORD, CA., 94305 - U.S.A.

E. GANDIN
UNIVERSITE' DE L'ETAT A LIEGE
INSTITUT DE PHYSIQUE (B5)
LABORATOIRE DE PHYSIQUE GENERALE
SART TILMAN, 4000 LIEGE 1
BELGIUM

O.F. GENCELI
ISTAMBUL TEKNIK UNIVERSITESI
MAKINA FAKULTESI
GUMUSSUYU
ISTAMBUL   TURKEY

E. GIORGETTI
C/O I.R.O.E. - C.N.R.
VIA PANCIATICHI, 64
50127 FIRENZE - ITALY

G. GIRONI
VALFIVRE S.P.A.
VIA PANCIATICHI, 70
50127 FIRENZE - ITALY

W.H. GLENN
UNITED TECHNOLOGIES RES. CENTER
EAST HARTFORD, CONNECTICUT 06108
U.S.A.

G. GROSCH
MBB
DEPT. LKE 326
PF. 80 11 60
D-8000 MUENCHEN 80 - F.R.G.

A. HARMER
BATTELLE, GENEVA RESEARCH CENTER
OPTICS & ELECTRONICS GROUP
7 ROUTE DE DRIZE
1227 CAROUGE, GENEVA, SWITZERLAND

S. HUARD
INSTITUT D'OPTIQUE THEORIQUE
ET APPLIQUEE
BAT.503, CENTRE UNIV. D'ORSAY
B.P. 43
91406 ORSAY CEDEX   FRANCE

A. IARIA
COMANDO CORPO TECNICO ESERCITO
VIA DELLA BATTERIA NOMENTANA 51
00162 ROMA - ITALY

J.T. IVES
UNIVERSITY OF UTAH
2220 E. 4800 S. NO 417
SALT LAKE CITY, UTAH, 84117
                          U.S.A.

D.A. JACKSON
PHYSICS DEPT.
UNIVERSITY OF KENT
CANTERBURY CT 27NR   U.K.

A. LEITE
DEPT. OF PHYSICS
UNIVERSITY OF PORTO
P. GOMES TEIXEIRA
4000 PORTO - PORTUGAL

H.J. KALINOWSKI
UNIVERSIDADE FEDERAL FLUMINENSE
AITEROL - 24100 - BRAZIL

J.P. LOEWENAU
B.M.W. AG
PETUELRING 130, BMW HAUS
D-8000 MUENCHEN 40 - F.R.G.

R. TH. KERSTEN
SCHOTT GLASWERKE
CENTRAL RESEARCH,   ZFG-1/KE
HATTENBERGSTR. 10
6500 MAINZ - F.R.G.

J.M. MACKINTOSH
OPTOELECTRONICS GROUP
DEPT.OF ELECTRONIC & ELECTRICAL
ENGINEERING, STRATHCLYDE UNIV.
GEORGE STREET
GLASGOW G1 1XW   U.K.

G. KERVERN
THOMSON-SINTRA
ACTIVITES SOUS-MARINES
ROUTE DE SAINTE-ANNE
DU PORTZIC
29601 BREST CEDEX   FRANCE

V. MAGNANI
C/O UNIVERSITA' DI PAVIA
VIA ABBIATEGRASSO, 209
27100 PAVIA - ITALY

R. KIST
FRAUENHOFER-INSTITUT FUER
PHYSIKALISCHE MESSTECHNIK
HEIDENHOFSTR. 8
D-7800 FREIBURG   F.R.G.

G. MARGHERI
C/O I.R.O.E. - C.N.R.
VIA PANCIATICHI, 64
50127 FIRENZE - ITALY

G. KNOLL
FRAUNHOFER-INSTITUT FUER
PHYSIKALISCHE MESSTECHNIK
HEIDENHOFSTRASSE 8
D-7800 FEIBURG   F.R.G.

M. MARINELLI
II UNIVERSITA' DI ROMA
VIA ORAZIO RAIMONDO
00173 ROMA - ITALY

R.P.H. KOOYMAN
TECHNISCHE HOGESCHOOL TWENTE
TWENTE UNIVERSITY OF TECHNOLOGY
DEPARTMENT OF APPLIED PHYSICS
P.O. BOX 217,   7500 AE ENSCHEDE
THE NETHERLANDS

S. MARTELLUCCI
II UNIVERSITA' DI ROMA
VIA ORAZIO RAIMONDO
00173 ROMA - ITALY

K.H. LAN
DEPARTMENT OF OPTICAL INSTRUMENT
ENGINEERING
UNIVERSITY OF ZHEJIANG
HANGZHOU,   CHINA

M. MARTINELLI
CISE S.P.A.
P.O. BOX 12081
20090 SEGRATE (MILANO) - ITALY

H.C. LEFEVRE
THOMSON-CSF
CENTRAL RESEARCH LABORATORY
DOMAIN DE CORBEVILLE - B.P. 10
91401 ORSAY CEDEX - FRANCE

E. MASETTI
LAB. FILM SOTTILI - ENEA -
00060 CASACCIA (ROMA) ITALY

F. MAYSTRE
UNIVERSITE' DE NEUCHATEL
INSTITUT DE MICROTECHNIQUE
RUE A.L. BREGUET 2
2000 NEUCHATEL 7 SWITZERLAND

R.S. MEDLOCK
BROWN BOVERI   KENT PLC.
BISCOT ROAD
LUTON, BEDFORDSHIRE LU3 1AL U.K.

A. MENCAGLIA
C/O I.R.O.E. - C.N.R.
VIA PANCIATICHI, 64
50127 FIRENZE - ITALY

R. MERCURI
C/O IESS
VIA APPIANO, 53
00136 ROMA - .ITALY

B. MICHAUX
DRDI - CIREIA
ELF AQUITAINE
TOUR ELF CEDEX 45
92078 PARIS LA DEFENSE, FRANCE

A.G. MIGNANI
I.R.O.E. - C.N.R.
VIA PANCIATICHI, 64
50127 FIRENZE - ITALY

T. NAKAYAMA
MITSUBISHI ELECTRIC CORPORATION
CENTRAL RESEARCH LABORATORY
8-1-1 TSUKAGUCHI-HONMACHI
AMAGASAKY, HYOGO 661 - JAPAN

A. NANNINI
CENTRO "E. PIAGGIO"
UNIVERSITA' DI PISA
FACOLTA' DI INGEGNERIA
VIA DIOTISALVI, 2
56100 PISA - ITALY

K. OGAN
THE PERKIN-ELMER CORPORATION
INSTRUMENT GROUP
761 MAIN AVENUE
NORWALK, CT., 06859-0111, U.S.A.

T. OKOSHI
UNIVERSITY OF TOKYO
DEP.ELECTRONIC ENGINEERING
7-3-1 HONGO,BUNKYO-KU
TOKYO, 113 - JAPAN

A. PANSINI
C/O UNIVERSITA' DI BARI
INGEGNERIA ELETTRONICA
VIA RE DAVID, 200
70100 BARI - ITALY

D.N. PAYNE
UNIVERSITY OF SOUTHAMPTON
DEPARTMENT OF ELECTRONICS AND
INFORMATION ENGINEERING
HIGHFIELD, SOUTHAMPTON, HANTS
S09 5NH - U.K.

H. POISEL
MBB-GMBH - DEPT. AE343
P.O. BOX 80 11 49
D-8000 MUENCHEN 80 F.R.G.

L. PRIANO
UNIVERSITA' DI BARI
VIA RE DAVID, 200
70100 BARI - ITALY

G.C. RIGHINI
GRUPPO NAZ. ELETTRONICA
QUANTISTICA E PLASMI - CNR
C/O IROE - CNR
VIA PANCIATICHI, 64
50127 FIRENZE - ITALY

A.J. ROGERS
DEPT.OF ELECTRONIC & ELECTRICAL
ENGINEERING, UNIV. OF LONDON
KING'S COLLEGE LONDON, STRAND
LONDON WC2R 2LS - U.K.

P. ROTH
CSEM S.A.
GUIDED WAVE OPTICS DEPT.
MALADIERE 71
CH-2000 NEUCHATEL 7 SWITZERLAND

V. RUSSO
I.R.O.E. - C.N.R.
VIA PANCIATICHI, 64
50127 FIRENZE - ITALY

G. SCELSI
C/O S. RIVA SANSEVERINO
DIP. DI INGEGNERIA ELETTRICA
UNIVERSITA' DEGLI STUDI
VIALE DELLE SCIENZE
90128 PALERMO - ITALY

A.M. SCHEGGI
I.R.O.E. - C.N.R.
VIA PANCIATICHI, 64
50127 FIRENZE - ITALY

J.O. SCHMIDT
RADIOMETER A/S
ANALYTICAL INSTRUMENTS DIVISION
SENSOR DEVELOPMENT GROUP
EMDRUPVEJ 72
DK-2400 COPENHAGEN NV DENMARK

J.E. SCHROEDER
ELECTROMAGNETICS INSTITUTE
TECHNICAL UNIVERSITY OF DENMARK
BUILDING 348
DK-2800 LYNGBY - DENMARK

O. SCHWELB
DEPT.OF ELECTRICAL ENGINEERING
CONCORDIA UNIVERSITY
1455 DE MAISONNEUVE BLVD. WEST
MONTREAL, QUE., H3G 1M8, CANADA

P. SCHWEIZER
SPECTEC S.A.
14, AVENUE SAINT-AUGUSTIN
IMPASSE LUCIEN GASTAUD
06200 NICE - FRANCE

P. SIXT
CSEM S.A.
GUIDED WAVE OPTICS DEPT.
MALADIERE 71
CH-2000 NEUCHATEL 7 SWITZERLAND

V. SOCHOR
FACULTY OF NUCLEAR & APPL.PHYS.
BREHOVA 7
115 19 PRAGUE 1, CZECHOSLOVAKIA

H. SOHLSTROM
THE ROYAL INST. OF TECHNOLOGY
DEPT. INSTRUMENTATION LAB.
S-100 44 STOCKHOLM - SWEDEN

P.T. SQUIRE
UNIVERSITY OF BATH
SCHOOL OF PHYSICS
CLAVERTON DOWN
BATH BA2 7AY U.K.

A. TADEUSIAK
TELECOMMUNICATION INSTITUTE
UL. NOWOWIEJSKA 15/19
00665 WARSAW - POLAND

G. TAITI
C/O I.R.O.E. - C.N.R.
VIA PANCIATICHI, 64
50127 FIRENZE - ITALY

T. TAMBOSSO
UNIVERSITA' DI PAVIA
VIA ABBIATEGRASSO, 209
27100 PAVIA - ITALY

S. TAMMELA
TECHNICAL RESEARCH CENTRE OF
FINLAND, SEMICONDUCTOR LAB.
OTAKAARI 7 B
SF-02150 ESPOO FINLAND

S. TRIGARI
I.R.O.E. - C.N.R.
VIA PANCIATICHI, 64
50127 FIRENZE - ITALY

R. ULRICH
TECHNISCHE UNIV.HAMBURG-HARBURG
HARBURGER SCHLOSSTRASSE 20
POSTFACH 901403
2100 HAMBURG 90 - F.R.G.

T. VAN ECK
UNIVERSITY OF CALIF., SAN DIEGO
DEPT. EE/COMPUTER SCIENCE, C-014
LA JOLLA, CA., 92093 - U.S.A.

PER VASE
INSTITUTE OF MANUFACTURING ENG.
LAB.OF INDUSTRIAL ACOUSTICS
TECHNICAL UNIVERSITY OF DENMARK
BLDG. 352
DK-2800 LYNGBY - DENMARK

L. VICARI
ISTITUTO DI FISICA
FACOLTA' DI INGEGNERIA
P.LE V. TECCHIO, 80
80125 NAPOLI - ITALY

G.VON TRENTINI
SIEMENS AG
SI KOMP V LWL
RUPERT MAYER STR. 44
D-8000 MUNICH 70   -   F.R.G.

R.A. WERNER
INNOVAL
P.O. BOX 6935
8023 ZURICH - SWITZERLAND

E. WINTNER
INSTITUT FUER ALLGEMEINE
ELEKTROTECHNIK U.ELEKTRONIK
TECHNICAL UNIVERSITY OF VIENNA
GUSSHAUSSTRASSE 27-29
A-1040 WIEN,   AUSTRIA

W. WODRICH
SCHOTT GLASWERKE
OPTICS DEPARTMENT
HATTENBERGSTRASSE 10
D-6500 MAINZ 1 - F.R.G.

H. WOELFELSCHNEIDER
FRAUNHOFER-INSTITUT FUER
PHYSIKALISCHE MESSTECHNIK
HEIDENHOFSTRASSE 8
D-7800 FREIBURG   F.R.G.

R. WU
LABORATORY FOR SURFACE DEVICES
SEMICONDUCTOR INSTITUT OF
CHINESE ACADEMY OF SCIENCES
CHINA